An Introduction to Environmental Systems

An Introduction to Environmental Systems

George Dury

Emeritus Professor of Geography and Geology,
University of Wisconsin — Madison

HEINEMANN
LONDON AND EXETER, NEW HAMPSHIRE

Published in the U.S.A. 1981 by
Heinemann Educational Books Inc.
4 Front Street, Exeter, New Hampshire 03833

Published in Great Britain 1981 by
Heinemann Educational Books Ltd
22 Bedford Square, London WC1B 3HH

Library of Congress C.I.P. Data

Dury, George H.   1916-
    An Introduction to Environmental Systems.

Includes Bibliographies and Index.
    1. Earth Sciences   2. Physical Geography. I. Title
QE26.2.D87   551   80–29151

ISBN 0–435–08001–6 (U.S.)
ISBN 0–435–3526–28 (U.K.)

Printed in the United States of America

Contents

Acknowledgements

For general discussion and for instruction and information on specific topics, I am grateful to Barbara Bell, W.A.R. Brinkmann, R.J. Chorley, F.D. Hole, J.C. Knox, L.B. Leopold, T. Langford-Smith, T.R. Vale, and G. Wahba. Critical eyes were cast on the manuscript, and helpful suggestions made, by various staff members of my publishers, and by Muriel Dury.

Particular help in obtaining photographs came from G.M. Habermann, T. Langford-Smith, R.J. O'Brien, and Élie Weeks. Among other individuals kindly providing pictures are R.J. Blacker, W.B. Denton, P.D. Feldman, P.J. Gersmehl, E.J. Hickin, S.A. Schumm, A.D. Short, Sigidur Thórarinnson, W.F. Tanner, and S.C. Zoltai.

Official agencies supplying photographs and giving permission for their use include the Australia Information Service, the Bahamas Ministry of Tourism, the Canada Department of Energy, Mines, and Resources, the Canadian Forces, the Canadian Government Office of Tourism, the Canadian Government Travel Bureau, the Division of National Mapping (Australia), Kitt Peak National Observatory, LANDSAT, the Long Beach Department of Oil Properties, NASA, the New South Wales Crown Lands Office, the New Zealand High Commissioner, and the U.S. Department of Agriculture. Private organizations giving permission for the use of their photographs are *The Capital Times* (Madison, Wisconsin), and Ward's Natural Science Establishment. John Wiley & Sons., Inc., permitted the use, in slightly adapted form, of a diagram in a book published by them. K.R. Walker, L.F. Laporte, and the editors of *American Scientist* were all involved in the correspondence that produced kind permission for the adaptation, in the present book, of diagrams by K.R. Walker and of some joint text. Yvon le Maho, in addition to supplying the photograph of emperor penguins, commented constructively on the adaptation of his own text to the purposes of the final passages in Chapter Twenty-one. References to the publications containing the closely adapted material appear either in the relevant captions or in the Instructor's Manual.

Where previously published data have been analysed, re-analysed, or otherwise considerably manipulated, original authorship is indicated throughout. Unaccredited photographs are my own. For credited photographs, copyright remains with the agencies and individuals named in the captions.

A final word of thanks goes to A.M. Davis for field survey, and to R.J. O'Brien for help with cartographic search.

Introduction

At any given point in history, half of all scientists who had ever lived were alive still. This is still true: half of all scientists who have ever lived are alive today. The consequence is the phenomenon called the *information explosion* or the *knowledge explosion*. The stock of available information, in a particular variety of environmental science, is doubling every ten or twenty years. Traditional scientific disciplines are becoming filled to bursting.

At the same time, each of us feels an instinctive need to develop some kind of world picture. One may suppose that the development of such a picture would be easier today than in earlier centuries, because of the contemporary speed of communication and travel. Not so, however; the very lack of information in past times eased the process of putting everything into a single frame. The eighteenth century is often called the Age of Reason because of the great revival and spread of interest in philosophical systems. What was then called 'natural history' embraced the whole environment of man. Much of it remained still obscure. The known part was so limited that it could readily be comprehended by a single person, and fitted into an all-embracing philosophical system.

Sciences of all kinds developed vastly during the nineteenth century. It was then that geology, oceanography, meteorology and climatology were founded, in forms still easily recognisable today despite the subsequent revolutions that have occurred in all four. It was still possible in the middle of the nineteenth century for a scientist to be competent in three distinct kinds of science. Even if the stock of *useful* information has been doubling only every fifty years, the stock for 1850 was only one-quarter the stock for 1950 and about one-sixth the stock for 1975. If the increase continues there will be eight times as much information for the year 2000 as in 1850. Let us assume that theoretical advances are only half as rapid as the generation of useful information. Even on that assumption, the theory bank for a given science would have been in 1850 only about two-fifths of what it was in 1975, and about one-third of what it will be in the year 2000.

Small wonder, then, that the later part of the nineteenth century and a great deal of the twentieth brought ever-increasing specialisation. Sciences of all kinds subdivided. Increasing fragmentation made it ever more difficult to communicate, not only within sciences, but even more, between sciences.

Three kinds of reaction to the situation are obviously possible. One is the creation of specialised professional societies, wherein specialists talk to other specialists of the same kind. A second is the attempt to communicate to the general public, by presenting scientific results in terms that can be generally understood. A third is the compression of as much material as possible into introductory texts. The fourth is drastic simplification.

All scientists belong to professional societies, which exist for the exchange of results and ideas. Some scientists, and presumably all scientific journalists, are interested in general communication. Some textbook authors indeed do try to include as much material as possible. If the trend continues, then some introductory texts in the year 2000 will be 8–10 cm thick. This outcome could particularly be expected in environmental science, because all the component sciences are experiencing their own knowledge explosions.

The way toward drastic simplification is indicated, however dimly, by the fact that much scientific work is done by teams. Simply because research is complex, more than one worker is often needed for a single project. In environmental science, the workers are apt to be drawn from different kinds of science. But because they can work together, they obviously have some interests in common. To give an example: it is easy to imagine that a soil scientist could combine profitably with a climatologist interested in former climates, in a joint study of ancient soils (paleosols). The team could readily be expanded to include a botanist, who would be responsible for the investigation of former plant life, and a geologist, who would identify the origin of rock materials included in the paleosols. Their various interests differ, but not without considerable overlap.

Now, this circumstance in turn suggests that they may share a common theoretical basis. It may at first glance seem highly unlikely that workers on soils, climates, plants and rocks should have theory in common, in addition to interests. However, the whole of environmental science has a common basis in systems theory. Application of this theory ensures great economy of effort in the study of the environment. The systems idea provides a complete frame of reference for the physical, biological and social worlds which we inhabit. It applies not only to our own planet earth, but to the entire universe. It works just as well for a galaxy as for a single grain of sand. It links instinct to the computer.

The systems idea will be summarised and illustrated in Chapter 1. It will be developed and applied in the remainder of the book.

Because our language is linear, there is no choice but to arrange the material presented in a sequence of chapters. The sequence adopted begins with a summary and illustration of the systems idea. After setting the planet Earth in the general context of the solar system, it then deals with the cascades of rock materials and water, moving next to surface forms of the land masses and their margins. An outline of the solar energy cascade introduces process-response systems of the atmosphere and the oceans, after which come an enquiry into the control of landscapes by climate and discussion of step-functional change in atmospheric behaviour. The nature and operation of ecosystems, in increasing order of complexity, leads to reviews of existing and of past and future ecosystems. Detailed scrutiny of the cascades of rock material and of radiometric dating is relegated to an appendix, while a second appendix reviews the analytical techniques (mainly graphical) that are used in the main text.

The indicated sequence deals first with physical objects and materials – rocks, rivers, landforms, beaches, glaciers – and next with weather and climate, which are undoubtedly experienced, but which can only in part be observed in terms of physical objects. On the other hand, the early appearance in the discussion of surface water means that the solar energy cascade must be, for the time being, taken for granted. This fact can scarcely be thought undesirable, because treatment of a system in cascading terms must mean that a great deal is taken for granted anyway. At the same time, numbers of users of this book may well prefer to deal with the solar energy cascade and process-response systems of the air and ocean, before attacking rocks and landforms.

Similarly, there is nothing to obstruct any reader who wishes from considering the details of the rock material cascades (Chapter 24) at the same time that the cascades themselves are examined, or from taking in the details of the solar energy cascade at the same time that this cascade is considered in broad outline. The two chapters which really stand apart are Chapter 1, in which the systems idea is presented as a whole, and Chapter 25 which recapitulates the analytical techniques that are employed. Chapter 1 is meant as a necessary introduction to the main text, to be read at the very outset, and to be referred to as frequently as needed on subsequent occasions. Chapter 25 is designed to enable readers to check rapidly on various types of relationships among physical variables, without the need to hunt back through the main text in order to locate the first example of a given relationship.

1 The Systems Idea

Thinking about the environment in terms of systems is likely to go through four stages. In the first stage we recognise systems as systems, without even thinking about it, and certainly without using the word *system*. In the very beginning, in fact, we operate by instinct rather than by conscious thought. The second stage introduces the definition of *system*. Because definitions in environmental science, as in science of all kinds, need to be precise, the definition of *system* must be formal. Because the systems idea is so widely applicable, the definition must also be very general. A highly formal, very general definition may at first seem difficult to apply to actual cases – even though when we realise what constitutes a system we start to perceive systems all about us. In this stage, the systems idea may appear at the same time self-evident and elusive.

In the third stage of thinking, examples multiply. Various kinds of system are identified. We recognise the general applicability of the systems idea. We see that it provides for any desired level of simplicity or complexity in our treatment of the real world. We also discover that many natural systems tend to be self-regulating. In addition, we come to suspect that the barriers between one kind of science and another are largely artificial. The systems idea works for all.

The fourth stage of thinking involves the analysis and classification of environmental systems, the construction of systems models and the use of selected models to predict how a system will behave. The construction of models is the process of abstraction. It provides us with generalisations. According to the degree of complexity of a system, the abstraction process can range from very simple to very difficult. Systems analysis ranges in the same way. Once we have discovered how a given system works, we can forecast its responses to various kinds of effects, including effects induced by man. Forecasting usually involves simulation, both because we need answers more rapidly than real systems can supply them, and because experiments on real environmental systems would be very risky. For analysis and simulation of highly complex systems, computers are necessary. On the other hand, the working of simple systems can easily be simulated by means such as the throwing of dice or the turning of cards.

1.1 Recognising systems without thinking about them

We all live in, and with, environmental systems. Indeed, each of us *is* a system. The dictionary tells us (although we shall need a revised definition presently) that a system is a complex whole, a set of connected parts, or an organised group of material objects. Usually we take for granted the systems that we meet every day. Clocks and cars for instance are physical systems. Cats and people are life systems, often called biosystems. We accept familiar systems as working wholes, without dividing them into component parts for analysis, studying their internal connections or bothering about their organisation.

Each of us knows what a clock or a car is, but very few of us could write out a list of components. Fewer still habitually think of a cat, or a person, in terms of component parts, even though we may be generally aware of the skeleton which provides the essential framework. How many bones are there in the human body? And how many hairs are needed to clothe a cat?

Even though we may well be conscious from time to time of some of our own component parts, we normally first pay attention to the whole. While I, the author–biosystem, am writing this paragraph, the sun is shining through my window. I do not take an inventory to find out if feet, legs, trunk, arms, hands and head are all separately warm. I satisfy myself with the total response of the whole system. I get the answer without needing to do the sum.

It is precisely in this way that the systems idea is highly economical. It permits us to accept the whole, whatever its complexity, without bothering about the parts – unless of course we wish to, or need to. The economy of approach resembles the difference between buying a car complete, and buying it one component at a time and then putting it all together. With biosystems in particular, it is impossible (except in science fiction) to produce a system by the assembly process.

To a great extent, we seem to begin instinctively with the entire system, and to consider its parts later, if indeed we ever consider them at all. Working instinctively means working at the unconscious level. Take for example the way in which a very small child develops the concept *cat*. Right from the start, the animal is accepted as a complete system – and a working system at that – named perhaps Cat, Kitty, or Puss. Acceptance occurs, long before the child bothers its mind about the differences between one kind of system and another. Examples could be the differences between animals that walk on four legs (cats, dogs) and animals that walk on two (birds, people). And very importantly, the child distinguishes the cat from the rest of the real world. The cat–biosystem has boundaries.

We can assume that, in time, the infant comes to see that the animal is furry, with a head at one end and a tail at the other. Eventually, it may also realise that the legs are four in number. It will certainly learn that the cat will sometimes purr, sometimes meow, and sometimes scratch. Although not all the observed facts are likely to be put into words, the following ideas have in fact been formulated:

THING = CAT, distinct from other things, including other cats;
HEAD, TAIL, FOUR LEGS, BODY = components;
HEAD AT ONE END, TAIL AT THE OTHER, LEGS ON THE CORNERS = systematic arrangement of the components;
FURRINESS, PURRING, MEOWING, SCRATCHING = characteristics.

The complex whole has been partly elaborated into a set of connected parts. Typical characteristics, including modes of behaviour, have been identified. Only two further mental steps are needed, in order to advance thinking to the scientific level. One step is very short: the other is very long.

1.2 Defining systems in general

The long step just referred to is the step between unconscious (instinctive) and conscious (rational) thought. This is the step from a child's apprehension of its immediate world to a scientist's formal description and analysis of the universe, its parts and their workings. This step is what science is all about. Many of us for many purposes operate at a kind of halfway mark. Thus, while few of us will study anatomy, we do recognise a general similarity between people and cats, in that both groups possess

backbones. We learn about other kinds of animals with backbones, many of which we shall never see. In this way, we come by rational means to know that there is such a thing as the vertebrate kingdom, which includes all backboned animals, but our knowledge remains very vague as to details. At the same time, we have moved out of the realm of pure instinct or completely unconscious thought.

The really long step from instinct to completely rational analysis is made simple by the short step, which we now take. It consists merely in stating the definition of a system:

A SYSTEM IS A STRUCTURED SET OF OBJECTS (that is, components), OR A STRUCTURED SET OF ATTRIBUTES (characteristics), OR A STRUCTURED SET OF OBJECTS AND ATTRIBUTES COMBINED TOGETHER.

The use of the term *set* merely indicates that the system has boundaries. It can be distinguished from the rest of the real world and from other systems of the same kind. We could, for example, regard our own skins as biosystem boundaries. The description *structured* only means that the system possesses internal order. As the dictionary says, the components are connected together, or at least are arranged in some kind of pattern. A heap of loose components, which if assembled together would make up a car, is not a system. It lacks structure.

What we call *characteristics* in everyday speech are the *attributes* of a system. They include appearance and behaviour, including sensory behaviour where appropriate. Clocks, cars, cats and people possess their own characteristic appearances. Clocks respond to winding by storing energy in their springs. Cars respond to pressure on the brake or accelerator pedal. Different behaviour patterns among different clocks and different cars appear in differing capacities to keep time accurately, or differences in mileage rates. Sensory responses in biosystems are highly familiar in everyday life. Cats respond to the smell of mice. The author–biosystem, in typing this text, and the reader–biosystem, in reading it, make use of the visual sensors which we call eyes.

However, *attributes* in the context of systems has to be understood somewhat more narrowly than *characteristics* in everyday speech. An attribute needs to be measurable; and two observers undertaking measurement ought to come to more or less the same answer. It is possible for instance to measure colour by referring to a standard colour scale. But it is not possible to measure the results of value-judgement. In the words of the ancient saying, beauty is in the eye of the beholder. Thus, we understand *components of a system* to be physical objects, and *attributes* to be non-physical, but rationally definable, characteristics. A language, regarded as a system, includes no physical components. It does, however, possess structure, without which it would have no meaning. Its attributes are its vocabulary and its grammar.

1.3 Systems all around us

Many readers will by now be multiplying examples of systems for themselves – bicycles, forests, farms, supermarkets, industrial plants, libraries, schools, hospitals, deserts, oceans, rivers, storms, tribes and families – the list is as boundless as the universe itself. Each of the listed items is distinct from other items. It is distinct from other items in the same class, and from the rest of the real world. It has boundaries which define it as a set. It possesses internal order – it is structured. It possesses attributes – it displays typical kinds of appearance and/or typical modes of behaviour.

But merely to recognise systems for what they are is not enough. We shall need to give attention to system functions, system states, the balance between input and output, patterns of behaviour both in space and in time and to different kinds of system structure.

1.3a System functions

Environmental systems interact with their surroundings in three main ways. We recognise three kinds of functional relationship: isolated, open with respect to the transfer of energy alone, and open with respect to the transfer both of matter and of energy.

Recalling that systems have boundaries, we define *isolated systems* as having boundaries that are not crossed either by energy or by matter. On the most gigantic scale possible, our universe seems to be an isolated system. So far as we can tell, no matter and no energy comes into the universe from outside, and no matter and no energy escapes. It is, of course, very difficult to imagine the meaning of *outside* in application to the universe, just as it is difficult to imagine that the boundary of the universe ends at nothing. On less than the universal scale, isolated systems are rare to absent. We might perhaps look on a time capsule as an isolated system. This is the container that is sometimes placed in the foundation of a major building, holding news clippings relating to the time of burial. It is meant to provide for some future generation a summary of recent events and prevailing circumstances. The capsule is fossilised. It is meant to be unaffected by heat, moisture and bacterial attack until it is opened at some future date. No matter leaves, no matter enters. No energy

leaves, no energy enters. No information leaves, no information enters. The system is entirely static.

Suppose, however, that the time capsule contains a fully wound clock. The stored energy of the clock spring is sealed in. But when the spring has expended all the stored energy, the clock stops working. Its energy content has been dissipated. Here is an illustration of the eventual fate of all isolated systems. They run down to a stop.

Systems which receive inputs of energy but not of matter, or which make outputs of energy but not of matter, are exemplified by our planet Earth in relation to the rest of the solar system. Solar energy comes in, earth radiation goes out. This example is only approximately correct when it is examined very closely, because meteorites and their dust do enter the earth–atmosphere system from space. It seems possible, and even likely, that some earth dust and some earth gas escapes into space, although on this topic nothing or little is known. However, the total exchange in comparison with the earth's bulk is so small that we can neglect it. We can look on the earth–atmosphere system as essentially *closed with respect to matter*, but as *open with respect to energy*.

On a minute scale, a light bulb works in the same way. It has a well defined boundary in its glass container. It is structured: the component parts of container, filament, fill of gas and plug or screw are systematically connected together. It receives energy in the form of electric current and radiates energy in the form of visible light. Energy, both incoming and outgoing, crosses the boundaries of the system. Matter does not.

Other examples of systems which receive and give out energy but not matter are the earth–atmosphere system, in which water is constantly recycled through the vapour, liquid and solid (ice) states, and the earth's crust, where rocks are being constantly broken down and reformed. The energy for the recycling process is supplied by the sun, by gravity and by radioaction in the earth's interior. It is partly dissipated within the systems, partly radiated away into space. The water in the one case and the rock material in the other, however, are retained within the system.

Systems *open to the transfer both of energy and of matter* are prime concerns in much of environmental science. Biosystems are invariably open. They require energy in order to operate; they require nutrients in order to exist; they have waste products to dispose of. A forest takes in energy, mainly from the sun. It gains mass as its plants and animal inhabitants grow. It radiates energy away into the atmosphere. It loses mass as plants and inhabitants die, and as mineral and organic materials are carried away by water.

Many physical systems in the environment are

similarly open. A rainstorm takes in energy, mainly from a heated land surface or from a warm sea. It takes in moisture, mainly from underlying water. It releases energy when its water vapour condenses into liquid, and drops moisture on the land or water beneath.

1.3b System states

This topic mainly concerns the balance or imbalance between input and output. It also involves the way in which many natural systems tend to be self-regulating. It leads directly on to the pattern of system behaviour through time.

As already noticed, the boundaries of isolated systems are not crossed, either by energy or by matter. There are no inputs and no outputs. Any energy that an isolated system contains to begin with becomes evenly dissipated throughout the system. The final condition is one of *static equilibrium*.

Systems which receive inputs, and make outputs, of energy, of matter, or of both, must either succeed or fail to strike a balance between output and input. If output and input match, the system is said to be in *dynamic equilibrium*. A system in dynamic equilibrium undergoes no long-term structural change.

Once again, a forest provides a useful example. It tends to grow beyond its existing boundaries as seeds are carried away by birds and by the winds. But consider a forest on the side of a mountain range, where the climate on the summits is cold and snowy, while the climate on the lower slopes is dry. Beyond a certain upper limit (the upper treeline) the forest is prevented from spreading because of low temperatures which kill any seedlings. Beyond a certain lower limit (the lower treeline) seedlings are killed by drought. At the same time, the forest is able to resist invasion by the mountain pastures from above, or by the dry scrub from below, simply because forest trees are better adapted to the conditions of the forest belt than are mountain grasses or dryland bushes. So long as climate remains stable, the forest limits also remain stable, despite the fact that the forest would spread if it could, and that grass and scrub would invade it if they were able. The tendencies of spread and invasion cancel one another out along each treeline. Input and output are balanced. The system is in dynamic equilibrium.

At the same time, it changes with the change of seasons. Its trees, animals, insects and bacteria have seasonal cycles of activity, numbers and reproduction. The seasonal fluctuations within the general condition of stability permit us to describe the system as being in a *steady state*.

If inputs and ouputs fail to balance, the system structure must change. So must its pattern of be-

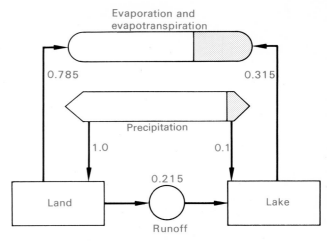

Fig. 1.1 Stabilisation of the level of a closed lake by negative feedback. Proportions reckoned for precipitation on land = 1.0. Based on data for The Salt Lake, N.S.W., Australia.

haviour. A system where serious imbalance occurs is said to be in *disequilibrium*. We need to say *serious imbalance* because many systems are adapted to absorb minor disturbances or to fluctuate considerably through time, without undergoing permanent disruption.

The transition from dynamic equilibrium with little or no fluctuation, through dynamic equilibrium with considerable fluctuation, to complete disequilibrium can be illustrated by the behaviour of three different kinds of lake. The illustration will include notice of the important *feedback principle*. The first lake is contained in an enclosed desert basin. The second possesses a natural outlet. The third lake is wholly artificial, contained behind a man-made dam. The central interest in all three cases will be the water budget.

The lake in the desert basin is fed by runoff from the land, and by rain falling directly on the water surface. As water is added, the lake surface rises. As the area increases, so does evaporation from the surface. When evaporation is equal to rainfall on the lake plus runoff from the surrounding land, the lake level and its surface area become stabilised (Fig. 1.1). A balance, the balance of dynamic equilibrium, has been struck between input (supply of water) and output (loss by evaporation). Moreover, *any increase in supply causes an increase in loss*. The system is self-regulating. Any disturbance caused by an increase or decrease in input is countered by a matching increase or decrease in output. The principle here at work is that of *negative feedback*. It is negative feedback which allows very many environmental systems to avoid self-destruction.

The lake with a natural outlet exists in a region of humid climate. There is enough water to keep the lake filled, and also to keep the outlet running. However, input varies from one season to another,

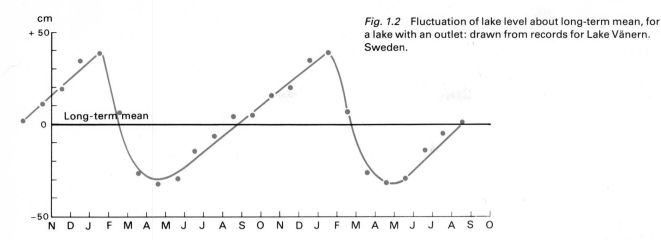

Fig. 1.2 Fluctuation of lake level about long-term mean, for a lake with an outlet: drawn from records for Lake Vänern. Sweden.

being least in late summer and autumn and greatest in late winter and early spring, when snow is melting off the drainage basin and the moist soil can absorb little or no water. The lake level therefore fluctuates from season to season (Fig. 1.2). The outlet is too small to discharge additional water immediately. Nevertheless, negative feedback is still at work. As the lake level rises, discharge through the outlet increases. Fluctuations in lake level are damped out to some extent. Also, in the illustration selected, the high level and low level of one year are closely matched in the next and succeeding years. Lake level fluctuates only within closely fixed limits. This system also is in dynamic equilibrium.

The man-made lake receives vast amounts of floodwater from a rare storm. Its dam is over-topped. Part of the crest of the dam is carried away. Water pours through the breach, enlarging it. The more the breach is enlarged, the more water rushes

through, and the greater the enlargement. Here is *positive feedback* at work. We are dealing with a runaway situation, which will not be resolved until the lake has been completely drained. The system has passed from one equilibrium situation (full) to another (empty).

1.3c Patterns of behaviour through time

Very many environmental systems exhibit cyclic patterns of behaviour. They respond for example to seasonal changes in climate. These changes in turn depend on the height of the sun in the sky at noon, and on the varying length of day. A graph of the height of the noon-day sun in middle latitudes forms a *sine wave* (Fig. 1.3). But the air does not heat up as rapidly as the noon-day sun rises higher, or as rapidly as the length of day increases. There is a certain *lag time* between input and response, as the diagram shows. In the same way, it takes time for rainwater from a heavy storm to become concentrated as floodwater in a valley bottom (Fig. 1.4). In dealing with the pattern of systems behaviour over time, we need to allow for lag time, rhythms, or jumps.

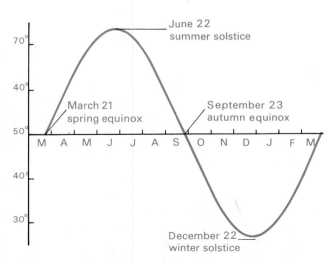

Fig. 1.3 Angular height of midday sun at 40°N latitude. For geometry of a sine wave, see Fig. 6.4.

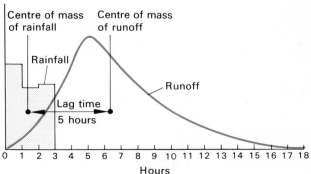

Fig. 1.4 Lag time between rainfall and runoff: based on a real example. Lag could be measured between respective peaks, but is usually, as here, measured between centres of mass.

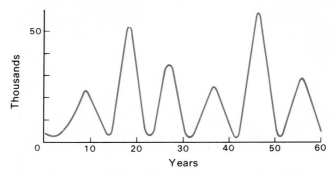

Fig. 1.5 Medium-term fluctuations in numbers of a wildlife species: based on records of lynx pelts taken in northern Canada.

Strictly speaking, an instant match of output with input should occur with the speed of light, the greatest speed which we can imagine. In actual practice, we are satisfied to recognise *short-term change*, which takes less time than one rhythm of the system. The time between peaks of the sun's height and peaks of the length of daylight is one year. The lag between maximum height of the sun and maximum length of day on the one hand, and maximum temperature on the other, is a matter of a few days to a few weeks.

Some environmental systems exhibit *medium-term change*. Examples are provided by certain wildlife populations, which undergo seasonal rhythms, but which also fluctuate in number through periods longer than one season (Fig. 1.5). In any given year, numbers are greatest in spring and early summer, during and immediately after the breeding season. They decline through later months, as the animals are eaten or starved. But superimposed on the seasonal rise and fall in numbers is the *cycle of*

abundance. For some years, the maximum number exceeds the maximum number of the previous year. Afterwards, the maximum number declines, then recovers again. We are dealing now with two cycles, one limited to the length of a single year, and another spanning several years. Only by taking medium-term change into account can we determine whether the population total tends to be stable, or whether it is tending to increase or to decrease.

Investigation of *long-term change* of behaviour poses serious problems. Many of these problems cannot easily be resolved. Whereas short-term change and medium-term change typically involve negative feedback, it may be very hard to choose between negative and positive feedback in the long term, or to say whether or not change is reversible. Consider change in climate. The earth's atmosphere cooled off through about 1°C between 1940 and 1970. Was the cooling part of a progressive trend – in which event it will lead to continental glaciation – or was it merely a swing of a long cycle? During about the same period, the lower atmosphere became increasingly dusty. Will it go on becoming dustier, or can the trend be reversed? Mankind is polluting the world ocean. Will pollution continue far enough to destroy marine life? We can only guess at answers to these questions: we do not know.

Long-term change has a way of coming on gradually, and then happening all of a sudden. Some earthquakes occur precisely in this way. Pressure on the two sides of a crack in the earth's crust builds up steadily, until the sudden moment of release (Fig. 1.6). The build-up may take a century or more: the release can be over in seconds. The shift from one state to another is abrupt.

We are dealing here, not with the smooth passage

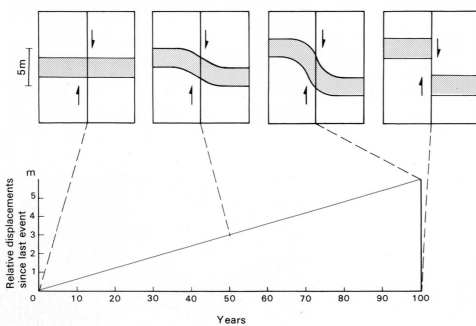

Fig. 1.6 Abrupt response of a tear-fault to continuous stress: upper panels are diagrammatic maps. Based on the behaviour of the San Andreas Fault, California.

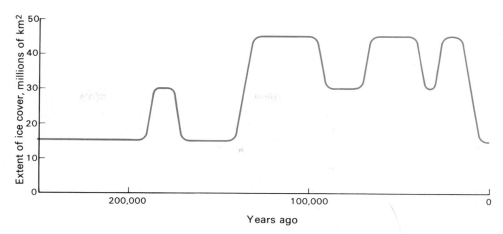

Fig. 1.7 Step-functional change in the extent of continental glaciers. Step functions of two different orders of magnitude are shown.

from the peak of a cycle to a trough, as in Fig. 1.3, but with a sudden jump. Some environmental systems switch rapidly from one equilibrium state to another. Continental glaciers do. The lag time, considerable in human terms, is very short on the geologic time-scale. The glaciers alternate between being vastly expanded and greatly diminished. A graph of their extent against time (Fig. 1.7) displays sudden jumps. It is the graph of a *step function*.

When a step function occurs, we can safely conclude that the system is in disequilibrium while the step is being made. We can also conclude that positive feedback is being applied. The step is halted only when negative feedback sets in once more. The step begins when the condition of the system passes some kind of *threshold*. To appeal again to some of the examples already used: the threshold of behaviour of the man-made dam occurs when the water overflows the crest and the wall is breached; a threshold in behaviour of a fault line occurs when the built-up pressure can no longer be resisted. A threshold in the behaviour of a wildlife population occurs when there is a population explosion, such as introduces a plague of locusts, or when the population is reduced below a critical level, so that extinction follows. For environmental purposes, the crossing of a threshold of behaviour, and the onset of positive feedback, is often followed by disaster.

1.3d System structures

Because we shall be dealing with environmental systems that are open to the transfer of matter, energy, or both, and because our interest in state is limited to the condition of equilibrium or disequilibrium, as the case may be, we need a classification scheme independent both of function and of state. This is provided by system structure.

A complete classification of natural systems would include individual life forms – single cells, plants,

animals – plus human social systems. However, a general view of the environment can dispense with the details of biology as well as sociology, anthropology and political science. Our scope includes the earth–atmosphere system, its workings and its evolution, plus the assemblages of life forms which the earth supports. It will be enough to deal, in increasing order of complexity, with morphological systems, cascading systems, process-response systems, control systems and ecosystems.

Morphological systems are defined in terms of their *internal geometry* – the number, size, shape and linkages of the components. We could put this statement the other way round, and say that, if we concentrate on the internal geometry of a system, we are treating it in morphological terms.

Think of the familiar honeycomb (Fig. 1.8). At first glance we take the pattern to be hexagonal. It is certainly repetitive. Its components (cells) are almost identical in size. The spatial relationship of a given cell to its immediate neighbours is constant

Fig. 1.8 A worker bee feeding nectar into a completed cell (Paul Popper).

throughout the set. When we take a closer look, we see that the cells are cylindrical, not hexagonal, inside. Honeybees are genetically programmed to build cylindrical tubes. The mode of arrangement that bees actually use ensures that the greatest possible number of cells can be packed into a given space, at the least cost in beeswax. For this morphological system, we have identified not only the shape of the components, but also their relationship to one another – in this case, an economical packing pattern.

A map of a stream network also represents a morphological system. The actual design of a network can vary widely from one stream system to another (Fig. 1.9), but the internal geometry can always be defined. That is to say, each unit of the system can be classified in relation to all other units. The spatial linkages already exist, on the ground and on the map.

Many of the mineral substances which make up the earth's crust will form crystals if they are free to do so. The crystals assume distinctive geometric shapes (Fig. 1.10). These shapes in turn are controlled by the arrangement of atoms within the crystal.

Morphological systems of the environment, then, are *identifiable shapes and patterns* – the shapes and patterns of crystals, joints in rock (Fig. 1.11), rock structures (Fig. 1.12), stream and glacier systems, shorelines (Fig. 1.13), entire landscapes (Fig. 1.14), and entire continents (Fig. 1.15).

Cascading systems receive and make *complex inputs* and *complex outputs* of matter, energy, or both. In dealing with them, we can work either in very general terms or in great detail, according to choice and need. The most familiar energy cascade is that of solar radiation. For very many purposes, we do not need to recall that the incoming radiation is highly complex, extending on the two sides of visible

Fig. 1.9   Some highly generalised, but also widely recognisable, patterns of stream network. See also Fig. 7.1.

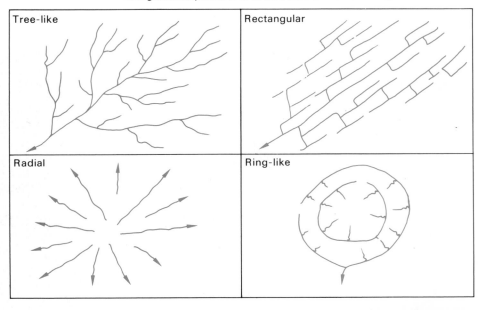

Fig. 1.10   Repetitive geometry of mineral crystals (scale in cm) (R. O'Brien).

Fig. 1.11   Approximately rectangular pattern of joints in limestone bedrock, West Yorkshire, England (J.K. St. Joseph).

Fig 1.12 (above)   Fold patterns displayed in cliffs, Welsh coast, Pembrokeshire (J.K. St. Joseph).

Fig 1.13 (below)   Cape-and-bay shoreline, with beaches forming approximately circular arcs; Aberdeenshire, Scotland (J.K. St. Joseph).

Fig. 1.14   About 14 000 km² of California and the adjacent Pacific Ocean. Water surfaces in the extreme west show up black. Rectangular patterns in the NE are farmland in the Great Valley. In between, hill country runs through the area from NW to SE (LANDSAT imagery).

light into the infra-red and ultra-violet ranges. For very many purposes, again, we do not need to recall that radiation from the earth into space takes place in another band of wavelength. What we are primarily interested in is that output and input, however complex either may be, are broadly balanced. If more energy came in than went out, we should all fry. If more went out than came in, we should all freeze.

But there is more to the situation than this. An increase in input would actually cause an increase in output, just as an increase of electric current through a burner on the stove causes the burner to become warmer. Cascading systems in the environment commonly include some kind of *regulator*. They also

commonly provide for *storage*. Some incoming solar energy has been stored in the earth, notably in the form of coal. There is some delay between total input and total output. Therefore, in assessing the behaviour of cascading systems, we are concerned with the *rate of throughput*.

Cascading systems of the environment include, among others, the incoming cascade of solar radiation, the outgoing cascade of radiation from the earth, the cascade of precipitation on the land, the cascade of rock waste moving downslope toward river channels, and the cascade of water moving through channel systems toward the sea. Here is a prime example of the effect of viewpoint. When interest is limited to internal geometry, a stream net constitutes a morphological system. When interest is extended to the flow of water and sediment, we are dealing with a cascading system.

Process-response systems change their internal geometry and/or behaviour in response to cascading

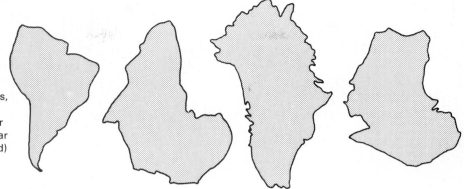

Fig.1.15 Continental outlines, drawn to varying scales and with varying orientation: pear (South America), inverted pear (Africa), thick pear (Greenland) and thick boomerang (Australia).

inputs. They involve at least one morphological system and at least one cascading system. The systems involved are linked together in some way, and may have components and operations in common.

Think back to the examples given earlier, of a lake in an enclosed desert basin, and a lake with an outlet in a region of humid climate. Each lake has area, depth and volume: it is for purely descriptive purposes a morphological system. However, it responds to changes in the cascades of input. If we make the convenient assumption that the incoming energy cascade is constant, then the lake in the enclosed desert basin will respond to changes in the incoming cascade of precipitation. The content of this cascade is common both to the falling rain and to the lake water. Increase of precipitation causes the lake level to rise, and the surface area to increase. *The geometry of the morphological system has altered, in response to a change in cascading input.* But the increase in area causes, as we have observed, an increase in total evaporation from the lake surface. Here is the *regulator* (negative feedback) at work. *Storage* is already effected, since there is water in the lake.

The lake in the region of humid climate also effects storage. It regulates its outflow by increasing discharge as the level rises, and decreasing it as the level falls. But it does *respond*, by changing its level and the area of its surface, to increases and decreases in the cascade of water from the sky or from the surrounding land.

Process-response study is highly important in the understanding of the environment. It deals with how things work. It also deals with the lag time (which we have previously met) between input and response, with the connection between process and form, and with thresholds (also previously met) beyond which an increase in input leads to a breakdown of system behaviour. There are presumably limits of tolerance for any system. Beyond a given limit of loading, a building will fall or a bridge will fail. Beyond a given

limit, an environmental system will break down. A wildlife population, reduced below a certain level, will become extinct. In process-response terms, we can say that the input of newborn creatures falls below the threshold necessary to maintain the system in working order. A drastic revision of the internal geometry of the system, expressed in numbers of population, follows. The total falls to zero.

In environmental science, we encounter process-response systems at every turn, once we go beyond the shape of morphological systems and the energy throughput of cascading systems. Morphological analysis is concerned with *what is there*. Analysis of cascading systems is concerned with *what happens*. Analysis of process-response systems is concerned with *how it happens*.

Control systems are process-response systems in which parts of the operation are controlled by an intelligence. Control systems vary greatly in scale. On a small scale, a farm is a control system. It receives cascading inputs of solar energy and precipitation. Its components react to changes in the incoming cascades – crops are watered by rain, the soil is warmed by the sun. But the cropping plan, the harvesting programme and the management of stock are controlled by the conscious decisions of the farmer. In some national systems, decisions about industrial production are highly dispersed: in others, they are highly centralised. In all cases, however, use of the cascading energy of the sun, the precipitation cascade, the surface-water cascade, and the cascade of energy from burning fuel is directed by human intelligence. We can summarise by saying that environmental management is concerned with the operation of control systems.

Ecosystems include sets of biosystems which interact with one another and with their immediate physical environment. Indeed, this immediate environment is part of the ecosystem. Thus, a forest ecosystem includes all the living organisms of the forest, from the largest trees down to one-celled microscopic bacteria, plus the soil in which most plants grow, the moisture that plants and animals take in and the special interior climate which a forest establishes for itself. Ecosystems are far more complex than the other kinds of system that have been reviewed so far, both because of their intricate internal linkages, and because of their frequently huge numbers of different kinds of component. The number of tree species in a tropical forest can run into thousands. Nobody knows how many, or how many kinds, of bacteria are at work in the soil.

We can isolate for study particular aspects of an ecosystem. Isolation is performed especially for the supply and consumption of nutrients, because, immediately above the level of extraction of chemicals from the soil, the inhabitants of an ecosystem rely largely on one another for food. Common representations of the food web are made to end with man and/or other higher life forms (Fig. 1.16). It is obvious however that the representation is incomplete. Dead human bodies are in their own turn consumed, their ingredients being taken into the recycling sequence.

Analysis of ecosystems is especially concerned with the interdependence of life forms on one another, with the relationship of particular life forms to the whole system and with ecosystem stability.

Interdependence is close and direct, where one species is a predator species and another provides the prey. Equally close and direct dependence connects plant-eating creatures with the regeneration and growth of appropriate plants. More distant connections may be very difficult to make out, especially since we, as representatives of mankind, tend to dislike and to kill plants and animals that seem offensive to us as human beings. However, if we can rid ourselves of the idea that all other living forms have some kind of duty to oblige us, and accept the proposition that they too are life forms,

we shall be able to look at their functions in the ecosystems of which they form part. Recent studies suggest that, of all the known kinds of animal life, only the female mosquito – which transmits malaria – may have no ecologically beneficial function. The male mosquito, it is suspected, may pollinate some species of flower. All other life forms that have been studied in detail perform functions which are of service to some other life forms. In this way, the living components of an ecosystem serve part of the total needs of the system, without necessarily benefiting themselves except in being alive. Furthermore, many components of an ecosystem are competing with other components for what the system can provide. The competition among plants for light, air and water is at least as vigorous as the competition among animals for food.

The presence and persistence of a given life form in an ecosystem implies that its numbers and distribution are more or less stable for the time being. So long as no sudden or drastic change takes place, we can view each species as occupying its own proper place in the total system. This place is its *ecological niche*. The number of niches in a given ecosystem is a measure of the complexity of the system. It ranges from enormous in the humid tropics to close to zero in polar regions.

Occupation of ecological niches seems obviously to be exchangeable. The North American grey squirrel, introduced into Britain during the nineteenth century, has proved competitively superior to the native red squirrel. The red squirrel is all but extinct. We conclude, therefore, that in the ecosystems of Britain there is only one squirrel niche. It does not matter to the ecosystem which type of squirrel occupies the niche, although it obviously must matter to the two squirrel species.

In westernised societies, organic garbage is dis-

Fig. 1.16 Diagrammatic layout of the food chain for a rocky intertidal shore: very freely adapted from the data of J.M. Emlen.

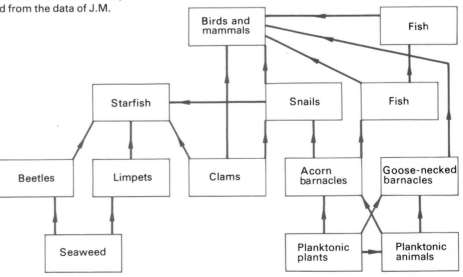

posed of by collectors, or by grinding machines which deliver waste to sewers. Elsewhere, garbage and dead animals are devoured by such scavenging birds as ravens, kites, buzzards and vultures. The scavenger niche exists. In the first case, its filling is thought socially necessary. It benefits the system, by effecting disposal and by supporting collectors and the manufacturers of garbage grinders. In the second case, it benefits the system by effecting disposal and by feeding the scavengers.

Because of the internal competition which they involve, ecosystems are unlikely to be static. However, unless complete breakdown occurs, they persist in conditions of dynamic equilibrium. Such conditions contain *displacement, speciation* and *radiation* within tolerable limits, and always imply *adaptation*.

Displacement has already been illustrated by the exchange of one occupant of an ecological niche for another. Long-term study of a forest ecosystem may well reveal a slow change in character, as one tree species is displaced by another. The whole record of fossil life demonstrates in some instances, and strongly implies in others, the process of displacement. From about 225 to 75 million years ago, the dominant animals were reptiles. Among the reptiles, dinosaurs were supreme. They attained lengths up to 30 or more metres and weights up to 50 tonnes. Today, the largest known reptiles are crocodiles and their relatives, including the largest land lizards, running to lengths of only about 3 m. Mammals have taken over. Today's mammals include the whales, among which are found the largest animals that have ever lived.

Speciation is the process of evolution at work. A genus splits itself into species, the species split into sub-species, and sub-species develop into varieties. When the new differences become sufficiently marked, a variety is promoted to the sub-species status, a sub-species becomes a species, and a species becomes a new genus. Life forms are multiplied by the process of speciation, but are simultaneously reduced in number by the process of *extinction*. Newly-evolved forms tend to spread through the ecosystem which has produced them: spreading is the process of *radiation*. Speciation and radiation tend to increase the complexity of a given ecosystem, extinction to reduce it. Generally speaking, the more complex an ecosystem, the more readily it can resist disturbance.

In all cases, we can expect life forms to be more or less closely adapted to the rest of the ecosystem – or at least, to those components and attributes of the system which are immediately important. Plants and animals in dry climates cannot find an ecological niche unless they are adapted to aridity: by some means or other, they must conserve water. Camels and cacti alike contain storage mechanisms. Plants and animals in cold climates must be adapted to frigidity. Typically, they end or subdue their life cycles during winter, and/or have developed means of conserving heat.

Man has proved to be a great displacer of occupants of ecological niches, as is shown by his destruction (still in progress) of many animal forms. Against this, he has supplied some new niches, and has provided occupants for niches which were formerly vacant. Pigeons and starlings have adapted well to life in many western cities. The term *well* must be understood to apply to the outlook of the birds, rather than to the opinion of city people. Pigeons and starlings are genetically programmed to live in the open country. Both types, however, have learned to feed and roost in cities – that is, they have adapted to the occupation of ecological niches which did not exist before cities were built. It may be thought ironical that city people, while objecting to pigeons and starlings, are not at all offended by sparrows, which are also open-country birds, but which seem to have led the city invasions. Like pigeons and starlings, mice, rats, cockroaches and bedbugs have successfully adapted to the new ecosystems constructed by man. Their successful occupation of newly-provided niches illustrates the resilient character of ecosystems in general.

Opening of hitherto vacant niches is finely illustrated by the introduction of the prickly pear and the rabbit into Australia. Introduced as a garden plant, the prickly pear escaped and ran wild (Fig. 1.17), eventually taking over vast extents of country. The ecosystem contained no check: the ecological niche opened enormously wide. Only when a predator was

Fig. 1.17 The spike-protected prickly pear cactus: height of plant is 2 m or more (Paul Popper).

Fig. 1.18 Explosive invasion of a vacant ecological niche: rabbits in Australia, before the days of viral control (Paul Popper).

introduced, also artificially, was the plant brought under control.

The rabbit, also artificially introduced, also ran wild (Fig. 1.18). There were no effective predators. Here again, the ecological niche opened enormously wide. Only when a virus control was introduced could rabbit numbers be reduced below the disaster level.

We conclude, then, that the interrelationships and operations of ecosystems are at the same time highly complex and open to serious disturbance. Ecosystems aside from man sort out their internal workings by interdependence, competition and evolution. The whole fossil record, of life forms which have existed during the last 1000 million years or so, is one of displacement, in a given niche, of one life form by another. But ecosystems that include man typically involve deliberate interference. Here comes the choice: management or disruption. If we do not understand the system, we risk disrupting it. If we understand it, we can at least attempt management.

Chapter Summary

A *system* is a structured set of objects, a structured set of attributes, or a structured set of objects plus attributes.

Systems are classified by *function* as *closed, open* to the transfer of either matter or energy, and *open to the transfer of matter and energy both.* Environmental systems are typically open to the transfer of energy, but some are closed to the transfer of matter.

Completely closed systems are in, or will reach, a *state of static equilibrium.* Open systems where input is equalled by output + storage are in *dynamic equilibrium.* Open systems where there is a serious difference between input on the one hand and output plus storage on the other are in *disequilibrium.*

The *state* of an open system is affected by *feedback. Negative feedback* tends to preserve *equilibrium, positive feedback* to promote *disequilibrium.*

Patterns of *system behaviour through time* often include *cyclic fluctuation;* and cyclic fluctuation can frequently be graphed as a *sine wave.* Apparent long-term change cannot always be proved to be *progressive.* In other cases, long-term change involves *step-functional switch* from one equilibrium state to another. The step occurs when some *threshold* of resistance is crossed.

The most important kinds of system in environmental science, as classified by *structure,* are the following: *Morphological systems,* defined by their *internal geometry; Cascading systems,* with *complex inputs and outputs; Process-response systems,* which change form and/or behaviour in *response to cascading inputs; Control systems,* which are process-response systems controlled in part by some *intelligence; Ecosystems,* composed of *complexly interacting biosystems and physical systems.*

These include *ecological niches* for the included life forms. They require *adaptation.* They tend to increase in complexity by the processes of *speciation and radiation,* and to decrease in complexity on account of *extinction.* They involve both co-operation and competition.

2 Solar and Earth Morphological Systems

This book is concerned with the planet Earth, and especially with the earth's skin, surface and atmosphere. But the atmosphere, skin, and some aspects at least of the surface cannot be properly understood, or even described, unless the planet is considered in relation to the solar system as a whole. The space relationship of earth to sun is a primary control of weather and climate. The composition of the sun and of planets other than the earth tells us something about the composition and origin of the earth itself. In this chapter, the earth will be treated as a minor component of the solar morphological system.

We define this system in the first instance in terms of the sizes and shapes of the components, and of the distances of non-solar components from one another and from the sun. Although slight irregularities are common throughout, we can afford to overlook these in the general view. For instance, the earth's equatorial diameter is slightly longer than the polar diameter. Its orbit is slightly elliptical. It takes very slightly longer than one earth year (about 0.055% more) to complete one orbit – which is why we have leap years. But for present purposes we can regard the earth as a sphere, of radius about 6000 km, which orbits the sun on a circular path once every earth year. It contains only about 0.0025% (one forty-thousandth part) of the total mass of the solar system.

In actuality, 99.9% of the total mass of the system is contained in the sun. By earth standards, the sun, with a radius of about 0.7×10^6 km, is colossal. The sun is larger than a million million earths. The nine planets are satellites of the sun, held in orbit by solar gravity. In outward order, they are Mercury, Venus, Earth, Mars, Jupiter, Saturn, Uranus, Neptune and Pluto. Their radius of orbit ranges from about 60×10^6 km for Mercury, through about 150×10^6 km for Earth, to about 6000×10^6 km for Pluto.

Other components of the solar system include the satellites of individual planets, such as earth's moon; asteroids; comets; meteorites; dust; gas; and radiation.

The second stage in defining the morphology of the system is reached when we consider the interrelationship of orbits. The basis of reference here is the plane of the earth's orbit, usually called the plane of the ecliptic. All the planets but Pluto orbit more or less in this plane, a fact which suggests that they originated in a single way. Pluto's orbit is inclined to the plane of the ecliptic, but intercepts the orbit of Neptune.

The asteroids, ranging up to 400 km in radius, orbit the sun at a distance of about 400×10^6 km, between the orbits of Mars and Jupiter. The largest of them have radii about half as large as the radius of the earth's moon, which makes them distinctly larger than the smaller of the satellites of Jupiter and Saturn.

The comets, consisting of frozen gas, seem to be concentrated beyond the orbit of Pluto. If they were twice as far out as Pluto, they would still be only one five-hundredth of the distance to the limit of the sun's pull, which extends through at least 6×10^{12} km, a thousand times as far as from the sun to Pluto. Comets become visible when they pass into elliptical orbit round the sun (Fig. 2.1). Their so-

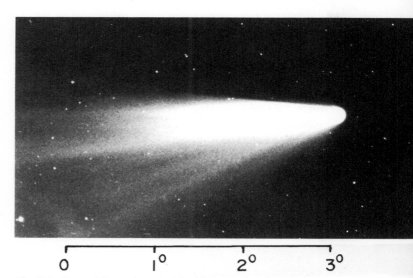

Fig. 2.1 Comet West, photographed from an Aerobee rocket on 5 March 1976. The Sun, off to the right, would subtend an angle of about O.6° of arc (P.D. Feldman).

Fig. 2.2 Results of meteor bombardment on the Moon and (inset) on Mars (Paul Popper).

Fig. 2.3 Galaxy in Canes Venatici (Kitt Peak National Observatory).

called tails are plumes of gas blasted away by solar radiation. If a comet is travelling away from the sun, its tail flows out ahead.

Meteorites probably originate as concentrations of dust, some of it perhaps trapped from beyond the solar system. Here on earth we call them meteors (shooting stars) when we see them burning in the friction of the earth's atmosphere, and meteorites when they land. They are typically either iron-rich or stony. The meteorite that formed Meteor Crater in Arizona may have weighed 100 000 tonnes when it landed. Planets and planet satellites with no or little atmosphere, and hence with no or little protection against meteorite bombardment, become heavily scarred if they themselves are rocky (Fig. 2.2).

2.1 The solar morphological system as an astronomical norm

When we view a galaxy in plan, it looks more or less circular or spiral (Fig. 2.3). Edge-on, however, it looks like a lens (Fig. 2.4). The components of a galactic lens swirl round their centre. Why this should be so remains obscure. However, a swirling motion seems in-built in the major concentrations of matter in space, whether it produces a spiral pattern or a circular pattern, as in the Milky Way galaxy to which our solar system belongs, and as in the solar system itself.

We can look on the solar system as a tiny galactic model. A galaxy consists of a concentration of star systems. A star system consists of the star and its satellites. In both cases, the origin seems to be a lens-like concentration of the gas hydrogen, the lightest element of all. By radioaction, heavier elements are produced from hydrogen. Gas is partly condensed into dust, and the dust is concentrated into individual components of the system.

2.1a A partially unfinished system?

The evolution of the solar system is not yet complete, for the central nuclear furnace of the sun has not yet expired. Presumably this furnace will long continue to convert hydrogen (at present 70% of its content) into helium (which is now 27%), and helium into heavier elements. But the possibility now to be discussed is that the system in its present form has in one respect failed to develop as far as it might have, and in another respect has changed since it came into being.

As a concentration of gas and dust, the system appears to have started its existence about 10 000 million years (10×10^9 years) ago. Rocky bodies, including the earth, the moon and stony meteorites, appeared about 4500 million years (4.5×10^9 years) ago.

The chief rocky bodies are the inner planets, Mercury–Venus–Earth–Mars. Those asteroids which are visible by telescope are also rocky. The outer planets, Jupiter through Neptune, are in considerable part gassy. Jupiter is not much denser than water, and Saturn is less dense.

Orbital radius increases fairly smoothly from Mercury to Mars, and again from Jupiter to Neptune (Fig. 2.5). As planets are added, orbital radius multiplies: the addition of one planet multiplies this radius by about 1.7 times. This is why a straight-line graph results from the plotting of a number of planets on a plain scale against orbital radius on a logarithmic scale. There is, however, a jump in the graph between Mars and Jupiter. If another planet were inserted here, the graph would run clear from Mercury to Neptune. It seems possible that the planet which is perhaps missing is represented by the asteroid belt. If so, it is represented by an orbiting array of small rocky bodies, in mid-range between the rocky planets and the gassy planets.

Fig. 2.4 Galaxies edge-on: the more slender is in Coma Berenices, the plumper is in Virgo (Kitt Peak National Observatory).

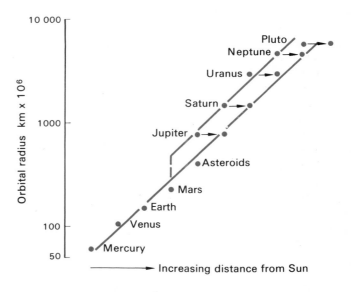

Fig. 2.5 Semi-logarithmic plot of orbital radius against order (= sequence) of distance from the sun.

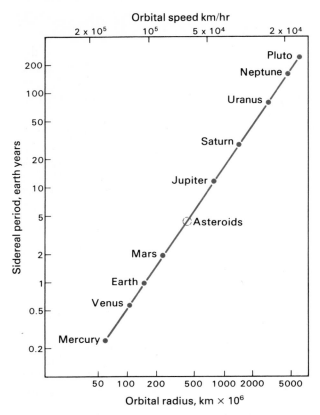

Fig. 2.6 Double-logarithmic plot of orbital radius against sidereal period (= time taken to complete one orbit): asteroid belt plotted as if it were a planet.

The same possibility emerges when orbital radius is considered in relation to time taken to complete one orbit. Regardless of differences in planetary density and mass, time taken to complete one orbit increases as orbital radius increases. In addition, speed along the orbital path decreases as radius of orbit increases. Mercury, the closest planet, goes round the sun once in every three earth months, travelling at about 175 000 km/hr. Earth, taking one year to orbit, travels at about 107 575 km/hr. Pluto, the most distant planet, takes about 250 earth years for a single orbit, at a speed of about 17 000 km/hr, some one-sixth the rate of Earth and one-tenth the rate of Mercury.

Now, when the time taken to complete one orbit is graphed against orbital radius (Fig. 2.6), an impressive regularity appears. Time taken to complete one orbit multiplies as radius of orbit multiplies. This is why the data are plotted in the diagram on a double-logarithmic scale. Time taken for one orbit multiplies about three times as fast as does length of orbital radius. It must follow that, as radius increases, speed divides. As radius multiplies, speed divides about 5.5 times as rapidly. A scale of speed has been added to the diagram.

The spacing on this graph between one planet and another, although not precisely regular, displays only two marked irregularities. One is the wide gap between Mars and Jupiter: the other is the narrow gap between Neptune and Pluto. If the asteroids had concentrated into a planet, the gap between Mars and Jupiter would have been halved, and the whole array from Mercury to Neptune would have been fairly regularly spaced throughout.

Pluto remains something of a mystery, as yet little known or understood. Its unusual position, small size, high density for its distance from the sun and its oblique orbit mark it down as problematical. One possibility is that it was once a satellite of Neptune (whose orbit Pluto intersects) which has been diverted into solar orbit by the gravitational pull of the sun.

Saturn can readily be suspected of being a partially unfinished subsystem. It has ten spherical satellites; but its icy or dusty rings seem potentially capable of having concentrated into three more. The rings lie in the orbital plane (i.e. the plane round Saturn) of the existing satellites. These rings could represent unfinished satellite bodies, just as the asteroids could represent an unfinished solar planet.

Of the bodies outside earth, we know most about the moon. This is the only outside body on which man has trod, and probably the only one on which man will ever tread. We can guess that the moon has condensed from orbiting gas and dust similar to the rings of Saturn, or from orbiting rocky bodies similar to the lesser asteroids. Alternatively, we can guess that it has been captured from some other part of the solar system, as the sun may have captured Pluto. However, simply because we know so much more about the moon than about other planets and their satellites, strong objections can be raised to both of the indicated alternatives. The moon's origin remains a puzzle.

2.1b Some leading attributes of the solar system

Certain attributes have already been noticed – in particular, the physical characteristics of the component bodies, and the speeds and directions of rotation of the planetary components. In contrast to the regularity observed in orbital relationships, there is no smooth progression, outward from the sun, in respect of planetary diameter, density, mass, number of planetary satellites, direction of rotation, period of rotation or angle of rotational tilt.

Rotation, with which this section is mainly concerned, is the spin of a planet on its own axis. In order to discuss this matter, we need to decide which direction, in the solar system, is up? On Earth, of course, *up* is away from the earth's centre. For the solar system as a whole, we assume that *up* is a direction perpendicular to the plane of the ecliptic, and that the up side is the side on which the earth's north pole is. By definition, the earth rotates on its spin axis once every earth day. This axis is tilted to the plane of the ecliptic at an angle of 66½°, a fact of fundamental importance in climate. The daily rotation is equally important in weather. Daily rotation causes sunrise and sunset. The tilt of the spin axis causes seasonal variation in time of sunrise, time of sunset and length of daylight.

Just as the earth orbits the sun, so does the moon orbit the earth. It would be more accurate to say that earth and moon circulate round a common centre. Because the earth's mass is eighty times that of the moon, this centre is beneath the earth's surface. The earth–moon system completes one circuit in twenty-eight days (as measured from Earth). The moon, spinning on its axis once in the same period, always presents the same side to the earth.

Because the earth spins on its axis faster than the earth–moon system revolves, the moon rises and sets. Because the plane of the revolving earth–moon system lies at about 5° to the plane of the ecliptic, the angle defined by sun, moon and earth varies through the lunar month. When earth is almost between sun and moon, maximum light is reflected back from moon to earth. The moon is full. When the moon is almost between sun and earth, we see only the shadow side of the moon. The moon is new. In between the two extremes appear the crescents of the waxing and waning moons. When sun, earth and moon are precisely aligned, eclipses happen. When the earth comes between sun and moon, the earth's shadow darkens out the moon in a lunar eclipse. When the moon intervenes between earth and sun, it blots out the sun in a solar eclipse.

2.1c Function and state of the earth–atmosphere system

Receiving radiation from outside its boundaries – mainly from the sun – and radiating energy back on different wavelengths, the earth–atmosphere system is open in respect of energy transfer. It is also open in respect of the transfer of matter. Although large meteorites mainly burn up in the atmosphere, between 100 and 200 a year reach the surface. Most are lost in the ocean. Very rarely, a large one strikes the ground. The addition of meteorite material to the earth's crust averages perhaps 2.5 tonnes per year. Since the atmosphere came into existence, total addition of meteoric material to the crust has been negligible. About meteoric addition of gas and dust to the atmosphere we know very little; and about loss of gas and dust from the earth–atmosphere system we know nothing. For all practical purposes,

the system is closed in respect of the transfer of matter.

The system is in a steady state, in the sense that input and output of energy balance one another out. If this were not so, the atmosphere would become progressively hotter or colder. But in this context, as in many others, we are obliged to consider the length of time for which this state is defined. As will be seen later, the state of the system undergoes step-functional jumps between one steady-state condition and another. In one state, illustrated by the present day, abundant land ice exists. In the other, illustrated by the major fraction of geologic time, land ice is rare to absent.

2.1d Special attributes of the planet Earth

The earth is special, in possessing an atmosphere composed dominantly of nitrogen and oxygen, and in having oceans. Because life on earth seems to have begun in the ocean, oceans are special indeed. Similarly, because most animal life forms depend on oxygen, the oxygen content of the atmosphere is also special.

Planetary atmospheres are thought to originate by means of the process of outgassing – the emission of gases from a planet's interior, as for instance by volcanic activity. Mercury is probably too close to the sun to be able to retain any atmospheric gases of its own. Mars does possess an atmosphere, composed mainly of carbon dioxide, but too thin to protect the planet from savage bombardment by meteorites. The thick and cloudy atmosphere of Venus, also composed mainly of carbon dioxide, attains temperatures of 450° at low levels, next to the planetary surface. No oceans exist on Mars, despite signs of water erosion on the Martian surface in the past. It seems impossible that oceans could exist on Venus, at low level atmospheric temperatures more than four times the boiling point of water on earth.

One might rely on the earth's oceans to remove carbon dioxide from the atmosphere. Nevertheless, the low carbon dioxide level in the air, and the high concentration of water in the ocean and of water vapour in the atmosphere, remain puzzling by comparison with the conditions on our nearest planetary neighbours, Mars and Venus.

2.2 Morphology and attributes of the earth's interior

The interior morphology of the earth deals with physical contrasts. The planet is composed of concentric shells of contrasting properties. These contrasting properties are themselves attributes. They include density, temperature, pressure and the density–elasticity relationship.

The earth is denser in its interior than at its surface. Astronomers can weigh our planet. Its weight is about 3.75×10^{21} tonnes, or 375, followed by nineteen zeroes, tonnes. But the total weight, only a tiny fraction of the weight of the solar system, is less interesting than the density. This, at 5.5 times the density of water, is the greatest in the entire solar system. Because the rocks at and near the surface average only about three times the density of water, the interior must be distinctly more dense.

Temperature increases from the outside inwards. We know nothing directly about temperatures in the deep interior, but observations in deep mines show conclusively that temperature in the skin increases with depth. We conclude that the interior of the earth is hot. This does not necessarily mean that it is all liquid. Melting-point rises as pressure rises. Earth materials can remain in the solid state under high pressure, far beyond temperatures where they would melt under normal air pressure.

Not even the deepest mines penetrate the earth's outer skin. For samples from beneath this skin we depend on volcanoes. Some of these bring up material from below the outer crust, indicating the chemical nature of the substance beneath.

Most of our information about the earth's inside comes from the analysis of earthquake waves. These are usually called seismic waves, the science of earthquakes being called seismology. Seismic waves come in four kinds. Excessively violent earthquakes cause the earth to vibrate like a bell, although the vibrations are outside the frequencies perceived by the human ear. Every ordinary earthquake sets off long waves (L), which travel round the earth's crust. Neither violent waves nor long waves tell us much about the interior. By contrast, the push waves (P) and the shake waves (S), set off by all earthquakes, tell us a great deal by their travel paths through the earth's body.

The push waves travel about twice as fast as the shake waves. For this reason, push waves are called primary, and the shake waves secondary. The difference in time of arrival at an observatory, between the primary-push-P waves, and the secondary-shake-S waves, identifies the distance between the observatory and the site of the earthquake.

More important for our purpose is the fact that P waves and S waves alike can be, and are, bent. They bend when they pass from one density layer to another. Bending is technically known as refraction. It happens because the speed of wave travel alters with a jump, when density abruptly changes. When a stick is thrust into still water, it looks bent. This is because the speed of light rays changes abruptly

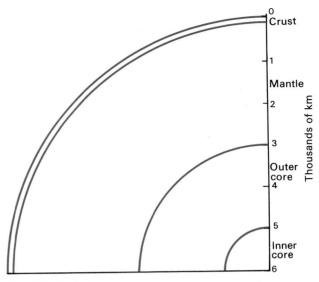

Fig. 2.7　Earth shells.

between the more dense medium (water) and the less dense medium (air). Try it for yourself.

Just so in the earth's interior. The speed of the P and S waves jumps forward, at certain intervals of depth. We infer that there are jumps of density. The behaviour of the earthquake waves is step-functional. Density, and speed of wave, increase suddenly at certain depths. We conclude that the earth consists of a system of concentric shells.

Two discontinuities in wave speed and in inferred density are of supreme importance. The first occurs at an average – although considerably variable – depth of about 16 km. It is called for short the Moho, after its Serbian discoverer Mohorovičič (pronounced, approximately, Mohorovitch). The Moho separates the earth's crust from the underlying mantle. The crust, accounting for only about 1% of the earth's substance, is rigid. In it, push waves travel at about 8 km/s.

The mantle extends from the Moho to the other major discontinuity at a depth of about 3000 km, halfway from the earth's surface to its centre (Fig. 2.7). It accounts for about two-thirds of the total substance of the earth. It transmits push waves mainly in the range 12 to 14 km/s. Wave speed increases inward, making minor jumps with increasing depth, and thus suggesting that the mantle can be subdivided into shells of different density. But the differences in density within the mantle are far less interesting than the fact that, although solid, the mantle subsystem includes currents. It may seem difficult to imagine that a solid body can flow; but, as will later appear, it is necessary to do so.

The major discontinuity at the base of the mantle separates the mantle from the earth's innermost component, the core. The core strongly bends the push waves, and drastically reduces their velocity

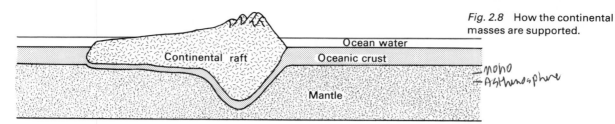

Fig. 2.8 How the continental masses are supported.

into the range 8 to 12 km/s. Shake waves completely vanish. In consequence, the core is regarded as liquid. Liquids will readily transmit push waves, as when a stone thrown into a pond raises spreading ripples, but are too weak to respond to shaking. There are probably differences between the outer core, 3000 to 5000 km down, and the inner core, 5000 to 6000 km down. Speed of push waves jumps a little in the inner core, which some workers think may be solid. But since it is shielded from shake waves by the liquid outer core, its true nature is difficult to determine.

In any event, conditions at great depths are difficult to imagine from the surface. Densities rise to four times those of surface rocks. Temperatures rise to 2500°C – more than 100 times normal air temperatures at the surface. Pressures rise to more than three million times air pressure at sea level. The best estimates are that the outer core at least is liquid, that its internal circulation produces the earth's magnetic field, and that the core as a whole is composed mainly of iron, with minor contents of nickel and of some combination of sulfur and silicon. The inferred total composition, which agrees with calculations of density, closely resembles the composition of iron-rich meteorites.

2.3 Large-scale morphology and major attributes of the earth's crust

The crust comes in two layers, respectively of oceanic and continental type.

Extending all across the ocean floors is oceanic crust, averaging about 8 km thick, and composed of crystalline rock rich in compounds of silica (SiO_2) with iron (Fe) and magnesium (Mg). Crust of this type is more than three times as dense as water. In its bulk chemistry it resembles the outer part of the mantle and the contents of stony meteorites. As a rock it belongs in the basalt family. Material of the same kind extends beneath the continents, but only as a layer which is mainly thin, with a depth of 1 or 2 km.

The continental crust, less than three times as dense as water, averages about 35 km thick. Because it contains materials which have been much redistributed, it is regionally far more variable than the oceanic crust. It is rich in silica and in compounds of

aluminium (Al) with silica, and also contains notable concentrations of carbonate ($CaCO_3$). In its bulk chemistry it strongly resembles the rock granite.

If the continental crust were spread evenly over the earth's surface, it would form a light outermost shell, about 12 km thick. The crust of oceanic type would then form a next inner and denser shell, about 5.5 km thick. But, by observation, the continental crust is not evenly spread. It is aggregated into the huge rafts which we recognise as continental landmasses.

These landmasses ride high because they are buoyed up by the mantle. The ocean floors stand low, because the combined density of ocean water and oceanic crust is greater than the density of continental crust. The principle involved here is the principle of isostasy–equipoise (Fig. 2.8).

2.3a Large-scale morphology of the continents

Despite their considerable morphological differences from one another, all continents possess certain essential components. For a given continent, these are the craton, at least one orogen, and the sedimentary cover (Fig. 2.9).

The craton consists of a rigid block of ancient rocks, 600 million to 4000 million years old. There is a tantalizing gap in the record, between the time of 4500 million years ago, when the earth came into being, and the age of the oldest known rocks. During the 500 million years in question there was probably a thin crust, now entirely vanished, on which volcanic and sedimentary deposits were laid down. This crust was invaded from below – and may have been entirely digested – by light melts migrating outward from the mantle. These melts froze into granites. Construction of the oldest parts of the cratons, mainly by the formation of granite, was completed by 2500 million years ago.

Between 2500 and 600 million years ago, the cratons grew. Belts of highly deformed rocks were welded on to their margins. We see them now as the roots of ancient mountain systems.

Orogens are belts of mountain-building. Obvious examples include the cordilleran mountain belts on the western side of the two Americas, and the Alpine–Himalayan belt which runs from west to east

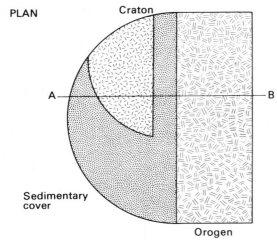

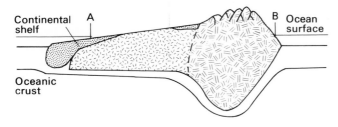

Fig. 2.9 Highly abstract model of a stereotype continent.

through Eurasia. Rocks involved in orogens range from sedimentary types which have been very little disturbed, through rocks contorted into complex fold patterns, to rocks which have been forced to crystallise under intense crustal pressure. There are also rocks formed of material brought up from great depth, including depths extending into the mantle. These are the rocks of cone-building volcanoes and of extensive floods of once-liquid basalt. In addition, the roots of orogens typically include large granitic masses.

This summary of the character of orogens, taken together with the description of the growth of cratons by means of welding-on at the margins, may seem at first to blur the distinction between orogens and cratons. In practice this is not so.

Beyond the date of 600 million years ago, we are dealing with mountain roots only. From 600 million years onwards, we are dealing with mountain systems where more than the roots may be preserved. The 600-million-year mark seems to have been set by a very extensive planation of previously existing continental relief. At the same time, in selecting this mark as a divide between the extension of cratons and the construction of mountains, we may be showing prejudice.

In considering the geologic record as a whole, we cannot avoid taking an intense interest in the de-

velopment of life forms, and in the ancestry of the plant and animal life which surrounds us, including of course our own ancestry. In the history of life on earth, there seems to be an abrupt break at about 600 million years ago. Before that time, the fossil record seems to consist mainly of bacteria and algae. At about 600 million years ago, however, the evolutionary process appears to have produced life forms that, however outlandish by the standards of today, are nevertheless recognisable as ancestral to existing forms. The jump seems, on present evidence, to have been quite sudden – that is to say, an evolutionary step function. It may however amount to nothing more than the development of hard parts, such as the horny covers of insect-like creatures, and the cases of shellfish, which could be preserved in sedimentary rocks.

The sedimentary cover is mentioned at this juncture, only for the sake of sketching in the full picture. Material stripped from cratons and orogens is redistributed over the continental surfaces, is laid down in the sea, or is fixed from seawater by sea-dwelling organisms. Sedimentary deposits are themselves subject to erosion and redistribution. The net result is that cratons and the older orogens are considerably obscured by sediments of younger age.

At the same time, the sedimentary cover is involved in a set of rock material cascades, which involve also the materials of the cratons and the orogens. The working of one cascade, discussed in the following chapter, explains why there are individual continental rafts, instead of a complete outer skin to the earth of continental material.

Chapter Summary

The *components of the solar morphological system* include the sun itself, planets, planetary satellites, asteroids, comets, meteorites, gases, dust, and radiation.

The internal geometry of the system is defined by the sizes of the components, by the distances of the other components from the sun, and by orbital paths. Orbital distance for the planets ranges from about 60×10^6 km for Mercury, through about 150×10^6 km for Earth, to about 6000×10^6 km for Pluto.

The solar system probably originated as a kind of minigalaxy, a whirling lens-like concentration of gas and dust. The major components appeared about 4500 million years ago.

Analysis of orbital speed and orbital radius suggests that *the asteroid belt may represent an unfinished planet,* and that the planet *Pluto may have been captured by the sun from a satellite orbit round Neptune.*

The *spin axes of the planets are inclined* to their plane of orbit. *For Earth, this fact is of fundamental climatic importance.*

The *earth–atmosphere system is open in respect of the transfer of energy*, but for practical purposes *closed in respect of the transfer of matter*. This system is *special, as including an atmosphere low in carbon dioxide, plus a world ocean*.

The *earth's interior is explored* with the aid of recorded *earthquake waves*. It includes a *core, liquid at least in its outer part*, probably of *iron plus some nickel and sulfur or silicon*. The *radius of the core is about 3000 km*. Outside the core comes the *mantle, which acts as if it were solid*, transmitting shake waves, but can *also act as if it flowed*.

The *earth's crust acts as if it were rigid*. It consists of *oceanic crust, chemically comparable* to the rock *basalt*, which is rich in *compounds of silica with iron and magnesium*, and the *continental crust* which is *rich in silica itself*, and in *compounds of silica with aluminium*. The *crust* is only *a thin skin*. The *continental parts* stand high because of their *low density*: they are *buoyed up by the mantle*.

The *continental rafts* consist of *cratons, orogens* and *sedimentary covers. Cratons, constructed between 4000 million and 600 million years ago*, include *extensive granites dating between 4000 and 2500 million years in age*, plus *welded-on edges dating from 2500 million to 600 million years old*.

Mountain belts, now very largely destroyed, *were added between 2500 and 600 million years ago. Younger belts in part survive*.

The break at about 600 million years ago may relate merely to a sudden elaboration of life forms.

Sedimentary covers considerably *obscure* the record of the *cratons* and the *older orogens*. But they form *integral parts of rock cascade systems*.

3 Rock Material Cascades: Igneous

Rock material cascades involve material which is physically and chemically complex. Rocks, by definition, are constituents of the earth's crust. They are broadly classified into igneous – solidified from a melt; sedimentary – derived from the breakdown of older rocks; and metamorphic – altered from their original characters. All rocks consist of rock-forming minerals, which are chemical substances, mainly compounds. In most natural circumstances, these substances assume crystal structures. Their component atoms arrange themselves in distinctive three-dimensional patterns (Chapter 25). If the crystals are free to grow, they assume distinctive crystalline shapes (Fig. 3.1).

In one sense, the rock material cascades begin with the original formation of an earth crust, and with the outgassing which, providing the original atmosphere and the original oceans, set in motion the forces of weathering and erosion. But subsequently, a great deal of rock material has been subjected to recycling. Because a recycling system is by definition closed in respect of the recycled material, it does not matter which point of entry is chosen to begin description (Fig. 3.2). On the other hand, sediments can be regarded as ultimately derived from igneous and metamorphic rocks, and metamorphic rocks as ultimately derived from earlier igneous and sedimentary rocks. Thus, igneous rocks provide a suitable starting-point for the rock cascades generally.

Some recycling systems merely propel their components along a closed path. A heating system using hot water acts as a carrier of heat. Applied heat causes the water to rise. When the heat is released, the cooled water returns to its starting point. The blood in the human body acts as a carrier of heat and oxygen. It moves outward from the heart through arteries and returns through veins. Neither in the hot water system nor in the human circulatory system does the recycled fluid undergo long-term or progressive change.

The closest parallel among environmental systems

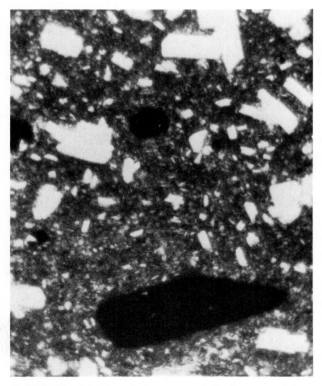

Fig. 3.1 Crystal shapes in igneous rock: the larger crystals are about 0.01 mm long (R. O'Brien).

Fig. 3.2 The rock recycling system, extremely simplified: compare Fig. 4.23.

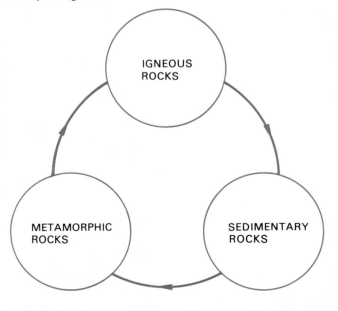

is the movement of water through the hydrologic cycle, or the transport of heat energy by air and ocean currents (Chapters 5, 16, 17). There is a quite close parallel among the rock material cascades, in the repeated re-use of sand grains to form new sandstones. The igneous cascade, however, resembles recycling systems in which the materials involved are restored to some former state.

Scrap iron is melted down to make new iron and steel; waste paper is used in the manufacture of fresh paper or paper board; recycling of glass bottles often involves crushing and melting. Parallels in the rock material cascade are the breakdown of sedimentary rocks to supply sediments of some contrasted physical character; the constant formation and destruction of ocean crust; and, at the extreme, the wholesale fusion of sedimentary or metamorphic rocks to supply new igneous melts.

3.1 The cascade of ocean crust

Ocean crust is created in mid-ocean. It forms the morphological systems of ocean floors. It is eventually destroyed, mainly near certain continental margins. Inputs are complex. They involve the separation of distinctive compounds from a freezing melt, in the form of crystalline assemblages. Outputs are also complex, in a reverse or negative sense. Ocean crust is destroyed by heating. It is returned to the molten state, wherein the chemical combinations of the frozen rock, and the crystal structures and shapes, no longer exist.

Whereas some portions of the continental crust are as much as 4000 million years old, and considerable portions are older than 2000 million years, ocean crust more than 250 million years old is a rarity. The rate of throughput for the ocean floors is clearly much greater than the corresponding rate for continental masses.

3.1a Creation of ocean crust

New ocean crust is created at mid-ocean rifts, narrow belts where the ocean floor is split apart (Fig. 3.3). Molten material from the outer mantle, rising into the rifts, freezes on to their margins in the form of new crust. Apart from combined oxygen, the added material consists of about one-half silicon, Si; one-fifth iron and magnesium together, Fe and Mg; one-sixth aluminium, Al; one-eighth calcium, sodium and potassium, Ca, Na and K; and a remaining small fraction of other substances.

Freezing at about 1250°C, the melt forms crystals. These are small, because cooling is rapid: there is no time for large crystals to grow (Fig. 3.4). About 80%

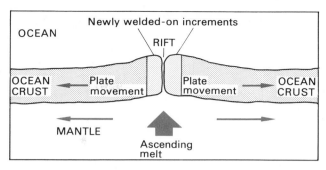

Fig. 3.3 Creation of ocean crust at a mid-ocean rift.

will be crystals of plagioclase feldspar (= skew-splitting common crystal), an aluminosilicate of sodium–calcium, and about 20% pyroxenes, which are complex aluminosilicates of calcium, iron and magnesium. The dark tints of the pyroxenes make the resulting rock look black. It is the rock basalt, of which ocean crust is composed.

For the purpose of considering the ocean crust cascade, it is enough to bear in mind that the concentration of iron and magnesium is far greater than in continental crust, and that oceanic crust is, in consequence, denser than continental crust.

Some of the crystals in the basalt melt are susceptible to earth magnetism. Still floating in the cooling mush, they are able to align themselves on the local magnetic direction, before the mush freezes into solid rock. They point to the north and south

Fig. 3.4 Tiny crystals in basalt: lengths up to 0.001 mm (R. O'Brien).

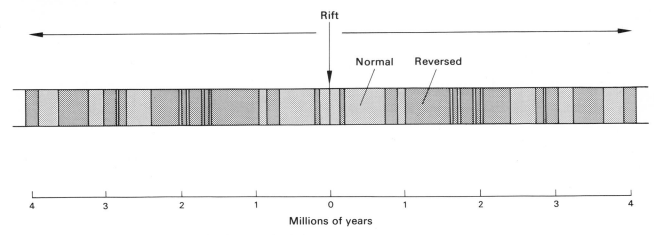

Fig. 3.5 Mirror-image array of normal and reversed polarity on the two sides of a mid-ocean rift.

magnetic poles. They also point down into the earth's body, at an angle depending on the magnetic latitude. The local magnetic directions are fixed into the rock when final freezing takes place. The rock possesses the attribute of remanent (= remaining, residual) magnetism.

3.1b The record of magnetic reversals

The remanent magnetism of most ocean crust formed during the last 750 000 years shows the north and south magnetic poles more or less where they are today. But ocean crust formed between 750 000 and 1 500 000 million years ago shows the north magnetic pole near the position which we call the south magnetic pole. We call the existing arrangement of magnetic poles normal, and the arrangement of 750 000 to 1 500 000 years ago reversed.

There have been repeated reversals of magnetic polarity in the geologic past. Hence, the magnetic record alongside a mid-ocean rift is one of alternating normal and reversed polarity. The array on one side of the rift is a mirror image of the array on the other (Fig. 3.5). Dating of the basalts by analysis of their radioactive components makes possible the calculation of the rate at which new crust is created – or, as it is usually put, the rate of ocean-floor spreading. Although there is some variation of rate from one area to another, a reasonable average is 2 cm/yr. This implies a widening of the average rift by 4 cm/yr. It also implies that the average slab of ocean crust could move sideways from the rift, through 4000 km in 200 million years. Here is the reason that ocean crust as a whole is so young. Before it can attain any great geological age, it has been destroyed.

The indications of the magnetic record are confirmed by the history of sedimentation in the ocean depths. Sediments derived from the world's lands lodge mainly close to the land borders, while lime-fixing marine organisms also live mainly close to land. In the North Pacific ocean in particular, density currents surge down submarine canyons and spread thin sheets of sand and mud over the ocean floor, but elsewhere the chief contribution is made by the remains of tiny marine organisms which live and die near the ocean surface. The hard parts of these organisms, consisting either of calcium carbonate or of silica, accumulate as the oozes of extensive ocean floors. In comparison with what happens near the margins of the lands, deposition of sediment in the open oceans is slow. Nevertheless, the fossil record is detailed enough to show that the oldest deposits are confined to the marginal areas of the oceans, whereas the central parts only contain material of lesser age (Fig. 3.6). The floors in mid-ocean did not exist when the older sediments were being laid down.

3.1c The drive of ocean-floor spreading

It is generally agreed that the primary drive of ocean-floor spreading is the ascent of convection currents in the mantle. The required energy is supplied partly by the earth's original heat, but chiefly by heat generated by radioactivity within the mantle. The convection currents are seen as forming a pattern of cells. Mid-ocean rifts occur at the top of rising cell walls. Not only is the ocean floor split open here: the rate of escape of heat through the crust is unusually high. But exactly how the spreading crust is moved remains uncertain.

The outward flow of currents in the mantle, away from the mid-ocean rifts, is inferred to average about 10 cm/yr, about five times the rate of crustal spreading. It might be thought that the crust is dragged along by friction. However, an alternative view is that the crust is actually sliding, being pulled by the weight of the descending limb where this sinks deeply down.

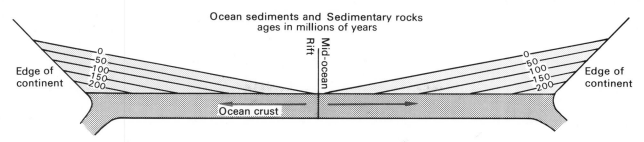

Fig. 3.6 Abstract model of the typical ages of sediment in an ocean basin.

3.1d Destruction of ocean crust

Ocean crust is destroyed in subduction zones, the counterparts of mid-ocean rifts. In a subduction zone, the edge of a plate composed of ocean crust plus the underlying rocky shell of the outermost mantle descends, either beneath the edge of a similar plate in mid-ocean, or beneath the margin of a continental mass (Fig. 3.7). Eventually the descending limb will be melted and reabsorbed in the molten outer mantle, except that a small fraction may go to supply the heat, and all or part of the molten rock, of volcanic chains.

3.1e Some consequences of morphology and of process-response

Subduction zones are expressed at the surface by ocean trenches (Fig. 3.7). Whereas ocean floors stand between 4 and 5 km below sea level, trenches descend to 10 km or more. A number of trenches, and the lines of volcanoes with which they are

associated, are arcuate in plan. We conclude that numbers of the descending limbs are planar: for a plane obliquely intersecting a sphere will trace an arc on the sphere's surface. The gentler the angle of descent, the tighter the arc of the ocean trench. A representative range of initial descent runs from 5 to 20° below the horizontal. However, arcs of volcanic islands typically lie within 400 to 500 km of the trenches where subduction begins. If the limbs merely bent down into descending planes, the volcanoes could be much farther distant. We infer that the limbs curve downward as they sink, with an average inclination of about 33°.

The paths of descending limbs are well documented by the records of deep earthquakes (Fig. 3.7). The place where an earthquake is set off is the focus. The foci of deep earthquakes in a subduction zone descend obliquely below the level of the ocean floor, along the path of the descending limb, which until it actually melts is liable to abrupt movements. The absence of earthquake foci below a maximum of some 700 km indicates the depth of final melting.

Fig. 3.7 Subduction zones in mid-ocean and at a continental margin.

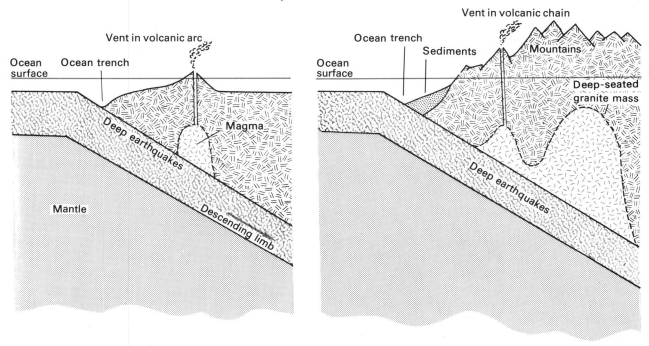

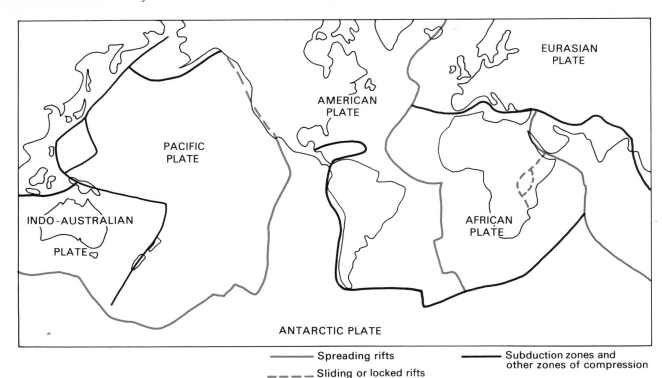

—————— Spreading rifts   —————— Subduction zones and
                                                                          other zones of compression
– – – – – Sliding or locked rifts

3.2 Plate tectonics

The combined distribution of earthquakes and subduction zones defines a morphological system of
crustal plates (Fig. 3.8). The plates move differentially. They expand at mid-ocean rifts, are consumed
in subduction zones, and slide against one another
elsewhere. The study of crustal plates and their
movements constitutes the science of plate tectonics.

Lines of sliding are transcurrent (side-slipping)
faults. They occur where two adjacent plates, instead of developing override and underride, rotate
in opposite directions. The Pacific plate is sliding
northward, relatively speaking, against the margin
of the North American plate. The resulting system
of transcurrent faults includes the notorious San
Andreas Fault of southern California, where a sudden movement in 1906 destroyed the city of San
Francisco. During the last 150 million years, slip on
the San Andreas Fault has amounted to 600 km, an
average of 0.4 cm/yr. But the rate has been increasing. The maximum slip in the 1906 earthquake was
about 6 m. Since then, the sides of the fault have
already bent as much as 3 m, indicating a relative
movement of some 4 cm/yr. An opposing movement
of 2 cm/yr on the two sides of the fault is by no means
excessive, in terms of the average rate of ocean-floor
spreading. The implications, however, are appalling.
It is impossible to believe that an earthquake shock,
broadly comparable to the shock of 1906, will not
take place. The longer the delay, the more severe
will be the event. On present showing, forces
enough to repeat the 1906 disaster will have built up

by about the end of the century.

Continental masses seem to play a rather passive
role in plate tectonics. They form light rafts with thin
underlayers of oceanic crust, appearing to be borne
along by the forces of spreading. Some pairs of
continental masses drift apart, other pairs collide.
As soon as reasonably good maps of the whole
Atlantic Ocean became available – that is, during
the 1500s – the rough jigsaw fit of the two sides of the
ocean was identified (Fig. 3.9). Hence the idea that
the two arrays of landmass had once been united.
Fits are best when the submerged edges of the
continental masses are used, instead of existing
shorelines; and it is necessary to exclude such geologic latecomers as the Caribbean Islands. Spreading
of the floor of the Atlantic Ocean accounts for the
separation of North America from Europe, and of
South America from Africa. Not only do the continental outlines match: distributions of rocks also fit
together, when the intervening ocean is closed up.

Reassembly of North America and Eurasia produces the former northern supercontinent of Laurasia. Reassembly of South America, Africa, India,
Australia and Antarctica produces the former southern supercontinent of Gondwanaland (Fig. 3.10).
During the last 150 million years or so, the chief
movements have been the sundering of the Atlantic
Ocean, plus the northward shift of all the southern
continents except Antarctica. Rates of northward
shift have been considerably variable. They have led
to the construction of island arcs in the Caribbean,
and to the considerable compression of the Mediterranean, with the Straits of Gibraltar being closed

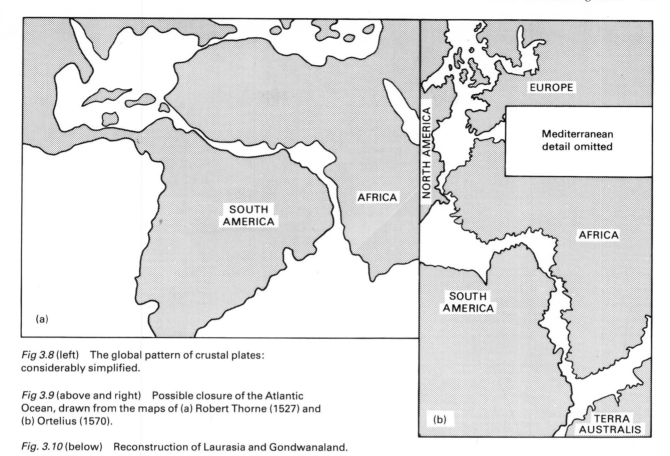

Fig 3.8 (left)   The global pattern of crustal plates: considerably simplified.

Fig 3.9 (above and right)   Possible closure of the Atlantic Ocean, drawn from the maps of (a) Robert Thorne (1527) and (b) Ortelius (1570).

Fig. 3.10 (below)   Reconstruction of Laurasia and Gondwanaland.

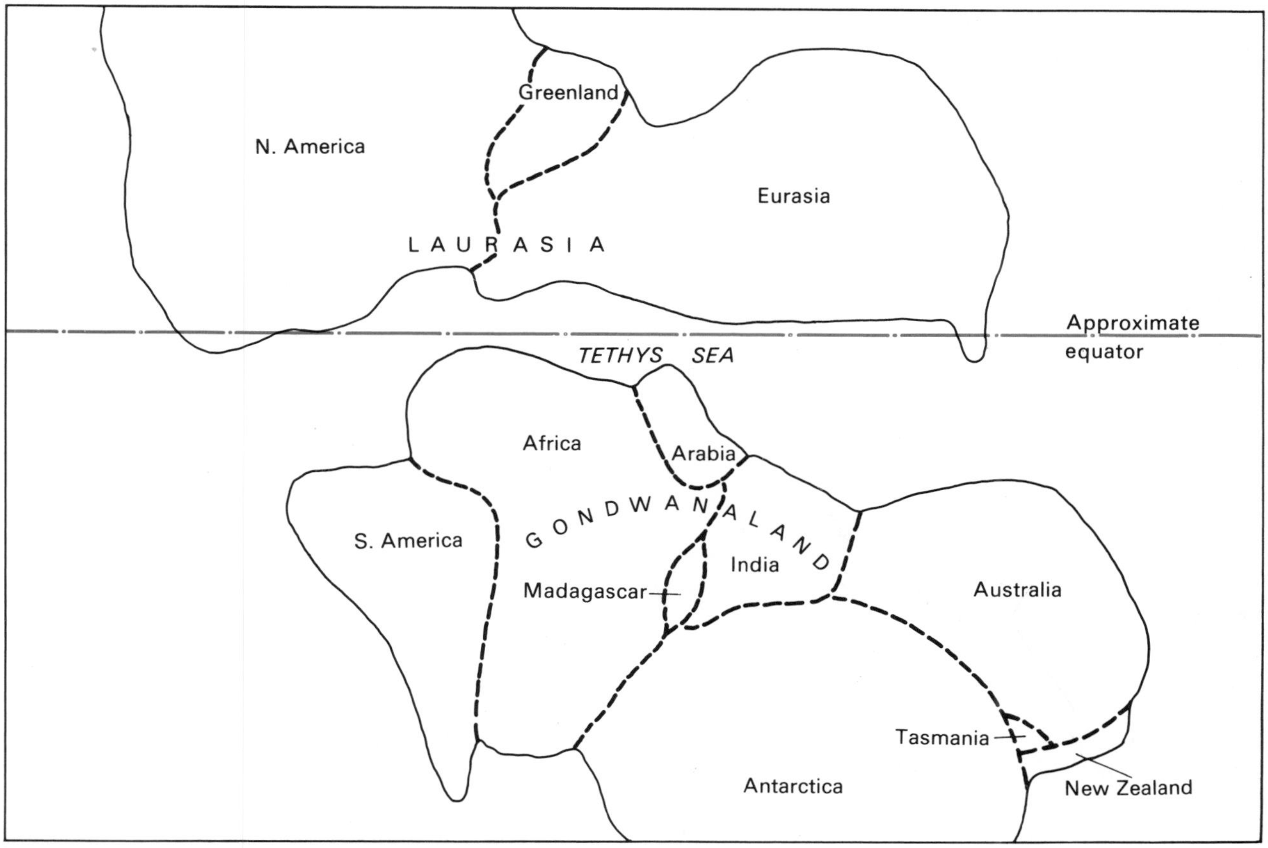

from time to time. The Caribbean area to a considerable degree, and the Mediterranean to some extent, represent the warm tropical sea of Tethys which once separated Gondwanaland from Laurasia. But in the existing Asian continent, this sea has been crushed out of existence. Its sediments stretch through the orogenic belt of Asia Minor and the Himalayas, and onward into China. Beneath the highlands of the Himalayas and Tibet, the Indian landmass is being forced beneath the Asian plate.

3.2a Changes in the pattern of plate movement

The spreading of ocean floors, away from mid-ocean rifts and toward the edges of continental rafts, can be taken as tending to perpetuate the contrast between deep oceans and high-standing lands. There is no mechanism whereby continental rock material can escape, in very large quantities, into the oceanic rock systems. But the assembly of the northern landmasses into Laurasia, the assembly of the southern landmasses into Gondwanaland, and the combination of Laurasia and Gondwanaland, about 150 million years ago, into the colossal supercontinent Pangea, demands something more. Specifically, it demands a pattern of convection cells in the mantle, which could concentrate continental crust into two single areas, or into one single area, of the earth's surface. Conversely, the breakup of supercontinents suggests a change in the convection pattern, capable of disrupting the initial concentration.

Changes in convection patterns are certainly suggested by Africa. To the northeast, the land of the east Mediterranean border is split by the Jordan Valley rift. The Arabian block has shifted away from Africa, letting ocean water into the Red Sea. Beneath the volcanic rocks of Ethiopia the rift system is obscured; but it persists, through 3000 km of length, through the rift valleys of eastern Africa. One may suspect that the African landmass is due to spread apart, the eastern portion moving out toward the Indian Ocean. But apart from a few volcanoes, the east African rift system seems to be inactive for the time being. It is as if a current system, tending at first to promote splitting, has ceased to function, or at least to function effectively.

Farther back in the past, we wonder about the landmasses ancestral to Pangea, Laurasia and Gondwanaland. It seems unlikely, on general grounds, that the only major disruption of supercontinents was confined to the last 150 million years: some 3000 million years of earth history are available for earlier events. Clues, hitherto faint, are available. Some parts of some continental masses are thought to display the suturing together of contrasted rafts of continental crust. Pangea, Laurasia and Gondwana-

land, then, may be merely the latest and the best known of a series, in the repeated formation and breakup of major continental rafts.

3.2b Reconstructing former continental positions

Remanent magnetism, already discussed in relation to ocean crust, can be analysed also for rocks of the continental masses. It is most easily treated in regard to igneous rocks, which have been injected into the crust in liquid form, or which have flowed over the surface as lava. As with ocean crust, local magnetic directions are frozen into the rocks so formed.

Most is known about former polar directions. The farther back that analysis goes, the greater the discrepancy of direction between one landmass and another. About 250 million years ago, the magnetic north pole for North America lay somewhere in the Gobi Desert of eastern Asia. The corresponding pole for Europe lay in the North Pacific, in the general vicinity of the end of the Aleutian arc. That for Australia was not far off Newfoundland. The only way to resolve the differences is to conclude that, 250 million years ago, the continents had different relative positions from those of today. Observations on former latitudes confirm that the continents have shifted, relative to their own magnetic poles.

It is in this way that the differential movement of the continents is traced, and that the breakup of the former supercontinents has been recorded. Future work of the same kind could establish earlier patterns of supercontinents, which resulted from earlier patterns of convection currents in the mantle.

Processes at work in the mantle are suspected of being responsible for certain major changes in the relative extent of land and sea. At the present day, any gently-sloping continental margins are submerged as continental shelves. During some past times, submergence was far more extensive. Marine sediments were widely deposited on what are now landmasses. Submergence seems to have peaked at fairly regular intervals, about 100, 300 and 500 million years ago. The earlier record is obscure. Also obscure are the mechanics of extensive submergence.

Discussion of this entire matter is complicated by evidence that our planet is slowly expanding. Astronomical observations show that the length of day is very slowly increasing – at an average of one second in 50 000 years. On the human time scale, the increase is negligible. On the scale of geologic time, its implications are great. It can be read as indicating an increase in the diameter of the earth, at a rate of about one millimetre a year. Studies of the daily growth lines of fossil corals produces confirmation. About 400 million years ago, there were 400 days in the earth year, each day about twenty-two hours long.

Although expansion should tend to increase the capacity of ocean basins, and thus cause landmasses to emerge, its effects in the actual record are largely or completely concealed by the effects of the more powerful processes inferred for the mantle.

3.3 The granite cascade

Granite is a coarse-grained igneous rock, formed from a melt with a surplus of silica. Its coarse texture results from slow freezing at depth. The rocks around and above the melt serve as insulation, through which heat leaks only very slowly. Freezing begins at temperatures in the range 600 to 300°C. A large granitic body may need 50 million years to cool down completely.

Aside from combined oxygen, a granite melt consists of about three-quarters silicon and one-eighth aluminium, with lesser quantities of sodium, potassium, calcium, iron and magnesium. As cooling proceeds, the constituents combine with one another in distinctive chemical assemblages. First comes plagioclase feldspar, which we have already met in basalt, and which in some granites is rare to absent. Next come the black micas (= biotite). These combine potassium and iron–magnesium in aluminosilicate compounds that include the OH ion. This is a kind of reduced form of water (H_2O), deprived of one hydrogen (H) fraction. Black mica crystallises as very dark flakes. It is responsible for the appearance of speckled granites.

After the black micas comes orthoclase feldspar (= square-splitting common crystal). Because potassium atoms are considerably larger than atoms of sodium and calcium, and because of a distinct difference of freezing points, orthoclase and plagioclase feldspars are not interchangeable. In some granites, the orthoclase crystals stand out very boldly (Fig. 3.11).

Next appears white mica (=muscovite), which like the black micas is always flaky. It is an aluminosilicate of potassium plus the OH ion. As in the feldspars, so in the micas: potassium compounds appear late in the sequence.

Finally, only surplus silica (SiO_2) remains. It fills in the spaces, after feldspars and micas have assumed their crystal shapes. But although the silica cannot take on crystal form, it can adopt crystal structure. It becomes the mineral quartz.

The bulk composition of the world's sedimentary rocks closely resembles the bulk composition of granite. Hence the conclusion that sedimentary

Fig.3.11 Coarse-grained granite: the large crystals are orthoclases, some showing two-stage growth (R. O'Brien).

rocks are actually derived from granite, and the further conclusion that remelting of sediments in bulk is capable of producing new granitic magma. Alternatively, some rocks at least can be reconverted to granite under great heat and great pressure, without actually being melted.

The granite masses observed at the surface of the crust today occupy two main types of setting. The oldest masses belong to the cratons, where they are associated with altered rocks of similar bulk composition. Their history presumably includes the separation of light crystalline materials from the mass of a heavier and molten earth body. They, and the associated altered rocks, are criss-crossed by faults and major joints, in square or diamond patterns.

Other granites occur in the roots of orogens. A representative length for a large mass of this kind is 1000 km. Width of the exposed surface depends on how much of the cover rocks have been stripped off. Width underground can range, like length, to 1000 km. Depth potentially ranges up to 50 km. These masses, in many cases, have melted their way across the surrounding rocks, providing the firmest evidence for the origin of granite melts by the fusion of

Fig. 3.12   The Thingvellir rift, Iceland: view is to the north; downthrow on the west side is about 40 m (S. Thorarinsson).

rocks of other kinds. In many places, loose pieces of the cover rocks are actually incorporated in the granites. The necessary heat is supplied by the pressure of opposing crustal plates, and by the driving of mountain roots deep into the high temperatures of the outer mantle. Somewhat exceptionally, a granite melt has shouldered the surrounding rocks aside, rather as if bubble gum had been forced sideways and had bulged into a compressed sandwich.

Granite, then, enters the rock system as a rock melt. It appears to be the ultimate source of land sediments and of near-land marine sediments. Recycling is effected along a variety of paths. The extreme path restores sedimentary and altered rocks to the molten or mush state. As with basalt, so with granite: freezing means the formation of the separate chemical compounds which we recognise as rock-forming minerals. Melting converts the distinctive compounds back to an undifferentiated melt. The paths between solid and melt will be explored further below.

3.4 Delivery of rock melt to the earth's surface

Rock melts are delivered to the surface indirectly, when igneous rocks frozen underground are exposed by erosion, and are delivered directly by volcanic eruptions. The form and behaviour of volcanoes varies a great deal, according to the setting with respect to plate boundaries, and according to the type of melt supplied. In actuality, type of melt and setting with respect to plate boundaries cannot be clearly separated from one another.

3.4a Basic volcanoes

These erupt basaltic lava. The volcanoes of Hawaii, thought to be located on a drifting hot spot in the floor of the Pacific, have constructed a colossal complex of lava domes. The summits rise 10 000 m above the main floor of the Pacific, as high as Mount Everest, supported by the underdriven edge of the Indian block, rises above sea level. The relative eastward drift of the hot spot has allowed the westernmost volcanoes to become extinct, while those at the eastern end remain active. Kaui, in the west, lies in erosional ruins. Beyond it, still further west, lie dead, submerged, and highly eroded volcanoes. Mauna Loa and Kilauea, in the east, are active today. Beyond them, still further east, are volcanic centres which still have to push their vents above sea level.

Individual eruptions in the Hawaiian Islands are quiet by volcanic standards. They include the play of lava fountains which rise as high as 300 m from the vents, the emission of rapidly-flowing basalt lava, and the discharge of large quantities of basaltic dust: but they are not explosive.

Equally quiet on the volcanic scale are fissure eruptions, which also discharge basalt lava. The most obvious locale for fissure eruptions is where the crust splits widely enough, and deeply enough, for basaltic melt to be released from below. The best-known fissure eruptions are those recorded for quite recent historical times from Iceland, where the mid-Atlantic Rift is exposed to view (Fig. 3.12). But fissure eruptions from mid-ocean ridges seem to be exceptional. The most far-reaching effects of fissure eruptions have been produced on land.

The continental crust in the Pacific Northwest of the U.S.A., in the northwest of the British area, in Siberia and in southern Brazil, has in past times split wide enough, and often enough, to allow basalt lava

Fig. 3.13 Flood basalts: eroded edges of individual flows, at least eight in number, exposed in Grand Coulee, Washington. Emplaced about 15 000 000 years ago, the basalts spread over an area of 2×10^5 km^2, covering the previous landscape to depths as great as 1500 m (V.R. Baker).

to flood the entire countryside (Fig. 3.13). The Snake–Columbia basalt floods of the U.S. Pacific Northwest cover 600 000 km^2 of country to an average depth of 0.5 km. The most extensive flows of all, however, are those of the Deccan of peninsular India, which have an estimated volume of 700 000 km^3.

We can guess that the crustal splitting reflected changes in the underlying systems of convection currents in the mantle, and that it went deep enough to allow mantle material to rise to the surface. It is also necessary to suppose that the rising magma was lighter than the mantle from which it originated, forming some kind of chemical separation.

The third type of basaltic volcano is that of oceanic island arcs (Fig. 3.7). The descending limb of a plate of ocean crust, melting in a subduction zone, can only supply basaltic magma at depth and basaltic lava at the surface. But basaltic volcanoes in island arcs are less common than volcanoes which erupt lavas of mixed composition.

3.4b Volcanoes with intermediate lava

Although some 90% of the igneous rocks of the world consist either of granite – silica-rich material – or of basalt – silica-deficient material – a prominent array of volcanoes erupts lava of intermediate composition. This lava is richer in silica than basalt but less rich than granite. It is called andesite, after the material erupted by the volcanoes of the Andes in

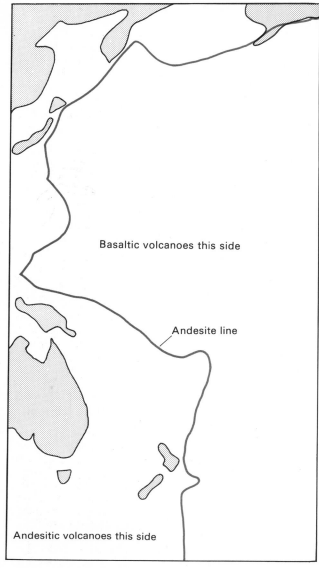

Fig. 3.14 The andesite line in the western Pacific.

Basaltic volcanoes this side

Andesite line

Andesitic volcanoes this side

South America. The volcanoes of the Pacific Ring of Fire are basaltic on the oceanic side, and andesitic toward and on the landmasses (Fig. 3.14). Andesitic melts are thought to result from a blending of basaltic melts with melts of rocks of continental origin.

Like basaltic volcanoes, andesitic volcanoes can erupt large amounts of dust. But, because their content of free silica lowers their freezing point, andesitic volcanoes are liable to plug their own vents, while sufficient pressure remains below to sustain further activity. Andesitic volcanoes, therefore, can be explosive. They can also discharge huge volumes of incandescent dust, such as destroyed the town of St. Pierre, on the French West Indian island of Martinique, in 1902. Thirty thousand people died in this calamity.

3.4c Volcanoes with acid lava

Like andesitic volcanoes, acid volcanoes are cone-builders. They too can discharge clouds of incandescent dust. But, with freezing points for their lava still lower than the freezing point for andesitic lava, they are especially liable to explosive activity. Their lava is extremely viscous and gassy, freezing at times into the rock pumice, in which drawn-out bubbles are clearly visible (Fig. 3.15). At the very extreme, acid volcanoes extrude masses of volcanic glass, as in the Puys group of the Central Plateau of France. From the composition of very acid lavas, it must be inferred that there is some connection with a body of granitic magma deep underground; but the means whereby granitic magma reaches the surface as very acid magma remains obscure.

3.4d Major delivery of acid igneous rock

Silica-rich magma, however, mainly fails to reach the surface of the ground. Instead, it freezes into deeply-buried granitic masses, entering the surface systems only when it is exposed by great uplift and deep erosion (Fig. 3.16). Only when erosion penetrates into the roots of an orogen, or attacks the deep foundations of a craton, is a major acid freeze exposed to recycling. But, as has already been seen, erosion has in fact gone far enough to feed granitic material back into the total rock cycle. However long the residence (= storage) times, the rock cascading systems are closed in respect of matter.

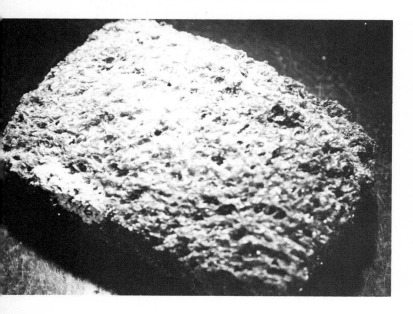

Fig. 3.15 Pumice: many bubbles are elongated from left to right (R. O'Brien).

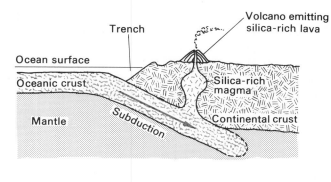

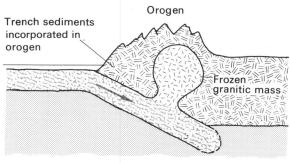

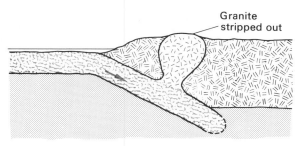

Fig. 3.16 Emplacement, uplift and erosion of a granitic mass.

Chapter Summary

Rock types are broadly classified as *igneous*, *sedimentary*, and *metamorphic* – frozen, deposited and altered. Rocks consist of *rock-forming minerals*, mainly chemical compounds, mainly with *crystal structures* and sometimes with *crystal shapes*. Rock material is subject to *constant recycling*.

Ocean crust is *created* at *mid-ocean rifts*. It is *consumed* in *subduction zones*. Mantle material *freezes* on to the *sides of mid-ocean rifts*, recording *magnetic directions*. It constitutes the rock *basalt*, which is *deficient in uncombined silica*.

By comparison with continental crust, *ocean crust is young*. It is remelted before it can attain great geologic age.

Spreading of the ocean floor away from mid-ocean rifts runs at about *2 cm/yr*. It transports the oceanic crust to *subduction zones*, which, marked by concentrations of *earthquakes* and by arcs of *volcanoes*, define the array of *crustal plates*. Some plates *slide* against one another along the lines of *transcurrent faults*. *Continental masses* appear to be *borne passively along* by the movement of plates, which is *driven* by *convection currents in the mantle*. *Former continental positions* can be determined from *former magnetic locations*. They indicate the assembly of the world's landmasses, *200 to 250 million years ago*, into *two supercontinents*, *Laurasia and Gondwanaland*, separated by the seaway *Tethys*, or into the *single supercontinent Pangea*.

Continental rocks are *similar in bulk composition* to the igneous rock *granite*, from which they are thought to be ultimately derived. Some granites appear to have originated by the *remelting of sedimentary and altered rocks*, others (in some parts of the cratons) to represent *original or at least early crust*.

Rock melt is *delivered* to the earth's surface either as the *lava of volcanoes* or as *frozen masses revealed by erosion*. The *chemistry of a melt* determines the *behaviour of volcanoes*. *Basaltic volcanoes* are *quiet* (= non-explosive). They occur at *hot spots* in the ocean floor, where continental crust is ripped by *deep fissures*, and in *some subduction zones*. Volcanoes erupting the *intermediate lava andesite* encircle the Pacific. They are *potentially explosive cone-builders*. Volcanoes erupting lava truly rich in silica are not particularly common. Most *silica-rich rock melt*, originating in part by the melting of existing rocks, becomes *granite*, *freezing at depth* and becoming *coarsely crystalline*, only *revealed at the surface* as a result of *deep erosion of a craton or an orogen*.

4 Rock Material Cascades: Sedimentary and Metamorphic

Igneous rocks crystallised from an original melt must be regarded as the original materials of the earth's crust. The existing array of igneous rocks, amounting to about two-thirds of the crust, includes rocks that have been re-melted and re-frozen. Basalts and related types account for some two-thirds of all igneous rocks, granites and related types for about another one-third. Only about 1% of the igneous rock content is left over for other rock types.

The one-third of the crust not composed of igneous rocks is made of sedimentary and metamorphic (altered) rocks. Metamorphic rocks, composing about a quarter of the entire crust, are more than three times as abundant as sedimentary rocks, which amount only to about one-twelfth of the whole.

4.1 The sedimentary cascade

Rock materials are stripped by erosion from the continental masses. They are transported by running water, by waves, wind and ice, and are redeposited in new places and in new forms.

Inputs to the sedimentary cascade are complex. They consist of rock waste, derived by a complex range of processes from pre-existing rocks. Transportation processes are complex, being highly variable through time, and often involving the storage of portions of the cascade. Outputs are also complex, resulting in an impressive array of end products. But in the broad view, the operation of the sedimentary cascade is simple. The path of input, throughput and output has already been summarised – weathering, erosion, and transportation and deposition (Fig. 4.1). Parts of the deposited output are cycled back into the crystalline rocks where the cascade begins.

The processes of rock destruction are the processes of weathering. Broken, rotted, or dissolved rock material moves downslope under the pull of gravity. Such material is rock waste. It ranges in coarseness and consistency from huge blocks to liquids. The net movement of rock waste is from the lands to the seas. But even when the cascade reaches the sea, the rock waste is not lost permanently to the lands. Sediment deposited on the sea bed, beneath the reach of wave action, goes to form new rock.

Fig. 4.1 Flow chart for sediments and solutes.

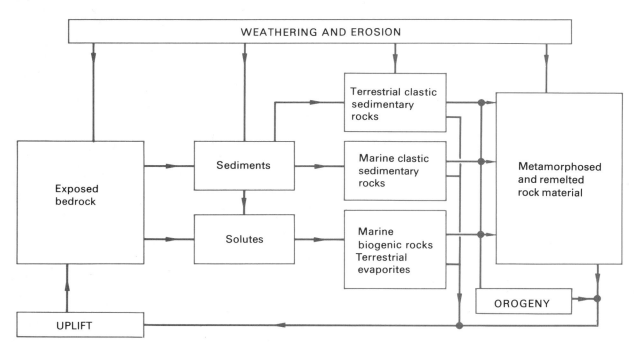

Dissolved materials are fixed into solid form by marine organisms, mainly close to the land margins. New rocks of all kinds, that are formed in the sea, will eventually be returned to the lands by sea-floor spreading, by orogeny, or by simple uplift. Some of the rock waste that enters storage on land can itself harden into new sedimentary rocks.

The relative importance of the components of the sedimentary cascade is roughly indicated by the total bulk of end products. About one-quarter of the world's sedimentary rocks by bulk are sandstones and related types. About one-half are shales. The remaining quarter consists chiefly of carbonate rocks, but there is a minor fraction of evaporites, which are precipitated from brines. In all cases, the attributes of sedimentary rocks tell a great deal about the environments of deposition. In many cases, they tell a great deal about the ecosystems of the past.

4.2 Setting the cascade in motion: rock rotting

Because the sedimentary cascade involves much recycling, and because the cascade as a whole can be viewed as a recycling system, it does not matter, logically speaking, which point of entry is chosen to begin a study. However, no sedimentary cascade operated at all, before atmosphere and oceans came into being. We begin, therefore, with the attack of weathering on igneous rocks.

The minerals in igneous rocks rot in direct order of crystallisation. That is to say, those which froze at the highest temperatures and pressures are the least stable at the low temperatures and pressures of the earth's surface. They are liable to attack by carbonated rainwater (rainwater + dissolved CO_2 from the atmosphere), by oxygen derived from rainwater, and by rainwater itself, which can enter into combination or can serve as a kind of catalyst for chemical reactions.

The plagioclase feldspars in granite are the first to go. Their combined sodium–calcium is lost in solution. It goes into soil moisture, or into rivers and thence into the sea. The remaining minerals combine with water (actually, with the OH ion) to form hydrated aluminosilicates. These are clay minerals. Black mica, if present, also decays early. Its magnesium is liable to early removal, but its iron can combine with oxygen (derived from water) in the stable compounds which we recognise as the familiar rust. Remaining materials again combine with OH to form clay minerals. However, the plagioclase feldspars and the black micas are not, on the whole, prominent constituents of granite. It is the range from orthoclase feldspar to quartz which most

Fig. 4.2 Rotten granite: the feldspars are breaking down, releasing quartz particles as a kind of coarse grit.

clearly illustrates contrasting resistance to chemical attack.

Within this range, orthoclase feldspar goes first (Fig. 4.2). Like the sodium–calcium of plagioclase feldspars, its potassium can go out in solution; but some can be caught up in the formation of clay minerals (Fig. 4.3). One of the chief products of the rotting of granite, then, is clay. The white micas are far more stable, in the electrochemical sense, than is orthoclase feldspar, and can remain more or less intact while the feldspar is breaking down wholesale. Thus, a sheet of rock waste produced by the chemical destruction of granite can contain abundant flakes of white mica. If the micas do eventually break down, the products are once again clay minerals.

Quartz, the most stable mineral of them all, can remain unchanged in a wide variety of climatic conditions. It comes out of the rotting rock as coarse grit with highly irregular grains. Here, nevertheless, is the ultimate origin of sandy beaches and of desert dunes.

Weathering of granite in humid tropical climates goes considerably further than the production of the familiar clays and sands of middle latitudes. We are thinking here of weathering in climates with mean annual temperatures of 18°C or more, and with at

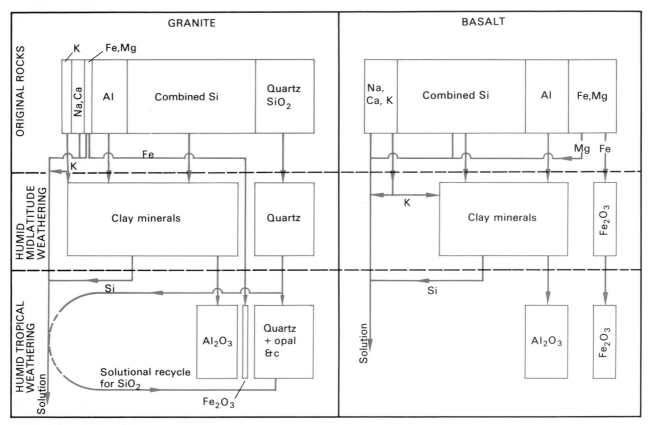

Fig. 4.3 Flow charts for the weathering of granite and basalt.

least 650 mm of precipitation a year, and often much more than this. In humid tropical conditions, the white micas are freely broken down into clays. Clay minerals are themselves liable to be broken down, their combined silica being removed. In conditions which are still not completely understood, the end product is a concentration either of the highly stable compound Al_2O_3, sesquioxide of aluminium, or of SiO_2 which has already been met as quartz. However, the processes of weathering in the humid tropics can include the mobilisation of quartz. Redeposition is common, producing quartz again, or the related (but hydrated) forms of opal and associated minerals. A provisional recycling loop for dissolved and reprecipitated silica has been inserted in the flow chart of Fig. 4.3.

The redeposited silica, if present, typically fills the pore spaces of a sandstone or conglomerate (= pebble-rock). Concentrated Al_2O_3 can form a surface crust. If rich enough, it becomes an economic deposit of the commercial mineral bauxite.

When basalts rot, two parallel series of reactions occur, one involving the plagioclase feldspars, the other involving silicates of iron–magnesium. The calcium-rich feldspars break down before their sodium-rich counterparts do. The variation in the rate of response among the iron–magnesium sili-

Fig. 4.4 Rotted columnar basalt, South Brazil: height of section about 15 m.

cates cannot be traced without a discussion of individual mineral chemistry, which has been deliberately excluded from this book. Suffice it to say that the iron–magnesium silicates are highly vulnerable, and in some cases extremely vulnerable, to weathering. The total effects, summarised in Fig. 4.3, are the formation of stable iron oxides and the production of clay minerals. As in the weathering of granite, sodium and calcium, magnesium, and some potassium are rapidly lost in solution. There is no free silica to provide quartz sand. Combined silica and aluminium unite to form clay minerals, while the considerable proportion of iron unites with oxygen, especially as the sesquioxide Fe_2O_3, the parallel compound to the already encountered sesquioxide of aluminium.

Again as with granite, intense weathering in the humid tropics can break down the clay minerals and release combined silica (Fig. 4.4). Thus, the end products of tropical weathering of iron-rich rocks are Fe_2O_3 and Al_2O_3, the respective sesquioxides of iron and aluminium. The iron compound, like the aluminium compound, is capable of being a commercially profitable metal ore.

4.3 Setting the cascade in motion: rock breaking

Rock rotting has been taken before rock breaking, because there are very few parts of the world where rocks are not subject to chemical attack. It is not possible to compare the total effect of rock rotting with the total effect of rock breaking. However, in certain settings, the effects of rock breaking are spectacular.

Frost in frost-prone climates splits off angular fragments from exposed rock faces. Water, percolating into the rock and freezing into ice, exerts insupportable pressures. Fragments pried off by the expansion of ice accumulate in banks below the eroding face. The generic name for accumulations of loose fragments is talus (Fig. 4.5). Banks of talus assume slopes equal to the angle of rest (= angle of sliding friction) of the loose material, commonly close to 32°. This, as will appear subsequently

Fig. 4.5 Talus banked against the foot of Mount Edith Cavell, Jasper National Park, Alberta, Canada (Canadian Government Travel Bureau).

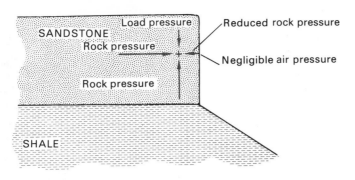

Fig. 4.6 Forces acting on a clifflike rock face.

(Chapter 9), turns out to be a kind of magic angle in slope evolution.

Frost action will continue to reduce loose blocks and particles of rock, down to about the silt grade of $\frac{1}{16}$ ($\frac{1}{2}^4$) mm diameter.

The largest individual masses supplied by rock breaking result from the relief of internal pressure. Just as the action of frost requires special conditions of climate, so does response to internal pressure require special conditions of rock exposure. The simplest condition is that where resistant rock is exposed on a hillside (Fig. 4.6). The external sideways pressure, the pressure of the atmosphere, can be neglected. The internal pressure, resulting from the original compaction of the rock, can be considerable. It can cause vertical slabs to peel off and fall. In this particular case, it can be very difficult to sepa-

rate the effects of pressure relief from those of frost.

Things are very different with the shedding of rock sheets (Fig. 4.7). Some rocky outcrops peel off along curved planes, leaving upstanding rounded hills. The mechanics involved are by no means certain. But it seems to be generally agreed that a sheet is detached by expansion. Erosion of covering rocks releases internal rock pressure. This, in turn, causes the affected rock to expand, and to rise in a surface sheet from the still sound rock beneath. Steeply rising hills and mountains produced by sheeting are most common on coarsely crystalline igneous rocks, but can also occur on sandstones and conglomerates.

4.3a Cascade components: rock fragments

Wherever rocks respond to weathering by breaking, loose pieces of rock are fed into the sedimentary cascade. To begin with they are usually angular, and often range considerably in size. They can for instance be as large as a house or as small as a golf ball.

They become rounded into boulders and gravel, mainly by the action of water (Fig. 4.8). Streams drag and bounce them against one another on channel beds. Waves on beaches pound them against one another, against the cliff, and on the sea bed. Fragments in transit get rapidly rounded off. Blocks too large to be moved have their angles battered

Fig. 4.7 Sheeting in granite, South Brazil: the main hillside is defined by a sheet surface, from which the overlying rock has slid; part of the sheet remains on the hilltop.

Fig. 4.8 Rounded riverbed pebbles exposed in borrow pit: Mulgrave River, Queensland, Australia (Australian Information Service: Bob Nicol).

away, and their bulk reduced, until they too can be fed into the subcascade of fragments in transit.

When eventually deposited, and cemented together (usually by calcium carbonate), gravels become the pebble-rock conglomerate (Figs. 4.1, 4.9).

4.3b Cascade components: sand

Sand consists, by definition, of small grains. We think of it as forming beaches, coastal and desert dunes, and the beds and banks of many rivers. Any fine-grained material will serve. Small crystals of gypsum ($CaSO_4 \cdot 2H_2O$) can be built into desert dunes. Some beaches, composed of grains of dark minerals, look black. Other beaches are not composed of rock waste in the strict sense, but of finely broken shell – shell hash. But most sand is quartz

Fig. 4.9 Gravel cemented into conglomerate: scale in cm (R. O'Brien).

Fig. 4.10 Sand grains photographed under the electron microscope; diameters about ⅕ to ½ mm (G.M. Habermann).

Fig. 4.11 Fine-grained sandstone, with the largest grains about 0.2 mm in diameter. Variation of tinting from white to grey results from photography in cross-polarised light, which makes cement filling the pore spaces appear black (R. O'Brien).

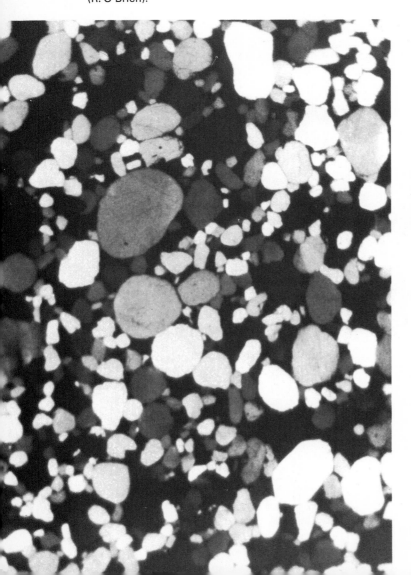

sand. We have already noticed its origin in the crystalline silica of granite. Throughout its history, which may be very many millions of years long, it retains its crystal structure, its chemical formula SiO_2, and its mineral name quartz.

The coarse grit weathered out of granite is chipped and rounded during carriage by wind or rivers, and is repeatedly beaten by waves on beaches. The grains become rounded (Fig. 4.10). A typical range of diameter is 2 mm to $\frac{1}{16}$ ($\frac{1}{2}^4$) mm. Above the 2 mm size we are likely to be dealing, not with individual grains, but with rock fragments. Below the 2 mm size a grain will rarely break, but it can be reduced by wear. At the $\frac{1}{16}$ mm size, more than 250 grains could be assembled in a flat sheet, on a surface 1 mm square.

The tint of grains of quartz sand ranges from clear to white, depending on the character of the original material. Quartz sand in the mass looks white, creamy, brown or red, depending on the amount of coating with iron oxide, and on the type of oxide.

Cemented together, sand grains make up the rock sandstone (Figs. 4.1, 4.11). The cement is often calcium carbonate, but cementing silica also occurs. It is quite common for some of the cementing silica to be taken up by the quartz grains themselves (Fig. 4.12). The regrowth partly compensates for the wear that has previously taken place.

In a real sense, all the quartz sand contained in the subcascade, between its release by the weathering of granite and its re-entry to the granite system by re-melting, can be regarded either as in transit or in storage. The length of storage varies from a few minutes or a few hours, as when beach surface sand comes to rest between waves or between tides, to hundreds of millions of years, as when ancient sandstones escape erosion. Desert sand may be liable to blow with every moderate wind, or may spend thousands of years in dead dunes. Considerable bulks of sand find their way into shallow sedimentary basins in continental interiors. The chief way in which quartz sand can be removed from the subcascade is however the alteration of sandstone by pressure (see below).

4.3c Cascade components: clays

The hydrated aluminosilicates of clay minerals crystallise into tiny flakes. We make a useful working distinction between the silt and clay grades. Silt particles range in diameter between $\frac{1}{16}$ and $\frac{1}{256}$ ($\frac{1}{2}^4$ and $\frac{1}{2}^8$) mm. They are prominent contributors to the dust of dry country. Raised by hooves or vehicles, they settle quickly down again unless the wind is strong. Clay particles are less than $\frac{1}{256}$ mm in diameter. More than 60 000 of them could be spread, without overlapping, on a square of 1 mm side. They can travel far.

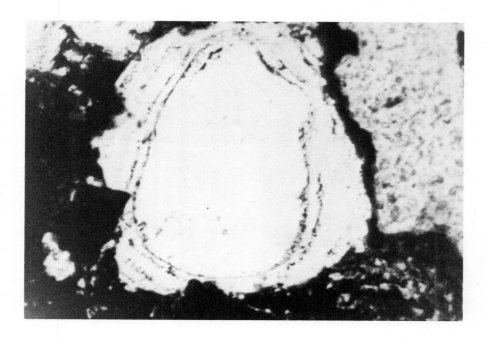

Fig. 4.12 (above)   Intermittent re-growth of quartz grains, recorded by dusting at intervals with iron minerals: diameter of grains is about 0.2 mm.

Fig. 4.13 (right)   Time taken to fall through one metre, in water and in air, by particles of various sizes. The distance/time ratio is the settling velocity.

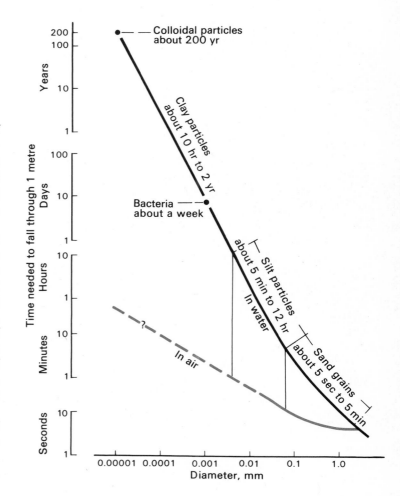

Fig. 4.13 graphs the time taken for particles of different sizes to fall through one metre. For convenience, it is plotted against logarithmic scales on both axes. These scales are used merely to extend the graph at the lower end, where it would otherwise be unreadable, and to compress it at the upper end, where it would otherwise shoot far off the paper.

In still water, sand falls in seconds to minutes, silt in minutes to hours, and clay particles fall in hours to years. The required time is of course less in air than in water. Not much is known of the behaviour of silt and clay particles in still air. In any event, the most notable travels are accomplished in turbulent air. Winds whip up silt from the low-flow channels of meltwater streams, depositing it as loess on the nearby land. Clay-sized dust from the interior of Australia reaches New Zealand, 3500 km distant. Similar dust from North Africa pollutes the air over the Caribbean, 10 000 km away.

Clay deposits form wherever mud settles out of water – in lakes, on river beds, on floodplains, in deltas, on muddy shorelines, and on the sea beds. Bedded clay is given the rock name shale (Fig. 4.1).

Fig. 4.14 Horizontal bedding: the dark beds are shale, which can itself be parted into very thin layers.

Shales accumulate most thickly on the floors of deep seas close to land margins. They can be almost as finely leaved as a telephone directory. Each thin bed is the product of one surge of muddy bottom current, set off at the margin of a continental block, either by the simple weight of accumulating mud, or by an earthquake. Any contained mica flakes, settling flat, help to accentuate the bedding (Fig. 4.14).

4.3d Cascade components: carbonates

Carbonate rocks consist essentially of calcium carbonate, $CaCO_3$. Their calcium is derived ultimately from the calcium-rich group of igneous rocks; it makes up about 3½% of the earth's crust. The oxygen comes partly from weathered rocks, and partly from the atmosphere and the oceans. In water and air, oxygen is abundantly combined with carbon as carbon dioxide, CO_2. The reactions involved are more elaborate than the simple $Ca + O + CO_2 \rightarrow CaCO_3$, but this is what they amount to in the end.

Carbonate rocks originate, in very large part, in the sea. Shellfish, corals, and a range of minute marine organisms use calcium carbonate to construct their hard parts. Where carbonate rocks have come into being, they can consist of consolidated shell banks, self-cemented lime muds, reef coral, or self-cemented shell hash (Figs. 4.1, 4.15). An abundant rain of the hard parts of tiny lime-using organisms produces the pure soft limestones known as chalks.

Some algae fix calcium carbonate, which in the fossil form preserves algal structures, hundreds of millions of years after the algae themselves died.

All these rock types, originating with construction by living creatures, are called biogenic. In special cases, calcium carbonate can be precipitated directly. In the lime-rich, warm water of the Gulf of Carpentaria, precipitated carbonate forms little round pellets. These are called ooliths – literally, eggstones – because they look like fish eggs. Like other accumulations of calcium carbonate, ooliths can be cemented into rock.

Limestone rock represents calcium carbonate in storage. For the time being, the material has been shunted out of the recycling system. Like other types of sedimentary rock, carbonate rocks can be uplifted as new land, and subjected to the weathering which promotes new inputs into the carbonate sub-cascade. But the weathering of carbonate rocks has attracted special attention, because of the storage subsystems which it often entails.

Although limestone is not directly soluble, it can be attacked by a combination of water and carbon dioxide. The water is supplied by rain, carbon dioxide by the air and the soil. Once again, the reactions are more elaborate than a simple equation would suggest, but they amount to $CaCO_3 + H_2O + CO_2 \rightarrow Ca(HCO_3)_2$, calcium bicarbonate, which is soluble. Underground solution of limestone produces cavities, including enormous caverns. Evaporation of lime-charged waters underground puts carbonate back into storage, in the form of the fantastic decorations which many caves contain (Fig. 4.16).

Fig. 4.15 (above)   Shelly limestone, with a matrix of lime mud (R. O'Brien).

Fig. 4.16 (right)   Cave decorations: stalactites hang down, stalagmites stand up. Buchan Caves, Victoria, Australia (Australian Information Service: Terry Rowe).

4.3e Cascade components: salts

These soluble materials are compounds of the metals sodium, calcium, potassium and magnesium. These metals are removed early and rapidly during the chemical weathering of igneous rocks; and they are also readily mobilised during the weathering of such other rocks in which they occur. They reach rivers and lakes, but principally the sea, as dissolved compounds.

They become rock-formers when they are precipitated out (Fig. 4.1). In order for precipitation to occur, the brines involved must become concentrated. Concentration is brought about by an excess of evaporation over inflow. Hence the general rock name evaporites for precipitated chemical compounds.

More than eighty chemical types of evaporite are known. The great bulk, however, consists either of sodium chloride or of calcium sulfates. We taste sodium chloride (NaCl) in sea water. We eat it as common salt. Its rock name is rock salt. Calcium sulfate comes either in the dehydrated form, $CaSO_4$, with the rock and mineral name anhydrite, or in the hydrated form, $CaSO_4 \cdot 2H_2O$, with the rock and mineral name gypsum. All three rock types are commercially important, as also are the much less abundant complex compounds of potassium. Rock salt is used in the chemical industry, sulfates are used in the making of plaster, and potassium salts form one of the main bases of the fertiliser industry.

Although $1\,km^3$ of average sea water contains thirty million tonnes of sodium chloride and $5\,km^3$ contains the equivalent of the world's annual consumption of salt, the concentration is low. Average sea water contains about 30‰ (= parts per thousand) of dissolved sodium chloride, and about 35‰ of dissolved salts of all kinds. To promote precipitation, the concentration must rise to at least 80‰; and, in some brines, the main precipitation does not occur until the concentration is well over 100‰. For instance, readings above 220‰ come from the Dead Sea, and the Great Salt Lake reaches 275‰ at low stages.

High concentrations can be attained only in enclosed, or almost enclosed, water bodies in regions of dry climate. Favourable settings include desert lakes, desert-bordered seas cut off from the main ocean by earth movement, and lagoons with shallow sills upon desert coasts. The Mediterranean Sea has been completely enclosed from time to time in the geologic past, as the African block thrust northward and shut the Straits of Gibraltar. Evaporation from the Mediterranean far exceeds inflow from the surrounding lands. Longterm closure of the Straits therefore results in the precipitation of salts on the sea bed.

It seems likely that the chief salt deposits of past ages were formed in subsiding basins, which connected to the open ocean across shallow barriers. The obvious present-day analog is the Gulf of Karabogaz, on the eastern side of the Caspian Sea. The Caspian is itself a sea closed in by earth movement. It last connected to the open ocean about 2 500 000 years ago. The input of fresh water by the Volga and Ural Rivers keeps the average salinity of the Caspian down to about one-third that of sea water, but concentrations rise to more than 300‰ in the desert-bordered Gulf of Karabogaz, where gypsum is being precipitated.

Because evaporites have been precipitated from solution, it follows that they can readily be redissolved if undersaturated waters can get at them. Those formed in past ages have been preserved by burial, typically under desert sediments.

4.3f Cascade components: carbon in plants

Plants use light energy to break the electrochemical bonds of water and carbon dioxide. Carbon, oxygen and hydrogen form carbohydrate compounds – the starches and sugars of cellulose. In a kind of two-stage reaction, some of the hydrogen of dissociated water is used by the plants, oxygen being released to the air. Here is the origin of the atmospheric oxygen, on which all our lives depend. Then, surplus hydrogen combines with surplus oxygen from carbon dioxide as water, which is again released to the air.

More energy is needed to break substances down than to build up new ones. The electrochemical charges of the breaking substances have to be overcome, while upbuilding uses charges already present. The difference is stored as chemical energy in the substance of the plants, especially in their carbon content. This is why wood and coal will burn.

Storage of plant material varies from as little as a year to very many millions. Plant material in a tropical forest undergoes continuous recycling. Certain cool moist climates are not too cold to prevent plant growth, but are cold enough to check bacterial decay: hence, on favourable sites, the growth of peat bogs. Most of these, being distinctly acid, help to keep the rate of decay down; and their rate of growth far exceeds the rate of chemical–bacterial destruction. In warm climates, something more is necessary. The most favourable site for the storage of plant material here is a swampy, peaty, intermittently sinking delta. The loading of sediment on to the delta causes the intermittent sinking. Each downward movement allows sediment to spread over the swamp and to bury the swamp. Many of the world's major coal deposits originated in this manner (Fig. 4.1). Others began in cool climates as peat bogs, now also buried.

4.3g Cascade components: hydrocarbons

Although hydrocarbons (= compounds of hydrogen and carbon) typically occur as gases or liquids, they do belong in the sedimentary cascade. As natural gas and mineral oils, they are contained in the pore spaces of sedimentary rocks. About 60% comes from the pore spaces of sandstones, about 40% from the less regular small cavities of limestones.

Whereas fossil plants in coals testify abundantly to the environment of formation, the precise origin of hydrocarbons remains uncertain. The best bet appears to be the fatty or the starchy/sugary content of the single-celled plants and animals that live close to the sea surface. But entry into storage requires, in almost every case, not only burial but an environment low in oxygen. The exception is the generation of the gas methane, which is produced by bacterial attack on vegetable matter. Even then, underground storage of methane, CH_4, requires burial. Other natural hydrocarbons have not only been buried (Fig. 4.1), but have been protected from bacterial decomposition by lack of oxygen.

The most likely environments of storage seem to be muddy beds of seas, lakes, or enclosed lagoons, offshore of muddy deltas. Enclosed basins naturally become oxygen-deficient. Fine-grained sediment is best fitted to protect organic remains from rapid decomposition. A prominent part may have been taken, in the formation of some deposits, by surging bottom currents of muddy water.

Since natural gas and mineral oil are so highly volatile, it may seem surprising that they do not readily escape. Some deposits do. Seepages of gas have been burning in Iraq for thousands of years. Seepages of much denser material form lakes of pitch, famous for their content of animal fossils. The hydrocarbons of the world's producing oilfields are contained in porous sandstones and limestones, which are sealed down by impermeable shales above. Oil shales, as yet little exploited, contain their hydrocarbons in pore spaces too small to permit migration.

4.4 Storage time for carbons and hydrocarbons

Oil deposits were accumulating more than 400 million years ago. Coal deposits began to accumulate later, when, in the neighbourhood of 350 million

years ago, forests first became extensive. The conversion of organic substances to mineral oil seems to take about 5000 years. The formation of coal takes longer. Coal composed of the debris of tropical forests, and 350 million years old, is the familiar black fuel, but can be sulfurous in some areas. Coal 250 million years old is sulfurous and smoky. Coal 100 to 50 million years old is soft. It has been incompletely converted from peat to rock.

There seems to be little pattern through time in the formation of oil and gas deposits, unless a pattern can be disentangled from the formation of the necessary enclosed basins. The formation of coal, on the other hand, seems to have been highly irregular through time, depending on the vast spread of delta swamps in warm humid climates, and of peat bogs in cool humid climates.

The tempo of extraction from storage is alarming. Depending on who is calculating the estimates, the world's reserves of oil and natural gas will become exhausted, or will near exhaustion, by the year 2000 or the year 2050. Discovery of additional reserves will put the exhaustion date forward. Nevertheless, even if we fix on 2069 as the precise exhaustion date, 200 years of usage since the sinking of the first commercial well will suffice to deplete the store of 400 million years. The current rate of depletion is two million times as great as the accumulation rate.

For coals, the prospect is mathematically brighter. The world's known coal reserves are enough for some centuries at present rates of consumption. But, just as oil reserves include oil shales, so do coal reserves include low-grade and dirty coals. If we reckon that the world's known coal reserves could last 350 years, we are still reckoning on a rate of consumption a million times as great as the rate of formation.

4.5 Throughput in the sedimentary cascade

Estimates of delivery of sediment to the sea by the world's rivers deal with suspended load (= silt and clay grades) and with dissolved load. Of the annual total of some 17 500 million tonnes, about three-quarters is suspended load. By comparison with the indicated total, the sum of coarser material carried by rivers, of material torn from the lands by waves or released by cliff fall, and of material carried in by glaciers is negligible.

Estimates based on present-day observations are however inflated by the effects of man's activities. In the absence of man, the annual delivery of sediment would be about 10 000 million tonnes a year. This rate cannot be taken as an average for geologic time, for the extents and the average heights of the

Fig. 4.17 Distribution of basic
igneous rocks, acid igneous
rocks, metamorphic rocks and
sedimentary rocks, in the crust
as a whole, and in the oceanic
and continental portions.
Sedimentary rocks in the subset
of oceanic crust should be
understood as merely resting
on the basic igneous crust
(compare Fig. 3.6).

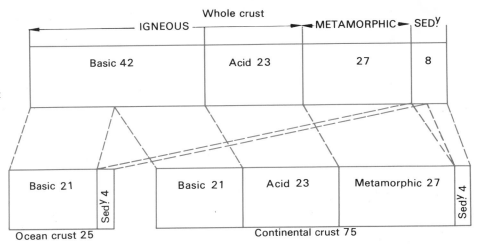

Fig. 4.17 Distribution of basic
igneous rocks, acid igneous
rocks, metamorphic rocks and
sedimentary rocks, in the crust
as a whole, and in the oceanic
and continental portions.
Sedimentary rocks in the subset
of oceanic crust should be
understood as merely resting
on the basic igneous crust
(compare Fig. 3.6).

continents have varied greatly in the long term. Nevertheless, the value does help to explain why some very old sedimentary rocks have escaped destruction. At 10 000 million tonnes a year, it would take 300 million years to put the existing continental masses, aside from their thin basaltic underlayers, through the sedimentary cascade.

4.6 The metamorphic cascade

Metamorphic rocks are rocks that have been drastically altered from their original characters. Alteration has been effected by pressure, by heat, or by the addition or removal of substances carried in gases or solutions.

Metamorphic rocks belong wholly to the continental masses, of which they constitute more than one-third. They are, in fact, rather more abundant in the continental blocks than either silica-rich or silica-poor igneous rocks, taken separately (Fig. 4.17). Since metamorphic rocks are usually far more durable than the original unaltered rocks, we conclude that metamorphism (= drastic alteration) is an important process whereby rock material is placed in storage.

About 80% of metamorphic rocks are gneisses (pronounced *nices*). Gneiss is a miner's term, mean-

Fig. 4.18 Gneiss: bulk chemistry and mineralogy resemble those of granite, but the different minerals have been largely segregated (R. O'Brien).

ing nest: many metamorphic rocks contain metal ores, the eggs in the rock nest. Gneisses are coarsely foliated. Their mineral constituents are roughly arranged in bands (Fig. 4.18). Gneisses amount to about one-fifth of the total crust, and to rather less than one-third of continental crust.

Schists (pronounced *shists*) are finely foliated. Their mineral constituents are arranged in thin wavy sheets (Fig. 4.19). They amount to about 15% of all metamorphic rocks. They are roughly as abundant as either sedimentary rocks on land, or sedimentary rocks and sediments in the ocean basins. The remaining 5% of metamorphic rocks is composed of quartzites, which are altered sandstones, or of marbles, which are altered carbonate rocks.

4.6a Alteration by heat

Lavas, flowing over the surface of the ground, can reach temperatures of 1000°C or more. They bake the rocks over which they flow. The processes involved are strictly comparable to those of kilns in which bricks or pottery are baked. The minerals of the baked materials recrystallise under the application of heat. The general effect is one of hardening. Similar effects are produced along the margins of small magma bodies that are injected into surrounding rocks.

Large-scale alteration by heat occurs at and near the margins of large magma bodies. With few excep-

Fig. 4.19 Schist: the rock tends to split along slightly wavy surfaces (R. O'Brien).

tions, these bodies consist of silica-rich magma which will eventually freeze into granite. Starting temperatures for freezing run between 600 and 1000°C, depending on depth and water content. The least possible effect of the slow escape of heat into the surrounding rocks is to bake these into hardness, just as clay is baked into brick in a furnace. Some rocks crystallise, or re-crystallise, completely.

The effects of simple heat, however, are in many ways much less impressive than the effects of the leftovers. Surplus magmatic liquid, surplus gases, and surplus solutions in water come off the freezing magma body, roughly in that order. They commonly include sulfur, which combines as sulfides with copper, tin, lead, zinc and silver. We are dealing here with a prominent group of metal ores. In addition, special minerals are formed, within a radius of 5 to 25 km of the margin of the magma. These include the gemstones emerald, ruby, tourmaline and garnet. The intensity of alteration, and the type of special mineral formed, changes with distance away from the magma.

4.6b Alteration by pressure

Whereas alteration by the heat and emanations that escape from a lava body is local, pressure can work on a regional scale. It is provided by the opposing movement of crustal plates, especially in orogens.

The effects of pressure vary, according to the intensity of the pressure applied. As with heating, the least possible effect is an increase in rock durability, which frequently means plain hardening.

Sandstone under pressure becomes quartzite. Its component grains pack completely together (Fig. 4.20). Although the original bedding may well be preserved, quartzite does not weather back into the original grains. It is extremely hard and highly durable in a wide range of chemical environments. Only the lattice of jointing – itself induced by pressure – allows quartzite to be attacked by weathering.

Carbonate rocks can be partly crystalline before alteration begins. Indeed, the fossil hard parts of organisms normally possess crystalline structure. Under pressure, and also under heat, the carbonate crystallises completely; and the rock becomes compact. Limestone turns into marble. A broken face of marble will twinkle when it is turned in the light, as individual crystals reflect into the observer's eye.

No new minerals are formed during the alteration of sandstone to quartzite, nor during the alteration of pure limestone to marble. Impurities in limestone can lead to highly decorative effects in marble; and alteration of limestone with a magnesium content (dolomitic limestone, chemically $CaMg(CO_3)_2$) can

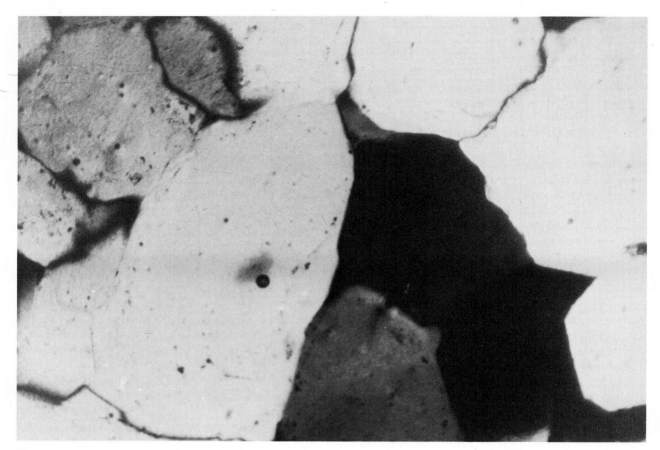

Fig. 4.20 Sandstone grains packed completely together in quartzite: mean diameter of grains is about 0.2 mm. Variation in tinting results from photography in cross-polarised light.

produce new minerals. But the most dramatic mineral changes occur during the powerful alteration of shale.

Moderate pressure causes shale to become hard. Hardened shale breaks into thick coarse slabs. More intense pressure, but still far less than maximum pressure, causes the contained mica flakes to align themselves at right angles to the direction of pressure, and also to start to grow. The rock cleaves (= splits) along parallel planes into thin slabs. It has become slate.

Still more intense pressure causes still further growth by the micas, and forces the parting surfaces to become wavy. The rock has become mica-schist (Fig. 4.21). Highly compressed mica-schists typically contain garnets, already noticed as one possible effect of heat metamorphism. They are complex silicates or aluminosilicates of magnesium, calcium, iron and manganese. Perhaps more than any other type of mineral, garnets are excellent indicators of intense metamorphism.

Basalts and allied igneous rocks also turn into schists under intense pressure. The flaky minerals developed in this group of schists are not micas, but the aluminosilicates of iron–magnesium which give basalt its rich dark tints in the first place.

4.4c Conversion to and from granite

Rocks which begin by being crystalline, and composed of contrasted kinds of mineral, can be made to flow under intense pressure. At the simplest, their minerals separate into coarse bands, which are alternately dark and light. The resulting rocks are gneisses.

It may be imagined that gneisses generally record the intense pressure metamorphism of granite, which has caused bands of feldspar, mica and quartz to separate out. Some gneisses have in fact developed in precisely this way; but they are not very common.

At the other extreme, the wholesale re-melting of assorted sedimentary rocks – or, more probably, metamorphic rocks – deep within an orogen is capable of producing new granitic magma.

In between these extremes, of pressure on the one hand and heat on the other, comes alteration by a combination of heat too little to cause complete melting, and pressure too little to produce wholesale flow-banding. The resulting rock is a very common type of gneiss, called migmatite (= mixed rock), which looks like older metamorphic rock invaded by newer granite. Migmatites are very common among the metamorphic rocks of cratons.

Fig. 4.22 summarises the location, on the combined scale of temperature and of depth/pressure, of basaltic and allied melts, granitic melts, and migma-

Fig. 4.21 (above)   Mica-schist with garnets (scale in cm) (R. O'Brien).

Fig. 4.22 (below)   Relationship between temperature and depth (as a surrogate for pressure) for major rock types: basic igneous rocks at the surface and at shallow depth omitted.

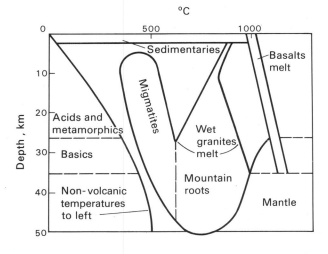

migmatites run below the temperatures for granite melts. These require something of the order of 800°C. Migmatites formed at temperatures of the order of 500°C. They can be regarded, therefore, as incompletely converted to granite.

One further consideration remains. Migmatites are richer in feldspars than would be possible if they had merely been altered from earlier rocks. We are bound to conclude that the raw material of feldspars has been brought in, either as liquids or gases, and has replaced part of the original rock substance. The addition of the feldspars makes it possible for the mixing process to convert any common rock type back to granite proper, given sufficient heat and pressure.

4.7 Repeated metamorphism

There is nothing to prevent a rock that has been metamorphosed once from being metamorphosed again. Just as sand grains can be involved in repeated recycling, so can metamorphic rocks be involved in repeated episodes of alteration. The disentangling of a metamorphic sequence is a matter of careful and highly technical detective work. It can reveal the record of repeated mountain-building. By its means, among others, the very early history of the movement of crustal plates is being unravelled.

tites. In all cases, the temperatures required are greater than the temperatures derived for non-volcanic regions. The melting temperatures increase with depth and pressure; and those for granite magma increase as water content decreases. But former temperatures determined for the cratonic

4.8 The complete rock cycling system

It will be clear by this time that the possible paths of rock material through the various systems of storage and recycling can range from very simple to highly elaborate. Fig. 4.23, itself a simplification, diagrammatically represents the interlock and interplay of flow through the complete rock recycling system.

Chapter Summary

The *sedimentary cascade* involves considerable *storage* and *recycling*. At the present rate of erosion, aside from the effect of man, the *continental rock masses* could be *put through the cascade in 300 million years*.

The *minerals of igneous rocks rot in direct order of crystallisation*. Feldspars and *micas* go to form *clay minerals*, *quartz* supplies *quartz sand*. Rocks *break* under the influence of *frost* and of *relief of pressure*.

Rock *fragments* become *rounded in transit*. *Cemented* together, rounded fragments make up *conglomerate*. *Sand grains* also become *rounded in transit*. They range mainly between *2 and ¹⁄₁₆ mm in diameter*. *Cemented* together, they compose the rock *sandstone*. *Clay minerals* are *hydrated aluminosilicates*. In the range ¹⁄₁₆ to ¹⁄₂₅₆ mm, the particles are called *silt* particles; *below ¹⁄₂₅₆*, they are *clay particles*. *Bedded deposits* of clay form the rock *shale*. *Carbonate rocks* consist chemically of *calcium carbonate* or of a *double carbonate* of *calcium* and *magnesium*. When *cemented*, carbonate material is the rock *limestone* or *dolomite*. The carbonate is initially *fixed by organisms*, mainly in *shallow seas*.

About *one-quarter* of the *world's sedimentary rocks* are *sandstones*. About *one-half* are *shales*. Most of the *remainder* consists of *carbonate rocks*, but a *small fraction* is *evaporites*, mainly *sodium chloride* and *calcium sulfates*. Evaporites are formed by *precipitation from brines*.

Plant carbon is stored as *solid fossil fuels*. *Organic hydrocarbons* are stored as *mineral oil* and *natural gas*. *Depletion rates* run at about *one million times storage rate* for *solid fuels*, and *two million* for *oil and gas*.

Metamorphic rocks have been altered by *heat*, by *pressure*, or by *both*. In some cases, *materials* have been *added or subtracted* by the action of *gases* and *solutions*. Metamorphic rocks make up *more than one-quarter of the entire crust*, and *more than one-third of the continental crust*. About *80%* of them are *gneisses*, which are *coarsely foliated*; *15%* are *schists*; and the *remainder* are *quartzites and marbles*.

Granitic magma underground *freezes* at *600 to 1000°C*. The released heat *bakes* the *surrounding rocks*, causes *crystallisation* or *re-crystallisation*, and promotes the formation of *special minerals*, including some *gemstones*. Surplus magma, gases and liquids produce materials which include *sulfide ores*.

Sandstone under pressure becomes *quartzite*; *limestone* becomes *marble*. Depending on the intensity of pressure, *shale* becomes *slate*, and slate eventually becomes *mica-schist*, often containing *garnets*. *Other schists* result from the *intense compression of basalts* and allied rocks. The iron–magnesium aluminosilicates of these rocks develop *flaky shapes*.

Gneisses can be formed by the intense compression of granite, just as granitic magma can be formed by the wholesale melting of sedimentary or metamorphic rocks. Very many *gneisses*, however, have formed at *pressures too low to* induce *complete foliation*, and at *temperatures too low* to induce *complete remelting*. Such gneisses are *migmatites*: they look like older metamorphic rock invaded by newer granite. Developed at *temperatures* of the order of *500°C*, they record the *addition of feldspar-building substances*, and represent the *incomplete conversion* of earlier rocks *to granite*.

The analysis of *repeated metamorphism* reveals the history of *repeated mountain-building*, and is helping to sort out the *early movements of crustal plates*.

Fig. 4.23 The rock recycling system.

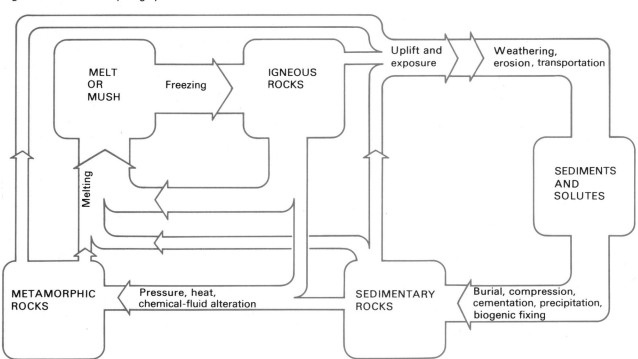

5 The Hydrologic Cascade: Storage and Recyling

Hydrology is the science of water. The present chapter deals with that cascading natural system which experiences a water throughput. The reason why the chapter title cannot be The Water Cascade, in parallel with the Rock Material Cascades of Chapters 3 and 4, will be self-evident.

This chapter will pay considerable attention to rates of throughput for the various subsystems of the total water system. It will therefore be concerned with the regulators that are characteristic of cascading systems in general, and also with the storage times that result from the operation of the regulators. Regulators of the hydrologic subcascades are far better understood than are the regulators of the rock material subcascades, in the sense that they can be the more easily measured. Because every output from any given subcascade of water constitutes an input into the next following subcascade, we shall be dealing throughout with the recycling process.

We tend to think of water in vague terms of rain and rivers – if not, indeed, in terms of rain and reservoirs. Rain nourishes growing plants. Reservoirs, whether man-made on the surface or provided underground by nature, supply water for domestic and industrial use. But the amount of water cycled through rainstorms and pipes, vital though it is to man's life on earth, is only a tiny fraction of the water budget of the earth.

5.1 The water budget

The water budget is the total of the world's water, distributed among various containing subsystems. It includes water in the air, on the land, and in the oceans, in the three possible – vapour, liquid, and solid (= frozen) – states. By comparison with the total mass of the earth, the sum of water is extremely small – only about one part in four thousand five hundred, 1×2^{-4}. Biological water, contained in the living substances of plants and animals, and making up a large fraction of our own bodies, amounts to so little as to be negligible in any general sum.

For all its limited bulk, water nevertheless covers 71% of the surface of the planet. Without it, weather as we know it would not occur: nor would the life forms with which we are familiar – including ourselves – exist.

Fig. 5.1 summarises the distribution of water

Fig 5.1 Distribution of water among subcascades: the right-hand block relates to water other than oceanic water.

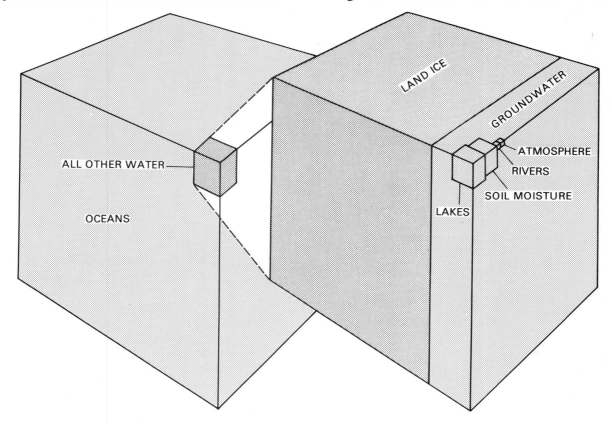

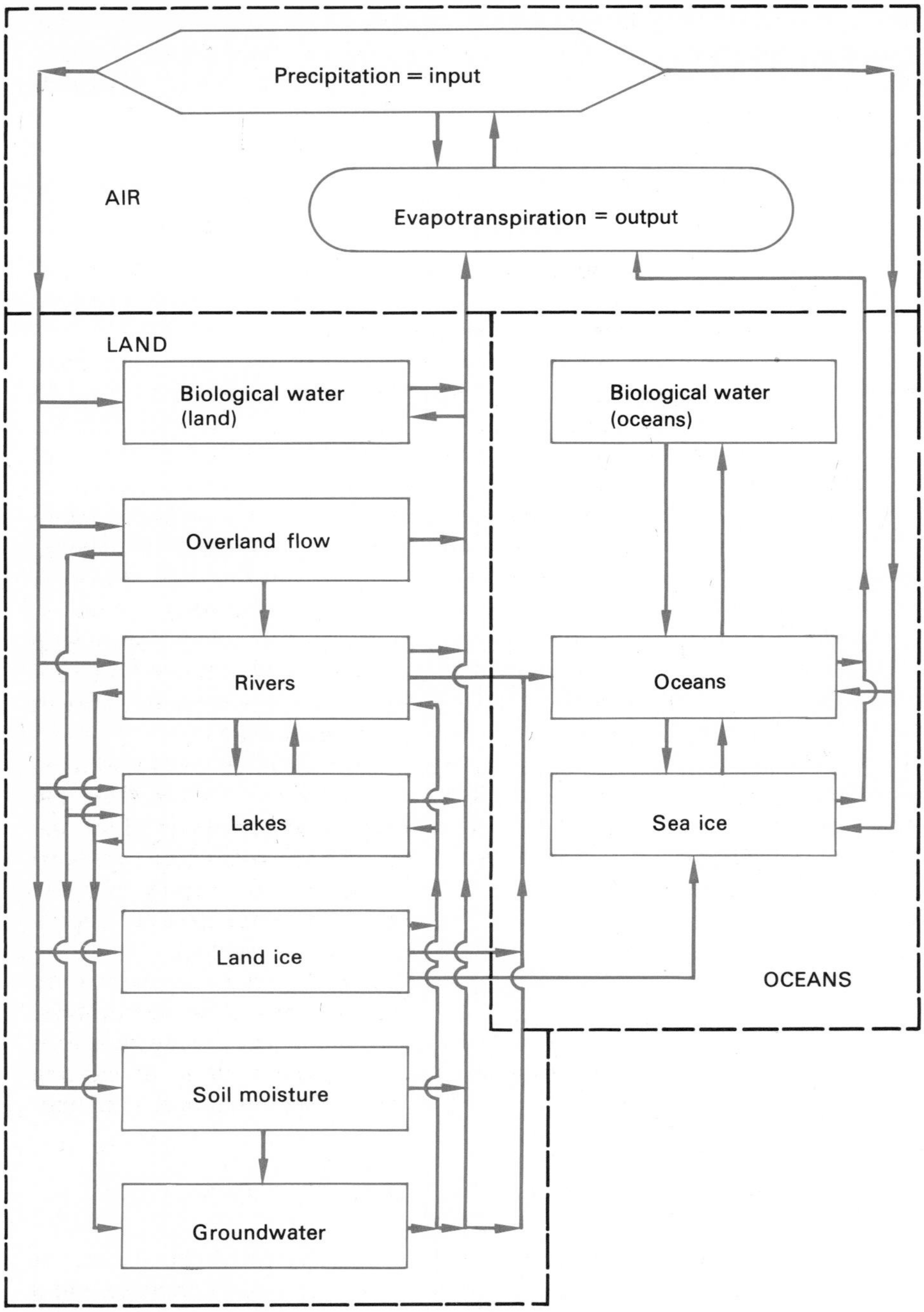

Fig. 5.2 Flow chart for the hydrologic cascade.

AIR

Precipitation = input

Evapotranspiration = output

LAND

Biological water (land)

Overland flow

Rivers

Lakes

Land ice

Soil moisture

Groundwater

Biological water (oceans)

Oceans

Sea ice

OCEANS

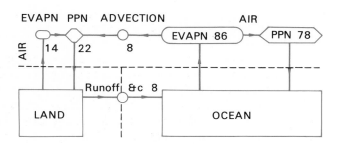

Fig. 5.3 Flow chart for cyclic water.

EVAPN   PPN   ADVECTION   AIR

AIR   14   22   8   EVAPN 86   PPN 78

Runoff &c 8

LAND   OCEAN

among the various subcascades of the total hydrologic cascade. Fig. 5.2 is a flow chart for the total cascade. Its implications will be explored throughout this chapter. The gross disproportions of content between one subcascade and another make it impossible to draw the diagram to scale.

5.2 Cyclic water

Although the flow lines in Fig. 5.2 involve only about 0.04% of the water budget (one part in three thousand: 1×3^{-4}), they stand for all the world's rivers, for world precipitation, and for world evaporation. The water involved in the recycling process at any given time is called cyclic water. Its distribution and generalised flow paths are illustrated in Fig. 5.3.

Precipitation on ocean surfaces amounts to 78% of cyclic water. Evaporation from oceans amounts to 86%. The difference is taken up by net advection (= balance of sideways transportation) of water vapour from over the oceans to over the lands. The output of water aloft, from oceans to lands, is matched by an input from lands to oceans at the surface, in the form of runoff. Over the lands themselves, there is an exchange of water between air and ground. Evaporation delivers moisture into the air; but the loss can be looked on as completely compensated by precipitation back to the ground. About two-thirds of precipitation overland originates by evaporation on land, the other third coming in from the oceans.

5.3 The ocean subsystem

This is a subsystem, only in the sense that it is fractionally less than the complete system. Some 97.5% of the combined total of liquid water, ice, and water vapour consists at present of ocean water. There is some exchange between ocean water and sea ice, and some leakage into the oceans by groundwater. There is, however, only one output of any importance, that of evaporation; and there are only two important inputs, those of precipitation (about $9/10$) and of runoff from the lands (about $1/10$).

Residence time for a system is duration of storage in that system. For the oceans, we calculate residence time by dividing loss rate into volume, arriving at an answer of about 3000 years. This value relates to existing climatic conditions. It is also an average for the ocean basins in their entirety. Only water at or very close to the surface can become involved in the evaporation process. Water at depth, and particularly bottom water (see Chapter 17), becomes stored for very long periods. Little is known of the rate of travel of bottom water, and

little, in consequence, of its age; but one may reasonably infer that some bottom water was generated many thousands, and possibly many hundreds of thousands, of years ago.

5.4 The land ice subsystem

Land ice contains about four-fifths of the water budget, exclusive of the content of the oceans. Residence time varies greatly.

Snow falling on a valley glacier may be melted within a day or a week. Snow taken into the ice of a valley glacier may reappear as water, where the glacier melts, in a few tens to a few hundreds of years. The longest storage occurs in the central parts of the ice caps of Greenland and Antarctica, where very great ages have been directly measured. Residence time here may amount to hundreds of thousands, and possibly even millions, of years. When the present rate of loss, by means of melting and the calving-off of icebergs, is compared with the volume of existing ice, an average residence time of about 10 000 years can be computed.

This value, which once again relates to existing climatic conditions, is distinctly longer than the average residence time for ocean water. Like the ocean depths, the deep central parts of ice caps have effectively nothing to do with the supply of cyclic water. In fact, the output of water from ice caps is only about $1/2$% of the cyclic water total.

Storage in land ice and storage in the oceans are interchangeable. If the ice caps grow significantly, sea level falls: if the ice caps waste away, sea level rises (Chapter 12). Complete melting of existing land ice would add about 2% to total ocean volume.

5.5 Groundwater subsystems

Groundwater is water contained in rocks, as opposed to contained in the soil. It sums to about one-fifth of the water budget, exclusive of the contents of the oceans. It will be discussed in some detail, because groundwater subsystems readily and clearly illustrate leading modes of systems behaviour.

5.5a Attributes of aquifers

An aquifer is a water-bearing rock formation. It is usually understood more narrowly, as a water-bearing formation from which water flows naturally or can be artificially extracted. The attributes of an aquifer, in addition to the contained water, include porosity and permeability. Porosity is the ratio of voids (= spaces) to total rock volume, usually expressed as a percentage. Identical spheres with hexagonal packing give a porosity of 26: just over a quarter of total volume consists of pore space. Permeability is the ability of an aquifer to deliver water. This ability depends on porosity, on pore size, on water pressure inside the aquifer, and on the slope of the groundwater surface towards the escape point. Whereas porosity can be cited for whole classes of aquifer, permeability needs to be calculated for each individual case. Nevertheless, some rough generalizations are possible. Permeability is expressed as delivery in a fixed time through a cross-section of fixed size – for instance, as m^3/day/ m^2, cubic metres per day through a cross-section of 1 square metre.

Every water well and water borehole taps an aquifer. The significance of aquifers and groundwater is well illustrated in the U.S.A., where about one-third of all consumed water comes from groundwater reservoirs. Surface water is in some ways easier to manage, requiring damming in contrast to pumping; but groundwater is likely to be purer than surface water, having been filtered by its aquifer.

5.5b Contrasts in porosity and permeability

The porosity of unconsolidated clay can exceed 50; but, because the pore spaces are microscopically small, permeability is effectively zero. It runs as low as $0.000005\ m^3$/day/m^2, at which rate it would take 500 years for a section of 1 square metre to deliver 1 cubic metre of water. Shales, typically only one-tenth as porous and one-tenth as permeable as unconsolidated clays, are even less promising as aquifers. They are chiefly important in groundwater behaviour as seals to permeable rock formations.

Most igneous and metamorphic rocks, with porosities below 1%, and also with low permeabilities, are useless as aquifers. Only if they are highly fractured do they become capable of receiving, transmitting and delivering water in quantity. The main exceptions are deeply rotted granite, which can supply shallow wells, and slaggy basalt, which can absorb heavy rainfall and supply copious springs (Fig. 5.4).

Among sedimentary rocks other than shale, limestones tend not to be especially porous. Perhaps an average of 10% of total volume will be pore space; and a representative limestone may deliver water at the rate of $0.25\ m^3$/day/m^2. Limestones as aquifers, however, function less by virtue of the porosity and permeability of the actual rock, than by virtue of the opening of underground cavities, which range from colossal cavern systems with underground lakes and rivers in them, to systems of opened joints and bedding-planes. It is these latter which enable the chalk of southeast England to be an aquifer of prime importance. Elsewhere, some limestone terrains include openings that discharge full-blown rivers (Fig. 5.5). Sandstones and conglomerates range in porosity up to 25%, and in permeability up to $5\ m^3$/day/ m^2. A water-bearing sandstone can easily prove a better aquifer than a water-bearing but cavity-free limestone, the more so since limestone waters are invariably hard.

Fig. 5.4   Part of the Thousand Springs complex in the Snake River Canyon, Idaho (V. R. Baker).

Fig. 5.5 Full-blown river issuing from Rainbow Spring, Florida.

Solid-rock aquifers are preferred over loose-deposit aquifers, because the former are the less liable to pollution. On the other hand, loose sands and gravels are the most highly porous, and also the most highly permeable, of any type of aquifer. Their porosities run generally above 25. Loose sands can deliver up to $500 \, m^3/day/m^2$, loose gravels up to $50\,000 \, m^3$.

The numbers in the permeability values are identical with the speed of flow in m/day. Very highly permeable loose gravels can generate flows of about 0.5 m/sec, about one-half or one-quarter the velocity typical of river channels full to the tops of the banks. A representative range of flow velocity for groundwater runs from 2 m/day to 2 m/yr: for most spring systems, there is a carry-over of storage, at least from one year to the next, on account of the lag time

between water input and water output. Residence time can be measured in days, months, or years; or for some sealed-down aquifers where water is stored deep underground, it can run into tens of thousands of years.

5.5c Groundwater subsystems, negative feedback and equilibrium

In order to discuss this topic, it is necessary to define the term water table. This term has two senses: the sense meant is always clear from the context. In one sense, water table connotes the groundwater in total: wells and boreholes are said to tap the water table. In the other sense, water table connotes the surface of the saturated rock underground. It is this second sense that applies in what now follows.

Consider an aquifer on a plateau top (Fig. 5.6). It leaks out at the edges through springs. The regulator

Fig. 5.6 A leaking aquifer on a plateau top: the situation illustrated is that of a perched water table.

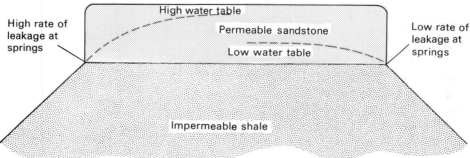

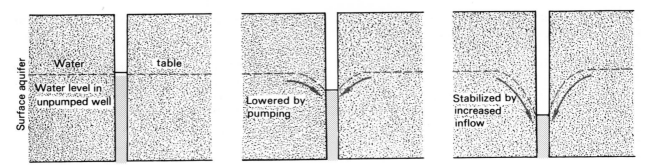

Fig. 5.7 Stabilisation, by negative feedback, of the water level in a pumped well.

of permeability (rate of throughput) ensures that leakage is not immediate. Hence, where winters are cool and rainy, the water table rises during the winter season. Its very rise, promoting an increased gradient down to the outlets, promotes increased flow through the leaks. Dry warm summers reduce or eliminate input. At the same time, the gradient toward the springs is reduced, so that leakage diminishes. Increased input increases output; decreased input diminishes output. Negative feedback is at work. Also, because the system oscillates between maximum height of water table and maximum spring flow on the one hand, and minimum height and minimum flow on the other, it is in dynamic equilibrium.

Negative feedback also operates, within limits, when water is pumped from an aquifer (Fig. 5.7). By lowering the water table (surface of saturation), pumping increases the gradient of the table down toward the bottom of the well, and thus increases the rate of flow into the well, up to the limit set by the permeability of the aquifer.

A particularly engaging equilibrium situation is represented by a situation of balance between fresh and salt water. Fresh water, being the less dense, occupies the lesser volume per unit mass. Therefore, it becomes possible for an offshore island to contain fresh water, which nourishes freshwater plants. This is why trees can grow on offshore islands in the sea (Fig. 5.8).

5.5d Groundwater subsystems, disequilibrium and positive feedback

Groundwater reservoirs from which water is extracted faster than it comes in are, by definition, in disequilibrium. The disequilibrium condition can potentially be corrected, at least in some cases, by artificial recharge, whereby sand-filtered water is fed back into an aquifer. On balance, artificial recharge has so far proved grossly incapable of counterbalancing extraction.

Over-extraction is typical of the use of artesian reserves, which are reserves confined under pressure. The aquifer is sealed in by impermeable rocks, both above and below (Fig. 5.9). It supplies artesian springs and wells (named after the French province

Fig. 5.8 The freshwater lens: lens depth is about forty times the difference between the level of the sea surface and the level of the water table. This kind of situation is abundantly illustrated by the offshore islands of the Gulf side of the Florida peninsula.

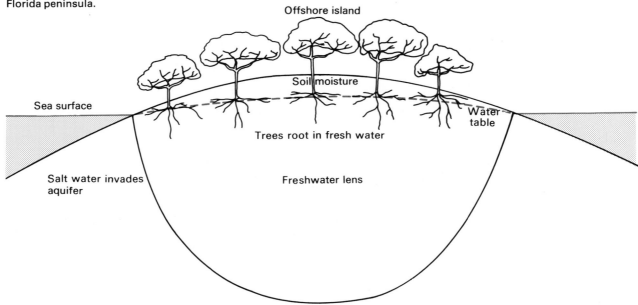

Artois), which flow at the surface. The chalk formation of southeast England formerly contained enough water, under enough pressure, to activate the Trafalgar Square fountains in the centre of London. Today, these same fountains use water recycled through pumps in their own ponds. The artesian table has been depressed, on account of progressive extraction, well below the surface of the ground. Similar histories apply to the artesian water tables of Australia, which is more dependent than

any other continent upon artesian supplies. Spanning the borders of useful lands and lands useless without groundwater (Fig. 5.10), Australia's artesian reservoirs have been, and are still being, vastly exploited. So long as the rate of extraction continues to exceed the rate of recharge, the reserve stock of water must continue to decrease, and the groundwater subsystems must continue to be in a state of disequilibrium.

The most highly disconcerting disequilibrium

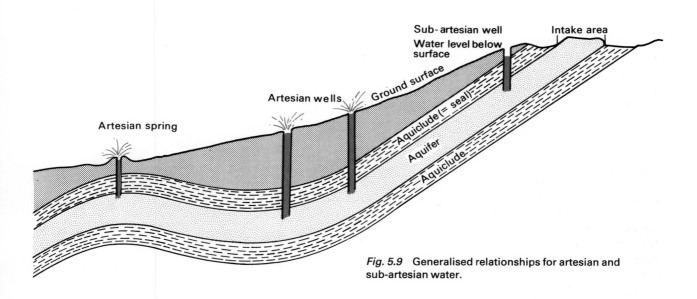

Fig. 5.9 Generalised relationships for artesian and sub-artesian water.

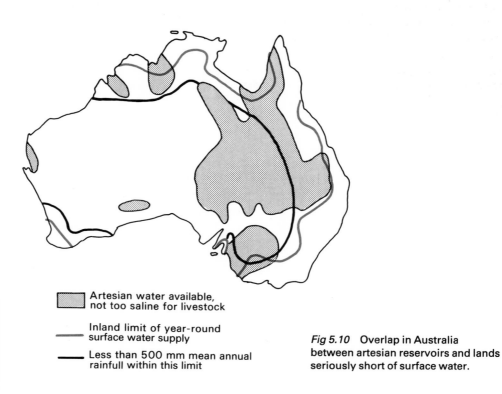

Artesian water available, not too saline for livestock

Inland limit of year-round surface water supply

Less than 500 mm mean annual rainfull within this limit

Fig 5.10 Overlap in Australia between artesian reservoirs and lands seriously short of surface water.

TOTAL SUBSIDENCE
1928 TO 1977

Fig. 5.11 Subsidence related to the extraction of petroleum, Long Beach area, California: values in feet. Subsidence was most rapid in the early 1950s (photograph and annotation courtesy of Long Beach Department of Oil Properties).

effect is the irreversible ruin of an aquifer. Porosities significantly above 25 imply either an open packing pattern, or a highly variable fragment/particle size, or both. If water is rapidly withdrawn from an aquifer composed of poorly-packed and/or variably sized fragments/particles, the packing pattern may change abruptly, greatly reducing the porosity. Over-pumping has caused some aquifers to collapse. They can never be restored. It could well be asked which is the more serious damage, the immediate response of subsidence, or the long-term consequence of aquifer destruction. Identical considerations apply to the extraction of petroleum from certain fields (Fig. 5.11).

Examples of subsidence caused by the over-extraction of groundwater include the cases of Mexico City, which has sunk through 7 m in ninety years; Shanghai, where resultant flooding of the waterfront has required extensive diking, plus the injection of water down boreholes; and Venice, where settling still continues at about 1 cm/yr, and where flooding during high tides is becoming increasingly destructive.

5.6 Lake subsystems

Lakes contain only about 0.7% of the water not contained in ocean basins. Somewhat more than half of this fraction is held in lakes with outlets (= freshwater lakes), and somewhat less than half in saltwater lakes (= lakes without outlets: closed lakes). About three-quarters of the total volume of freshwater lakes is accounted for by Lake Baikal in Asia, the large lakes (including the Great Lakes) of

North America, and the large lakes of Africa. The world's enclosed lakes are dominated by the Caspian Sea, which contains three-quarters of their total volume.

Residence time is increased by lack of an outlet, with evaporation from the lake surface the only means of depletion. But because most enclosed lakes exist in dry climates, evaporation loss is often considerable. The Caspian Sea, with a surface area of 435 200 km^2 and a net evaporation loss (= total evaporation minus rainfall on the water surface) of more than 850 mm/yr, evaporates water at the rate of some 12 000 m^3/sec, the equivalent of the discharge of a distinctly large river. Mean residence time is rather more than 200 years.

Conversely, residence time is reduced by discharge through an outlet. It increases in drained lakes with the volume of the lake basin. For the Great Lakes system of North America, it averages about a century, approaching two centuries for Lake Superior, and averaging about ninety years for Lake Erie. This last value is alarming, in the context of the serious pollution that Lake Erie has already undergone. Because water at depth records far longer than average residence time, the cleanup process now in hand could require several centuries to become really effective.

5.6a Equilibrium behaviour and step-functional behaviour

Lakes, both closed and drained, have already been used to illustrate states of dynamic equilibrium that involve negative feedback (Figs. 1.1, 1.2, and text). The general equilibrium equation is

   Output = Input

which tells us nothing about the nature of inputs and outputs. For a closed lake, the equation becomes

   $E_l = P_l + I$

where E_l is evaporation from the lake surface, P_l is precipitation on that surface, and I is inflow from the surrounding land area. The fact that lakes in many desert areas lasted long enough, and stood long enough at or near fixed levels, to cut beaches and to precipitate rims of evaporites, demonstrates that the equilibrium situation formerly obtained. The fact that the former lakes have now vanished, in response to an increase of dryness (Chapter 18), demonstrates a step-functional response. Either the climate is wet enough to sustain lakes, or it is so dry that the water level falls drastically, and many lakes dry up altogether.

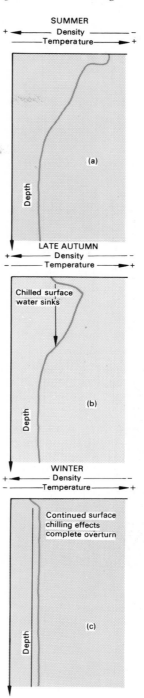

Fig. 5.12   The mechanism of lake overturn.

Step-functional behaviour is also illustrated by the seasonal overturn which affects many mid-latitude lakes as the winter season comes on. During summer, the surface water becomes warm and stays warm. In a small lake, one can dive through a shallow skin of warm water into much colder water beneath. The warm surface water stays on top, because its density is reduced by heating (Fig. 5.12). As winter draws on, the surface water cools off.

When it becomes colder than the water below, it also becomes more dense. It therefore sinks. Here is the convection process at work. Continued chilling of surface water, and continued convectional sinking, promote a complete overturning of the lake water. The surface-to-depth exchange occupies only a fraction of the year. In terms of the annual calendar, it is abrupt. It constitutes a step function.

5.7 Soil moisture subsystems

Soil moisture involves two components of the hydrologic cascade – water that is merely passing through, and water that is retained – and this makes the soil variably moist. Water merely passing through drains downward, under humid climates, into the groundwater table. In dry climates it is liable to be drawn back toward the surface, and to be evaporated there.

The total of soil moisture is estimated at only 0.44% – say, one two-hundredth part – of the water not contained in the oceans. Even at that, it is far more abundant than the water contained in river channels. It is of supreme biological importance.

The moisture contained in the soil is withdrawn from storage by evaporation from the soil surface, and by transpiration by plants. Plants use moisture to carry nutrient materials up from the soil, and transpire the moisture into the atmosphere. The combination process of evaporation and transpiration is evapotranspiration.

Actual evapotranspiration is measured loss rate. Potential evapotranspiration is the loss rate that could occur if the ground carried the maximum possible plant cover, and if there were enough soil moisture to provide for maximum possible evaporation and maximum possible transpiration. The difference between actual and potential evapotranspiration is what irrigation could usefully supply.

Potential evapotranspiration varies with temperature – that is, effectively with input of solar radiation, about two-fifths of which is used in the transpiration process. It also varies with the varying demands of different kinds of plant cover. The kinds of magnitudes involved are illustrated by the values of 500 mm/yr of rain-equivalent for short grass in southern England, 600 mm/yr for grain crops in the same area, and 800 m/yr for alfalfa in central California. Here lies the snag. A rough average for potential storage in the soil is 100 mm of rain-equivalent. If the demand during the growing season exceeds storage capacity, three possibilities are open: storage can be replenished by summer rain; irrigation water can be supplied; or, plant growth can be checked by lack of water – and, at the extreme, crops will pass their permanent wilting point, and die.

Residence time of soil moisture is obviously short. If we assume seven months for the growing season,

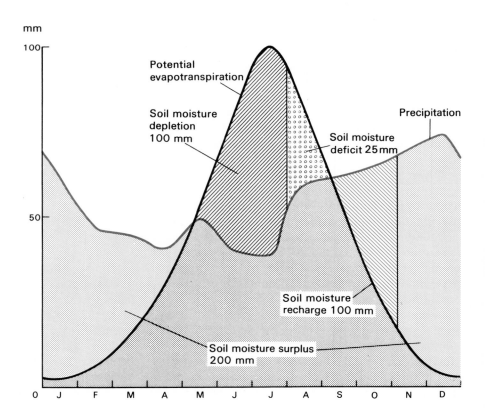

Fig. 5.13 Interrelationship of soil moisture supply and soil moisture demand: the diagram relates to a humid midlatitude climate (see also Chapter 15).

an effective demand equivalent to 700 mm of rain, and a storage capacity of 100 mm, we arrive at a residence time of one month.

For individual areas, the demand and supply relationship can be computed (Fig. 5.13). Because we are dealing with a cyclic situation, it does not matter where the cycle is entered: however, it is convenient to enter where demand (= potential evapotranspiration) first exceeds supply during the passage of the spring season. Growing plants tap the soil moisture store. As the warm months advance, soil moisture is withdrawn. The difference between potential demand and actual supply is a soil moisture deficit. As air temperatures decline through late summer, the demand declines. Eventually, the supply rate comes to exceed the demand rate. There is a surplus of supply, taken up in the first instance by the recharge of soil moisture, and then by the copious filling of river channels. Here is the physical basis of the description, February fill-dike.

5.8 Water storage in rivers

Some time must elapse between the time when water enters a river channel and the time of discharge into the sea. We can calculate the order of time involved, from speed of flow and length of river. The answer turns out to be hours, or even minutes, at least, and a matter of days at most. Water flowing at the modest average rate of 3 m/sec would take only about twenty days to traverse the entire channel length of the world's longest rivers.

River channels contain only about 0.005% of the water not contained in the oceans – that is, only about 0.00013 of the total water budget. The typically brief residence time, plus the tiny fraction of total water involved, may seem to make it surprising that so many rivers can sustain flow throughout the year. Delivery of groundwater, with its much slower throughput than that of channel water, is what mainly enables rivers to sustain perennial flow.

Rivers have already been specified as effecting the return flow to the oceans, of the net input aloft from oceans to continental areas. They have also been identified as great shapers of the landscape. Their minute total content of water at any given time in no way mitigates their significance in the cycling of cyclic water, in the cycling of sediments, or in the shaping of the surface of the ground.

5.9 Overland flow

Just as there is a lag between delivery of water to rivers and delivery of river water to the sea, so there is a lag between the fall of rain on the ground, and the delivery of overland flow to rivers. In cold climates, the lag between precipitation and runoff is still further increased, by the storage of precipitated moisture in the form of snow.

It is not possible to estimate the total storage of liquid water, at any given time, in overland systems. The speed of runoff from land surfaces can be crucially important in the promotion of flood peaks, and also in the rate of sediment delivery to river channels. Residence time in overland flow systems is increased by density of vegetation cover, and especially by forest cover. Residence time is decreased by anything that increases speed of runoff, especially by bareness of the land surface – and, on this count, by stripping associated with construction, and by the paving-over of urban areas. Two extreme possibilities arise here. A stripped surface ensures maximum potential erosion, plus high potential speed of runoff: construction works often involve very high rates of erosion. Complete paving-over reduces erosion to zero, but maximises speed of runoff. It may well prove impossible to provide drainage systems that can cope with heavy storm runoff from a completely paved urban surface.

5.10 Storage in the atmosphere

About as much water is contained in the atmosphere at any one time as is contained in the world's river channels. The mean daily rate of exchange between the atmosphere and the land–ocean surface is roughly 5 mm, with 2.5 mm leaving the atmosphere for the surface, and an equal amount leaving the surface for the atmosphere. Calculation from this rate shows that average residence time in the atmosphere is about ten days.

It is particularly in the earth–air exchange that inputs and outputs are complex, and that the cascading nature of the subsystems can be recognised. Still more obvious, and centrally important to man's activities, are the chief processes by which the exchange is effected: evaporation from the surface, and precipitation of rain or snow back to the surface, and especially to the surface of the land (Chapter 16).

5.11 Biological storage

The total storage of water in living organisms is extremely small, even on the scale of the right-hand cube in Fig. 5.1. The total throughput is a different matter. The amount of water annually released by transpiring plants may well be equal to the amount discharged by the world's rivers. Complexity of a given biological subcascade varies with the different ways in which organisms manage their inputs and outputs of water. Mankind, for instance, emits water in sweat, in urine, and in vapour form in outgoing breath. Residence time for water in biological storage also varies much, ranging from hours for many warm-blooded animals to seasons for plants, while water permanently stored in tissues is released only on the organism's death.

Chapter Summary

Water amounts to a very *tiny fraction* of the *mass of our planet,* but nevertheless *covers 71% of the surface.*

Cyclic water is *exchanged among oceans, atmosphere and lands.* Amounting to some *0.04% of the total water budget,* it includes all *rivers, rain, water in the air* and much *soil moisture* and *groundwater.* The water cycle involves a *net transfer from oceans to lands aloft,* and a return of *runoff to the oceans.*

Ocean water sums to about *97.5% of the water budget.* *Inputs* are about *⁹/₁₀ precipitation* and *¹/₁₀ runoff. Average residence time* is about *3000 years.*

Land ice contains about *⅘ of the water not held in the oceans. Residence time varies greatly,* possibly reaching *millions of years* in the *centres of ice caps.* Storage in land ice and in the oceans is *interchangeable.*

Groundwater reservoirs contain about *⅕ of the water not contained in oceans.* Groundwater is held in *aquifers,* which are rocks *porous* enough to contain water and *permeable* enough both to take it in and to let it out. Most aquifers in actual use are *alluvial sands and gravels, sandstones,* and *limestones.* Groundwater tables are *liable to negative feedback,* increasing spring flow in response to rainfall, reducing it in response to drought.

Disequilibrium results from over-extraction. *Artesian reservoirs,* where the *aquifer* is *confined* both *above and below,* are especially liable to over-use. Whether artesian or not, an *aquifer is ruined, if over-extraction changes the packing pattern of the rock,* in such a way as to promote a drastic decrease in porosity. Changes of the packing pattern can be accompanied by disastrous *subsidence at the surface.*

Lakes contain about *0.7% of the water not in ocean basins.* They are sub-classified into *lakes with outlets* (fresh) and *lakes without outlets* (salt). *Residence time varies greatly,* increasing with size of lake and with lack of outlet. *Closed lakes standing at a given level* illustrate the *equilibrium situation.* If they *dry up,* they are acting *step-functionally. Seasonal overturn,* experienced by many lakes in middle latitudes, is also a *step function.*

Soil moisture includes *water passing down* to the groundwater table, and *water taken into the* substance of the *soil.* It amounts to *0.44% of water not in the oceans.* It is liable to *withdrawal by evaporation* and by *transpiration* through plants. *Mean residence time* is perhaps on the order of a *month.*

Potential evapotranspiration is the *expectable loss* for *maximal plant cover* and complete *satisfaction of demand.* In many regions it runs far *higher,* in the latter part of the warm season, *than supply.* Measurement of potential evapotranspiration can be made to indicate *potential irrigation demand.*

Storage in rivers involves only about *0.005% of the water not in the oceans. Residence time* is *days* at most. Leakage from *groundwater* tables is important in sustaining *perennial flow.*

Speed of overland flow increases on *cleared* and *paved surfaces.* It can mean serious problems of erosion of construction sites, or of storm drainage for built-up cities.

The *atmosphere* contains *about as much water as rivers* do. Mean *residence time* is about *ten days.* The total is less immediately important to man than are the exchange processes.

Biological water amounts to an *extremely small total. Residence time* also *varies greatly,* from minutes to a lifetime.

6 The Hydrologic Cascade: Patterns in Time

Leading patterns of systems behaviour in time were introduced in general terms in Chapter 1. Some have just reappeared, in their hydrologic connections, in the immediately preceding chapter. Further examples will be needed later, especially in the form of seasonal climatic rhythms and the step functions of climatic change.

Patterns of systems behaviour in time are by no means confined to the hydrologic cascade. It is however for hydrologic systems that behaviour in time has been intensively studied. Study has been directed both at determining what happens, and at predicting happenings in the future. This chapter will discuss two main methods by which time series can be analysed. One method depends on the recognition of cycles. The other method assumes – at least to begin with – that variation in the system under study is completely random through time. This method will be needed for parts of the discussion in Chapters 7 and 8.

We need to know about annual regimes of precipitation, about the rhythm of the water need of growing plants, and about streamflow. We need to define the odds for heavy rain in single storms, and of rainfall amount in a given month or a given year. We need to know about the prospect of high floods or prolonged drought. For parts of the Great Lakes area, where high water levels are accompanied by erosion of the shoreline, we need to know about the rise and fall of the water surface. In all cases, if some kind of order can be brought into the observed time series, we can at least look at the prospect of forecasting odds for the future.

To anticipate: evidence will be brought that attempts to bring order into time series can be guaranteed success, so long as the analyst does not bother about whether the results are real or not. Evidence will also be brought that attempts to forecast for the future are apt to be bedevilled by progressive or step-functional change.

6.1 The search for order

Just as we seem instinctively to accept dynamic systems as working wholes, so we seem instinctively to search for order in their patterns of behaviour over time. Until system components are assembled into a pattern, they remain merely confusing. In the same way, until the attributes of behaviour over time are analysed into a pattern, they tell us nothing. In the nothing state, systems behaviour may even seem offensive. Think of disorderly behaviour on the part of an individual human biosystem: such behaviour can lead to a court of law, or to an institution for the insane.

6.1a Reduction to order by main force

A crude kind of order can be imposed on a time series by the calculation of long-term means. All variation about the mean is, by definition, suppressed when the mean is calculated. That is to say, all noise (= unexplained variation) is eliminated. Variation through the day, from one day to another, from one month or one season to another, and from one year to another, is all averaged out.

6.1b Some results of long-term averaging

Moscow records a long-term mean annual precipitation of 630 mm, an average of 52.5 mm per month. The monthly average conceals the difference between the wettest month, July, with a long-term average of 76 mm, and the driest month, March, with a corresponding value of 28 mm. The mean annual value nevertheless has a certain worth. Moscow is obviously wetter than the desert station of Iquique in Chile, where the mean annual precipita-

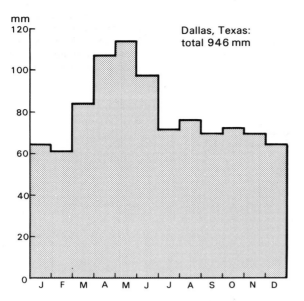

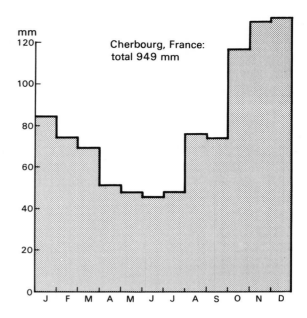

Fig. 6.1   Graphs of mean monthly precipitation for two stations with almost identical annual means, but with contrasted regimes.

tion is only 2 mm. Completely rainless years are frequent at Iquique: completely rainless months are very frequent indeed. Moscow on the other hand seldom experiences a rainless month, and never a rainless year.

Dallas, Texas, records a mean annual precipitation of 946 mm. Cherbourg, in northern France, records 949 mm. The two totals are, for all practical purposes, identical. But the seasonal regimes stand in strong contrast to one another (Fig. 6.1). At Dallas, mean monthly precipitation is concentrated in spring, with a peak in May. The concentration at Cherbourg comes in winter, peaking in December.

Long-term means of streamflow, although useful for reservoir operations, have little to no value for other purposes. The Chatooga River at Clayton, Georgia, recorded for an eighteen year period a mean discharge of 16.5 m³/sec. The low extreme discharge hidden in this mean was 2.5 m³/sec, about 0.15 of the mean value. The high extreme was 820 m³/sec, about fifty times the mean. Still greater contrasts can be expected in dry climates. In a fourteen year period, the Cimarron River at Guy, New Mexico, recorded a mean flow of 0.4 m³/sec. The low extreme was zero flow, infinitely smaller than the mean. The high extreme was 240 m³/sec, 600 times as great as the mean.

The two records, of fourteen and eighteen years, are quite short. Longer records could be expected to increase the spread between extremes.

Long-term monthly means are less crude than long-term annual means, to the extent that they define annual regimes. The long-term monthly mean levels of Lake Michigan at Milwaukee range be-

tween 15 cm on either side of the long-term annual mean (Fig. 6.2). The extreme highs for the period of record rise to about 90 cm above the long-term mean, while the extreme lows fall to about 80 cm below. But despite the considerable noise that is suppressed by the monthly means, these means do indicate a likelihood of high lake levels in mid-year, and of low levels at the year's turn.

6.2 The yearly time series

Fig. 6.3a is a graph of high extreme monthly mean levels for Lake Michigan at Milwaukee. Each point represents the highest monthly mean level for one year in the series. Distinct peaks and troughs occur, as they do on many other yearly series. If the array could be analysed into some kind of pattern, then future high levels could be predicted.

It is in connection with annual series that cyclic analysis has been mainly tried. There is a strong temptation, in this case, to suppose that the rise in peak levels from 1940 to 1952, and the subsequent descent to 1960, represent some kind of wave. The depths of troughs in the graph become less from 1934 to 1950, and then increase again. Here also is a suggestion of a wave.

6.2a Smoothing

The suggestion is reinforced when the record is smoothed. Fig. 6.3b superimposes a graph of five-year means on the graph of the actual record. Smoothing is a perfectly legitimate exercise. It is

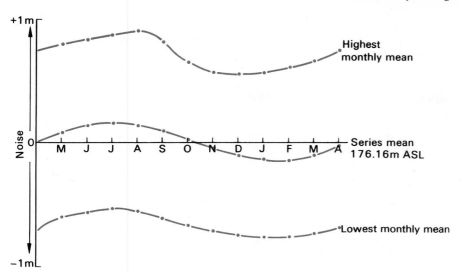

Fig. 6.2 Long-term mean levels of Lake Michigan at Milwaukee, referred to the series mean: monthly highs, monthly lows and monthly means.

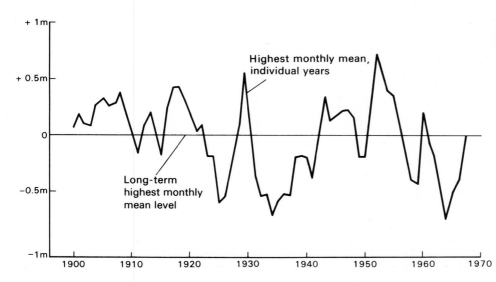

Fig. 6.3a High extreme monthly mean levels of Lake Michigan at Milwaukee, 1910–70: basis of reference is long-term mean monthly high level.

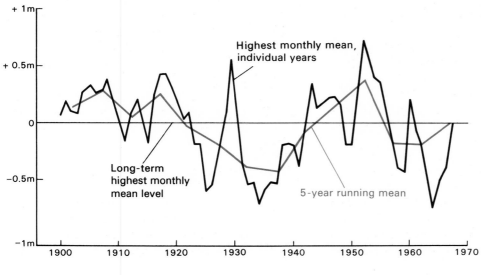

Fig. 6.3b Five year running mean superimposed on the graph of Fig. 6.3a.

designed to suppress the noise of short-term variation. In this case, the graph of five-year means displays two troughs, one in the middle 1930s and one at about 1960, separated by a peak at about 1950. Although no strong single peak occurs in the early part of the graph, the five-year means do run consistently above the long-term mean, for the interval 1900–21.

6.2b Sine wave analysis

An immediate temptation at this point is to suppose that we are looking, not only at a wave pattern, but at a pattern of sine waves. The graph of five-year means seems to resemble, in a general way, the graph of sun height in Fig. 1.3, or the graph of mean monthly lake levels in Fig. 6.2. As shown in exaggerated form in Fig. 6.4, the rise and fall of mean monthly levels describe an almost perfect sine wave.

The diagram also shows how a sine wave is generated graphically. In a right-angled triangle, the sine of an angle is the length of the perpendicular opposite that angle, divided by the length of the hypotenuse. For a graphically generated wave, the length of the hypotenuse = length of the radius of the arc = a constant. Thus, the perpendiculars are proportional to the sines.

Sine waves actually do occur, and occur often, in the behaviour of natural systems. Partly for this reason, and partly perhaps because they are easy to calculate, they have commanded much attention in cyclic analysis and in attempts at prediction. Long before the days of electronic computers, notable success was achieved in the mechanical prediction of the rise and fall of tides. The machines integrated sine waves of differing length and height. Fig. 6.5 illustrates the combination of two wave trains, one, with a 50-year periodicity, and amplitude (= distance from baseline to peak or trough) of 0.3 m, the

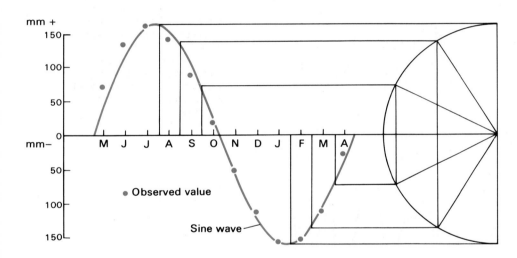

Fig. 6.4 Long-term mean monthly levels of Lake Michigan at Milwaukee, as in Fig. 6.2, fitted to a sine wave: fit is about 98% perfect.

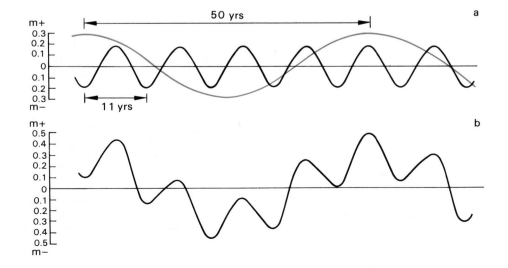

Fig. 6.5 Combination of wave trains of different periodicities and somewhat different amplitudes: (a) the two separate trains; (b) the trains combined.

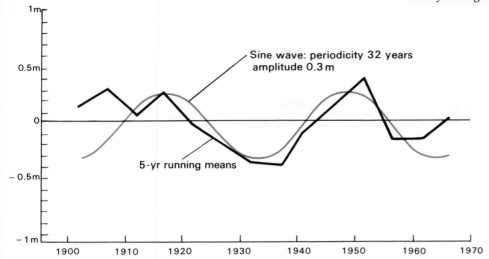

Fig. 6.6 Five-year running means of mean level of Lake Michigan at Milwaukee, compared to sine wave with periodicity of thirty-two years and amplitude of 0.3 m: basis of reference is long-term mean level.

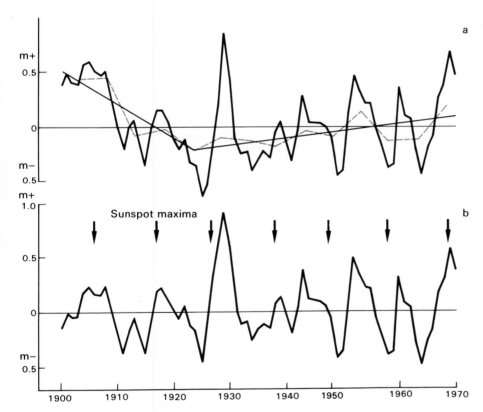

Fig. 6.7 Residuals from Fig. 6.6, after allowance for the supposed thirty-two year cycle: (a) residuals (solid peaked line), five-year running mean of residuals (dashed line), and possible progressive trends (solid straight lines); (b) residuals after supposed thirty-two year cycle and progressive trends, compared with incidence of sunspot maxima.

other with eleven-year periodicity and amplitude of 0.2 m. The combined graph bears some kind of general resemblance to the graph of lake levels in Fig. 6.3a.

Tide-predicting machines dealt, as computers to-day deal, with more than two wave trains at a time. At least six are usually needed: and the most elaborate machine handled more than sixty. The graph of Fig. 6.3a could, in fact, be resolved into a combination of sine waves: but whether or not the results would mean anything real is another matter.

Let us however for the sake of discussion assume that the trend of the five-year means in Fig. 6.3b does reflect a sine-wave variation. A thirty-two-year

periodicity, with peaks in 1918 and 1950, and with an amplitude of 0.3 m, produces a quite close resemblance to the smoothed graph of five-year means, from about 1910 to about 1964 (Fig. 6.6).

6.2c Residuals

The next step is to see what happens when the effect of the thirty-two-year cycle is removed. The variation about the line of the thirty-two-year cycle is noise in relation to that cycle. It consists of residuals – amounts remaining, plus or minus, when the cycle has been taken off. Fig. 6.7 graphs the residuals in this case.

6.3 Progressive variation

The residuals can be smoothed in the same way that the original data were smoothed. Five-year means are also plotted in Fig. 6.7a. They quite strongly suggest a decline in high lake levels until about 1903, followed by a slow rise thereafter. If the suggestion is correct, then we are probably dealing with progressive variation, as shown by the straight-line graphs inserted in the diagram.

Progressive linear variation in lake level can certainly happen. Fig. 6.8 plots the variation in the annual mean level of Lake Nyasa for a fifty-two-year period. Except for an apparent jump in level of about 1 m between 1916 and 1917, the trends seem remarkably clear – a steady slow fall from the beginning of the record until 1916, a steady and somewhat faster rise from 1920 to 1930, then a steady swift rise to 1937, a steady swift fall to 1943, and thereafter a renewed steady swift rise.

Progressive variation, being by definition non-cyclic, is not popular with the devotees of cyclic analysis. Indeed, it is not likely to be popular with anyone trying to make predictions for the future, because its direction, rate, and typically sudden changes of direction and rate are not predictable at all. Let us however, again for the sake of argument, accept as real the progressive trends marked in Fig. 6.7a. The residuals after the thirty-two-year cycle and the progressive trends are graphed in Fig. 6.7b. All suggestion of a long cyclic swing has vanished. The graph displays seven major peaks, divided by intervening troughs. The mean peak height, 0.45 m, is not markedly different from the mean trough depth, 0.4 m. The intervals from peak to peak are, in order, 12, 11, 14, 9, 8, and 9 years.

6.4 The sunspot cycle

It has long been known that the number of sunspots is far greater in some years than in others. A generalised graph of sunspot numbers strongly resembles a sine wave, with maxima at an average interval of 11.1 years. A particular interval may be as short as 8 or as long as 16 years. The intervals between the peaks in Fig. 6.7b, therefore, fall within the possible range for the intervals between sunspot maxima. Moreover, the average interval between pairs of peaks in Fig. 6.7b is 10.5 years, identical with the average interval between pairs of sunspot maxima in the same period.

Even though the possible relationship between the sunspot cycle and the hydrologic cascades is not understood, that cycle has attracted a great deal of notice, simply because it is real. Unfortunately, efforts to relate lake levels or other hydrologic variations to the sunspot cycle usually end in disappointment. One well-known early study dealt with the annual mean level of Lake Victoria. From 1899 to 1922, the variation of lake level closely matched the variation in numbers of sunspots, doing so through two complete waves (Fig. 6.9). If variation in lake level is assumed to result – in however complex a way – from variation in the sun's effect as expressed by sunspot number, then 75% of the variation in lake level is explained by variation in sunspot number. Here would seem to be a reliable means of predicting changes in lake level, since sunspot number can be predicted. But after 1902, the relationship between sunspot number and lake level breaks down completely. There is a precisely zero relationship between the two variables, from 1903 on.

Fig. 6.8 Progressive trends in the level of Lake Nyasa, East Africa: reference is to an arbitrary datum.

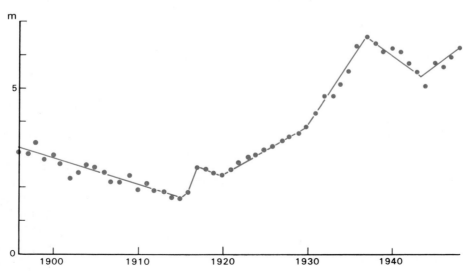

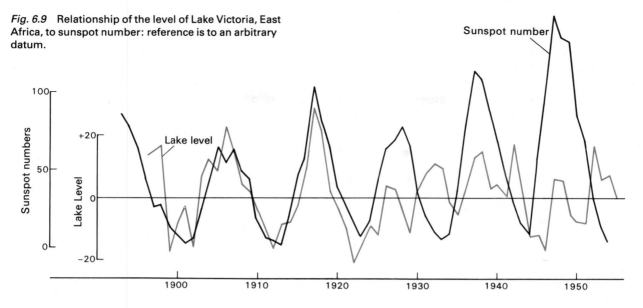

Fig. 6.9 Relationship of the level of Lake Victoria, East Africa, to sunspot number: reference is to an arbitrary datum.

For Lake Michigan levels, there does at first seem to be some connection between the incidence of sunspot maxima and peak residuals after the thirty-two year cycle and the progressive trends have been discounted. One would however have to suppose that some peaks occur close to sunspot maxima, while others lag behind. A lag time of one or two years is to be expected in the response of this particular lake system; but then all peaks ought to lag equally. These peaks do not. The double sunspot maxima of 1905–7 are well matched by the peak of residual level; but the sunspot maximum of 1938 is not followed by a major peak in residual level until 1943. The sunspot maximum of 1947–8 nearly coincided with a low in the lake record, while the maximum of 1958 coincided exactly with a low.

6.5 General conclusion on cyclic analysis

Many records are too short to justify the identification of cycles thirty or more years in length. Even if the sunspot cycle would hold good in some instances, predictions based upon it have too often failed to support its use. Moreover, a great deal of noise remains, after definable cycles have been taken out. To reduce an hydrologic time series to a combination of sine waves, while completely possible in practice, may not be warranted on logical grounds. An artificial cyclic description of past events may be quite untrustworthy as a guide to the future. Attractive though the idea of cyclic analysis and prediction undoubtedly is, it will not appear reliable until at least some cycles have been shown to affect a number of hydrologic sub-cascades in the same region. We could reasonably expect a cyclic effect in the flow of the Thames to be reflected also in the flow of the Severn and the Trent. A cycle in the level of one of the Great Lakes should be identifiable for the major rivers in the drainage area, and also for rivers in neighbouring basins. And even if the required matches can be made, cyclic prediction will remain suspect, until the causes of cycles can be determined.

6.6 Non-cyclic analysis

This form of analysis relies on the calculation of odds. Crudely described, the basic principle is that records exist to be broken. However great a past flood, an even greater flood will happen in the future. However severe a past drought, an even more severe drought will occur in years to come. However much rain has fallen in a given month or a given year, a future month or year will prove even wetter. However great the damage inflicted by a tornado or a hurricane, a future tornado or hurricane will prove even worse. Non-cyclic analysis concentrates especially on events of great magnitude. For events of given magnitude, it calculates probabilities: hence the title, magnitude–frequency analysis.

Consider the annual peak discharges on the Fraser River at Hope, British Columbia (Fig. 6.10). Like the graph of high levels on Lake Michigan, this diagram at first sight suggests a cyclic variation, with peaks at about 1920 and 1956 – although an analysis into sine waves could not possibly account for the extreme peak in the period of record, in 1948. When the annual peak flows are listed in order of magnitude, they can be analysed for probability. The

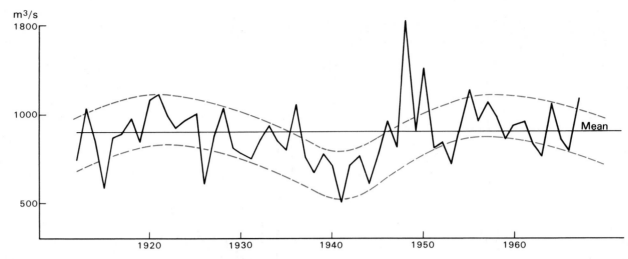

Fig. 6.10 Peak discharges on the Fraser River at Hope, British Columbia: one peak per year. Dashed lines of quasi-envelope curves illustrate the temptation to engage in cyclic analysis.

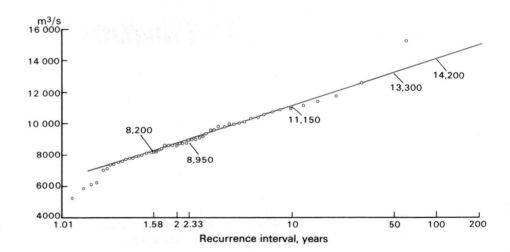

Fig. 6.11 Magnitude–frequency graph for the peak flows used in Fig. 6.10.

required technique is very simple. The frequency of a discharge of given magnitude is called the recurrence interval. This is computed as 1 + the number of years in the record, divided by the ranking order of a given discharge:

$$r.i. = (n + 1)/r,$$

where *r.i.* is recurrence interval, *n* is the length of record, in years, and *r* is rank order.

Table 6.1 lists the annual peak discharges of the Fraser River as they happened, from 1950 through 1967: a short series is used here for the sake of brief illustration of the method. In the third column of the table, the discharges are rearranged in ranking order: that order is listed in the fourth column. The fifth column lists the recurrence intervals, computed

according to the above equation, for the ranked discharges.

The series of peak flows, one flow per year, is the annual flood series. The peaks are called floods, whether or not they cause inundation of the valley bottom. Recurrence intervals have been computed for the complete series, 1912–67. The observed floods are plotted against their computed recurrence intervals in Fig. 6.11.

The graph paper used here is a special kind, called Gumbel paper, after its inventor. The vertical scale is plain (arithmetic). The horizontal scale resembles, but is not identical with, a logarithmic scale. It is so designed that if the observations conform strictly to a certain kind of probability, then the plotted points will fall on a straight line. The points for annual floods on the Fraser River, between the recurrence intervals of 1.1 and 28.5 years, do almost precisely that. We can confidently draw in the straight line by eye.

6.6a Some particular recurrence intervals

The graph cuts the ten-year mark at $11\,150\,\text{m}^3/\text{sec}$. This is the magnitude of the ten-year flood, which is written as q_{10} (annual series). The q stands for discharge – magnitude in cubic metres per second. The ten-year flood is the discharge that can be expected, as an annual maximum, ten times in a century. The long-term average spacing of ten-year floods is obviously ten years; but we do better to say ten times a century, in order to avoid all suggestion of regular spacing. There is nothing to prevent two or more ten-year floods from coming in successive years. Another way of putting things is to say that the odds against a ten-year flood in any given year are ten to one.

It is a property of the statistical design that the mean annual flood has a recurrence interval of 2.33 years. The magnitude read against this recurrence interval in Fig. 6.11 is $8950\,\text{m}^3/\text{sec}$. The arithmetical average for the entire series is somewhat lower, at $8856\,\text{m}^3/\text{sec}$. But the actual record includes four low peaks that plot well below the line of the graph, and one highest peak that plots well above. When these values are taken out, the arithmetical average becomes $8975\,\text{m}^3/\text{sec}$, only about 0.25% away from the graphically-derived value.

Table 6.1 Sample of the annual Flood Series for the Fraser River at Hope, British Columbia (discharges in m^3/sec)

Year	Annual floods Recorded	Annual floods Relisted	Rank order	Recurrence interval, years
1950	12 500	12 500	1	19
1951	8 040	11 600	2	8.5
1952	8 330	11 300	3	6.3
1953	7 220	10 800	4	4.2
1954	9 060	10 500	5	3.8
1955	11 300	9 770	6	3.2
1956	9 690	9 690	7	2.7
1957	10 500	9 520	8	2.4
1958	9 770	9 350	9	2.1
1959	8 740	9 060	10	1.9
1960	9 350	8 740	11	1.73
1961	9 520	8 580	12	1.58
1962	8 210	8 330	13	1.46
1963	7 700	8 210	14	1.36
1964	11 600	8 040	15	1.27
1965	8 580	7 900	16	1.19
1966	7 900	7 700	17	1.12
1967	10 800	7 020	18	1.06

It is also a property of the statistical design that the median annual flood has a recurrence interval of two years. In the long-term average, half of the annual floods should exceed the median value, half should fail to reach it. The two-year recurrence interval is the appropriate one: in the long-term average, the median value is equalled or exceeded every other year. The constant 1 in the calculating equation is used to ensure that the median annual flood does in fact have a recurrence interval of exactly two years.

The floods of chief interest in the behaviour of rivers through time are the most probable annual flood, and floods of high recurrence interval and therefore great magnitude. The most probable annual flood has a recurrence interval of 1.58 years: it is written $q_{1.58}$ (annual series). It is to this discharge, at the most probable annual flood, that stream channels tend to be adjusted. If we used a more elaborate time series, taking account of more than one flood peak a year, we should find that the recurrence interval of the most probable annual flood is precisely one year.

6.6b Some usual problems of analysis and forecasting

Short series of records may well, by pure chance, include events of low probability. Even the record for the Fraser River, which at fifty-six years is of quite respectable length, includes events that are not likely to happen in an average fifty-six year period. Such events are represented by the very low flows at the lower tail of the series, and by the record discharge in 1948 of $15\,200\,\text{m}^3/\text{sec}$. The graph suggests that the recurrence interval of so high a discharge is about 200 years. The event just happened to fall within the fifty-six years of record.

Not many stream records are as yet long enough to permit the definition, with any degree of confidence, of the 50-year and 100-year floods. On the other hand, it is floods of this range which must be considered for many purposes of planning. The odds against a 50-year flood are 50 to 1 in any given year; those against the 100-year flood are 100 to 1. But in the long run, one 100-year flood and two 50-year floods can be expected in the average century. If necessary, a magnitude of $13\,300\,\text{m}^3/\text{sec}$ for the 50-year flood could be read from the graph. Really firm information, on the other hand, is confined to recurrence intervals from about thirty years downward. It could prove especially risky to fix on $14\,200\,\text{m}^3/\text{sec}$ as the magnitude of the 100-year flood.

Fig. 6.12 graphs an unusually long annual flood series – that of the Danube at Orsova in Romania. Extending from 1840 through 1968, the series con-

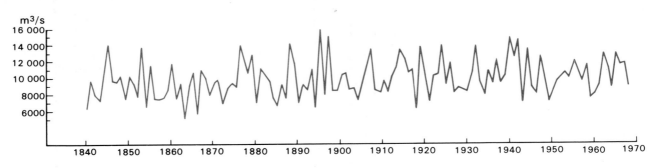

Fig. 6.12   Annual flood peaks on the Danube River at
Orsova, Romania.

Fig. 6.13   Magnitude–frequency graph for the peak flows
used in Fig. 6.12.

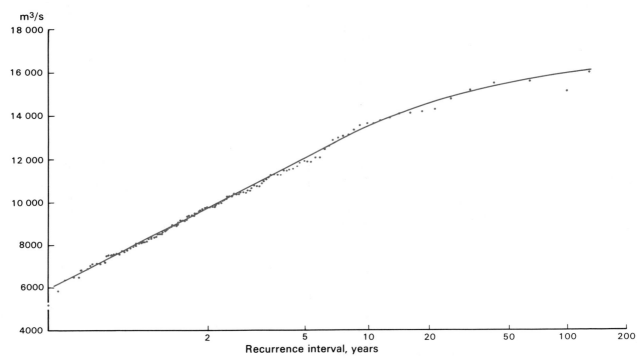

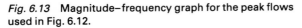

tained flood listings for 129 years. When recurrence intervals are computed, and the points plotted up, a straight-line graph appears up to the recurrence interval of about ten years, but then the band of the scatter bends over (Fig. 6.13). It seems possible that we are dealing with a variant type of probability, where an increase of recurrence interval brings a lesser proportionate increase in discharge than occurs for instance on the Fraser.

Roughly speaking, an annual series should for preference be twice as long as the recurrence interval used in predicting flood probability. To predict a 50-year event, we need a 100-year series. To predict a 100-year event, we need a 200-year series – if we can get it. The 129-year record for the Danube permits us to predict with some confidence for recurrence intervals from about 65 years downward.

The second highest point on the graph is plotted at the recurrence interval of 65 years exactly, the third is plotted at 43.3 years, and the fourth at 32.5 years. It seems really likely that the curve-over of the graph indicates the true behaviour of this annual flood series.

A similar conclusion can be drawn, for similar reasons, about the highest mean monthly levels on Lake Michigan (Fig. 6.14). The very lowest record, and the one highest record of all, plot away from the generalised graph; but it is reasonable to take the mark of 0.8 m above long-term mean level as the 50-year monthly mean high.

An exceptionally long record exists for outflow from Lake Vänern in Sweden (Fig. 6.15). Annual mean discharges are used here, as opposed to extreme high flows, in order to show that magnitude

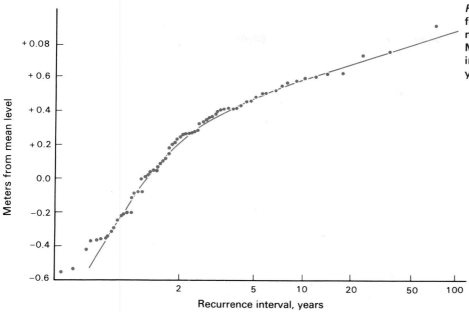

Fig. 6.14 Magnitude–frequency graph of the highest monthly mean levels of Lake Michigan at Milwaukee, as used in Fig. 6.3: one extreme per year.

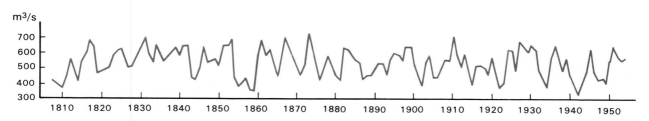

Fig. 6.15 Graph of annual mean discharges, Lake Vänern, Sweden.

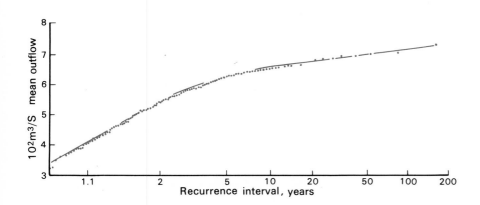

Fig. 6.16 Magnitude–frequency graph of the discharges used in Fig. 6.15.

–frequency analysis works for events other than extreme events. The magnitude–frequency graph appears in Fig. 6.16. Once again, the graph is curved over. We are dealing here, not only with a type of probability that brings down very high extremes, but also with negative feedback. A rise in lake level increases discharge through the lake's outlet: the greater the rise in level, the greater the increase in outflow. The greater the increase in outflow, the greater the feedback which reduces very high lake

levels. Both for Lake Vänern and for the Great Lakes system, the feedback relationship is very simple. The connection between lake level and outflow is linear (Fig. 6.17).

No such simple influence of negative feedback can be easily imagined for the discharges of the Danube at Orsova. We seem obliged to conclude that the slow increase in magnitude through the range of very high frequency at that station is in some way related to the low probability of very high runoff. But not all

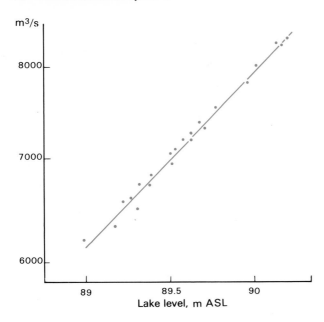

Fig. 6.17 Linear relationship of lake level and discharge, Lake Michigan: more than 99% of variation in discharge is explained by variation in level.

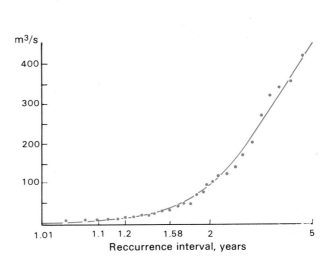

Fig. 6.18 Magnitude–frequency graph for annual peak discharges on the Wollombi Brook, N.S.W., Australia.

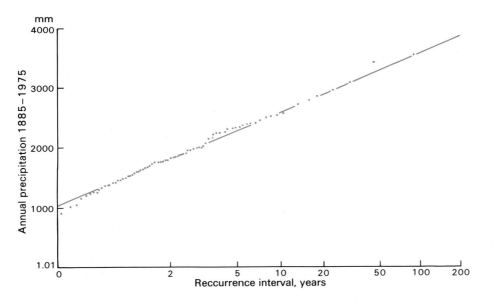

Fig. 6.19 Magnitude–frequency graph of annual rainfall at Sydney, N.S.W., Australia: the plot strongly suggests a straight-line graph, but see Fig. 6.20 and accompanying text.

rivers act either in this way, by producing a curved-down graph, or like the Fraser River, in producing a straight-line graph. For some rivers, the graph of magnitude–frequency is curved upward (Fig. 6.18). Such a graph represents a third type of probability, where, as frequency diminishes, magnitude rapidly increases. There is presumably some cutoff point, beyond which an increase in magnitude of discharge becomes physically impossible. This point comes where the water necessary to increase streamflow still further simply cannot be carried in the air.

Magnitude–frequency analysis can be applied to rainfall totals in individual years or in individual months. Fig. 6.19 is a magnitude–frequency graph of annual rainfall totals at Sydney, Australia. In the ninety-one years of record, 1885–1975, the extreme high total was 3477 mm. Plotted at the computed recurrence interval of ninety-two years, this value lies exactly on the line that generalises the rest of the scatter. The indicated 100-year total is about 3500 mm.

But when the individual annual totals are plotted as a time series (Fig. 6.20), it looks very much as if we are dealing with something more than purely

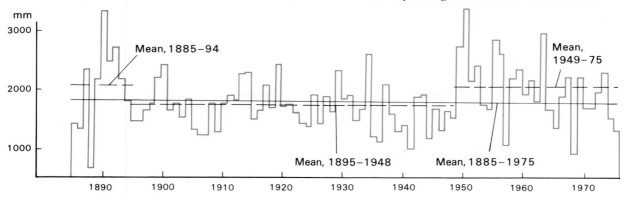

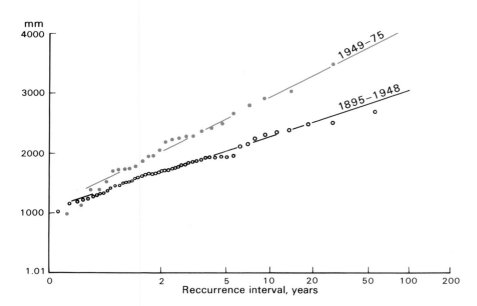

Fig. 6.20 Annual rainfall at Sydney plotted as a time series: differences of means between the different blocks of years are statistically significant.

Fig. 6.21 Magnitude–frequency graphs of annual rain at Sydney, plotted for two separate blocks of years in accordance with Fig. 6.20: contrast Fig. 6.19.

random variation. For the brief recorded interval 1885–94, the annual totals include the extreme low of record, plus the second greatest recorded high. The annual mean for this interval is just short of 2100 mm. For the interval 1895–1948, the annual mean is 1750 mm. Few peaks rise far above this value, few troughs dip far below. From 1949 onwards the series once more becomes highly variable, with some very high peaks and some very low troughs. The average for this final interval is 2100 mm per year.

6.6c Step-functional variation

It seems highly probable that we are here dealing with step function. Before 1895, variability was great but the annual mean ran high. From 1895 to 1948, variability and the interval mean were both low. From 1949 onwards, high variability and a high annual mean have reappeared.

When Gumbel graphs are separately drawn, for the annual rainfall totals 1895–1948 and 1949–75, the results are those shown in Fig. 6.21. The 100-year annual rainfall for the former interval is only 3050 mm, which has been exceeded more than once in the actual record. The corresponding total for 1949–75 is 4100 mm, 25% as great again, which has not yet been logged. If existing trends continue, such a total will be observed.

In a similar way, magnitude–frequency analysis for annual floods on the Fraser River at Hope sorts itself out into two series (Fig. 6.22). For 1925–46, the mean annual flood is 8200 m^3/sec, and the projected fifty year flood is 12 600 m^3/sec. For 1946–67, the mean annual flood is 9600 m^3/sec, and the projected fifty-year flood is 14 800 m^3/sec. If a step function has actually occurred, then the existing river needs a channel nearly one-fifth as large again (9600/8200) as it needed in earlier years. The implications for bank erosion are obvious: the river will enlarge its channel by cutting its banks back.

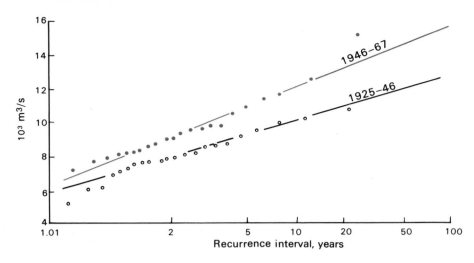

Fig. 6.22 Corresponding graphs for the Fraser River: contrast Fig. 6.11.

Step-functional behaviour of hydrologic systems is at least as difficult to cope with as is progressive variation. Like progressive variation, step-functional variation fails to respond to cyclic analysis. It too involves typically sudden changes in rate and direction – and also in level, as in the mean level of annual rainfall at Sydney. But if step functions are the moves that hydrologic systems make, they are what we have to live with. The matter of predicting steps is something altogether different.

Chapter Summary

Observed *hydrologic systems may be analysed for cyclic and non-cyclic patterns of behaviour over time.* Analysis determines *what happens*, and encourages *attempts to predict for the future.* As yet, however, *future prediction* promises to be only *partly successful*, on account of *progressive variation* and *step functions.*

Long-term means impose a sort of *crude order by main force.* They *suppress* all the *noise* of variation about the mean value. Stations with identical mean annual rainfall can have highly contrasted annual rainfall regimes.

Annual regimes are *defined by monthly mean values.* These values *suppress* the *noise of variation within the month.*

Cyclic analysis has been *mainly applied to annual series. Many* annual series, such as those of lake levels and peak streamflow, *suggest sine wave variation.* Sine waves are *also suggested by smoothed records*, where for instance the data are averaged over five-year periods. Smoothed records *suppress* the noise of *short-term variation.* Sine waves actually do occur in the behaviour through time of many hydrologic subsystems: they have accordingly commanded much attention in the analysis of system

behaviour and in attempts at prediction for the future. *Tides* can be interpreted, and *forecast, as a combination of sine waves.*

Subtraction of sine wave variation *can suggest progressive variation. What is left over* after sine wave variation is subtracted *is the residuals. Some residuals* undoubtedly *show progressive variation*, as for the water level of Lake Victoria.

Many efforts have been made *to relate cyclic variation* to the *sunspot cycle* of about eleven years. They have so far proved *unsuccessful.*

Non-cyclic analysis relies on the *calculation of odds.* Magnitude of an event (for example, a stream discharge at a given station) is related to time, in terms of probability.

The *10-year flood* (= annual flood peak) has a $1/10$ *probability* in a given year. The *100-year flood* has a $1/100$ *probability. The probability converts to the recurrence interval* = average spacing. In the long term, the 10-year flood can be expected ten times in a century, and the 100-year flood once. The *mean annual flood* has a recurrence interval of *2.33 years*, on the series where only one peak flow a year is considered. The *most probable annual flood* has a recurrence interval of *1.58 years*, but *if lesser flow peaks* are also *taken in*, it can be seen to tend to occur *once a year.* It is to discharge at the most probable annual flood that *stream channels tend to be adjusted.*

Streamflow analysis is commonly undertaken with the aid of *Gumbel graphs.* Crudely put, the statistical design states that *records exist to be broken. Three types of probability* apply. If the *graph curves down in its upper end*, some effect of *negative feedback is at work*, as in lake outflow, tending to suppress the rise in level. *If the graph curves upward, interpretation can be difficult. If the graph is straight, then the variable* (rainfall, streamflow, lake level) *seems to behave randomly* through time.

However, *some time series can be dissected into intervals.* Magnitude–frequency graphs drawn for discrete intervals produce *contrasted results*, and also *contrasted predictions* for what is likely to happen in the future. The examples used suggest *step functions*, as opposed to cyclic or progressive systems behaviour over time.

7 Surface Morphological Systems: Stream Nets, Stream Slopes and Drainage Basins

Streams are fed by water that rises in springs, or that runs off the surface of the ground after rain falls, or that comes from melting ice or snow. The relative contributions vary greatly from one area to another. In high mountains with glaciers, streams begin with the flow of meltwater. In regions with a winter cover of snow, spring meltwater soon reaches the stream channels. In deserts and in areas of impermeable rock, springs contribute nothing. In some areas of limestone country and in some basalt terrains, water coming out of the ground is far more abundant than anything running directly over the surface (Figs. 5.4, 5.5). In all cases, however, all that is needed to form a stream is enough water to cut a channel. The study of rivers is the study of channelled surface flow.

For practical purposes we make a rough distinction between perennial streams and ephemeral streams. By strict definition, perennial streams ought never to stop flowing. In practice, some of them dry into stagnant pools in occasional times of drought. In climates with hot dry summers but rainy winters, there is a well-defined season of low streamflow, which at the extreme involves a complete drying-up. But streams in really dry climates are truly ephemeral. They flow at high irregular intervals, and then only for a short time. There may be years between two occasions when the channel of an ephemeral stream actually carries water.

This and the following chapter will be concerned primarily with perennial streams. Much of what will be said however – especially what is said about network geometry – applies to ephemeral streams also.

7.1 Network pattern

It is important to distinguish between network pattern and network geometry. The study of network pattern is concerned exclusively with the way that networks are related to rock structure, whereas network geometry deals exclusively with the internal relationships of the network itself.

The ways in which networks become adjusted (or for that matter, maladjusted) to structure will be discussed in Chapter 10. For the time being it is enough to recognise certain common types of pattern. Something here depends on the scale of obser-

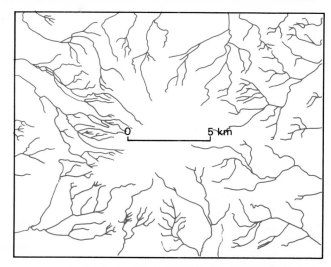

Fig. 7.1a Radial drainage on a volcanic cone. (Mt. Rainier, Washington).

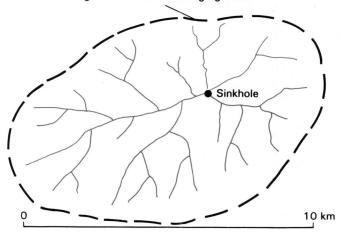

Fig. 7.1b Inward-radial drainage in limestone country (synthesised from actual examples).

vation. On the local scale come the outward–radial streams developed on the slopes of a volcanic cone (Fig. 7.1a). The opposite case is that of inward-radial streams, that drain a solutional hollow in limestone country and finally vanish underground down the sink at the centre (Fig. 7.1b). The parallel

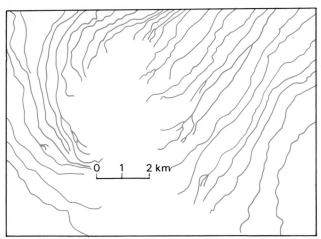

Fig. 7.1c Parallel drainage (Antelope Peak, Arizona).

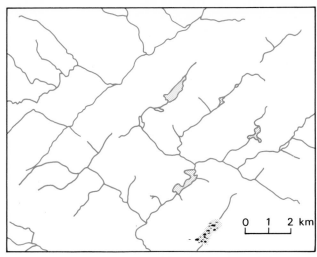

Fig. 7.1e Rectangular drainage (Elizabethtown, N.Y.).

streams of Fig. 7.1c flow down a uniform slope: such streams occur on artificial banks, on the sides of artificial cuts, and on tidal flats. In all these three local cases, ground slope is the main control over stream direction.

Other patterns appear on the regional scale. The dendritic (= tree-like) network of Fig. 7.1d has developed with only general control by slope and no control by structure. This is the kind of pattern that streams would develop if they could. But where structural control is imposed, the network cannot become tree-like. The highly angular (rectangular) stream net of Fig. 7.1e is adjusted to the cross-faults of part of a craton. The land surface as a whole is subdued, but the rocks are strong. Lines of faulting provide the only lines of weakness. The streams have concentrated themselves upon these lines. A

structural dome involving alternating strong and weak formations of sedimentary rock becomes etched into a system of ridges and valleys which forms rings in the plan view. Rivers are mainly contained in the ring-valleys, forming a pattern of ring drainage (Fig. 7.1f).

The trellis drainage pattern often developed on eroded folded rocks is somewhat difficult to understand at first sight: it is, however, well-defined in nature (Fig. 7.1g). The main streams run along the fold axes. Tributaries come down the flanks of the ridges of resistant rock. Finally, the really patternless (= deranged) drainage where the landscape has been blanketed by glacial deposits, and where the network has not had time to organise itself properly, stands in direct contrast to all the patterns distinguished so far (Fig. 7.1h).

As stated, stream nets tend to grow like trees, if they are free to do so. We can look on structure-guided patterns as trees trained, by force, into artificial patterns. Network geometry is independent of network pattern. It works for all properly organised networks, regardless of the pattern displayed.

Fig. 7.1d Dendritic drainage (Effingham area, Illinois).

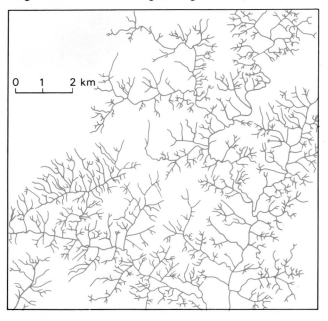

Fig. 7.1f Ring drainage (Maverick Springs, Wyoming).

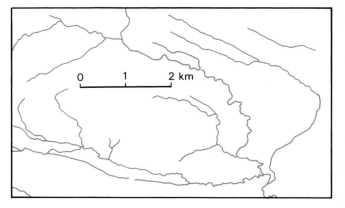

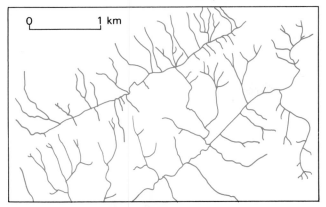

Fig. 7.1g Trellis drainage (Ewing, Kentucky–Virginia border).

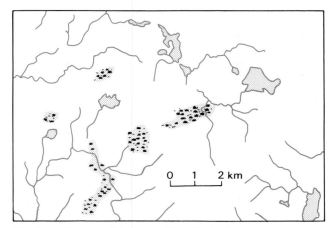

Fig. 7.1h Deranged drainage (New Hampshire).

7.2 Network geometry

Like trees, streams send out branches. Indeed, we refer to a trunk stream or to a main stream, and give the name branch or fork to principal tributaries. But the range of names for the component parts of trees is limited. It is also imprecise. The words stem,

trunk, branch, fork, bough and twig more or less exhaust the available choice – unless we wish to compare the root system with a delta. How for instance can we distinguish between a branch and a bough? For the analysis of stream network geometry, something more elaborate is needed. That more elaborate something works also for trees, as it works for networks with well-defined patterns, such as those that have been illustrated.

7.2a Stream order

The basic concept is the concept of stream order. Order indicates the ranking position of a given stream in the network that contains it. Classification of streams by order permits the components of a network to be identified by rank.

The concept is extremely simple. A headwater stream with no tributaries belongs to the lowest rank of all, the first order. When two first-order streams unite, they produce a stream of the second order. When two second-order streams unite, they produce a stream of the third order, and so on (Fig. 7.2a). The attribute of order defines the position of a stream in the network hierarchy (= ranking system). It shows how a particular stream is related to the total network: that is, how the components of the network system are structured together.

Promotion up the hierarchy occurs only when two streams of equal order unite. A stream of the fourth order can receive any number of streams of the third, second and first orders, without being promoted to the fifth order. Only when another fourth-order stream comes in is the combined stream promoted (Fig. 7.2b).

The practice of labelling streams by rank order is the practice of stream ordering – or, as it is usually called, the practice of ordering-up the network. Stream ordering has two purposes: to provide a basis of comparison, and to permit the analysis of network geometry.

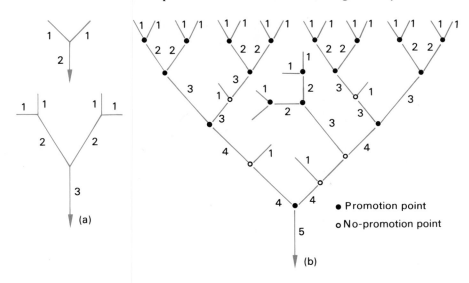

Fig. 7.2 Stream order: (a) the simplest case of continuous promotion; (b) combined promotion and non-promotion.

• Promotion point
○ No-promotion point

7.2b Stream number

For a given network (= basin), stream number is the number of streams per order. For the first order, it is the total of first-order streams; for the second order, it is the total of second-order streams; and so on. Number must obviously increase as order decreases. There must be at least twice as many first-order streams as second-order streams, because a second-order stream branches into two first-order streams. Similarly, a third-order stream splits into two second-order streams. But because low-order streams can come in, without causing high-order streams to be promoted in rank, the ratio between number of streams in a given order (say, fourth) and number in the next higher order (in this case fifth), is commonly greater than two.

Fig. 7.3 portrays an ordered-up network. Numbers are listed against orders in Table 7.1. The table also gives values for stream length and for bifurcation ratio, which are explained below.

Table 7.1   Data for the Stream Network in Fig. 7.3

Stream order	Stream number	Stream length km	Bifurcation ratio
1	131	0.36	
			3.97
2	33	0.71	
			4.13
3	8	1.67	
			4.00
4	2	4.08	
			mean 4.03

7.2c Relationship between stream number and stream order

When number is plotted against order on a semi-logarithmic graph, the plotted points usually lie on, or close to, a straight line. In this case (Fig. 7.4) the fit to a straight line is nearly perfect. As order increases by addition (1+1 = 2, 2+1 = 3, 3+1 = 4), number decreases by division. The dividing factor between one order and the next order is the bifurcation ratio. For the streams in the table, 131 in the first order divided by 33 in the second order = 3.97, and so on. The mean bifurcation ratio for the network is slightly greater than 4.0: that is to say, for every fourth-order stream there are four third-order streams; for every third-order stream there are four second-order streams; and for every second-order stream there are four first-order streams. As already seen, the lowest possible bifurcation ratio is 2.0 – there must be at least two first-order streams for

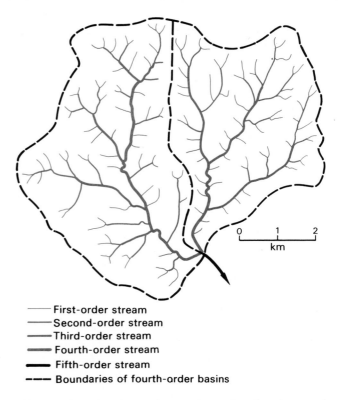

- —— First-order stream
- —— Second-order stream
- —— Third-order stream
- —— Fourth-order stream
- —— Fifth-order stream
- - - - Boundaries of fourth-order basins

Fig. 7.3   An ordered-up real network: two fourth-order basins.

every second-order stream. Many real networks have bifurcation ratios of about 3.0 to 4.0, because low-order tributaries come in without promoting the high-order trunks.

The kind of relationship illustrated in Fig. 7.4 applies to the vast majority of natural stream nets, regardless of the type of network pattern illustrated in Fig. 7.1. It simply expresses the way in which streams tend, and tend very strongly, to develop their network geometry. Network geometry is normally systematic, in the sense of revealing an orderly relationship among the various system components.

7.2d Stream length

Stream length is the average length of stream per order. It is the total length of all streams in the order, divided by the number of streams in that order. Determination of stream length is inevitably tedious. For the 131 first-order streams in Fig. 7.3, individual measurements on each component produced a total of 47.01 km of length, and the indicated average of 0.36 km. Assume the bifurcation ratio of 4.0; then, for a system containing one tenth-order stream, there will be four ninth-order streams, 16 eight-order streams ... and so on, up to 262 144 first-order streams, and 349 524 streams in the total system. For mean length per order to be determined, every single stream must be measured. Measurements have, however, been made on very many networks. The results are not only what might be expected in a general way, but are once again very orderly.

As one might guess, mean length increases as order increases (Table 7.1). The trunk of a tree is longer than a twig. The main stem of a river is longer than an unbranched headwater. Once again, a systematic relationship appears. Fig. 7.4, already used to show the relationship between order and number, also shows the relationship between order and length. Another straight-line graph appears. The fit of the points to the straight line is only slightly less than perfect. Length (= mean length per order) multiplies as order adds. In this particular case, the multiplying factor for length is about 2.25 – second-order streams average about 2.25 times as long as first-order streams, and so on. The multiplying factor for length, like the bifurcation ratio, can vary from one network to another; but, like the number –order relationship, the number–length relationship holds good, regardless of network pattern. We are dealing with a sweeping generalisation.

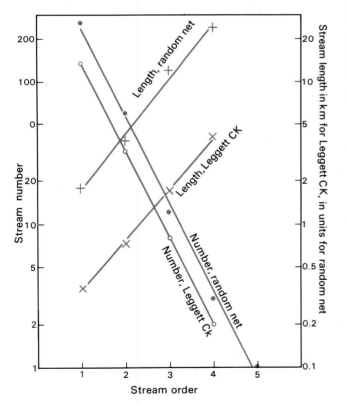

Fig. 7.4 Plot of stream number and stream length (vertical scale: logarithmic) against stream order (horizontal scale: arithmetic) for the networks shown in Figs. 7.3 and 7.5. Straight lines fit the sets of data points with better than 99% accuracy.

7.3 Network geometry and a growth law

The type of relationship where one variable changes by multiplication or division, while the other variable changes by addition or subtraction, is called exponential growth or exponential decay (see Chapter 25). It is especially familiar in the contexts of compound interest and the population explosion.

Money invested at 6% compound interest will slightly more than double itself every twelve years. Every 100 units invested will produce 200 units in twelve years, 400 units in twenty-four years, and 800 units in thirty-six years. The rate of increase remains constant, but the absolute amount added each year increases.

Similarly with population totals. A growth rate of 25% per decade – well inside observed limits for some countries – implies a growth rate of more than 2% a year. At an even 2% a year, a population total would increase by more than 2½ times in 50 years and by more than 7 times in 100 years. Numbers multiply as time adds.

Comparable results appear in any instance of exponential growth. Consider the legend of the origin of chess. The legend tells that an emperor in ancient China, tiring of war (or possibly having conquered all available foes), wished for a war game. A court sage invented chess – the earliest known war game of any complexity. The emperor, delighted with the campaign of the chessboard, offered a reward up to half the content of the imperial treasury. The inventor politely declined. He asked only for one grain of rice for the first square on the chessboard, two grains for the second square, four grains for the third square, eight grains for the fourth square, and so on, doubling up each time to the sixty-fourth square.

The emperor, as pleased with the modest price as with the game itself, agreed at once. Now try it on a graph. A graph drawn on plain paper cannot cope with the huge totals per square that soon begin to appear. A graph drawn on semi-logarithmic paper can: it forms a straight line. The horizontal scale is numbered at uniform intervals for the squares on the board, 1, 2, 3, 4, ..., 64. The vertical scale is numbered at uniform intervals for the numbers of grains per square, $1, 2, 2^2, 2^3, ..., 2^{63}$. Incidentally, the total of 2^{63} grains on the final square is itself greater than all the rice that exists in the world today.

The branching system of a tree can be ordered, numbered, and measured for length, just as a stream net can. The results produce diagrams strictly similar to Fig. 7.4. Analysis of the branching tubes and tubules in a human lung again produces the same kind of result. The systems of human arteries and veins do the same kind of thing – although most here

First order
Second order
Third order
Fourth order
Fifth order

Fig. 7.5   A stream network synthesised by random means.

is known about the ratios of diameter. When an artery subdivides, or when veins join, the diameter of the larger tube (one order up) is precisely twice the diameter of the two smaller tubes (one order down). In all these and many other contexts, it seems that nature – whatever nature may be – having once hit on a good idea, uses it over and over.

7.4 Network geometry and random variation

Branching stream nets can be simulated by random methods. Simulated nets display the same kind of network geometry that is observed for real streams in nature.

Randomly branching networks are most easily simulated with the aid of a computer, but good results can be obtained by means of card-flipping, coin-tossing, or the throwing of dice. The network in Fig. 7.5 was generated by card-flipping: club = move

one step forward, diamond = move left, heart = move right, spade = move back. A few ground rules are needed. The array has defined boundaries, which no stream except the main trunk is allowed to cross. Impossible instructions are ignored: alternatively, the instruction is obeyed at the first possible point. The pack of cards is thoroughly shuffled after each run through. Nine hundred and fifty flips generated the random network in Fig. 7.5. The data for the simulated network are listed in Table 7.2

The results of ordering, numbering, and measuring for length are very similar to those obtained for the real network. In fact, the real network in this case produces rather better fits to straight-line graphs than does the random network (Fig. 7.4). As in the real network, so in the simulated network, stream number divides as stream order adds: the graph of number is straight, descending to the right. Moreover, its slope is very similar to that of the graph for the real network: the close correspondence would be expected, from the near identity of the two mean bifurcation ratios. Again, the graph of length for the simulated network is straight, ascending to the right: length multiplies as order adds. The two graphs of length are almost parallel: the rates of increase of length, as order adds on, are almost identical. If one arbitrary unit (= step) in the simulated network is taken to represent ⅙ km, then the mean length for fourth-order simulated streams is 4.0 km, while that for simulated first-order streams is 0.31 km. Compare these values with those listed in Table 7.1.

We draw the indicated and acceptable conclusion that network geometry, as observed in nature, and as imitated by simulation models, results from random branching.

7.5 Network geometry and channel dimensions

Channel dimensions include width, depth, cross-sectional area and downstream slope. These can obviously be called attributes of channel systems. But the study of channel dimensions had begun, long before the systems idea was in use. The general term adopted for dimensional attributes, plus discharge of water and sediment, plus channel form in plan, is channel characteristics. It is far too late in the piece to replace the word characteristics by the word attributes. Thus, characteristics = attributes; and dimensional characteristics form a subset of characteristics in general.

Channel dimensions can be studied in relation to network geometry. Numbers of studies are in fact on record. They tend to show that width, mean depth,

Table 7.2   Data for the Simulated Network in Fig. 7.5

Stream order	Stream number	Stream length (arbitrary units)	Bifurcation ratio
1	262	1.85	
			4.38
2	60	3.53	
			5.00
3	12	12.25	
			4.00
4	3	24.0	
			3.00
5	1	incomplete	
			mean 4.10

and cross-sectional area all increase exponentially as order increases. As order is added, width, mean depth and cross-sectional area multiply. As order increases, slope decreases exponentially: as order adds, slope divides.

But the investigation of channel dimensions in relation to stream order has been largely short-circuited. It is the flowing water that shapes the channel, determining width, depth, cross-sectional area, and slope. Where information on discharge exists, channel dimensions are studied in relation to discharge (Chapter 8). Where information on discharge does not exist, channel dimensions are studied in relation to drainage area. Drainage area provides an implied substitute for discharge. Position in the network involves a further substitution – network position is substituted for drainage area.

Nevertheless, the relationships can be instructive. For a theoretical perfect network, channel width should increase by a factor of 1.75, for each promotion up the network hierarchy. Mean depth should increase by a factor of 1.44. Cross-sectional area should increase by a factor of $1.75 \times 1.44 = 2.52$.

7.5a Network geometry and channel slope

Channel slope is in some ways a different proposition from width, depth and cross-sectional area. Like these, it often demands measurement in the field, as opposed to measurement on maps or air photographs. Where measurements have been made, they tend to show that slope decreases as order increases; also, that the rate of decrease becomes less as order continues to increase. But slope studies were well advanced, long before the idea of stream order had been established.

To begin with, slope was related to distance along

the channel. More precisely, channels were measured for height above sea level in relation to distance. A plot of height against distance is a long-profile (Fig. 7.6a). Long-profiles tend to be concave-up, and to decrease in gradient from source to mouth. They also tend to be exponential, forming approximately straight lines when plotted on semi-logarithmic paper (Fig. 7.6b).

Now, by definition, the uppermost part of a long-profile belongs to a first-order headstream. Equally, the lowermost part belongs to the highest-order stream in the network. A first-order stream drains only a very small area. Its channel-forming discharge is the least possible. A great river in its seaward reaches, draining a vast area, collecting an enormous channel-forming discharge, and ranking at the top of its network hierarchy, needs a huge channel.

7.5b Channel size, channel slope, and flow velocity

A very small (= first-order) channel, exerting maximum friction on the flowing water, needs a steep slope to keep the water moving. In a very large channel, the water in the vast midstream is not affected by friction on the bed and banks. The slope

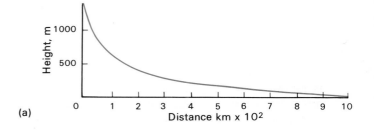

(a)

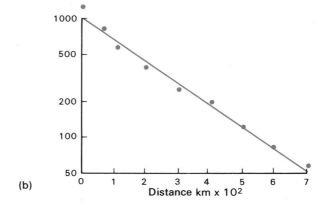

(b)

Fig. 7.6   Channel slope: (a) arithmetic plot of height–distance relationship; (b) semi-logarithmic plot of the same data. Measurements taken on a real stream. Fit of straight line to data points in Fig. 7.6b is better than 99% accurate.

can be very gentle, yet still permit flow velocity to be maintained.

Two things need to be stated at once. One is, that the idea of streams roaring tumultuously down mountainsides, but flowing sluggishly across the contrasted plain below, has no basis in observed fact. The other is, that we need a basis of comparison. This basis is provided by discharge and velocity at the most probable annual peak of discharge (Chapter 6). This discharge has been recorded as the discharge to which channel dimensions tend to adjust. The adjustments include adjustments of channel slope.

Flow velocity in real streams is affected by so many variables that it can be impossible to make out any pattern of variation along the length of a given stream. Another way of putting this is to say that it is usually impossible to prove that velocity at $q_{1.58}$ is any different in the headwater reaches from what it is in the middle or downstream reaches.

Theory predicts that, on an ideal stream, velocity should increase very slowly in the downstream direction. We do in fact have records of quite modest velocities, of the order of 1.2 m/s, on high mountain streams, where slopes run at about 40 m/km, and of a velocity in excess of 2 m/s on the lower Amazon, where slope is of the order of 1 cm/km, only one four-thousandth of the slope of the high mountain streams.

7.6 Stream slope: random variation or compromise?

Just as stream nets can be simulated by random means, so can long-profiles. Any simple technique of simulation produces profiles that descend in steps;

but when the steps are generalised by a smooth curve, the familiar concave-up profile comes into being.

There is however another way of looking at long-profiles. It can be shown that, if the work done by the stream were equally distributed along the profile, slope would be inversely proportional to the square root of discharge. That is, slope would decrease in proportion to the increase in the square root of discharge. If the work were so distributed that its total for the stream became the least possible, then slope would decrease in inverse proportion to discharge: a doubling of discharge would halve the slope.

The two extreme statements are written

$$S \propto q^{-0.5}$$

and

$$S \propto q^{-1.0}$$

– equations which merely express by symbols what has already been written in words.

The profile where slope is proportional to the −0.5 power of discharge is much less strongly concave than the one where slope is proportional to the −1.0 power of discharge. Fig. 7.7 illustrates the difference. It is drawn for realistic assumptions about channel length, drainage area, discharge, and total fall from mountains to the sea.

Most real long-profiles fall somewhere between the two extremes. They seem to represent compromises between the tendency for work to be distributed uniformly along the stream, and for the tendency for work to sum to the least possible total. The intermediate profile shown in the diagram, where slope is proportional to the −0.74 power of discharge ($S \propto q^{-0.74}$) is that of an ideal stream.

Fig. 7.7 Typical long-profile as a compromise between equal-work and least-work situations.

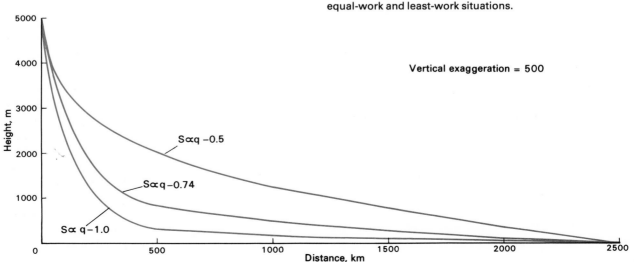

Vertical exaggeration = 500

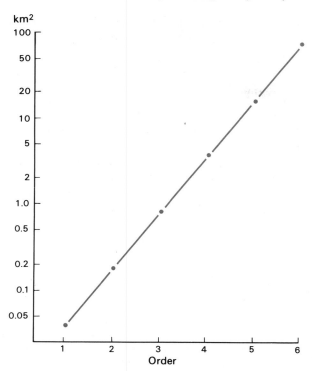

Fig. 7.8 Relationship between drainage area and stream order, for a sixth-order basin in S.W. Wisconsin: analysis involves more than 10 000 first-order streams.

7.7 Drainage basins

Basin order provides a basis for comparison, both within and between basins. Such attributes as channel dimensions and valley-side slope can be compared between basins of the same order in different networks, or between basins of different order in the same network. Like channel dimensions, basin area within a given network tends to increase by multiplication as order increases by addition (Fig. 7.8). In this particular example, the multiplying factor is about 4.5; the average area of a second-order basin is 4.5 times the average area for a first-order basin, and so on up the hierarchy.

7.7a Basin order and process-response

We come now to a slight qualification of the earlier comments about the substitution of stream order for drainage area, and of drainage area for discharge. In many of the mid-latitude networks that have been studied so far, there is a break in process-response at about the fourth, fifth, or sixth stream (and basin) order. Where the break occurs in a particular network depends to some extent on how accurately the first-order streams have been identified. If numbers of them have been missed, the break will come between third and fourth, or fourth and fifth, order. If most or all have been counted in, then the break

will come between the fifth and sixth, or between sixth and seventh, order.

The break can be represented by a slight mismatch, for instance of the two parts of a graph of channel size or channel slope. It probably results from the impact of single storms. In the low-order streams below the break, flood waves crest and fall very rapidly. In the higher-order streams above the break, flood waves are slower to rise and also slow to fall. Fig. 7.9 graphs the response to the flood rains of Hurricane Agnes (1972) on the part of two contrasted streams. Skippack Creek, with a drainage basin of 139 km² and an order certainly no higher than six, reached its flood crest and then subsided within a matter of hours. At peak flow, it discharged 2.2 m³/s for every km² of basin. The Schuylkill River, of eighth order or thereabouts, and with a basin of 4900 km², spread its flood wave over three days. The peak flow was only 0.6 m³/s for every km² of basin.

It is obvious that the outlet stream of a large network will integrate the effects produced on individual feeders. As drainage area increases, the time for a flood to pass is increased, and the flood peak is flattened. However, there does seem to be a little more to the matter than the general influence of increasing drainage area.

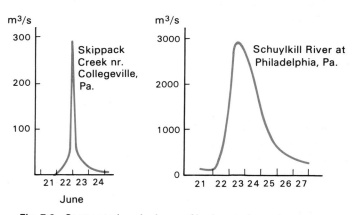

Fig. 7.9 Contrasted peakedness of hydrographs on two streams of contrasted drainage area, for the Hurricane Agnes floods of 1972: note the difference of vertical scale.

Although individual rainstorms are rarely confined to a single basin, low-order basins have a greater chance of being swamped in a given storm than have high-order basins. The sixth-order basins in Fig. 7.8 have an average area of 78 km². The predicted area for a seventh-order basin in the same network is about 350 km², and that for an eighth-order basin is nearly 1600 km². Where thunderstorm rain is at all important in supplying water to the stream channels, the break in process-response seems to come somewhere near the point where basin area resembles the average extent of an individual rainstorm.

7.7b Drainage area and stream length

The comparison between extending drainage networks and growing trees also needs some qualification. Trees grow up into the air, but also outward from their trunks. Stream nets extend themselves upslope. But only one line of the network can grow directly upslope. For the main branches from this trunk, there must be a cross-slope component. The odds that a network will grow upslope seem to be greater than the odds that it will extend itself by sideways branching. Extension upslope seems likely to be easier and more rapid than extension sideways.

If a drainage basin grows sideways as fast as it grows headward, then the relationship between trunk stream length and drainage area must always be

$$L \propto A_d^{0.5}$$

where L is trunk stream length and A_d is drainage area. Stream length is proportional to the square root of drainage area.

Consider the abstract model of a basin in Fig. 7.10, where the basin boundaries form a square and the trunk channel traces a diagonal. If the side of the square is 10 km long, the total area will be 100 km^2 and the diagonal will be 14.142 km ($= \sqrt{10^2+10^2}$). If the basin grows, keeping its same shape, until one side is 1000 km long, then the area is 10 000 km^2 and the diagonal is 141.42 km long ($= \sqrt{100^2+100^2}$). The area has multiplied itself by 100, while the diagonal has only multiplied itself by 10 – the square root of 100: $\sqrt{100}$, or $100^{0.5}$.

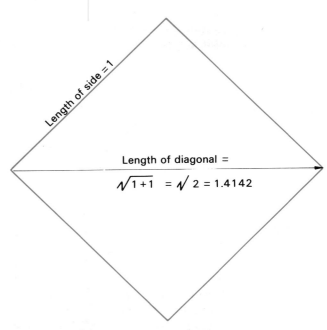

Fig. 7.10 Abstract model of the situation where trunk stream length and basin area increase at the same rate.

The relationship holds for any shape of basin. If a basin expands as fast as the trunk stream grows headward, then the length of the trunk stream is proportional to $A_d^{0.5}$. If the basin expands faster than the trunk stream grows headward, then the length of the trunk stream is proportional to less than the 0.5 power of A_d. If the trunk stream grows headward faster than the basin expands, then the trunk length is proportional to more than the 0.5 power of A_d. This last situation is what we commonly find in the real world. Specifically, length tends to increase with the 0.6 power of basin size:

$$L \propto A_d^{0.6}$$

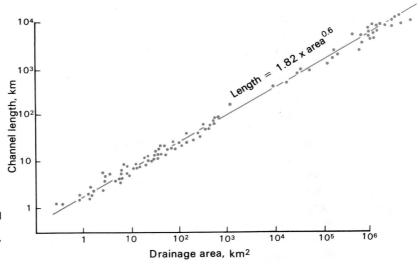

Fig. 7.11 The real-world situation: basins tend to lengthen faster than they expand.

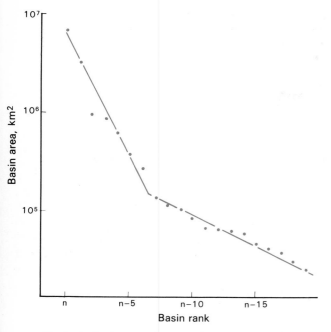

Fig. 7.12 Basin area plotted against basin-size rank, for large rivers in South America: the rate of increase in area with increase in rank rises abruptly, somewhere above $10^5 \, km^2$ of area.

7.7c Major river basins of the real world

For any continental landmass (Antarctica of course excepted), the size relationship between one basin and another turns out to be as systematic as it does for basins of different order in a single network. It is impossible however to start at first-order basins and work up through the hierarchy. Even if full coverage of maps or air photographs were available, the task of ordering and numbering for a whole continent would simply be too great.

Instead, basins are ranked in order of size, starting with the largest. This basin is given the order n. The second largest basin gets the order $n-1$, the third largest gets the order $n-2$, and so on. When basin area is plotted against order, quite simple graphs come into being. They are of the same general kind as those obtained for individual networks: area multiplies as order adds.

Because its drainage systems have been little changed by glaciers, South America will be used as an example (Fig. 7.12). For the seven largest basins, one step up the rank order means, on the average, multiplication of drainage area by 1.86 times. In the lower part of the range, a similar step means multiplication of drainage area by a factor averaging 1.15.

While it is easy to comprehend the systematic relationships among basins of different order in a single network, the systematic relationships among basins of different rivers of a whole continent is not easy to explain. We can only say that many systems in the natural world simply do sort themselves out into simple arrays of rank order, and that it is with the perception of some particular array that scientific enquiry often begins.

Chapter Summary

The *study of rivers* is the study of *channelled surface flow*.

It is important to distinguish between *network pattern* and *network geometry*. Network patterns are *affected by slope and by rock structure*. Common patterns are *radial* and *parallel*, affected by slope; *dendritic*, little affected by structure; *rectangular*, *ring*, and *trellis*, strongly influenced by various kinds of geologic structure; and *deranged*, where rivers have not reorganised themselves after glaciation.

Network geometry deals with the *internal relationships of a network*. The basic concept is that of *stream order*. A *first-order stream has no tributaries*. Where *two first-order streams join*, they *form a second-order stream*. Two second-order streams join to *form a third-order stream*, and so on. *For a given order* in a given network, *stream number is the number of streams in that order*. *Stream number divides as order is added on*. The dividing factor is the *bifurcation ratio*.

Stream length is the *average length* of streams *in a given order*. It *multiplies as order is added on*.

The relationships identified illustrate a *common mode of growth or decay – exponential growth or decay*. Other examples are the increase of money at compound interest, and the population explosion.

Branching stream nets can be simulated by random methods. Simulated nets display the same kind of number–order and length–order relationships that real nets do. The inference is that *real nets result from random branching*.

Channel dimensions tend to change exponentially as order increases; but the study of dimensions in relation to order has *largely been by-passed by their study in relation to discharge*. Network position is a substitute for drainage area, which itself is a substitute for discharge.

Channel slope has often been *studied in relation to distance* along the channel; a *graph of height against distance* is the *long-profile*. Long-profiles *tend to be concave-up*, with slope decreasing in the downstream direction. The *first-order headstream* at the top of the profile *needs a steep slope*, because its drainage area and channel are small and friction runs high. The *highest-order trunk* at the mouth *needs only a low slope*, because its drainage area and channel are large and friction is reduced.

There is often *too much noise on real streams* to reveal any tendency for *velocity*, at $q_{1.58}$, *to vary along the length of the stream*. Theory predicts that, *on an ideal stream, velocity* at $q_{1.58}$ *should increase slowly* in the downstream direction.

Many *real profiles* can be regarded as compromises between an *equal-work profile*, which is *less strongly concave* than a typical real profile, and a *least-work profile*, which is *more strongly concave* than a typical real profile.

Basin area in a given network increases by multiplication as order increases by addition. Channels in small basins rise rapidly in flood, and also rapidly subside. In large basins, flood peaks are spread and suppressed. There is often a *slight break in network geometry at the fourth to sixth order*, probably related to the relationship between size of basin and extent of single rainstorms.

Very many networks have grown headward somewhat *faster than they have widened*. A common situation is that where *length of trunk stream is proportional to the 0.6 power of drainage area*.

For the continental landmasses, information on *drainage area* is available only *for the highest-order streams*. Area plots against rank order in a *semilogarithmic graph*, similar to graphs for single networks, although there can be a distinct break in the multiplying factor within the total array.

8 Surface Morphological Systems: Stream Channels

Certain types of stream channel occur again and again. We recognise a particular type by its form in plan – its plan geometry. The primary basis of classification is morphological.

This chapter will not attempt a full classification. Two very common types of channel will be examined in some detail, while two other types will be treated summarily. In order to go beyond mere description, we need to take account of characteristics other than channel form in plan. These will include the width, depth, cross-sectional area and slope which were encountered in the previous chapter, plus width/depth ratio, bed form, flow velocity, discharge of water, and roughness (= total friction). For purposes of explanation, we shall need to consider channels as process-response systems. We shall also need a look at channel systems states, which is most conveniently taken at the very outset.

8.1 Negative feedback, steady state and dynamic equilibrium

Channel response to changes of input has been best studied in respect of single channels, as opposed to the multiple channels which some rivers form. The studies in question have mostly concerned channels with mobile beds – beds shaped in loose material which can be alternately scoured and re-deposited.

Channels with mobile beds are liable to short-term variations – variations spread over a few hours or a few days, or alternating from season to season. At the simplest, the bed is scoured as a flood peak rises, and is infilled as the peak subsides. Water and sediment entering the upstream end of a channel reach are eventually delivered through the downstream end. In the medium-term view, output matches input.

Some set of mechanisms is evidently at work, which tends to restore the dimensions of the channel to some central set of values. Although there is some time lag between input and response, or between input and output, we can regard these mechanisms as promoting negative feedback. A channel enlarged by scour on the rising flood will be larger than usual when discharge starts to fall. In consequence, velocity through the cross-section on the falling flood will be low. Low velocity will promote deposition of sediment, and the restoration of the channel section to something like its size and shape before the flood came through.

Although this description of what actually happens is much simplified, it does serve to show negative feedback at work. The very enlargement of the channel on the rising flood promotes infilling on the falling flood. The reversion to some central set of size and shape illustrates the steady-state condition. Were it not for negative feedback, steady state could not be maintained. The system would run wild (= fall into disequilibrium). This is precisely what happens when a gully cycle starts.

In the very long term, as the landmass is worn down, a river must reduce its slope. The total fall between the highest summits and the sea becomes less. Reduction of stream slope must mean net erosion of the channel bed. But the changes involved here are very slow in comparison with the short-term changes that occur between one flood peak and another. Maintaining its systematic identity through time, the channel system is said to be in dynamic equilibrium.

8.2 Channel plan geometry: meandering

Meandering streams are streams with single channels that wind from side to side (Fig. 8.1). The name comes from a river in Asia Minor, known to the ancient Greeks as the Maeander, and to modern Turks as the Menderes. The English name is derived from the ancient Greek form. One meander consists of one loop. Thus, two loops make up one wavelength (Fig. 8.2).

Many meandering streams approach dynamic equilibrium. The outside curves of the channel banks are liable to erosion, but deposition on the inside curves keeps pace with erosion on the outside curves, so that bedwidth is maintained. In addition,

Fig. 8.1   Meanders on Endrick Water, Dunbartonshire–
Stirling border, Scotland (J.K. St. Joseph).

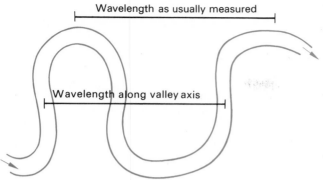

Fig. 8.2 Measurement of meander wavelength.

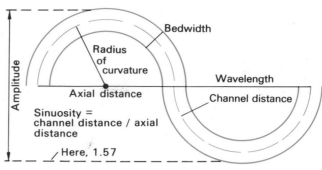

Fig. 8.3 Complete meander plan geometry.

the meander waves tend to sweep (= migrate) down-valley, but they can do this without losing their identity. The channel and the wave trains change their positions in space, but continue in being, preserving their plan geometry. The rate of erosion of outside curves varies much from one river to another, as also does the rate of downstream sweep. Some bends on the Rio Grande, which marks the boundary between Mexico and the U.S.A., have averaged a sweep rate of 0.2 km/yr for short periods.

Morphologically, meandering channels are treated in terms of bedwidth, meander wavelength and radius of curvature, which are of considerable interest; of sinuosity, which is of some interest; and of amplitude, which is scarcely interesting at all and which will not be discussed (Fig. 8.3).

Bedwidth is the width of channel measured between banktops. Wavelength is most easily measured from the extremity of one loop to the extremity of the next. Radius of curvature is measured for the arc of a circle drawn in mid-channel. Sinuosity is the ratio between distance along the channel and the straight-line (= axial) distance. There can be no fixed rule about how sinuous a channel has to be, before it is classed as truly meandering; sinuosities of 1.5 and upwards are widespread in nature. If the meander arcs were perfectly semicircular, then the sinuosity would be $\pi/(2\times\text{diameter}) = 1.57$.

8.2a Meanders on floodplains and in bedrock

The most familiar meanders are those which occur on floodplains. These are alluvial meanders. A typical floodplain is the creation of a meandering stream. At the outsides of the meander bends, the valley sides are trimmed back. The alluvium (= material deposited by the stream) of the floodplain has been carried down by the stream itself. It enters storage in the floodplain, either by settling out of

floodwater when the valley floor is inundated, or by lodging on the channel banks on the insides of meander bends. Deposits at these inside positions are called point bars. If the river constructs banks along the edges of the channel, these are called natural levees. Some rivers build levees, others do not.

The floodplain alluvium – general spread, point bars, levees and all – is subject to constant reworking as the channel shifts its position and as the meanders sweep down-valley.

Floodplains are liable to inundation, simply because the channel is adapted to accommodate the most probable peak flow ($q_{1.58}$: Chapter 6). Higher flows occur from time to time. They spill out of the channel as floodwater (Fig. 8.4).

Fig. 8.4 Inundation of the floodplain: the Severn in Gloucestershire, England (J.K. St. Joseph).

Fig. 8.5 (above) Meanders cut into bedrock, on the lower Wye, Welsh Border (J.K. St. Joseph).

Fig. 8.6a (below) Distorted meanders on Crooked River, Florida (U.S. Dept. of Agriculture).

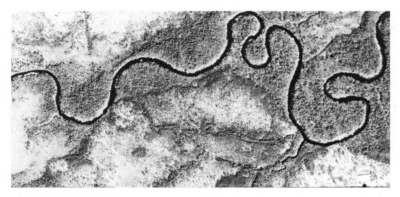

In striking constrast to streams on floodplains, some meandering streams have cut their bends deeply into bedrock (Fig. 8.5). Because such bends are typically enlarged as they are cut down, they are called ingrown. But apart from the fact that ingrown meanders are by definition not contained on a floodplain, there is nothing to choose between them and alluvial meanders. Some ingrown meanders are, it is true, highly distorted by the influence of structures in bedrock, but the average meander geometry is the same for both sets of bends. Apparent exceptions to this statement will be dealt with in Chapter 18.

8.2b The meander plan

There is a strong family resemblance between any one set of meanders and any other set, regardless of where they occur, and regardless also of scale. Admittedly, some meander trains are smooth and regular, while others – including some on floodplains – are highly distorted (Fig. 8.6a). At the extreme, distortion can mean that one meander is short-circuited by the next (Fig. 8.6b); for alluvial meanders, this in turn means that the short-circuited loop is liable to have its ends sealed by alluvium, becoming first a cutoff (= curved lake), then a

Fig. 8.6b (above)   Cut-off alluvial meander on the Tay, Perthshire, Scotland (oblique view: J.K. St. Joseph); (below) cut-off bedrock meander (actually a valley meander) on the Kickapoo, Wisconsin (vertical view: U.S. Dept. of Agriculture).

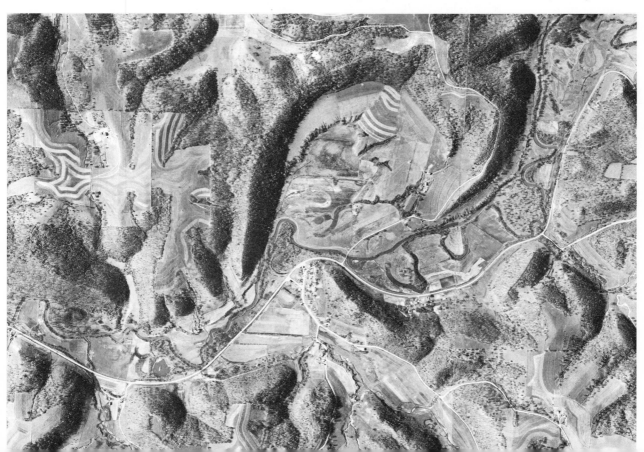

Fig. 8.7   Meander plan simulated by a bent wire cable (R. O'Brien).

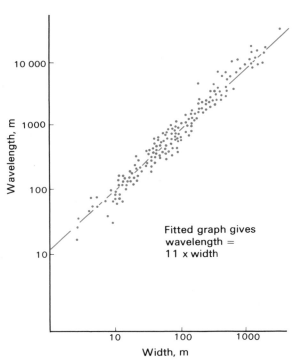

Fitted graph gives
wavelength =
11 x width

Fig. 8.9   Wavelength/width ratio for a large sample of natural channels.

Fig. 8.8   The most expensive meander simulation ever: train wreck near Greenville, South Carolina, 31 May 1965 (Leon E. Carnes).

swamp, and then after complete infilling an unrecognizable part of the floodplain. But distorted and cutoff meanders are in a sense only a distraction from the general case. They result from accidental irregularities in the resistance of floodplain alluvium, or from the structural guidance, already mentioned, of meanders ingrown into bedrock.

Fig. 8.7 presents a simulated meander bend. Its close resemblance to the real bends of Fig. 8.1 is undeniable. The simulation experiment uses a spring of thick wire cable. On a far larger scale, Fig. 8.8 presents a series of simulated bends. The spring here is formed by a railroad train of flatcars, loaded with bundles of rail track. A bulldozer on the rail line caused the flatcar train to derail. The three hauling locomotives came to a stop, but the middle and rear of the train kept coming on. The massive spring was deformed into a wave pattern, quite clearly simulating that of a meandering stream channel.

Both experimental springs were free to distribute their stresses along their length. It follows that the distribution of stress was evened out along the length involved. We are here dealing with an equal-work situation. If work had not been equally distributed, the bend patterns would have been distorted. We are also dealing with a least-work situation. If a channel is to be curved at all, then the standard meander plan enables the river to do least work in turning.

Another reason why meanders tend to look much the same, wherever they are found, is that their proportions do not vary much. It is unusual to find meander wavelength less than eight times bedwidth, or more than twelve times bedwidth. As a result, a

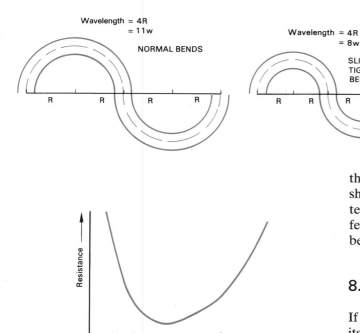

Fig. 8.10 Relationship of radius/width ratio to resistance to flow round the bend.

very large meander on a very large channel looks like an enlargement of a small meander on a small channel. The best central value for the wavelength/width ratio is 11 (Fig. 8.9).

The underlying explanation is that very tight bends (say, with a wavelength/width ratio of 5) and very open bends (say, with a wavelength/width ratio of 20) would increase resistance to flow round the bend. Resistance is least when radius of curvature is between two and three bedwidths (Fig. 8.10).

As the diagram shows, one wavelength is roughly equal to 4×radius of curvature. We should therefore expect wavelength to equal between 4×2 and 4×3 bedwidths, = 8 to 12 bedwidths. Above and below

these values, extra resistance to flow round the bend should be taken up by erosion or deposition that tend to restore the usual proportions. Negative feedback once more operates to control systems behavior.

8.2c Why rivers meander at all

If a meandering channel has a sinuosity of 1.5, then its slope is 1/1.5 = 0.67 of what it would be, if the channel went straight downslope. At first sight, we might be tempted to guess that meandering is merely a means of reducing channel slope. But floodplain slopes at least are determined by the rivers themselves. It seems highly unlikely that a river would construct a floodplain slope that is too great for the channel-forming discharge, and then correct it by becoming sinuous.

Many straight channels prove to be unstable, both in nature and in the laboratory. They develop sequences of pool (= deep) and riffle (= shallow). Pools or riffles are spaced at about 5 to 6 bedwidths apart (Fig. 8.11a). Once a pool-and-riffle sequence has been established, many channels develop side-to-side swing. Alternate pools migrate to opposite sides. Riffles remain at the crossings between bends (Fig. 8.11b). The instability of straight channels suggests that, with meandering streams, we may be

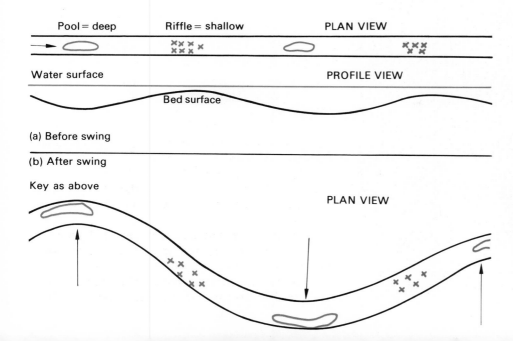

Fig. 8.11 Relationship of pool-and-riffle sequence to channel plan: (a) ratio of spacing to width in a straight channel; (b) location with respect to bends in a sinuous channel.

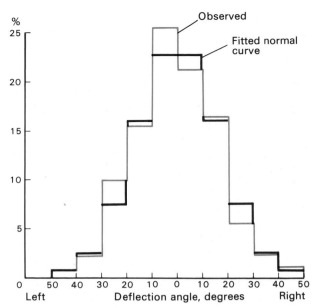

Fig. 8.12 Frequency distribution of deflection angles, from measurements at 300-m intervals on six meanders and 18 km of channel length on the Red River of the South, U.S.A. Fit of observed to ideal distribution is 97.5% accurate.

dealing with some kind of wave phenomenon, similar in at least some respects to the raising of wind waves on a water surface, or to the formation of corrugations when traffic passes over a dirt road.

But it seems still more likely that the meandering channel plan can be explained in terms of probability and of random variation. As already seen, the standard meandering plan allows a river to do the least work in turning. On this count, it can be looked on as a most probable plan, if the channel is to deviate from straightness at all.

Now, if a single stream channel is to be distinctly other than straight, then the meandering plan minimises changes in channel direction. In view of the very considerable changes in direction that actually occur on a meandering channel, this statement may at first seem surprising. But when changes in direction are graphed by frequency, they concentrate themselves about the central value of zero (Fig. 8.12). The frequency distribution approximates the familiar bell-shaped curve. A reasonable conclusion is that the changes of direction are randomly distributed about the zero mark, and that the meander plan, like the geometry of stream networks, can be regarded as the random result of process-response. The random inputs in this situation are presumably random deviations from straightness.

8.3 Channel plan geometry: braiding

Braided streams are those which, at low stages, exhibit multiple interlacing channels (Fig. 8.13). The channels are separated by bars (= banks, low islands) of gravel. Each bar is called a braid.

Fig. 8.13 Braiding on the meltwater stream of the Tasman Glacier, South Island, New Zealand (New Zealand High Commissioner).

Fig. 8.14 The Waimakariri River, South Island, New Zealand: (a) braided channel sunk into gravels of outwash plain; (b) immediately adjoining reach, in upstream view – channel meandering through bedrock outcrop.

If a braided stream is cutting down, some braids will dry out enough to be colonised and fixed by vegetation, but unfixed braids are liable to be eroded (or entirely destroyed) at stages of low flow, and to be rebuilt at stages of high flow. Wherever it is unfixed, the stream bed is mobile. It is also coarse.

There is a faint comparison here with meandering streams, where pools are scoured out, and riffles are built up, at stages of high flow. But there is no fixed position for a given braid in a given braided channel.

The essential difference between a single channel with meanders and a multiple channel with braids (at the low-flow stage) resides in the width/depth ratio. The depth value used in the width/depth ratio is the mean channel depth at bankfull flow. On small single channels the width/depth ratio runs low – say, from about 5/1 to about 10/1. On very large single channels, it is distinctly higher – say, of the order of 50/1. But the depths involved in this last value are considerable, of the order of 20 m. On braided channels, discharge for discharge, the width/depth ratio runs much higher than on single channels, reaching 500/1 or more on moderately large braided streams.

The great increase of width, and the corresponding decrease in depth, on braided streams, ensures that maximum velocities on braided streams occur close to the channel bed. Although these velocities can be expected somewhere in midstream, and thus far from the banks, the shallowness of channels with a high width/depth ratio means not only that maximum velocities will occur close to the channel bed, but also that the velocity will decline very rapidly down to the bed of the channel.

This amounts to saying that shear on the bed is strong – powerful friction is transmitted from the flowing water. The bed is deformed, but not into

pool-and-riffle. Instead, randomly spaced banks of gravel are constructed. These banks are the braids of a braided stream. They control the pattern of splitting and rejoining of channels at the low-water stage (Fig. 8.13). Simulation analysis shows that the pattern of splitting and reunion of low-stage channels, like the pattern of braid position, is random. Once again, we are faced with random variation in natural systems.

8.3a Possible explanations of high width/depth ratios

Random variation cannot, however, explain the high width/depth ratios which are necessary to begin with. A partial explanation is that rivers which mainly carry bedload, as opposed to fine sediment that can travel in suspension (compare Fig. 4.13), tend to have wide and shallow channels – provided that the banks are weak. Numerous braided streams occur on the Canterbury Plain of South Island, New Zealand, where the influence of bank strength is very easy to demonstrate.

These streams are working mainly in outwash gravel, carried by meltwater from the deposits of former glaciers. The gravel provides the coarse outwash from which braids are built. Although this material can sustain steep slopes at the water's edge, it is completely uncemented and easily eroded.

Now, the gravel has been spread over an earlier landscape, that contained upstanding hills cut in resistant bedrock. In places, the braided streams, in cutting down, have encountered the bedrock hills. Fig. 8.14 shows what happens in such a case.

Fig. 8.14a is a photograph of the Waimakiriri, where the channel is formed exclusively in gravel. The stream is braided. Fig. 8.14b is a photograph of the same stream, just a little downvalley, where the channel has cut down into a buried bedrock hill. The channel narrows into a single unit – and also de-

scribes one meander loop, complete with point bar. Just as much water and gravel are going through the meander as come down the braided reach upstream. The only difference between the two reaches is strength of channel banks. Downstream of the gorge in bedrock, and in outwash gravel again, the river reverts to braiding. With bank strength as the only process variable, we are dealing with two highly contrasted situations of process-response.

8.3b Braided channels and sandbed channels

For braiding to occur, it still seems necessary for bedload to be really coarse. Coarse bedload is most abundantly supplied by glacial outwash. Some streams in regions of semi-arid or arid climate transport bedloads of sand. Like the gravel loads of braided streams, abundant bedloads of sand promote the development of high width/depth ratios (Fig. 8.15). But sandbed channels tend not to braid. Their cross-sections are often almost rectangular. Even when sinuous in plan, they may show no sign, or only extremely faint signs, of developing sequences of pool-and-riffle. In a way, they are intermediate between meandering channels and braided channels.

8.4 Channel plan geometry: anabranching

An anabranch is a channel that diverges from a trunk channel, to reunite with the same trunk, or to joint some other trunk, after a considerable distance (Fig. 8.16). Since this distance may run to 100 km or more, we are really dealing with network geometry rather than with channel geometry. On the other hand, we are also dealing with channel response to a particular kind of situation.

This situation is that of very low downstream and cross-stream slopes. Anabranches are named from the Riverine Plain of inland southeastern Australia,

Fig. 8.15 Sandbed channel near Denver, Colorado, in the dry season. Fence posts on the right stand on top of the near bank. View is upstream.

where up to 385 000 km^2 of country has been overspread with alluvium. Mainly fine-grained, this alluvium provides a downstream slope of about 10 cm/km – about 1 in 10 000. Cross-slopes are effectively zero. In addition, many streams of the Plain are levee-builders. When an anabranch takes off from a trunk, reunion may be impossible over a very long distance.

8.5 Channel plan geometry: deltaic

A delta is a concentration of sediment at a river mouth, with splitting channels on the surface (Fig. 8.17). The name comes from the ancient Greek capital letter △ delta, which the plan outline of the Nile delta roughly resembles.

Deltas form where a stream, entering a lake or an ocean, deposits more sediment than waves of currents can remove. The lengthening of a channel across the sediment means that cross-slopes are reduced. During times of flood, when the channel banks are overtopped, some of the water is apt to take a short cut. Repeated channel splitting produces the typical morphological system of distributaries.

Deltas are process-response systems, the process being sedimentation. Responses include not only splitting of channels, but, for all great deltas, subsidence. Under its sedimentary load, the Mississippi delta has sunk 3000 m in the last fifteen million years. Negative feedback is once again at work.

Fig. 8.16 Anabranch: based on the actual pattern of part of the Riverine Plain, inland SE Australia.

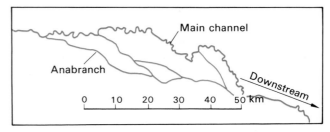

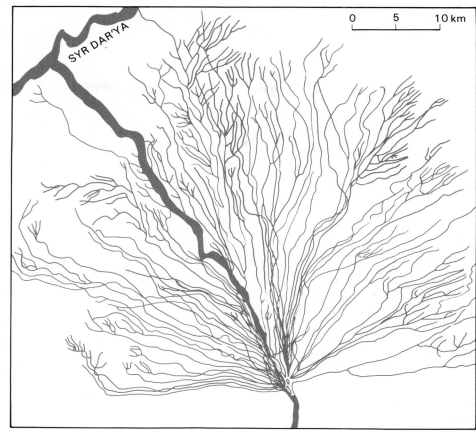

Fig. 8.17 Radial splitting of a tributary of the Syr Dar'ya, USSR: although the existing pattern is that of an alluvial fan or inland delta, it is precisely comparable to that of deltas built into water.

Deposition, causing the delta to subside, reduces seaward growth much below the extent it would reach if there were no subsidence.

8.6 Stream channels and hydraulic geometry

Hydraulic geometry analyses the relationship between the dimensional characteristics of channels on the one hand, and discharge on the other.

Discharge at any given cross-section varies through time (compare Fig. 7.9). As discharge increases from low stage up to the bankfull stage, width of water surface and mean water depth both increase. Cross-sectional area, the product of width and mean depth, increases also. The increase of cross-sectional area brings about an increase in velocity, because it involves a decrease in average friction. But variations at a given cross-section through time, called at-a-station variations, are in many respects far less important and far less interesting than variations along the length of the channel. These latter variations are called downstream variations.

The basis of comparison for downstream variations is the most probable annual flood – the peak flow with the recurrence interval of 1.58 years on the annual series, and of 1.0 years on the series that takes in more than one peak a year (Chapter 6). It is discharge at the most probable annual flood that fills a channel up to the banktop level, if the stream is neither rapidly downcutting nor rapidly infilling its channel (Fig. 8.18).

Fig. 8.18 Observed and model log-normal frequency of bankfull discharge. Fit is 95% accurate.

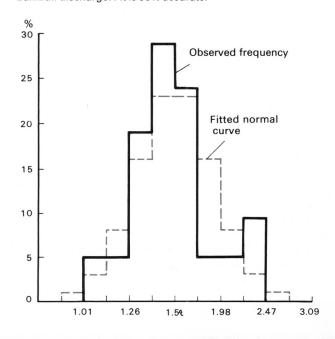

8.6a Power-functional relationships between channel characteristics and discharge

The first characteristics to need attention are width, (mean) depth, and flow velocity. If width and depth are known for a given discharge, velocity can be calculated. In practice, things usually work the other way round. Width, depth, and velocity are measured, and multiplied together to give discharge. Writing q for discharge, w for width, d for mean depth, and v for velocity, we get

$$q = wdv$$

which merely states that discharge is the product of width, depth, and velocity. If width and depth are measured in metres, and velocity is measured in m/s, then discharge will be calculated in m^3/s. The equation is of the same kind as that used to calculate the volume of a prism:

$$\text{volume} = \text{area of base } (wd) \times \text{length } (v).$$

Whether for at-a-station variation, or for the downstream variation which is under discussion, width, depth and velocity turn out to be power functions of discharge. That is, width multiplies as discharge multiplies, but not at the same rate (Fig. 8.19). The same applies to depth and velocity.

Power functions are plotted on double logarithmic paper. Recall that semi-logarithmic paper is used to plot exponential growth, where one variable multiplies as another adds (Chapter 7), and that the multiplying variable is plotted against the logarithmic scale. Where both variables increase by multiplication, a double logarithmic scale is needed, as in Fig. 8.19 (see also Chapter 25).

The theoretical relationships between width, depth, and velocity on the one hand, and discharge on the other, are for downstream variation

$$w \propto q^{0.55}$$

$$d \propto q^{0.36}$$

and

$$v \propto q^{0.09}$$

(width, mean depth and velocity vary, respectively, with the 0.55, 0.36, and the 0.09 power of discharge). These theoretical relationships are very well borne out by field observation. The power of velocity, 0.09, that is theoretically associated with the downstream increase in discharge, defines the slow rate of increase that was referred to in the last chapter.

Because meander wavelength is a linear function (= multiple) of bedwidth, it follows also that

$$l \propto q^{0.55}$$

where l is wavelength (meander wavelength varies with the 0.55 power of discharge).

Power functions also connect channel slope and roughness (= total friction) with discharge. The slope relationship in the ideal case has already been met in the previous chapter, namely,

$$S \propto q^{-0.74}$$

where S is slope. The corresponding ideal relationship for roughness, n, is

$$n \propto q^{-0.21}$$

(slope and roughness vary, respectively, with the -0.74 and the -0.21 power of discharge).

Fig. 8.20 summarises, for the ideal case, the implications of these various power-function relationships. Wavelength (if meanders are present) and bedwidth increase quite steeply as discharge increases in the downstream direction. Depth also increases considerably, but more slowly than does bedwidth. Velocity, as already noticed, increases very slowly. Slope decreases very rapidly as discharge increases, while roughness decreases at a moderate rate. Although many large river systems possess low-order tributaries with stone-strewn and therefore rough channels, in contrast to high-order outlets with smooth channels lined by silt, clay and fine sand, the downstream decrease in roughness is in fact chiefly a function of increasing channel size.

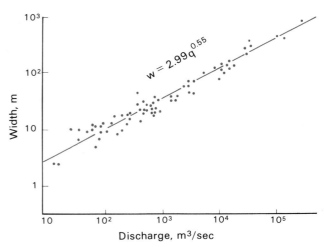

Fig. 8.19 Power-functional relationship of channel width to discharge at most probable annual flood. Fit is better than 95% accurate.

8.6b Interchangeability of channel characteristics

To a very considerable extent, channel characteristics are interchangeable. A variation in one of them will set off a compensating variation in one or more of the others.

So much, indeed, has already been stated, with reference to the difference between single (meandering) and braided channels, where an increase in width is offset by a decrease in depth. Furthermore, a braided channel is rougher than a meandering channel. Just as the flowing water applies friction to the channel bed, so does the channel bed apply friction to the flowing water. For an equal discharge, the braided channel is far less efficient than the meandering channel.

The lack of efficiency is apt to be compensated by an increase in slope. For an equal discharge, the slope of a braided stream is likely to be about half as great again as that of a meandering stream. The increase in slope offsets the increase in roughness.

It is easy to imagine all kinds of combinations of change among width, depth, roughness, slope, and velocity. One more example may suffice. Assuming, as before, a constant discharge, also assume that roughness is constant. Then, a decrease in slope must decrease velocity. In consequence, the cross-sectional area of the channel must increase, because the water is going through more slowly than before. Width, depth, or both, must increase.

Channel dimensions, channel roughness, and flow velocity can be regarded as representing compromises among a whole series of opposing, counteracting, compensating and interchangeable influences. A change in one characteristic can be expected to provoke changes in others. The compromise situation has been widely overlooked by those who have undertaken to alter natural channels.

8.6c Interchangeability and interference by man

Attempts to alter natural channels have commonly been applied to only one, or only a few, characteristics. The results have frequently proved disastrous. What were meant to be control systems turned into process-response systems in disequilibrium.

A stream reach is likely to have a preferred channel size – the central value of cross-section to which the channel tends to revert after scour or fill. A change in discharge will alter the preferred channel size, as is well shown by numerous streams in the English Lowlands. For long centuries, water has been diverted from the natural channels into the offtakes for water mills. The natural channels have become somewhat infilled. Conversely, if discharge remains the same, channel size tends to remain the same, as it has done on the Huang Ho River of northern China.

There, on a vast alluvial plain, the natural levees of the Huang Ho have been strengthened and raised, during a period of at least 2000 years. The river has responded by infilling its bed, restoring the cross-section to the natural size. As a result, in many reaches the bed of the stream lies higher than the

Fig. 8.20 Graphical summary of ideal power-functional relationships to discharge, along the length of the channel, of width, depth, velocity, wavelength, slope and friction.

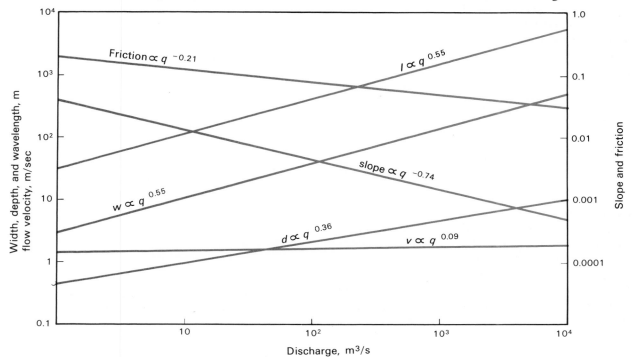

surrounding countryside. When the levees are over-topped and breached – as, inevitably, they are from time to time – the whole river can pour out.

An extent of 140 000 km^2 is permanently endangered. Nearly all of it is in farmland. A really serious flood blankets great areas with sediment, adding deaths by starvation to deaths by drowning and disease. The Huang Ho flood of 1931 was the most disastrous in known history, causing deaths estimated at four million.

In some agricultural areas, including the U.S. Midwest, gullying of farmland has been savage. There seem to be two main causes of man-induced gullying. One is the clearance of forest from valley sides, the result being to reduce the lag time between rainfall and flow in channels, and to raise the peaks of channelled flow. The channels have responded to the increased flow by reducing their slopes. A suddenly downcut low-order channel occupies the bottom of a gully.

A more complicated case is the gullying induced by the enlargement of channels of, say, the third or fourth order. With enlargement goes straightening, the idea being to contain and deliver water that otherwise would inundate the valley bottom. An enlarged artificial channel can indeed contain more water than a smaller natural channel, and thus mitigate flooding at the outset. It also enables water to move downstream faster than in the smaller natural channel. But, because the artificial channel is straight, it has a steeper slope than the winding natural channel that it replaces: it makes the same fall in a shorter distance. On the other hand, because it has been enlarged and carries more water than before, it needs a reduced, not an increased, slope. On both counts, its new slope is unduly great.

Several results now follow. Of the sundry dimensional characteristics, slope is the most easily adjustable. The new channel will have its slope reduced by erosion at the upper end, and by deposition, either at the downstream end or beyond the enlarged reach. Where deposition occurs, choking of the channel will increase the flood risk, not reduce it. Where erosion takes place, it will spread up the tributaries. Deepening themselves first where they enter the eroded main channel, these will in time deepen themselves throughout their lengths, and may well be able to extend themselves upslope of their original heads. The activated tributaries cut gullies.

The rock waste removed from the gullies will all be carried to the main channel. This will certainly be further choked by sediment at its downstream end, and may end up by being choked throughout its length. In this way, the attempt to improve a channel by enlarging and straightening it proves not only self-defeating, but actually brings about the un-wanted consequences of increased flood risk, in addition to the wave of gullying which has spread up the tributaries. The artificial increase in channel size and slope has provoked a response of adjustment throughout the system. Only if close-spaced check-dams are inserted in the enlarged channel, forcing the stream to descend in gentle steps, can results of the kind described be prevented.

To begin with, positive feedback comes into play: enlargement and straightening of the main channel throws the whole system into disequilibrium. Later, negative feedback sets in, and a new state of equilibrium is established. The very process of gullying set off by alterations to the main channel tends, in time, to restore that channel to something like its former size and slope. A possible, however mournful, conclusion could be that rivers are best left to themselves.

Chapter Summary

Stream channel patterns are *classified by plan geometry*. However, it is *difficult to concentrate solely on morphological systems* and to *ignore system state* or *process-response behaviour*.

Negative feedback is illustrated by the tendency of many channels to scour on a rising flood, and to fill on a falling flood, with the *cross-section varying about some central value*. This tendency exemplifies the *steady-state condition*. *In the long term, channel systems* maintain their identity as the landmass is reduced: they *remain in dynamic equilibrium*.

Meandering streams have sinuous channels. Meanders *on floodplains rework their own alluvium, eroding outer banks* and *constructing point bars on inner banks*. The *downvalley migration* of meanders is known as *sweep*. Meanders *in bedrock* are commonly *ingrown*, having enlarged themselves as they have cut down.

The typical *meander form can be simulated by bent springs*, indicating an *equal-work situation*. Another reason for the family resemblance among meanders is that the *ratio of bedwidth to radius of curvature* normally *varies within* quite *narrow limits*. The typical *meander form minimises the work* done by the river in turning; it *minimises resistance to flow* round the bend, and thus represents a *least-work situation*.

The cause of *meandering* can be *considered in more than one way*. Many *straight channels*, both in nature and in the laboratory, *are unstable*. They develop *pools and riffles*, each type of feature *spaced at about 5 to 6 bedwidths* apart. *Side-to-side swing* produces meandering channels. But because the *meandering plan* enables the river to do the least work in turning, it can be looked on as the *most probable plan*, if the channel is to be other than straight. Also, the meandering plan *minimises changes in channel direction*. The meander plan can therefore be regarded as the *result of random variation*.

Braided streams at low stage have *multiple gravel bars* and *multiple interlacing channels*. They carry abundant coarse bedload. They possess *very high width/depth ratios*.

Braiding results from strong *shear on the bed*, which follows from the high width/depth ratios. These ratios result, at least to some extent, from *weak banks*. *Sandbed channels*, often with roughly rectangular cross-sections, and also with high width/depth ratios, seem to be *intermediate between meandering and braided channels*.

Anabranching channels split and rejoin over large distances, where downstream slopes are slight, levees are common, and cross-valley slopes are effectively zero.

Deltaic channels split into *distributaries*.

The *hydraulic geometry* of stream channels concerns the relationship between *channel dimensions and discharge. Channel dimensions are power functions of discharge*: their values multiply as the value of discharge multiplies, but at a different rate. Theoretical and empirical (= observed) rates of comparative multiplication agree well.

Channel characteristics are largely *interchangeable* in nature. A change in one characteristic, such as channel pattern, can readily be offset by an opposite change in another, such as channel slope. *Human interference*, however, has often *concerned fewer than the minimum* required number of characteristics.

Raising of natural levees can cause a stream to *infill* its channel so as to *restore the natural cross-section*, and thus to *increase the risk of flooding. Enlargement* and *straightening* of natural channels can *set off waves of gullying* in low-order tributaries.

9 Surface Morphological Systems: Slopes

Slopes on land surfaces are produced both by erosion and by deposition. They are chiefly considered in relation to valley sides and hillsides. The choice of term depends in part upon point of view, although we tend to use valley side (or valley wall) when the valley floor is narrow to absent and hillside (or hillslope) when the slope base lies far distant from a drainage line.

Erosion in this context refers chiefly to the erosion of bedrock. A usual assumption is that erosion of slopes is controlled by some limiting factor, such as wave attack on the margin of the land, or the downcutting of streams. Slope development through time involves the destruction of bedrock by weathering, and the progressive removal of weathering products. Where slope bases lie far distant from the nearest drainage line, it seems necessary to conclude that slope systems, once brought into being, begin an independent evolutionary life of their own. This is especially so in dry climates, where many slope systems cannot possibly be controlled by downcutting streams. For instance, downcutting cannot possibly apply, where the centre of a desert basin is being infilled by sediment. Depositional slopes occur where rock material is being deposited faster than it can be removed, as for example upon sheets and fans of talus on the flanks of mountains, or on the sides of volcanoes.

Three principal groups of workers are interested in slope form, slope erosion, and slope stability or instability. One group is composed of civil engineers, who deal with thresholds of slope failure, and with the potential effects of excavation or constructional loading. The engineering approach concerns process-response in individual situations. The second group includes those soil scientists and agricultural experts who deal with soil erosion and soil conservation. To some extent, this group looks on the fact of erosion as a cascading system. The third group consists of geoscientists who investigate the form of the ground. Members of this group, like civil engineers, are interested in slope failure; and, like soil scientists and agriculturalists, they are interested in soil erosion. In addition, they deal with the natural evolution of slopes through time. Common ground is supplied by considerations of slope angle, which figures prominently both in civil engineering and in soil conservation, and also – as will be seen – in geoscience. For this reason, the present chapter takes slope morphology as its starting-point. The discussion will however become progressively concerned with process-response.

9.1 A basic slope model

The simplest possible slope model is one in which slope angle is uniform throughout. Angle is measured either above the horizontal, when the measurement is taken from below, or below the horizontal when the measurement is taken from above. The one-component model could consist of a vertical slope, as on some coastal cliffs, or a less than vertical slope, as with some hillsides developed across rocks that cannot sustain vertical faces. Most slope systems however include more than one component.

Although it may appear a simple matter to construct a basic model with several components, in practice the setting up of any model with two or more components invites the outbreak of heated argument. There are at least four reasons why this is so. In the first place, the establishment of a basic model is an act of abstraction – that is to say, an act of generalisation. A generalised model, however widely applicable on the world scale, may differ strongly from the slope systems that are most familiar to us as individuals. We tend to regard the familiar as normal or standard. In the second place, even though we start with slope form, that form is shaped by slope process. In consequence, every morphological slope model is backed by a process-response model. Because very little is yet known about the relative importance of slope-forming processes, even morphological models can be highly controversial. Little, again, is known about the ways in which slope morphology is controlled by climate. In certain respects, it may not be controlled by climate at all. The debate for and against fundamental climatic influence constitutes the third reason for argument. And in the fourth place, the development of slopes through time is central to the whole concept of the evolution of the physical landscape. In this connection, two extreme opinions exist: that hillside slopes decline in slope angle through time, and that they do not decline. Any reader can be pardoned for thinking that these extremes match the

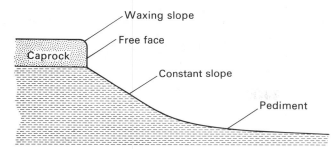

Fig. 9.1 The basic slope model and its four components.

Labels in figure: Waxing slope, Free face, Caprock, Constant slope, Pediment

controversy in *Gulliver's Travels*, between the faction that held that boiled eggs should be cracked at the little end, and the opposing faction that held that the eggs should be cracked at the big end. The argument does however involve a basic theoretical difference. We may perhaps add that the theoretical analysis of slope development through time often assumes a uniform type and strength for the bedrock across which a hillside is developed, whereas in the real world, both type and strength of bedrock frequently vary on a given slope,

The next least complex situation to that of uniform rock type and rock strength is the one where a hillside is developed across flat-lying sedimentary rocks, with a resistant formation of limestone or sandstone at the top, and a less resistant formation of shale beneath. The resistant formation is called the caprock. As an alternative to a sedimentary formation, it could consist of a sheet of igneous material. If the caprock is strong enough to sustain a vertical face at its edge, then the basic slope model (= slope morphological system) includes four components (Fig. 9.1).

9.1a The waxing slope

The horizontal surface on the hilltop does not form part of the hillside slope sequence. The typical curve-over at the edge of the summit does. This is the waxing slope, so called because, on a given vertical line, it increases in slope angle through time. As erosion causes the whole slope system to recede through the hillside, the low slope angle near the top of the waxing slope changes into a steeper angle in a lower position and nearer to the vertical face. At the edge of the hilltop, the bedrock is attacked by weathering, both from above and from the side. Hence the rounding-off of the waxing slope. Because the waxing slope is only a minor component in the total system of slope morphology, it has rarely been investigated in detail. It will be excluded from further discussion.

9.1b The free face

This is the vertical, cliff-like edge developed on the caprock. Erosion of the next lower slope component tends to undermine it, and to cause slabs to break away. In addition, the difference between the negligible pressure of the air outside, and the considerable built-in internal pressure of the caprock, favours the development (or at least the opening) of vertical joints, thus promoting the breaking-off of slabs and the maintenance of a vertical edge.

9.1c The constant slope

Much confusion and controversy surround the name of this component of the hillside slope system, which occurs next beneath the free face (Fig. 9.1). As originally proposed, the name stood for a slope component that remained straight in profile (= slope angle constant throughout), whether or not the slope angle changed through time. That is to say, the word constant merely referred to relationships in space. Then usage shifted, so that the term came to connote a slope component with a uniform angle that did not alter as the slope developed through time. The word constant had been transferred to time relationships. As will presently be seen, both modes of usage can be misleading about what happens in reality.

On numbers of rock types, the maximum angle for the constant slope is about 32° or 33°, very close to the angle of rest of rock debris. The value also resembles the angle of 33½° at which dry sand, on the lee side of a desert dune, begins to slip. Because many constant slopes are mantled only with a very thin veneer of rock waste, it seems as if the angle of rest of debris in some way controls the angle of the slope eroded across bedrock.

If there were no connection between angle of rest and angle of slope, we might expect constant slopes frequently to approach an angle of 45°. The forces acting on a rock fragment on an inclined surface can be classified as the slide force and the stick force. Both forces result from gravitational pull (Fig. 9.2).

Fig. 9.2 Slide force and stick force.

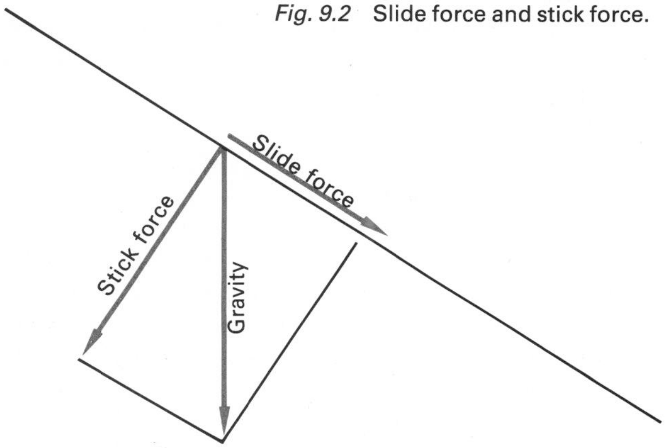

Labels in figure: Slide force, Stick force, Gravity

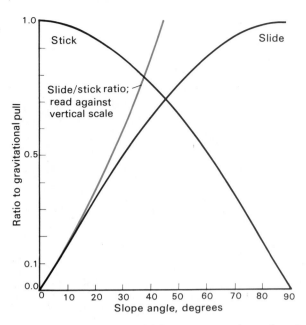

Fig. 9.3 Slide force and stick force, expressed as ratios of gravitational pull, in relation to slope angle; the slide:stick ratio, on a scale of 0.0 to 1.0, in relation to slope angle.

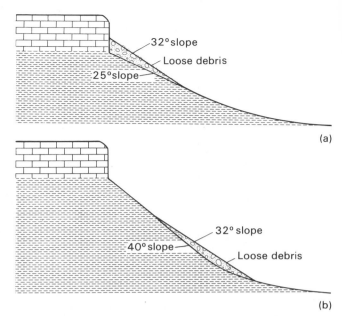

Fig. 9.4 Lodging of debris: (a) at the slope head; (b) at the slope foot.

The stick force acts at right angles to the bedrock surface, the slide force parallel to that surface. On a horizontal surface the slide force has a value of zero. Its value increases as slope angle increases. If we take the maximum possible value as 1.0, then this value is reached on a vertical slope (Fig. 9.3). The stick force has a value of 1.0 on a horizontal surface. The value decreases as slope angle increases, becoming zero on a vertical slope. Expectably, as shown in the diagram, the two forces are equal in value on a slope of 45°. But deficiencies of friction, and – in wet material – excess of lubrication, ensure that the rest angle of rock debris is typically below 45°. Rock fragments will move downslope at lesser angles than this.

In addition, by no means all constant slopes match the rest angle of debris that falls onto them. Some allowance must obviously be made for rock strength. If the constant slope develops an angle of, say, 25°, then falling debris will lodge at the top (Fig. 9.4a). If the angle of the constant slope is 40°, then the falling debris will reach the base of the slope and accumulate there (Fig. 9.4b). The end result – or better, final tendency – in both cases is to produce a surface slope of about 32°, but with the depositional portion distributed in contrasted ways.

9.1d The pediment

The fourth component in the basic slope model is the pediment (Fig. 9.1). This is a concave-up component, decreasing in slope angle in the downslope direction, again eroded across bedrock. It derives its name from an architectural feature (Fig. 9.5) – an ornamented gable end, or a similarly shaped feature on top of a pillared portico. Early English-speaking stonemasons are thought to have called the architectural feature a pyramid, with the pronunciation passing into pereminth and eventually into pediment. There are probably some cross-links with the independent words plinth and pedestal. But the pediments of architecture correspond to the constant slopes of landscape (Fig. 9.6). By some obscure linguistic shift, the name has been transferred to the slope component next below the constant slope. As so often happens, the linguistic process proves impossible to reverse. Constant slopes remain constant slopes, despite the unsatisfactory nature of their name, and pediments are concave-up instead of being straight in profile.

Where they have been surveyed in detail, pediments tend strongly to display profiles like those of ideal streams – that is, the forms of growth curves that resemble the compound interest graph. From the base of a pediment upward, slope angle multiplies as distance is added (Fig. 9.7).

There is no agreement whatever about the processes that shape pediments. We can probably assume, with safety, that the downslope change from the straight profile of a constant slope to the concave-up profile of a pediment goes with a change of process. The assumption seems inescapable where, as is often the case, the constant slope and the pediment are cut across the same type of rock. Beyond this, however, almost everything is obscure. Despite claims to the

Fig. 9.5 (above)   An architectural pediment on the campus
of the University of Wisconsin, Madison (G.M. Habermann).

• = Survey point

Free face: vertical

Constant slope: angle 19° 30′;
95% fit to straight profile

Pediment: 97% fit to
semilogarithmic profile

m altitude

50
40
30
20
10
0

0   100   200   300   400

m distance

Vertical exaggeration = 2.5

Fig. 9.6 (above)   Compare the profile slopes of the
architectural pediment in Fig. 9.5 with the (steeper) constant
slopes in this photograph: the term pediment, in geoscience,
has become shifted to the concave-upward slopes next
below (Wyoming Butte: S.A. Schumm).

Fig. 9.7 (left)   Semi-logarithmic curve fitted to the surveyed
profile of a pediment: Platte Mounds, SW Wisconsin (survey
by A.M. Davis).

Fig 9.8 Valley side (left) consisting almost entirely of constant slope: at the foot it runs straight into the stream channel, while the free face is represented only by a few minor crags near the top (near Mt Kosciusko, Australian Alps: Australian News and Information Bureau; W. Brindle).

contrary, pediments are not confined to dry climates. On the other hand, the processes that move rock waste down their slopes may differ from one climate to another. Workers in regions where the sub-surface is now, or formerly was, in a permanently frozen state recognise pediments shaped by the action of frost. In humid mid-latitudes, the mantle of rock waste on a pediment seems to be moved chiefly by creep (see below). In dry climates, the only available processes appear to be the downhill creep of individual rock fragments, the surface wash promoted by occasional rainstorms, and the flaking of bedrock from exposed surfaces. A common difficulty here is that many pediments in regions that are now arid are thickly mantled with coarse water-rolled gravel. An obvious possibility is that the pediments in question developed in climates very different from those of the present day.

9.1e The valley floor

Although not an essential part of the hillside slope system, the valley floor deserves consideration. Many hillside slope sequences descend to valley bottoms that are wider than simple floodplains. It must follow that the cascade of rock waste from the hillside is in some way removed or absorbed. Debris in small quantity comes off the waxing slope. More abundant debris is delivered by the free face to the constant slope. The combined total is moved down the constant slope, anything yielded by the weathering and erosion of this slope component being additional. The aggregated total is transported down the slope of the pediment, being increased by everything removed from the pediment surface itself. If a wide and flat valley floor exists, then some process must operate to move rock waste across it. Almost no investigations into this matter have been made. We seem obliged, in the meantime, to guess that creep is the chief responsible agency.

9.2 Variations on the basic slope model

Slopes in the real world can vary considerably from those of the basic model. In special circumstances, they can be produced by the clean stripping of some resistant rock formation: a vertical dyke of igneous rock can be stripped out of a cone of volcanic debris. Strictly speaking, its two sides constitute free faces. But variations on the basic model deal mainly with omission, repetition and elaboration of components. In addition, some slope morphological systems include deceptive components.

9.2a Omission of components

A vertical shoreline cliff usually rises to a waxing slope, but most of the system corresponds to the free face of inland slopes. On some valley sides, where rock type is uniform throughout and the valley floor is narrow to absent, the free face and the pediment may both be missing (Fig. 9.8). Almost all is constant slope. Especially in such situations, slope angles have been analysed for distribution. On a given rock type, the slope angle tends to be centred on a particular value (Fig. 9.9). The significance of particular central values is far from being understood, except for the values of about 45° and about 32–33°. Some researchers have concluded that slope angles of 43–45° develop on bedrock within which the pore water pressure is low; that angles of 33–38° typify loosely packed talus; that angles of 25–27° are usual on taluvial slopes; that sandy slopes often display angles of 19–21°, and that slopes cut across clays normally fall into the range of 8–11°. Taluvial slopes require definition. They are slopes formed on taluvium, a combination of rock rubble and sand, or on rock rubble and a sand–clay mixture. The rock rubble is talus. The sand or sand–clay mixture is colluvium – rock waste finer than the rock fragment grade, delivered downslope by wash and creep. The tal from talus and the uvium from colluvium combine to spell taluvium.

On flat-lying or gently inclined sedimentary rocks, where the caprock is too weak to sustain a vertical edge, the free face will be missing. Such is widely the case in the lowlands of western Europe, and especially of southeast England, where hillsides frequently include two components of constant slope, one much gentler than vertical at the edge of the caprock – say, with a slope angle of about 30° – and the other – say, with an angle of 10–12° – developed across still weaker rock beneath (Fig. 9.10). At the very extreme, free face and constant slope may both be omitted, the waxing slope merging directly into the pediment; but such apparent merging is probably most likely to occur with certain of the deceptive components discussed below.

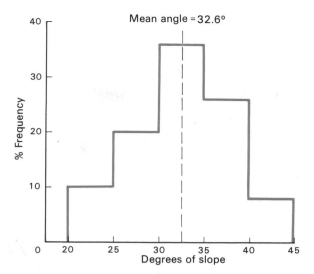

Fig. 9.9 Distribution of slope angle, on a single rock type and in a single valley system; values are close to normally distributed about the mean: from an actual example.

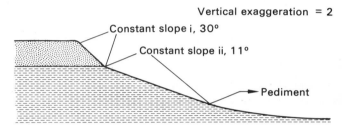

Fig. 9.10 Segmentation of the constant slope on sandstone and shale, with the free face omitted: based on typical conditions in parts of the Jurassic belt of the English Midlands.

9.2b Repetition of components

If alternating strong and weak rocks crop out on a slope, then the slope angle will vary repeatedly. The alternation of free faces and constant slopes is dramatically illustrated in the Colorado Canyon (Fig. 9.11), where the sequence descends through 1000 to 1500 metres. In some places the pediment

Fig. 9.11 Alternation of free faces and constant slopes in the Grand Canyon.

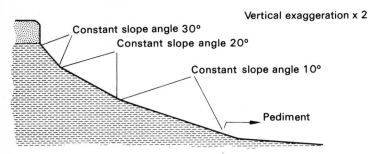

Fig. 9.12 Segmentation of the constant slope on a single rock formation: based on field surveys in the Jurassic belt of the English Midlands.

component is also present, but tends to merge rapidly into horizontal benches developed by the clean stripping of resistant rock formations.

9.2c Elaboration of components

Even when the constant slope is developed across a single rock type, its profile can be segmented. The constant slope is subdivided into subcomponents, each straight in profile throughout, but with the slope angle increasing upward through the sequence (Fig. 9.12). The reason seems to be that slope angle can be controlled, not so much by the inherent strength of the underlying bedrock, but by the cohesiveness of the overlying mantle of rock waste. It appears likely that this cohesiveness becomes progressively reduced, as weathering breaks down the constituents of the waste mantle; but also that response to progressive weakening does not change steadily, but changes by behavioural jumps. The

process involved here is called threshold–slope decline. It has been recognised so recently that we do not yet know whether it is general, or whether it can only operate on particular types of rock. A further possibility is that it can operate only in particular types of climate – those moist enough to promote the concentration of clays in the waste mantle.

A second main way in which the components of a slope morphological system can be elaborated consists in the elaboration of the basic model itself. Fig. 9.13 illustrates the nine-unit model, which relates slope morphology to slope-forming process. This model, although more complex, is not difficult to reconcile with the basic model. Its units 1 (part of) to 3 correspond to the waxing slope; unit 4 is equivalent to the free face; units 5 to 7 correspond to the constant slope and the pediment, although making more allowance for mass-movement (see below) than is made in the basic model. Units 8 and 9 in the nine-unit model are not considered in the basic model.

9.2d Evolution of slopes with stabilised bases

Up to this point, the discussion has dealt with slope systems that are retreating into the surviving high ground. Cliffs undercut by waves recede in the landward direction. Waxing slopes evolve, as free faces recede into the hillside; constant slopes recede above the heads of extending pediments; and pediments themselves bite deeper and deeper into bedrock as rock waste is removed from them and carried

Fig 9.13 Relationship of slope unit to geomorphic processes in the nine-unit model.

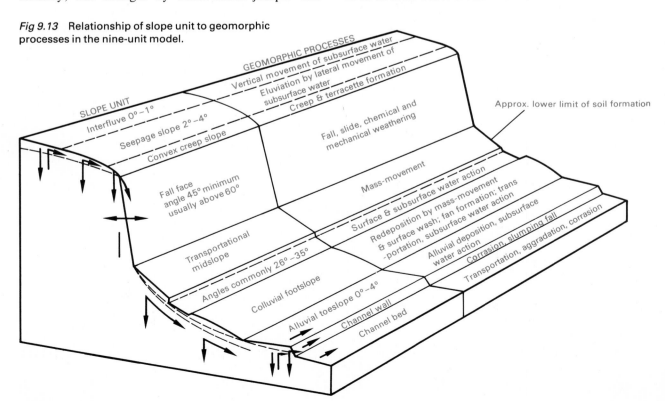

into valley bottoms or on to the floors of enclosed basins. But if a shoreline becomes stabilised at a particular level and in a particular position, or if a stream on a given drainage line ceases to cut down, then the future evolution of a slope system becomes independent of activation from below upwards. Some workers conclude that, in such conditions, slope angles will flatten through time, especially on single rock types where no caprock is present.

9.2e Deceptive slope components

When we look at slope morphological systems, we are looking at them as they exist today. They may, however, not have evolved in existing climates. The steep slopes of glacially eroded valleys present no problem. They can be recognised at once for what they are. They are being converted, however slowly on the human timescale, to slopes fashioned in non-glacial climates. By contrast, the slopes produced by the action of ground ice and the thaw/freeze process can be highly deceptive. They include the less than vertical clifftops along the English Channel, where ten or more metres of the chalk rock have been shattered by frost, the corresponding broad convexities on chalk scarps inland, and the surfaces of sludge that moved downslope under former periglacial climate (Fig. 9.14; see also Chapter 13, section 13.8).

A particularly striking case of the effect of ground ice upon slope morphology is that where a shale formation beneath a caprock of limestone or sandstone has been mobilised, and where the caprock itself has also been mobilised to some extent (Fig. 9.15). As now observed, the caprock is draped across the brows of hills, displaying surface slopes of up to 3°. These slopes are far too wide to be classed as waxing slopes. They result from the squeezing out of the underlying shales, and from the springing

Fig. 9.14 Anomalously large waxing slope, developed on chalk shattered to a depth of some 10 m in a former permafrost climate.

apart of the caprock by ground ice that formed along the bedding planes and in the perpendicular joints. The general name for the bending-over of the caprock is cambering. It can be accompanied by the arching-up of another resistant formation beneath valley bottoms, as illustrated in Fig. 9.15

9.3 Interplay of delivery down slopes and through channels

Implicit in most of what has been said so far is the notion that material delivered from one slope component above is transported eventually over the next slope component below. We imagine rock fragments falling off a free face to be delivered downhill along the constant slope. Rock waste coming off the constant slope is transported down the slope of the pediment. Even when the rest angle of falling debris is greater than the angle of a rock-cut constant slope,

Fig. 9.15 Cambering (shallow downfolding) and valley bulging (shallow upfolding); complex detailed effects (not illustrated) can be produced in the cambered flaps. Based ultimately on structural exploration in the Northamptonshire Ironstone Field.

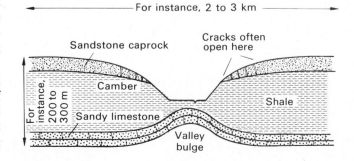

so that talus accumulates about the junction of constant slope and free face, newly-falling fragments slide down, and lodge on, the talus sheet. The same kind of thing happens about the foot of the constant slope, in cases where the rest angle of falling debris is less than the rock-cut angle of the constant slope. Numbers of theoretical analyses deal with the evolution and form of rock-cut surfaces where talus accumulates in sheets. No analytical result seems to match closely with field observations. Far more important for our immediate purpose, however, is that fact that a great deal of rock waste is delivered, not simply downslope, but through channels.

Riverborne sediment is not necessarily in question here. Steep mountain sides can develop runnels that discharge falling rock fragments from a point, so that the accumulation takes the form of a fan (Fig. 9.16). Debris fans can of course merge into a talus sheet, if they are large and closely spaced enough. But if the runnels deepen and lengthen into valleys, then they will come to be occupied by streams, and to discharge waterborne sediment rather than dry rock. The accumulations of rock waste are then called alluvial fans (Fig. 9.17). These fans have much in common with deltas built outward into lakes or seas. The stream works from side to side across them, bursting out from a newly raised portion to an older and lower portion, and raising it in turn.

So far, alluvial fans have been studied chiefly in relation to the steep mountainous rims of certain desert basins. Analysis shows that, up to areas of about 75 km^2, fan area is roughly equal to source area – the area of the mountain valley that supplies the fan sediment. Something obviously depends on the depth to which the supplying valley is cut; and somewhat surprisingly, above about 75 km^2, fan area possibly fails to increase as rapidly as source area does.

9.4 Slope-forming processes

The mechanical and chemical breakdown of rock materials has already been discussed (Chapter 4, sections 4.3, 4.5). We are concerned here with the processes that move rock waste downslope. These processes can be subclassified into surface wash, throughflow, and mass movement. Surface wash includes the action of rainsplash and that of flowing sheets of water on the surface during heavy rainstorms. Throughflow is the action of water in the sub-surface, where not only dissolved substances but also diffused clay particles can be moved downslope.

9.4a Surface wash

The direct effect of rainfall on bare soil is most obvious on ploughland, which is artificial, and in dry climates, where rain is rare. In extreme cases, as much as 90% of soil erosion may be triggered by rainsplash, which detaches soil particles and injects them into the surface sediment cascade. Erosion by unchanelled surface flow is controlled by a whole series of variables. The drier the soil, the more water it can soak in before it becomes saturated. But the speed of soaking varies widely, according to the coarseness of the pore system. Moreover, very intense rain can fall so rapidly that, on many soils, surface flow begins before saturation is complete. Again, the velocity of surface flow varies with the roughness of the surface, with vegetation cover, with surface slope, and with slope length. Nevertheless, equations have been developed that relate average annual soil loss (e.g. in tonnes/km^2) to a rainfall factor, a soil-erodibility factor, a combined factor of slope length and slope steepness, and to factors of cropping–management–conservation practice.

Understandably enough, the effects of rainsplash and of surface wash have been studied almost exclusively in relation to farmland, where the annual loss can range as high as 15 to 150 tonnes/ha, depending on cropping practice.

9.4b Throughflow

This process is the downhill movement of water in the subsurface. Although its velocity is typically only about one-thousandth that of surface flow, 20 cm/hr as against 20 000 cm/hr, throughflow can start before surface flow begins, and can continue after surface flow ceases. Very little is known about its quantitative importance, except that, in humid-tropical climates, it is capable of moving clay particles, even on very gentle slopes. Thus, clays can be removed from the waste mantle, even on the flanks of very shallow valleys in humid tropical climates, and can be concentrated in the valley floors. The throughflow process is conceivably far more important in such climates than is surface wash, except that in the wettest climates of all, surface wash may be dominant. The balance of effect between surface wash and throughflow seems to depend in part on the effectiveness of the forest cover in the filtering of rainfall. Trees blur the impact of storm rainfall upon the surface of the ground. Throughflow is specifically noticed in the nine-unit slope model (Fig. 9.13) for units 1, 2, 3, and 5, 6, 7.

Fig. 9.16 Talus fans: Wastwater Screes, Cumbria, England (J.K. St. Joseph). Contrast Fig. 4.5.

Fig. 9.17 An alluvial fan, slightly truncated by stream in the foreground (Burnaby River, British Columbia: E.J. Hickin).

Fig. 9.18 Badly gullied and abandoned farmland on the southern U.S. Plains in the 1930s (U.S. Dept. of Agriculture Soil Conservation Service). Incredibly enough, restoration measures on land such as this have proved widely effective.

9.4c Gullying

Gully erosion bedevilled the cotton planters of the American South: it has also ravaged farmlands elsewhere (Fig. 9.18). Gullying bridges the technical/artificial gap between slope erosion and channel erosion. Where enough water collects to promote flow in channels, a channel forms. As a sad agricultural history shows, in many countries, channels can form on, and can consume, ploughland. Collection of surface water can readily be promoted, in numbers of climatic and vegetational settings, by the clearing of land for cultivation, or even by the clear felling of a forest.

The gullying process often involves the collapse of gully banks, and thus promotes mass-movement. The deepening of gullies promotes the diversion of throughflowing water into surface channel systems.

The headward extension of gullies constitutes an extension of a channel network. In these various ways, gullying, whether induced by man's activities, or whether promoted entirely by non-human causes, occupies an intermediate position between the erosion of hillside slopes and the erosion of stream channels.

9.5 Mass movement

Although mass movement is only one of the sets of processes which force slopes, it has been hitherto the most closely studied of all. As a result, it can be discussed in such detail that it merits a section of its own. Moreover, some individual mass movements deliver impressive totals of material. Whereas the general balance among surface wash, throughflow, and gullying cannot yet be calculated, mass movement can easily be recognised, on the world scale, as dominating the transportation of rock waste from hillside into valley bottoms.

Mass movement includes all types of slope failure.

Failure can be sudden, as with the events that are loosely classified as landslides. It can by contrast be slow, as with creep. But its general effect is to deliver rock waste in the downslope direction. Expectably, engineering studies and hazards research are concerned mainly with sudden failures.

Fig. 9.19 provides a classification of type of slope failure by type of process and by type of material affected. As shown, the processes of sliding and flowing cannot always be separated where mud and debris are involved, and catastrophic debris flow displays very high speeds, comparable to those of rockfall. In general, however, mass movements can be satisfactorily described by reference to type of process and to type of material, and the process –material relationship can be related in turn to degree of moisture content, failure angle and velocity of movement.

9.5a Heave

The heave process, resulting from alternating expansion and contraction of loose rock fragments, from the swelling and shrinking of clays in consequence of wetting and drying, and from alternating freezing and thawing, works to move rock waste downslope. Material shifted upward shifts perpendicularly to the hillside slope. When it settles back, it settles vertically. The net movement is downhill. The effects of heaving, and the allied effect of simple gravitational pull, amount to creep and solifluction. Creep is the very slow downhill movement of unfrozen rock waste; solifluction involves thaw–freeze action.

Creep on constant slopes in dry climates affects individual rock fragments, which often lie thinly scattered on an eroded bedrock surface. Speed of creep in these conditions is controlled by slope angle – more precisely, the speed varies with the sine of the slope angle, the ratio between vertical height and horizontal distance. The few detailed studies that have been undertaken so far indicate creep speeds of about 50 mm/yr for slopes of 32–33°, and of about 10 mm/yr for slopes of 5°. At this latter rate it would take a hundred years for a rock particle to travel one metre in the downslope direction.

In regions of midlatitude humid climate, creep can affect the whole waste mantle down to a depth as great as 1 m, if the waste mantle consists at least in part of swelling clays. Movement is however typically concentrated at and near the surface, and is slow to imperceptible below a depth of 20 to 30 cm. Displacement at and near the surface is measured only in a few millimetres a year, while average displacement through the column may run only at 1 mm/yr.

Solifluction, the downslope movement of the ac-

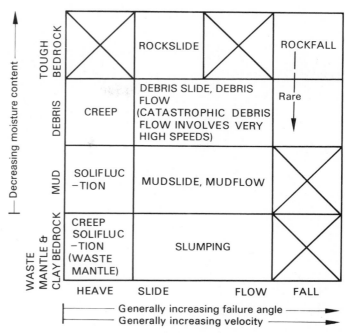

Fig. 9.19 An outline classification of mass movement.

tive layer of permafrost terrain (Chapter 13, section 13.7), is faster. A representative value for displacement at and near the surface is 20 mm/yr, associated with a representative average of 10 mm/yr for a 50 cm depth of the waste mantle.

Creep and solifluction are alike capable of producing minor forms of waves or steps on the surface of the ground. On moderate slopes, solifluction leads to the formation of turf-banked terraces (Fig. 13.23). Creep in regions of mid-latitude climates can produce terracettes (Fig. 9.20a), which roughly resembles staircases. Creep in regions of arid and semiarid climates can produce gilgai (named from an Australian aboriginal word), which vary widely in their surface expression, but which on air photographs are often expressed by a kind of fingerprint pattern that reflects the micro-distribution of desert plants (Fig. 9.20b).

9.5b Slide and flow

Although the vague term landslide is often applied to any kind of sudden slope failure, true sliding involves movement down a well-defined surface. Such a surface may be a bedding plane, an erosional surface, or a shear surface developed underground. Coarse debris can slide; but, as clay content and moisture content increase, it becomes increasingly probable that the moving material will flow rather than slide. As will appear shortly, dry material can flow in certain conditions.

Sliding over planes of bedding occurs most readily

Fig. 9.20a Rhythmic deformation of the waste mantle: terracettes on a valley side near Bakewell, Derbyshire, England (Paul Popper). Patches of especially deep shade indicate shallow sliding.

Fig. 9.20b Rhythmic deformation of the waste mantle: fingerprint pattern on vertical air photograph is produced by low scrub vegetation, which responds to minute environmental differences within the low flat waves of gilgai. Spacing of waves between crests is about 15 m, amplitude at the ground surface is 5 to 10 cm (New South Wales Department of Lands).

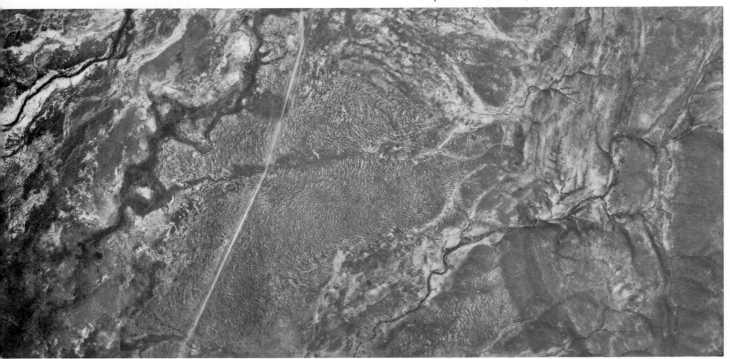

Fig. 9.21a Rockslide: the broken remains of a sheet, which slid off the face partly visible at the top of the photograph. (see also Fig. 4.7)

Fig. 9.21b The Gros Ventre debris slide of 1935, Wyoming, seen partly reclaimed by forest (Ward's Natural Science Establishment, Inc.).

where these planes are steeply inclined, and/or when they include shale partings between formations of limestone or sandstone, and/or when the planes are lubricated by water. The 1963 Vaiont Dam disaster in Italy was caused by a rockslide (Fig. 9.21a). Two weeks of heavy autumn rain wetted inclined planes of bedding, greatly reducing the stick component, and permitting 600 million tons of rock to slip into the reservoir. Although the dam itself held, water overtopped the crest of the dam by 100 m. With an initial velocity of 30 m/sec. about thirty times the speed of ordinary streamflow, a flood wave swept downvalley. Within fifteen minutes, more than 2500 people had been drowned.

Debris consists of weathered and broken rock, mixed with material in the sand–silt–clay grades. It can be supplied for example by the breakup of a sheet of bedrock that lifts from an expansion joint, and which in falling becomes mixed with its original cover of waste mantle. Where the waste mantle is

Fig. 9.22 The St. Jean-Vianney mudflow, Québec, 1971: part of the town, with a few houses still more or less intact, has been engulfed (Canadian Forces photograph).

thick enough, it can develop internal shear planes on which sliding takes place. A noteworthy debris slide occurred in 1925 in the valley of the Gros Ventre River, Wyoming (Fig. 9.21b). Debris loosened and weighted by heavy rains slid downhill, damming the valley bottom to a height of about 80 m. The lake that formed on the upstream side of the dam rose high enough to spill over in 1927. A rapidly eroded outlet channel discharged 55 million m³ of water in only five hours. Thick waste mantles on steep slopes are obviously slide-prone. The 1970 earthquake in Peru set off a whole series of debris slides in mountainous areas. One slide killed 25 000 people in and near the town of Yungay.

Debris clayey and wet enough to flow will spread out fanwise at the base of the slope, forming a deposit which generally resembles a delta. The most dramatic flows of all, however, are those of dry material. These are the catastrophic debris flows which, beginning with rockfall and often splashing at the slope base, will flow across flat valley bottoms and then suddenly settle. If the valley bottom is narrow enough, the flow will surge up against the opposite valley wall before falling back. Flow speeds across the valley floor can easily exceed 30 m/sec. The falling broken blocks become enveloped in a dense dust cloud which permits the whole turbulent mass to move as a fluid. Magnitudes of material involved are impressive. In a single catastrophic debris flow on earth, the magnitude is known to range up to 40 000 tonnes. Still greater magnitudes probably apply to events on the airless moon, while debris flows on the air-poor planet Mars can shift 100 000 tonnes a time.

Many river-damming mass movements in the Himalayan mountains are catastrophic debris streams. In 1893, the valley of a tributary of the Ganges was blocked through a length of 2000 m and to a height as great as 400 m. The newly-dammed lake held 450 million m³ of water. Although a warning system was set up, it proved inadequate. The dam was overtopped, and rapidly cut through, during a night of bad weather in 1894. The town of Srinagar, more than 100 km downstream, was wiped out by a flood wave 20 m high. The most horrifying of all events of this kind dates from 1840. The valley of the upper Indus had been blocked by a catastrophic debris stream, which also impounded a lake. When the landslide dam failed, the discharge of mud-laden water was great enough, 400 km downstream, to generate a wave front 30 m high and to wipe out a whole army in its riverside encampment.

A special type of debris flow is the lahar. This is a mixture of rock debris and water, discharged down a valley on the flank of a volcanic cone. The best known lahars are those of the Mount Rainier volcano in Washington State, in the Cascade Ranges of the U.S. Pacific Northwest. A major lahar can travel as far as 100 km downvalley from its source area. It can move with a 300 m wave front, and deposit as much as 30 m of sediment on the valley bottom. The Mount Rainier lahars are composed of warm and chemically weathering volcanic fragments, and of snow melted into water by escaping volcanic heat. Speed of flow is, for all practical purposes, unknown, but is certainly great.

Mudslide and mudflow, by definition, involve material in the silt and/or sand grade. Concentrations of such materials in slide-prone or flow-prone situations can constitute serious environmental hazards. Mudslides merge into debris slides, for instance on the flanks of glacial troughs that are plastered with glacial sediment. Mudflow, however, can occur on almost horizontal slopes. Notably in southwest Sweden and in the lower valley of the St. Lawrence River in Canada, clayey sediments accumulated in late-glacial times. These sediments, laid down in narrow arms of the sea, locked a great deal of water into their tiny pore spaces. Rebound of the land when the ice caps melted back has raised the valley floors above existing sea level. Rivers, cutting down through the clays, have developed steep banks. Here is a situation packed with danger. Sudden slope failure of a river bank in a particular place not only removes the support of sites upslope, but can also cause the clay in the subsurface to release its water and to turn into flowing slurry. A major slide of this kind occurred in 1971 on the Petit Bras River, a component of the Saguenay River drainage system in Quebec (Fig. 9.22). The material affected – 7 million m³ beneath an area of 270 000 m² – was mainly the debris of a 500-year-old and much larger slide. During the preceding twelve days, spring thawing of the ground had taken place, and more than 180 mm of rain had fallen, causing actual flooding in places. There had also been minor sliding along the river bank. Forty minutes after the main slide began, it reached the outskirts of the town of St. Jean-Vianney. Five minutes later forty houses had been carried away, and thirty-one people were dying or already dead. Liquefying completely, the moving mud reached speeds of 7 m/sec, comparable to the flow of a fast river: 80% of the mass reached the trunk channel of the Saguenay, almost 3 km distant.

9.5c Slumping

The slumping process starts with sliding and ends with flowing (Fig. 9.23). It is probably the most closely examined of all modes of sudden slope failure, because the form of the slide surface can be calculated by the methods of civil engineering. The slumped mass settles downward at its head, and spreads downward and somewhat sideways at its toe.

Fig. 9.23 Slumping of cliff on the Dorset coast, England. former cliff edge is now represented by step half-way down, while toe of slump spreads out at the bottom.

Sliding over a curved shear surface occurs above, slow flow on to lower ground occurs below. Hence the notation in Fig. 9.19, where slumping is made to bridge the separation of sliding from flowing.

9.5d Rockfall

Strictly speaking, rockfall includes the free-air descent of single fragments, say of pea size upwards. Such fragments shower down from many bare bedrock faces. But some falls mean the breaking-off of a whole spur of hillside, or even of the whole projection of a mountain (Fig. 9.24). The activating processes are the undercutting of the slope base and the internal pressure of the exposed rock. In suitable conditions, the falling material transforms itself into the catastrophic debris streams discussed above.

9.6 Triggers of rapid mass-movement

Preconditions for rapid mass-movement have been noted in the foregoing paragraphs – internal rock pressure, the opening of expansion joints, the pre-sence of rock material and rock waste that can be mobilised, and the existence of slopes steep enough for slide, flow, or fall to occur. Anything that acts to cause rock waste to accumulate on a slope, and anything that causes slopes to steepen, such as crustal uplift and downcutting by rivers, must tend to increase the probability of slope failure. The probability of failure is also increased by forest clearance on steep valley walls and on the flanks of mountains, and, in limited instances, by quarrying.

What process, or combination of processes, triggers a particular mass-movement varies both with type of physical setting and with the type of movement. In every case, however, a threshold of cohesion is crossed. Rockfall occurs when the internal cohesion of the bedrock fails to match gravitational pull. Rockslide occurs when the value of the stick component is reduced below the value of the slide component. Debris slide, debris flow and mudslide are similarly controlled. Even with the less

Fig. 9.24 Rockfall from the chalk cliff at Beachy Head, Sussex, England: fallen mass, spread as a dry rubbly fan on the foreshore, will soon be removed by waves (J.K. St. Joseph).

violent movement of slumping, the crossing of a threshold of cohesion permits shearing. Mudflow on gentle slopes also begins with the crossing of a threshold of cohesion at the initial breakaway; a second such threshold is crossed when slurrying begins.

The trigger for an individual movement, then, is some process that alters cohesion of the mass. Common triggers are earthquakes and heavy rainfall. Earthquake shocks are capable of loosening dry rock on steep slopes, and of setting debris and mud in motion. An earthquake in Guatemala in 1976 is estimated to have set off more than 10 000 mass-movements, most occurring on steep slopes, but some affecting gentle slopes (gradients of 1 in 3) where fine-grained material in the subsurface turned into slurry. Corresponding mudflows, prepared for by the uplift of the land and the downcutting of rivers, have been set off by a shock no greater than that of a starting train. Heavy rainfall, whether prolonged over days, or unusually intense in a single storm, is particularly effective with debris and mud. It not only weights the rock waste, but reduces its internal cohesion. In the case of rockslide, it lubricates the bedding planes that are also slide planes. Small wonder, then, that major slides characterise mountainous areas that are liable both to earthquake shock and to major rainstorms. The effects of earthquake shock in triggering mass-movement are perhaps best known for the Andes mountains of South America – a mountain system which is still growing. The effects of heavy rainfall are best known for the Himalayan mountains, also still growing, but especially liable to receive heavy downpours because of their location in the Indian monsoonal system.

9.7 Comparative effectiveness of slope processes

Comparison of effectiveness among surface wash, throughflow, and the various forms of mass movement is difficult. In part, this is because comparative data are few: and this fact has led to criticism of the nine-unit slope model (Fig. 9.13). Again, the results of field observation can be expressed in more than one way. Some data on velocity of movement of the waste mantle have already been given. But some investigations are concerned with the rate of recession of the constant slope, or with the rate of surface lowering, or with the rate of throughput of rock waste for a particular slope component.

Some workers have concluded that the rate of recession of the constant slope tends to vary with climate. Suggested representative intervals for recession through a horizontal distance of 1 metre are a

hundred years in humid-tropical climates; a thousand years in humid-midlatitude climates; ten thousand years in low-latitude semiarid climates; a hundred thousand years in cold high-latitude climates; and as much as ten million years in the really arid climates of the subtropical deserts. Here, in the almost entire absence of surface moisture, slope forms appear to change with infinitesimal speed – except that really arid climate may not prevail as long as ten million years. Moreover, recession rates vary wildly from one site to another. They can be as low as 1 metre in fifty thousand years in regions of humid midlatitude climate, fifty times as slow as the apparent representative value.

Rates of surface lowering by surface wash alone appear to vary to some extent with climate, ranging from about 0.005 mm/yr where wash produces little effect, to as high as 15 mm/yr under tropical rainforest. The main control over the rate of surface lowering in general, by the sum of all processes involved, seems however to depend on slope rather than on climate. On low to moderately sloping hill country, the total rate of lowering is perhaps 25 to 75 mm in a thousand years, at which rate it would take 50 to 60 million years to remove the existing continental masses. The rate increases where relief becomes steep. It rises perhaps to 75 to 750 mm in a thousand years. Even at the extreme, it would take 5 to 6 million years for a mountain belt to be torn down. And beyond this fact is the further consideration that rapid processes work only locally. Rapid mass-movements affect only small parts of the flanks of mountain belts.

9.7a Mass movement and continental denudation

We assess rates of continental denudation by means of analyses of the sediment and dissolved substances carried by major rivers. A discount of about half is necessary, to allow for the effect of mankind in accelerating erosion. The discounted values, by continents, are given in Table 9.1

Some major mass-movements displace far more material, in a single event, than is stripped off entire

Table 9.1 **Annual Delivery of Rock Material to the Sea, discounted for acceleration of erosion by human activities: million tonnes**

Asia	4957
South America	1413
Africa	1139
North and South America	1224
Europe	384
Australia	183

Table 9.2   Amounts (in millions of tonnes) Moved in selected events of Rapid Mass-movement

Name of event = location	Amount
Saidmarreh	37 5000
Flims	31 200
Engelberg	7 150
Pamir	5 200
Osceola	5 000

continents in the course of a year (Table 9.2). But we know very little about the total effect of mass-movement on the continental scale. We know even less about magnitude-frequency relationships, such as those discussed for rivers in Chapter 6, section 6.6, except in respect of limited areas. It has been shown, for example, that major slope failures triggered by unusually prolonged or unusually intense rainfall can be referred to precipitation events with recurrence intervals of 100 years of more.

On the other hand, we can reason that, if the long-term average delivery of rockwaste by mass-movement exceeded the average power of rivers to effect removal, many mountain valleys should be choked by debris. Some are; but by far the greater proportion is not. It must follow that, on the whole, streams can shift what slopes deliver. A puzzling observation here is that the material eroded from drainage basins at the present day seems in large part to go into storage. Streams of very low order may deliver as much as 90% of the material eroded off divides, but the fraction dwindles to 5%, on midlatitude streams, by the time that the drainage area has increased to 25 000 km². We seem bound to conclude that the sediment has lodged in the flood-plains of valley bottoms; and in certain cases, this is known to be true. But the known cases all relate to accelerated erosion caused by man. What happens, or would or could happen, in the absence of man, is completely unknown.

Chapter Summary

Slopes on land *are produced by erosion* of bedrock *and by* the *deposition* of sediment. *They are studied by engineers*, mainly for *slope failure*, by *soil scientists*, mainly for *erosion of farmland*, and by *geoscientists* who work on the *form of the ground*.

The basic slope model includes four components. The *waxing slope* is the curve-over at the edge of a hilltop. The *free face* is the cliff-like edge developed on the caprock. The *constant slope* is *straight in profile*, but *can be segmented*. A commonly occuring *maximum angle* is *32° to 33°*. This is *less than the 45° for unlubricated material*. The *rest angle* of debris *complicates matters*. The *pediment* is a *concave-up* compo-

nent. Its *profile* often *forms a growth curve*. Processes acting on pediments are controversial to unknown. *Something must* also act to *transport rock waste across valley floors*.

The basic model can be varied by the *omission of components*, the *repetition* of components, the *elaboration* of components, as with the *segmentation of the constant slope by threshold–slope decline*, and the elaboration of the model itself, as in the *nine-unit model*. This model has been challenged in respect of its allocation of slope-forming processes among the nine units. In addition, *some slope morphological systems include deceptive components*, especially in areas formerly, but not now, subjected to periglacial climate.

Theoretical analysis of debris-covered slopes and their evolution is *not often matched in the field*. A partial *exception concerns alluvial fans*, on the steep mountainous rims of desert basins. *Up to* areas of about *75 km², fan area* closely *matches supply area*.

Slope-forming processes, additional to rock weathering, *include surface wash, throughflow, and mass movement*. Surface wash can eventually lead to *gullying*. *Heave* results from *heating and cooling*, the *swelling and shrinking* of clays alternately wetted and dried out, and in appropriate climates from alternating *freezing and thawing*. The *downhill movement is expressed as solifluction* in periglacial climates, *and as creep* in other climates. Both solifluction and creep can produce minor step-like and wave-like forms in the surface of the ground. *Slide and flow affect* material through the complete range from *dry rock* through *debris* to *mud*, with catastrophic debris flow of dry material providing a special case. *Lahars are flows* of wet debris or mud *on the flanks of volcanoes. Certain clays slurry* when they start to move; they can flow down very gentle gradients. *Slumping involves sliding* in the upslope situation, *and highly viscous flow* in the downslope situation. *Rockfall is free-air descent*, but can promote the forming of catastrophic debris streams. *Immediate triggers of rapid mass-movement* include *earthquakes, intense rain*, and *prolonged heavy rain*.

Rates of *recession of constant slopes* perhaps range from *1 m/100 yr in humid-tropical climates* to *1 m/10 000 000 yr in arid climates*. Great local variations, however, seem to occur within particular types of climate.

Rates of lowering by surface wash possibly varies with climate, but *may be controlled* mainly *by slope angle*. The likely *range* seems to be *from* as low as *25 mm/1000 yr on moderate slopes*, to as much as *750 mm/1000 yr on steep slopes*.

Some mass-movements deliver more material than is stripped from a whole continent in a single year. But the *frequency* of major mass-movements is presumably *low*, because few valleys are completely choked with debris. All the same, some of the *sediment stripped* off the land *by soil erosion* typically *enters storage* somewhere in the drainage basin. The *delivery ratio* (supply divided by delivery) *decreases up the stream network hierarchy*, falling in mid-latitudes to as low as 5% at drainage areas of 25 000 km². All relevant studies relate to areas where man's activities have caused accelerated erosion. What would happen in the absence of man remains unknown.

10 Surface Morphology and Subsurface Structure

Slopes in the real world are developed on actual outcrops of bedrock, actual spreads of sediment, and actual masses of rock waste. Form can be intimately related to substance, with surface details reflecting the details of rock structure. Similarly with entire landscapes: the grain of structure can be powerfully reflected in the grain of relief.

During the fairly brief history of geoscience, approaches to the surface–structure relationship have undergone drastic changes of fashion. Two hundred years ago, it was widely believed that deep chasms were originally opened by some kind of crustal convulsion. In the minds of some authors, the belief extended to valleys in general. When no idea existed of the immensity of geologic time, mountains also were ascribed to sudden catastrophic events. This interpretation persists today, in the form of a vague folk-belief that all mountains are volcanic. In systems terms, a catastrophic interpretation of physical landscapes calls for instantaneous inputs of energy, and for instantaneous morphologic responses.

The principle of differential erosion became increasingly well understood during the nineteenth century. This principle holds that erosional forces will selectively attack weak rocks and weak structures. In consequence, strong rock units that are not traversed by planes of weakness come to stand in high relief. Weak rock units are differentially lowered by erosion; weak structures, such as faults and master joints, guide the direction of erosional attack. In systems terms, the principle of differential erosion accepts the sum of erosional processes as a black box or cascade, passing over the physics and chemistry of the processes themselves, and going directly to the structure–surface and strength–surface relationships.

In many instances, these relationships are easy to demonstrate at the component level – that is, at the level of individual landforms. As will be illustrated below, very many landforms can be directly and simply related to rock attitude and rock strength. This circumstance is the chief basis of the sub-science of photogeology, wherein stereoscopic air photographs are used to map the attitudes and thicknesses of rock units of varying strength.

In the closing years of the nineteenth century and during the first half of the twentieth, landform analysis became increasingly concerned with situations where surface does *not* match geologic structure. The underlying proposition was that, given enough time, the forces of erosion would reduce the relief of an entire continent, or of a considerable fraction of it, to an extremely subdued state. Analysts of the physical landscape concentrated on evidence of planation, as opposed to the signs of structural control.

Since about 1950, interest in the surface–structure relationship has been forcibly renewed, not only because of the development of photogeology, but also because of the growing attention paid to the additional subsciences of structural geomorphology and neotectonics. The former is concerned precisely with the influence of structure upon surface. The latter deals with the morphological influence of current and continuing crustal movement.

Furthermore, the theme of the evolution of landscapes through time, thrown into discredit by the excesses of the planation-fanciers, is now being restored to respectability. We may, without hesitation, pursue the general topic of the interplay of two systems and two sets of forces. These are the structural systems contained in bedrock and the morphological systems expressed in landscapes, the internal forces of uplift and the erosional forces of relief reduction.

This chapter will deal first with individual landforms and their structural bases, and then with a landscape where structure–surface relationships involve both match and mismatch. As will be made clear, by no means all of the obvious problems have yet been solved. As will also be made clear, a full understanding of the landscape involved, that of the folded Appalachians, requires far more information than can be read from maps or from air photographs.

Fig. 10.1 Geologic outliers, geomorphic residuals, in an Arizona landscape: the capping formation on the left and right is represented in the centre merely by pillars and slabs (Paul Popper). Compare Fig. 10.2.

10.1 The simplest possible case: horizontal strata

The simplest imaginable case of the structure–surface relationship is that where drainage systems cut down through horizontal sedimentary rocks of completely uniform composition. This case however is an extreme, attained only locally. The next most simple case, widely observable in the real world, is that where the rocks are still horizontal, but with a strong component overlying a weak one. The opposite case can also be imagined, where the weak component is uppermost; but it seems likely here that the weak cover will be rapidly stripped off, leaving the drainage to cut into the strong unit beneath.

Incision of drainage through a strong unit above, and into a weak unit beneath, leads in time to the elaboration and widening of valley floors, and to the progressive removal of the strong cover. Sooner or later, portions of the strong unit will be completely detached. Geologically they become outliers. Geomorphologically they are called residuals (Fig. 10.1). Although no specifications of size exist, a qualitative progression is recognized. The continuous spread of the original caprock is a structural plateau. A large detached mass, along with its underlay of protected weaker rock, forms a mesa. Next smaller than a mesa is a butte. The smallest outlier–residual, consisting of a single column, is a pinnacle (Fig. 10.2). Pinnacles are commonly bounded by next-neighbouring master joints in the caprock.

The sequence of plateau, mesa and butte can also develop on broad sheets of basalt, where the lava flowed over terrain of gentle relief, and where the underlying rocks are weaker than the solidified basalt.

Fig. 10.2 The temporal, spatial and formal succession from plateau to mesa to butte and pinnacle.

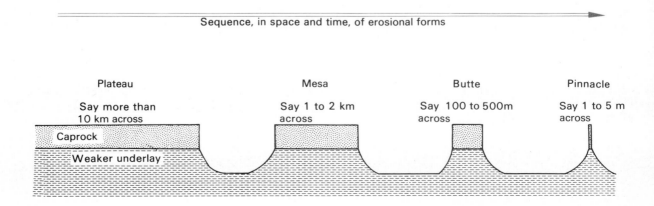

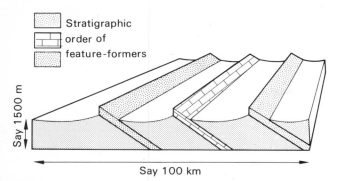

Stratigraphic
order of
feature-formers

Say 1500 m

Say 100 km

Fig. 10.3 A scarpland etched out of uniclinal strata.

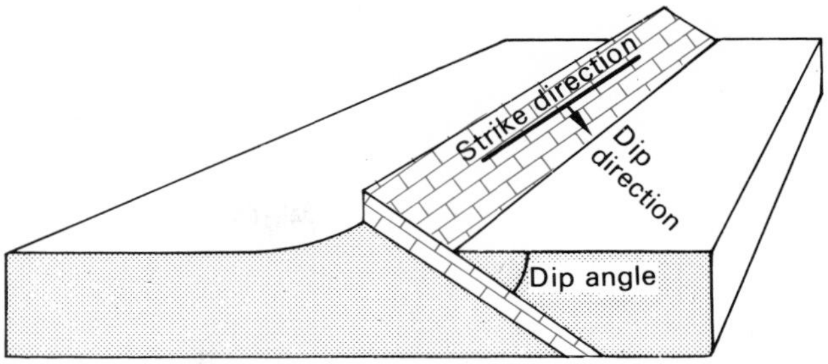

Fig. 10.4 Defining the attitude of bedrock: dip and strike.

10.2 Uniformly inclined strata

The simplest case of the differential erosion of inclined strata is that where the rocks have been gently and uniformly tilted. The typical alternation in a sedimentary sequence of sandstones and limestones with shales ensures differential rock strength. Leaving aside for the time being the connection between differential rock strength and differential development of drainage, we turn directly to the typical outcome of differential erosion – the excavation of lowland belts on the shales, and the etching of the strong units into relief (Fig. 10.3).

Instead of sustaining a plateau, a strong rock formation now rises in a belt of asymmetrical hill. The steep slope eroded through the thickness of the formation, together with a less steep extension across the weaker underlying rocks, is a scarp. The gentler slope on top of the strong formation is a backslope. Scarp + backslope = cuesta. The repetition of cuesta and lowland constitutes a scarpland.

10.2a Specification of attitude

Inclination of a bedding plane is measured in degrees below the horizontal. The measurement gives the dip. The term slope is reserved exclusively for descriptions of the ground surface. The direction at right angles to the dip is the strike. (Fig. 10.4). In a scarpland, the cuestas are developed along the strikes of strong formations, the lowland belts along the strikes of weak formations — hence the term strike vale.

10.3 Domes and basins

A dome, or a basin, can be either structural or topographic – that is, expressed in the attitudes of the rock units, or in the form of the surface. Structural and topographic expression do not necessarily coincide.

In a structural dome, dips radiate outward from the central part. The strike direction curves round the centre (Fig. 10.5a). When a structural dome formed in sedimentary rocks is eroded, the scarps

Fig. 10.5a Abstract field map of structural dome: rock formation shown where exposed.

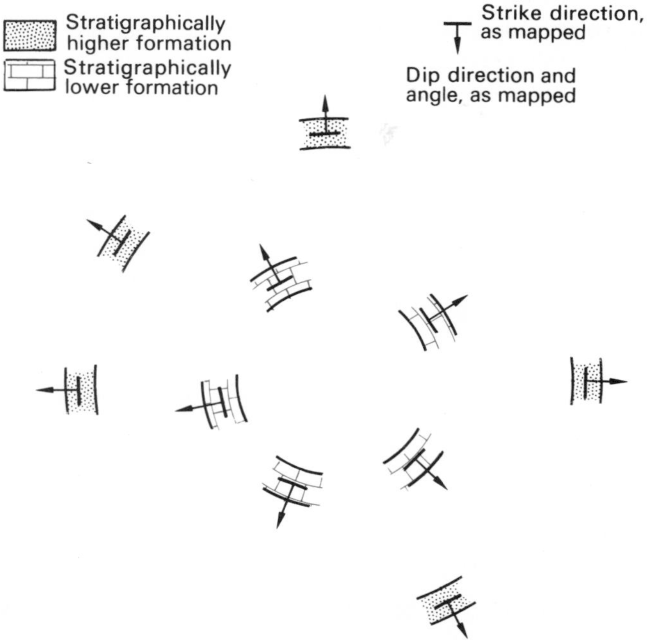

Stratigraphically
higher formation
Stratigraphically
lower formation

Strike direction,
as mapped

Dip direction and
angle, as mapped

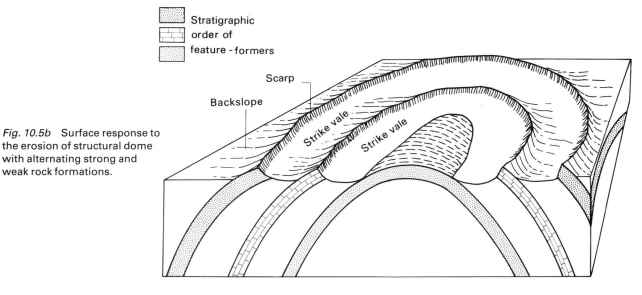

Stratigraphic
order of
feature - formers

Scarp

Backslope

Strike vale

Strike vale

Fig. 10.5b Surface response to
the erosion of structural dome
with alternating strong and
weak rock formations.

cut across strong rock formations will face inward
toward the centre. Cuestas on the strong formations,
and valleys on the weak formations, will curve round
in accordance with the curve of the strike direction
(Fig. 10.5b).

In a structural basin, as in a structural dome, the
strike direction curves round the centre; but the dips
run inward (Fig. 10.6a). If strong and weak rock
formations alternate through the succession, ring-
like valleys separated by cuestas will again develop,
but, in direct contrast to what happens on a dome,
the scarps will face outward (Fig. 10.6b).

10.4 Eroded folds

A section across the middle of a structural dome is,
in effect, a section across an upfold. Equally, a
section across the middle of a structural basin is
effectively a section across a downfold. Scarps face
in toward the axis of the upfold, outward from the
axis of the downfold.

10.4a Plunging folds

The structural corrugations of folds cannot continue
indefinitely. The folds eventually die away at their
ends. The fold axes are said to plunge. Plunging
folds can readily be simulated by the bending of an
elbow: look at the fold patterns produced in your
sleeve.

There is no rule about how asymmetrical a dome
or basin must be, before it is classed as a plunging
fold. Perhaps a 3:1 ratio between widths along and
across the structure might be a reasonable separa-

Stratigraphically
higher formation
Stratigraphically
lower formation

Strike direction,
as mapped

Dip direction and
angle, as mapped

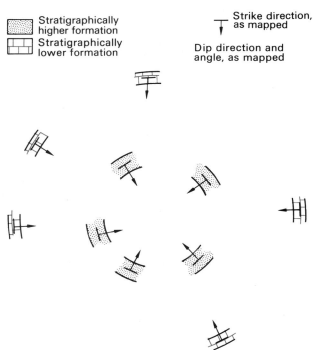

Fig. 10.6a Abstract field map of structural basin: rock
formations shown where exposed.

tion point. Some plunging folds attain a ratio of 20:1
or greater.

What has already been said of the dip–strike
relationships in domes and basins illuminates the
surface–structure relationships of plunging folds. At
the plunging end of an upfold, the dip and strike
directions suddenly swing round (Fig. 10.7a). Scarps
face inward. The two limbs of high ground developed on a single strong formation unite in the
nose of a fold, in a form somewhat like that of the
end of an upside-down canoe (Fig. 10.7b). Dip and
strike directions similarly swing round sharply at the

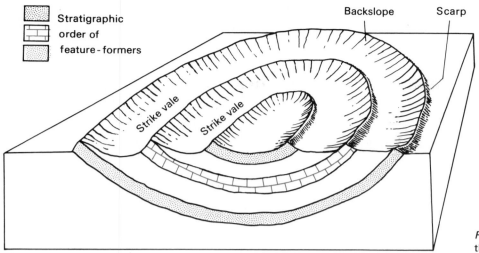

Stratigraphic
order of
feature-formers

Backslope   Scarp

Strike vale   Strike vale

Fig. 10.6b Surface response to
the erosion of a structural basin
with alternating strong and
weak rock formations.

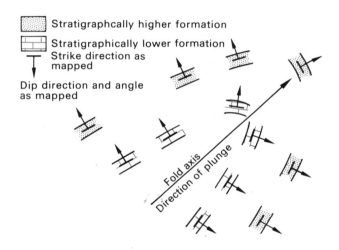

Stratigraphcally higher formation

Stratigraphically lower formation
Strike direction as
mapped

Dip direction and angle
as mapped

Fold axis
Direction of plunge

Fig. 10.7a Abstract field map of plunging upfold: rock
formations shown where exposed.

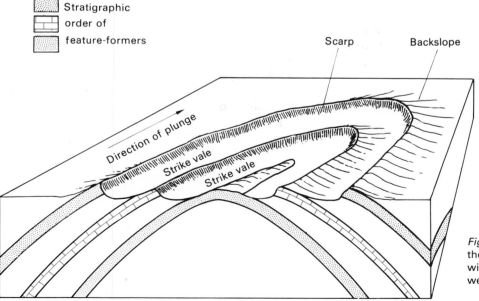

Stratigraphic
order of
feature-formers

Scarp   Backslope

Direction of plunge
Strike vale
Strike vale

Fig. 10.7b Surface response to
the erosion of a plunging upfold
with alternating strong and
weak rock formations.

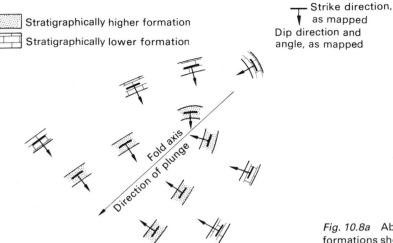

☐ Stratigraphically higher formation
☐ Stratigraphically lower formation

⊤ Strike direction, as mapped
⊤ Dip direction and angle, as mapped

Fold axis
Direction of plunge

Fig. 10.8a Abstract field map of plunging downfold: rock formations shown where exposed.

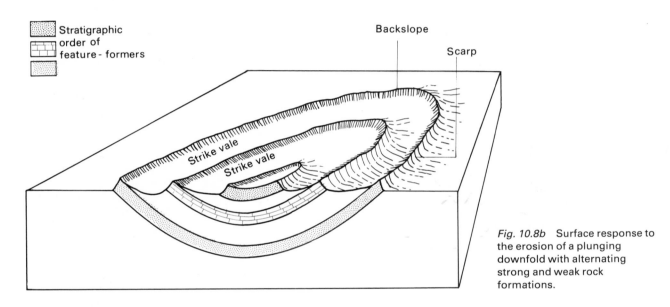

☐ Stratigraphic order of feature-formers

Backslope

Scarp

Strike vale

Strike vale

Fig. 10.8b Surface response to the erosion of a plunging downfold with alternating strong and weak rock formations.

plunging end of a downfold (Fig. 10.8a); but, because scarps face outward, the form of the ground here resembles the end of a canoe set right-side up (Fig. 10.8b).

10.5 Overfolding and thrusting

Folding obviously represents a response to lateral compression. If this compression is severe, and exerted mainly from one side of the fold belt, folds can be overturned (Fig. 10.9a). Still greater pressure can result in the thrusting of rock sheets over the adjacent foreland (Fig. 10.9b). Thrusting in the Swiss Alps has extended through as much as 40 km.

Considerable thrusting also typifies much of the Rockies.

While genuine thrust-faults exist, where the overlying rock has been pushed upward and forward, they fail to record displacements that match those of the Alps. For many years, the mechanics of large-scale thrusting remained obscure. Calculations of rock resistance produced insuperable values. Calculations based on the idea that the Alpine sheets had slid downslope, under the influence of gravity, produced impossibly high slope angles. The explanation lies in hydrostatic pressure of the water originally trapped in the overrunning sheets, or released by the processes of metamorphism. With values of hydrostatic pressure no greater than those observed in deep oil wells, rock sheets 5 km thick could glide down slopes as low as 2° of arc – that is, roughly 1 in 30, or 35 m/km.

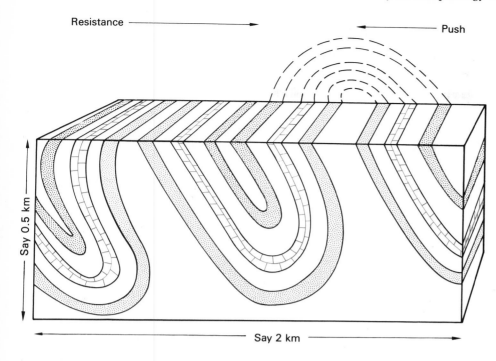

Resistance ⟶   ⟵ Push

Say 0.5 km

Say 2 km

Fig. 10.9a Overturned folds. Any case where the fold-plane axis is tilted through more than 90° from the horizontal is one of overturning. The likelihood of thrusting increases as the degree of overturning increases, but there seems to be a real difference between the high-angle thrusting associated with overturning, and the kind of thrusting illustrated in Fig. 10.9b.

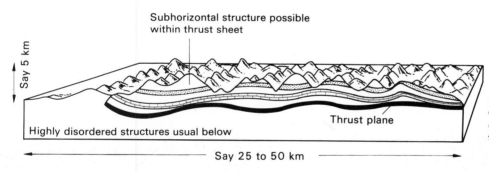

Subhorizontal structure possible within thrust sheet

Say 5 km

Highly disordered structures usual below

Thrust plane

Say 25 to 50 km

Fig. 10.9b Thrusting: note the difference of scale between the effects illustrated here, and those shown in Fig. 10.9a.

10.6 Faults

Faults are cracks in the earth's crust. Faults in the continental crust are typically dominated either by vertical or by lateral movement. Their expression at the surface depends on the local history of faulting, the local history of erosion, and on whether or not a fault brings together rock units of contrasted resistance to erosion. Fig. 10.10 summarises the leading types of morphological response to dominantly vertical fault movement.

Dominantly lateral movement has been intensively studied along the San Andreas Fault system in California, where displacement amounts to about 550 km in the last 150 million years. A particularly easy case is that of the Great Glen Fault in Scotland, where vertical displacement ranges up to about 2 km, but where horizontal displacement is about 100 km. Horizontal displacement is easy to measure on the Great Glen Fault, because the line of fracture cuts a granitic mass in two (Fig. 10.11).

10.7 Adjustment of drainage to structure

Differential development of low relief on weak rocks implies differential growth of stream systems. The process of progressive alignment of the larger streams on weak outcrops is the process of the adjustment of drainage to structure. A simple general model for the development of drainage on a structural dome will show how the process works.

The original structural doming is assumed to raise a topographic dome, on which radial streams

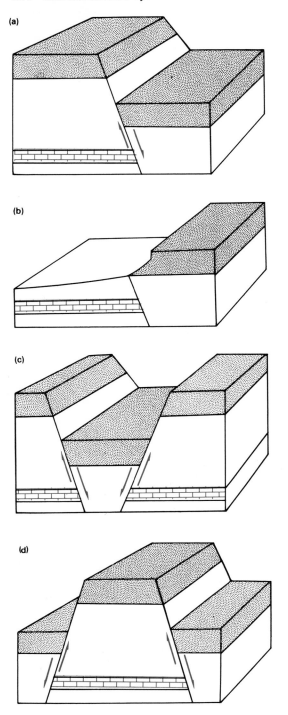

Fig. 10.10 Leading types of morphological response to dominantly vertical movement: horizontal strata, resistant formations tinted. Width of block base can be imagined as 10 km. In (a) high ground lies on the upthrow side; in (b) differential erosion has caused high ground to lie on the downthrow side, but further erosion may reverse the situation; the state (c) is that of a rift valley, where continued movement is likely to maintain the down-rifted strip as low ground (compare Fig. 3.12); but the upthrust slice in (d), probably reflecting a one-shot set of movements, could eventually become eroded into a depression, as can be seen by comparison with (b).

develop (Fig. 10.12a). When the strong caprock is breached, erosion rapidly attacks the weaker rocks beneath. The radial streams throw out branches that take over more and more of the total drainage network.

Now, the original radial streams are most unlikely to be equal in size. The larger of them will cut down faster than will the smaller, one important reason being that a large stream needs a gentler slope than does a small stream. Downcutting, particularly in the middle and lower reaches, serves to reduce channel slope. But the streams that cut down the most rapidly will be the first to breach the caprock, and the first to grow tributaries along the strikes of weak rocks. A particularly large stream can cut down so rapidly that a growing tributary will actually tap the channel of a neighbouring radial stream, in the process known as stream capture or stream piracy. It could equally well be called stream cannibalism, in which the stronger always dismembers the weaker. A stream that loses its headwaters by capture is more at a disadvantage than ever, whereas the captor stream, with its increased volume, will cut down still faster and still farther than before. On a structural dome in alternating strong and weak rock formations, the joint product of differential erosion and capture is a concentric arrangement of cuestas with infacing scarps, and of valleys cut in the weak units, the whole drained by streams with a dominantly ring-like pattern (Fig. 10.12b).

Fig. 10.13 presents an example from the real world, that of the Weald of SE England. A number of rivers do succeed in breaking through the infacing marginal cuesta, but well over half the total length of the higher-order streams is aligned on weak outcrops.

This particular region aptly illustrates the difficulties and the possibilities of reconstructing former drainage lines. It is tempting to match gaps in the marginal cuesta with apparently radial stream courses in the high ground of the centre. But where a gap floor now lies high above the surrounding low ground, amounting to no more than a slight depression in the crest of a line of hills, possibilities other than capture must be taken into account. For instance, two low-order headstreams, working into the hill belt from opposite sides along the line of a fault, could readily notch the crest.

Only when stream deposits can be traced through a gap that no longer carries a stream can capture be conclusively demonstrated. Evidence of the kind necessary is available at the northwestern rime of the Weald. The Blackwater river formerly passed through the gap at Farnham, but that part of its network upstream of the gap has been captured by the Wey. The diversion is recorded by the gravels of an earlier floodplain, which run downstream toward the gap and actually pass through it (Fig. 10.14).

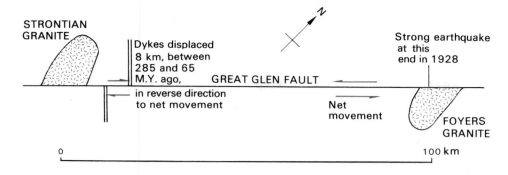

Fig. 10.11   Displacement on the line of the Great Glen Fault, Scotland: net left-lateral displacement is given by the spacing between the Strontian and Foyers granites, originally two halves of a single mass, but lesser right-lateral displacement is recorded by later dykes (basic data after W.Q. Kennedy, J.M. Speight, and J.G. Mitchell). The fault line is still active; the Inverness earthquake of 1928 shook the crust as far away as the English Midlands.

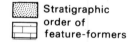

Stratigraphic order of feature-formers

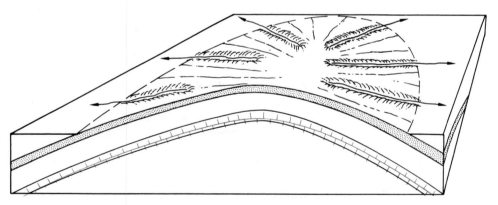

Fig. 10.12a   Radial streams developing on a rising structural and topographic dome.

Stratigraphic order of feature-formers

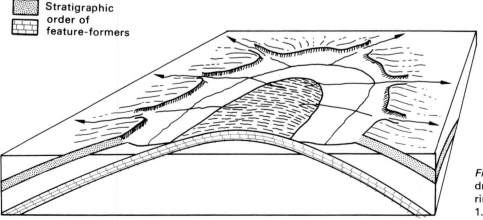

Fig. 10.12b   The original radial drainage largely taken over by ring drainage (compare Figs. 1.9 (part) and 7.1f).

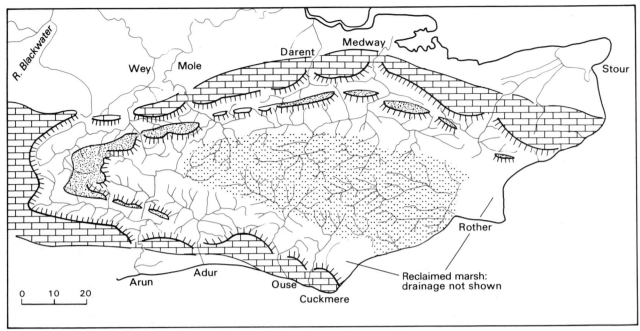

Legend:
- Cuesta-former: chalk
- Cuesta-former: sandstone
- Sandstone-based hills at the centre

Fig. 10.13 Drainage pattern, rock strength, and relief in the Weald of SE England.

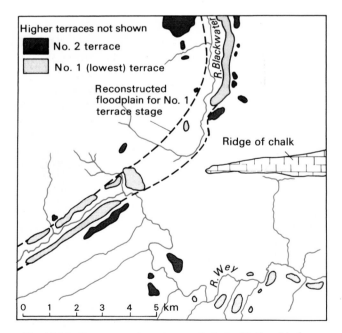

Higher terraces not shown
- No. 2 terrace
- No. 1 (lowest) terrace

Reconstructed floodplain for No. 1 terrace stage

Ridge of chalk

Fig. 10.14 Stream capture demonstrated with the aid of former trains of alluvium: diversion of what is now a head of the Wey from the Blackwater system, SE England.

10.7a The mechanics of adjustment

In what has been said so far, stream capture is seen to depend on differential downcutting on the part of neighbouring master streams. The invasion of one network by the feeders of another depends on a marked difference of level between the two trunk channels. However, something more needs to be said about the selective extension of streams on weak outcrops.

Where the weak rocks are shales, and the strong rocks are more or less permeable limestones and sandstones, springs and seepages are to be expected at hill margins along the base of the caprock. Now, springs can eat their way back into a hillside. The longer a spring-fed tributary becomes, the larger its basin. Again, the larger its basin, the greater the proportion of impermeable shale exposed in the basin floor. Both on account of rock strength and on account of water supply, streams growing along the strike of weak units are well endowed with powers of further extension.

Of the drainage patterns illustrated in Fig. 7.1, the ring-like drainage of Fig. 7.1f results from adjustment to dome structure in sedimentary rocks, as explained above. The trellis drainage of Fig. 7.1g results from

adjustment to fold structure – specifically, to the structure of plunging folds. The rectangular network in Fig. 7.1c, developed in response to the cross-faulting of crystalline rocks, shows that differential strength of rock types is not alone in guiding adjustment: patterns of structural weakness will serve equally well. Here again we may provisionally infer the operation of stream capture, wherein master streams working along the lines of major structural weakness progressively take over the drainage system.

10.8 Maladjusted drainage

Drainage is said to be maladjusted when it is markedly mismatched with geologic structure and with the pattern of relief. Maladjustment is brought about in two main ways – by the development of structures across drainage lines, and by the lowering of drainage lines on to a new set of structures.

If a fold or fault-block rises across the line of a stream, but not so quickly that the stream is diverted, then the stream can persist across the fold axis or the line of faulting (Fig. 10.15). The systems interplay involved here is called antecedence: the stream antedates (= is antecedent to) the structure. Numerous fault blocks in the Great Basin of Utah –Nevada appear to have arisen across stream courses, slowly enough for the streams to escape diversion. On a larger scale, the Indus, the Ganges and the Brahmaputra persist in their southward escape from the rising Himalayas, even though the

Fig. 10.15   Entry of the South Fork of the Humboldt River, Nevada, into a gorge through a fault-block: some miles away to either side, the mountain barrier dies out, with no obstacle to prevent the river from flowing round either end. The river, however, was here before the mountain grew.

Brahmaputra in particular flows for 1300 km along the direction of the mountain axis. Backing-up of sediments behind the rising fold belt is especially well marked on the Indus and the Brahmaputra.

If a drainage network, however well adjusted to the structures within which it is developed, is let down on to a completely different set of structures, it must be thrown completely out of adjustment. This is precisely what happens when the drainage cuts through a sedimentary cover and into the rocks of an underlying craton. The structural discontinuity between cover and craton is an unconformity – specifically, an unconformity of structure. In extreme cases, maladjustment of drainage may supply the only presumptive evidence that an unconformable cover ever existed. This subtopic will be further explored below.

It has been postulated, but on very little evidence and with no demonstration whatever, that adjusted drainage might lose its adjustment in the final phases of continental planation. The postulate seems to rely on the production of a deep mantle of rock waste, and seems also to imply a gentle tilting of the crust, sufficient to displace drainage lines from their earlier adjusted positions. Again, more of this follows below.

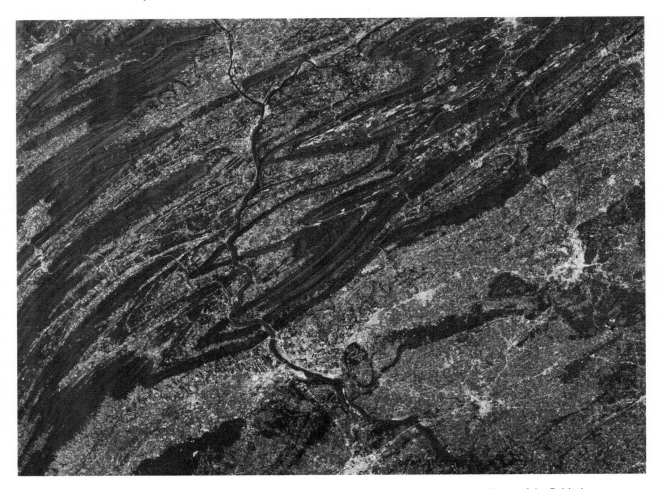

Fig. 10.16 Satellite picture of part of the Folded Appalachians (LANDSAT imagery).

10.9 The folded Appalachians

Fig. 10.16 is a satellite photograph of a portion of the Appalachian Mountains, measuring about 160 × 130 km. The zigzag pattern of dark bands is the system of tree-covered ridges developed on strong formations. The area will now be used in a case study of the interrelationships between surface and structure, drainage direction and structure, sedimentational history and subsidence, plate-tectonic movement and orogeny, between uplift and erosion, and among uplift, erosion, planation, transgression and drainage development.

10.9a Surface and structure

Fig. 10.17 identifies the chief feature-formers. Southeast of the belt of zigzag ridges occur two patches of crystalline and metamorphic rocks. The crystalline rocks are of Precambrian age, more than 600 million years old. The metamorphic rocks, now quartzites, originated as sandy sediment in Cambrian times, somewhere between 600 and 500 million

years ago (see Fig. 24.7). The sand was first converted to sandstone, mainly under its own weight, and then compressed by earth-movements into quartzite.

Within the fold belt, the chief feature-formers are the Tuscarora formation, the Pocono formation, and the Pottsville formation. The first of these is a sandstone, whose original sediments were deposited between 430 and 400 million years ago. The Pocono formation is also essentially sandstone, the Pottsville formation mainly so. Their sediments were deposited between 340 and 280 million years ago.

Several things can immediately be read off Fig. 10.17. The extensive outcrops of the older Tuscarora formation west of the Susquehanna, in contrast to dominantly Pocono and Pottsville outcrops east of the river, indicates generally greater uplift to the west than to the east. Generally older rocks are exposed to the west; on the east, the Tuscarora formation disappears under the Pocono, and that in turn to a considerable extent beneath the Pottsville.

Secondly, the sharp projections of the Pocono outcrop toward and across the line of the Susquehanna river delineate synclines that are plunging

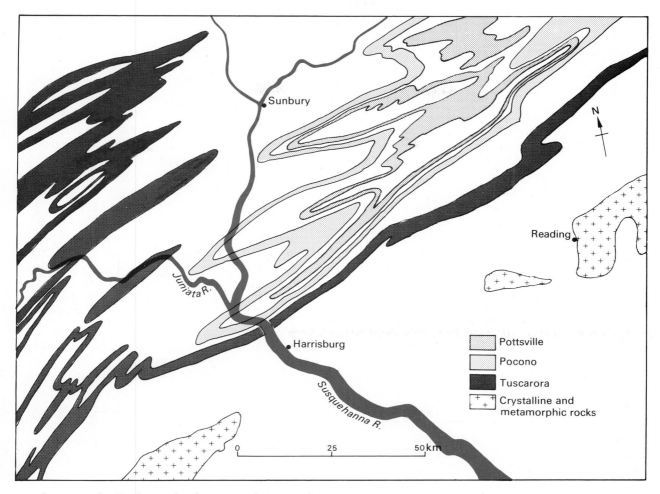

Fig. 10.17 Selective geologic map of the area photographed in Fig. 10.16.

northeastward. Each projection contains a sub-parallel projection of the Pottsville outcrop, which is younger than the Pocono. Outcrop angles directed toward the northeast correspond to the noses of plunging anticlines.

10.9b Drainage direction and structure

Only a few selected components of the existing drainage systems appear in Fig. 10.17. The sole intention is to illustrate the fact that the direction of the trunk Susquehanna river bears very little relationship to the pattern of fold structures and zigzag ridges. With modified emphasis, something of the same kind can be said about the direction of the Juniata river. There do exist reaches where the Juniata flows parallel to the lines of structure and relief, but in total it succeeds in cutting across the structural pattern. The Susquehanna is highly maladjusted to structure – a fact that raises far-reaching questions about the nature and origin of the surface on which it first commenced to flow. These questions however cannot be put, until the sedimentational and structural history of the area has been outlined.

10.9c Sedimentation

It was for the Appalachian area that the concept of geosynclinal sedimentation was first developed. A geosyncline is a major crustal sag. The Appalachian geosyncline was nearly 400 km wide and more than 3000 km long. When the former supercontinent of Pangea is reassembled (Chapter 3), the Appalachian geosyncline can be extended into South America and Eurasia. Eroded plunging folds recur in north-west France.

Deposition of sediment, beginning about 600 million years ago and continuing for about 300 million years, produced total thicknesses of sedimentary rocks as great as 12 km. The mean rate of net accumulation, 0.04 mm/yr, is well within conceivable bounds; but it conceals much recycling of material, connected especially with the elevation of parts of the geosynclinal belt to form dry land. For this reason, the net accumulation rate is no indication of the net rate of lowering by erosion.

Because the sediments are dominantly shallow-marine, and some are even deltaic, there can be no question of the filling of an initially deep trough. Calculations of loading and density show that if the total thickness resulted from a combination of initial sagging and subsequent sedimentary loading, the initial trough would have had to be 5 km deep. The shallow-water character of the sandstones, limestones and shales proves that it was not. On the other hand, sedimentary loading alone could not have forced the floor of the trough down, regardless of the height of the adjoining landmasses and of the abundance of the sediment delivered by them. Something more is required.

10.9d Geosynclinal sagging and plate-tectonic movement

A distinct possibility seems to be that the Appalachian geosyncline was formed in response to the opening of an ancient Atlantic Ocean, with the North American plate of the time moving away from the African and European plates. Not a great deal is yet known of events on the African side; but on the North American side, it is quite possible that the geosynclinal sag was a response to the cooling of the ocean floor, as it spread away from the mid-ocean ridge of the time. The greater the distance from the ridge where ocean crust is created, the cooler and therefore the denser the ocean crust. That is to say, the Appalachian geosyncline may have formed along the trailing edge of the North American continent.

Some support for this proposition can be read from the fact that former geosynclines, where they have been studied in detail, can often be subdivided into two parts. On the margins of the former landmass, sedimentary deposits of the continental slope define the miogeosyncline (mio approx. = lesser). Sediments of continental origin, banked against the descent from the continental slope to the ocean floor, compose part of the fill of the deeper eugeosyncline (eu approx. = greater, or complete). Eugeosynclinal sediments are typically intermingled with abundant volcanic materials; indeed, the prefixes mio and eu were originally applied precisely to connote the contrast in the degree of former volcanic activity.

Widespread volcanic activity might at first sight appear to contradict the idea of a cold subsiding slab of ocean crust. On the other hand, if the subduction process involved melting, it could have generated the magmas which poured as lavas through the vents of eugeosynclinal volcanoes.

The sedimentary rocks of the folded Appalachians originated as miogeosynclinal sediments; so did those of the Appalachian Plateau country next to the west.

10.9e Orogeny

Conversion of the geosynclinal fill to a belt of mountains – an orogen – calls for closure of the ancient Atlantic ocean basin. What contribution was made to the orogen by sediments derived from the ancient African continent is uncertain. The European extension of the orogen is well displayed in SW Britain and in NW France, whence the belt of compression and deformation loops eastward. Closure of the ancient basin is thought to have begun about 500 million years ago, and to have been completed about 375 million years ago. This date however does not mark the end of compression, which continued for another 100 million years or so.

While the shallow-water sedimentary rocks of the miogeocyncline were compressed into the fold structures illustrated in Figs. 10.16 and 10.17, and while the corresponding rocks farther west were gently warped, very severe deformation took place to the southeast. The Cambrian sandstones, metamorphosed into quartzites, were incorporated in a belt of intense deformation and alteration.

The resulting rocks and structures are not particularly well displayed in the area illustrated, although the two blocks of crystalline and metamorphic rocks supply part of the evidence of what happened. Farther to the southwest, the prominent Blue Ridge along the Appalachian flank, the tip of which appears in Fig. 10.17, is an upthrust slice of the Precambrian basement. To its northwest, sedimentary rocks of the miogeosyncline have been thrust over the more stable foreland. In New England, the metamorphic rocks of the most highly compressed part of the orogen, and the granitic bodies that invade them, are widely exposed.

10.9f Rates of uplift and rates of erosion

An uplift rate of 2 cm/yr is documented for some orogenic belts of the present day. This implies a thickening rate of 20 cm/yr, since for every 1 cm of uplift, another 9 cm of light mountain root must be driven down. Convergence of opposing plates moving at 10 cm/yr each is not theoretically impossible, but appears somewhat extreme. In actuality, the orogenic process, in the Appalachian geosyncline and elsewhere, seems to have been irregular. We may infer that the mean rate of uplift was distinctly less than 2 cm/yr. All the same, the span of time necessary to produce the system of plunging folds may have been surprisingly short. Assume that folding reduced a strip of crust from an original 400 km wide to an existing 100 km. At a rate of 2 cm/yr, the total work of compression could have been effected in only fifteen million years.

If compression of the sedimentary fill and the resulting isostatic rise were the only processes responsible for the rise of fold mountains, then thickening by a factor of four, from 12 to 48 km, would imply a mountain belt some 5½ km in average height, supported by a depth of about 42½ km of mountain roots. Allowing fifteen million years, the average rate of uplift becomes 0.035 cm/yr – once again, well within the bounds of possibility.

Modern rates of denudation – lowering of continental relief by the stripping off of solids and by the runoff of solutes – run upwards from about 0.0005 cm/yr. This value, which allows for the accelerating effect of human activities, applies however to whole continents rather than to regions, especially rather than to regions of currently active orogeny. For these latter, perhaps a fourfold increase in the denudation rate, to 0.02 cm/yr, might be in order. Our hypothetical mountain belt, roots and all, could be entirely destroyed at this rate of denudation in 240 million years. In point of fact, the existing Appalachians are modest in relief as orogens go, while gravity survey reveals that their roots are shallow.

But something additional is certainly involved. Not only the compression and uplift of the orogen, but also its reduction by erosion, have been spasmodic. Mountain-building has alternated with planation.

10.10 The planation controversy

The stratigraphic record – that is, the upward sequence of sedimentation through geologic time – contains major discontinuities. These correspond to erosional gaps in the depositional record. They are expressed in the field by unconformities, surfaces where later sediments rest, with strong structural contrast, on much older materials, including the crystalline rocks of the Precambrian basement. Because an unconformity, as observed in a limited field exposure, often appears smooth, the idea has grown that unconformities in general result from the erosional planation of a landmass, with subsequent marine invasion and renewed sedimentary deposition. Exceptions, where the discontinuity is highly irregular in relief, are known and recognised, but can be accommodated by the principle of differential erosion, as applied to the older rockmass. There is a side-controversy about the course and mechanics of widespread planation, which need not be explored here. Of more importance in the immediate context is the possible linkage of planation and renewed sedimentation to the drainage pattern of the folded Appalachians.

10.10a Major unconformities

The spasmodic and prolonged episodes of orogeny were accompanied and followed by severe and widespread erosion. By something like 200 million years ago, extensive planation had been accomplished, despite the fact that the orogen had not been completely destroyed. This particular planation is recorded in the Piedmont region southeast of the fold belt, where sediments of Triassic age (see Fig. 24.7) rest unconformably on Precambrian crystallines and on metamorphosed Cambrian sedimentary rocks.

The sub-Triassic unconformity is now preserved beneath down-faulted bodies of Triassic rocks. Presumably the erosional plane that it represents extended northeastward across the fold belt. Sheets of basalt buried within, or intruded into, the Trias combine with the record of faulting to herald the breakup of the landmass of the time, and a new opening of the Atlantic ocean basin.

A second, and more extensively preserved, unconformity underlies the Cretaceous sediments of the Coastal Plain region. The planation to which the sub-Cretaceous unconformity corresponds had been completed by about 100 million years ago. Widespread marine invasion, with its accompanying sedimentation, is known for the Cretaceous period from many parts of the world. For the Appalachian system, a crucial question is that of how far inland from the present shore the Cretaceous cover extended.

10.10b Major unconformities and the origin of drainage lines

If that cover did formerly extend over the fold belt, then such streams as the Susquehanna could have resulted from uplift. Although the mechanism of general uplift, long after the original orogeny, is obscure, the surviving Cretaceous rocks undeniably dip to the southeast. General uplift of the Appalachians, between say 65 and 40 million years ago, must be accepted as a fact. If there was a Cretaceous cover, then master streams flowing to the southeast would have formed upon it, as soon as the land rose above sea level. Whatever the structures of the cover, and however well adjusted streams may have been to those structures, the drainage would immediately have been thrown out of adjustment as soon as the cover was cut through. According to this interpretation, the Susquehanna is maladjusted, simply because it has cut down through an unconformity. Not enough time has since elapsed for readjustment to be accomplished.

The hypothetical extension of a former Cretaceous cover, far to the northwest of surviving outcrops on the Coastal Plain, fails however to stand up to testing. No Cretaceous outliers survive in the Appalachian region. Furthermore, part of the Cretaceous succession in the Coastal Plain is dominated by sediments derived from the crystalline and metamophosed rocks of the Piedmont; at the relevant time, these rocks must have been exposed on land. Again, the fraction of sediments derived from the folded Appalachians increases upward in the Cretaceous record. The fold belt must also have been included in a land area; the increase through time in the fraction of sediment derived from it relates, in all probability, to the headward extension of streams draining toward the southeast.

10.10c Erosional platforms

Observers have been repeatedly struck by the fact that, seen in perspective, the ridges of the fold belt appear to have remarkably even tops, with the largest and most prominent ridges in a given area rising to much the same height. In the area shown in Figs. 10.16 and 10.17, this height is between some 575 and 600 m. Numerous workers have claimed that the level and accordant summits can only be explained by general planation in some past time. A few have sought to identify the reconstructed erosional platform with the sub-Cretaceous unconformity; but, as has just been seen, such identification goes against the evidence of the sedimentary record. A few, again, have looked to marine planation, but this also is ruled out for the same reasons.

Now, between the ridges lie valley floors. Wherever these are long and wide, they tend strongly to be of very gentle relief, forming almost flat strips between the steeply-rising ridges. By extension of the idea of general planation at ridge-top level, the valley floors have been taken to record partial planation at lower levels. The number of steps recognised varies according to the worker involved, ranging from two to eighteen. In addition, because each lower step is considered to be younger than the next higher step, the whole sequence of descending erosional steps implies a step-functional fall in the relative level between ocean and land.

Obvious difficulties in all this stem from the observed rates of denudation. Lowered at a rate of 0.005 cm/yr, the ridges would be reduced in height by 50 m every million years. If, as has been claimed, the erosional plane represented by the ridgetops was perfected about 30 million years ago, the existing summits may be as much as 1500 m below the hypothetical erosional plane. The valley floors are lower still.

10.10d Other possible explanations

Some authors, seeking to explain the maladjustment of main drainage without recourse to an unconformable cover, have looked back to a landsurface as old as 250 million years, upon which to initiate the original master streams. Others have argued for complex sequences of stream capture, to reverse northwestward drainage into southeastern lines. Others again have rejected altogether the concept of planation, regarding the evenness accordance of ridge summits as representing a dynamic equilibrium between the rate of general denudation and the specific rate of the lowering of crests. The only possible conclusion is, that controversy continues. There is much about this region that we do not yet understand. But any new hypotheses advanced in the future must take plate-tectonic action into account. It is no longer possible to deal with the physical landscape in isolation.

Chapter Summary

Differential erosion results in the *selective lowering* of the ground surface on *weak rocks* and *weak structures*. *Resistant rocks* stand in *high relief*.

Erosion of *horizontal strata* produces the morphological sequence of *plateau, mesa, butte and pinnacle*. Erosion of gently *tilted strata* produces belts of *lowland on weak outcrops* and asymmetrical *cuestas on strong outcrops*. Erosion of a *structural dome* produces one or more *rings of infacing scarps;* erosion of a *structural basin* produces one or more *rings of outfacing scarps*. Similar relationships apply to folded strata; *scarps face in toward the axis of an upfold, out from the axis of a downfold*.

Plunging folds can be looked on as grossly elongated structural domes and basins. At the *tip of a plunging upfold, infacing scarps* suddenly *change direction*. At the *tip of a plunging downfold, outfacing scarps* similarly *change direction*.

Folding results from compression. Accordingly to the degree of compression, folds are *open, tight*, or completely *closed up*. Asymmetrical folding and *overturning* are widely known.

Thrusting through short distances carries one rock mass over another. Something *more than simple push is needed to explain thrusting through distances of tens of kilometres*, such as is observed in well-investigated orogens. This something is the *internal hydrostatic pressure* of water trapped in the rocks at the time of deposition, or released during metamorphism.

Faults are cracks in the crust. Movement along faults in continental crust is dominantly to wholly *vertical*, or dominantly to wholly *horizontal*, according to the particular fault in question. *Most faults* where movement is *dominantly vertical* are *tensional;* a minority, including thrusts, are compressional. The *effect of differential erosion* of faulted rocks *varies* with age and extent of faulting, history of erosion, and comparative rock strength on the two sides of a fault.

Differential erosion involves differential growth of stream systems. Stream nets extend themselves preferentially along the strike of weak rocks and along weak structures. Particularly *large,* or otherwise advantaged, *master streams* can *dismember their neighbours,* tapping the competitor trunk in the process of *stream capture.* Frequently suspected, and frequently appealed to in the reconstruction of landscape development, *capture can only be proved when stream deposits pass through the gap* from which drainage has been diverted.

However well adjusted it may be, *drainage can be thrown out of adjustment by the development of new structures* that cross drainage lines, and *by its cutting through an unconformity.* Little or nothing is known of the possibility that streams on an erosional plane could lose their adjustment.

The folded Appalachians provide the material for a *case study.* The axes of the tight plunging *folds run northeast-southwest.* The Susquehanna and other *trunk streams flow* generally *to the southeast, markedly out of adjustment* with the fold structures.

The Appalachian *orogen began* to form *with the development of a geosyncline.* The necessary crustal *sag quite possibly resulted from* the *subduction* of a cold limb of oceanic crust. *Sediments* in the geosyncline were *dominantly shallow-marine,* indicating that sedimentation kept pace with subsidence. The *rocks of the folded Appalachians* and of the plateaus next to the northwest originated as *miogeosynclinal* sedi-

ments. *Eugeosynclinal materials and parts of the Precambrian basement were highly compressed* when the geosyncline closed up; *thrusting* and *metamorphism* were accompanied by *granitic invasion* at depth. *Thickening* of the compressed mass *created mountains, supported* isostatically *by mountain roots.*

Compression, uplift, and planation have been spasmodic. The *sub-Triassic unconformity* is preserved beneath down-faulted Triassic sediments. *Faulting,* and the intrusion and extrusion of *basalt, record* a new *breakup of the supercontinent of the time.* Planation was again effected by Cretaceous times; Cretaceous rocks of the Coastal Plain rest on the *sub-Cretaceous unconformity.*

Suggestions that the Susquehanna and similar streams *originated on* an emerged and *unconformable Cretaceous cover are unwarranted.* By Cretaceous times, the Piedmont and the folded Appalachians were parts of a landmass.

The *even-crested ridges* of the folded Appalachians, along with the *general similarity of height* among the larger of them, have frequently been *taken as evidence for planation* – by most of the workers concerned, planation later than that of the sub-Cretaceous unconformity. *By extension, wide flat valley-floors* have been *taken as evidence of incomplete planation. Episodes of planation,* whether complete or partial, *range from two to eighteen,* according to the worker concerned. An *opposing view* is that evenness and accordance of the major ridges merely represents *dynamic equilibrium.*

11 Hydrologic and Morphologic Extremes: Deserts

Deserts represent hydrologic extremes, in the sense that they do not generate overflowing rivers. They also represent morphologic extremes, in the sense that they contain large expanses of bare rock or loose sand. Further types of extreme can be added. Deserts are too dry to sustain continuous plant covers: desert vegetation is discontinuous to absent. Plants and animals alike must be adapted to withstand great dryness, to take advantage of moisture whenever it becomes available, and to survive – if they can – rainless spells which may stretch through whole runs of years.

As hydrologic and morphologic extremes, deserts might suitably have been discussed immediately after stream nets, stream channels, or even slopes. Against this, it is rarely possible to investigate desert landscapes without reference to total landscape history, such as has been examined in a quite different context in Chapter 10. The former operation of landscape processes, moreover, can be treated in either the context of the physical landscape itself, or in the context of climatic shift (see Chapter 18, sections 18.5a and 18.5b, and Chapter 19, section 19.4b). As at several other junctures, we are up against the awkward fact that a sequence of chapters is by definition linear, whereas the structure of geoscience resembles a web rather than a chain.

11.1 Extent and location

Occupying some 30 000 000 km^2 of the world's land area, deserts account for about 20% of the total continental extent. They constitute responses to subsiding, warming, and therefore drying air. The subtropical deserts correspond to the permanent or semi-permanent high-pressure cells of the general circulation (Chapter 15). Extensions of the subtropical deserts occur in midlatitude regions of strong rainshadow. There, potentially rain-bearing winds are dried out, by mountainous relief in the upwind direction, or simply by distance of wind-travel from moisture-supplying oceans. Thus, the Atacama Desert of Peru and northern Chile, on the western – Pacific – side of the continent, but dominated by subsiding air, crosses the Andes into the lee side of the mountains to merge into the dry country of Patagonia in southern Argentina.

The Sahara is a subtropical desert. The Rajasthan Desert in the northwest of the Indian subcontinent is another, but with some rather special atmospheric attributes. Its subsiding air powers the Indian monsoonal system in the low-sun season. Heating in the high-sun season, although responsible for sucking in the inflowing and rain-bearing monsoon, rarely brings rain to the desert itself. Desert dust in the lower atmosphere reflects back so much radiation that the base of the air column persists in subsiding, and in warming and drying during its descent.

Potentially rain-bearing winds are drawn away from most of Arabia and much of Iran by the suck of the Indian monsoonal system. Inland, mainly at low levels but separated from the ocean by mountains, a vast spread of desert and semi-desert reaches 1500 km eastward from the shores of the Caspian Sea. North of the Himalayan barrier, the Tibetan Plateau is dry, cold and windy. Farther still into the interior of the Asian continent, the Takla Maklan Desert occupies the Tarim Basin, while the Gobi Desert overlaps broadly from northwest China into Mongolia. Desert vegetation in inner Asia extends beyond 42°N, the latitude of Rome or Philadelphia.

In the southern hemisphere, in addition to the Atacama Desert and its extension, desert topography occurs in Namibia, while semi-desert extends broadly through the Kalahari. Australia contains the world's greatest continental proportion of desert and semi-desert terrain. The Atacama, Namib and Australian deserts all represent responses to the subsidence of air in the major subtropical cells of high atmospheric pressure.

11.2 Desert rainfall

Because of the considerable extensions of rainshadow deserts into middle latitudes, it is not possible to set a precise value on the limiting rainfall for desert climate; allowance must also be made for temperature. Nevertheless, we may make the crude generalisation that mean annual precipitation on a great deal of the world's desert area runs at 250 mm or below. Averages as low as 12.5 mm are known. Averages, however, signify little. Spells of ten years with no rain at all can afflict not only deserts proper, but adjacent regions classed, on the basis of long-term averages, as semi-desert.

With scanty precipitations go low cloud cover, low relative humidity, and high rates of evaporation, Readings of relative humidity not far above zero have been taken in the Sahara. Potential evaporation in the centre of Australia has been calculated at about 2300 mm/yr, some twenty times as great as mean annual precipitation. The calculations have been confirmed by observations of the drying of Lake Eyre after a rare flood.

Notable exceptions to the rule of low cloud cover, low relative humidity, and high evaporation rates occur in the foggy deserts of Peru and Namibia. In both areas, heating overland causes air to be drawn in from over the ocean. But because the offshore waters are chilled by equatorward-moving currents, and by upwelling close inshore, the inflowing air is thermally stable. It condenses its moisture into fog but not into unstable rainclouds. Thus, despite high relative humidities, rain fails to fall. Only if the chill offshore water is overspread by a thin surface skin of warm water, as occurs from time to time off the Peruvian coast, does convectional rain break out.

11.3 Desert hydrology

Rivers that cross deserts, eventually reaching the sea, include the Colorado, the Indus, and the Nile. All three are nourished by precipitation in mountain areas.

The upper reaches to the Colorado system tap the Wasatch Mountains and the Southern Rockies, where a mean annual precipitation of 500 to 750 mm falls partly as winter snow. Released from storage by summer melting, the snow contributes markedly to the year-round flow of the river. Evaporation losses in transit are reduced by the canyon walls, which prevent sunlight from ever falling on some channel reaches.

The Indus is fed by the monsoon-drenched western Himalayas, where mean annual precipitation runs as high as 1500 to 2000 mm, and where great glaciers add their storage capacity to that of snowfields. Equatorial rains of about 1500 mm/yr, in the general vicinity of Lake Victoria, nourish the upper reaches of the White Nile system. The tributary

Sobat River draws on the Abyssinian highlands, as does the more voluminous Blue Nile. Rain in Abyssinia, up to 2000 mm/yr in the highest ground, is brought by a kind of bypass loop in the Indian monsoonal system.

Although scarcely touching desert proper, the River Niger describes a huge bend, about 750 km long, through country where mean annual precipitation runs below 500 mm, and where mean annual temperatures exceed 25°C. Its headstreams form in the Guinea Highlands, where monsoon-like rains in the high-sun season supply most of the 1500 to 2000 mm on mean annual precipitation. Penetrating into successively drier country, the Niger suffers marked evaporation losses. It also loses by percolation of water into its massive inland delta. Enough water remains, however, for the river to persist. It breaks out southward to the ocean, through mountains drained by rain-fed confluents.

These four rivers are exceptions to the general rule that desert rivers do not reach the sea. The Humboldt River finally dies away in Carson Sink, Nevada, when indeed it reaches that far. In Australia, the Diamantina River and Cooper's Creek are eventually lost in the flats of the Lake Eyre Basin. Not enough water comes down the Volga to force a connection between the Caspian and Black Seas. The Amur Darya and the Syr Darya vanish in the saltpans of the so-called Aral Sea. In the Tarim Basin of inner Asia, a whole group of rivers dry up and disappear; if they persist far enough, they finally evaporate in the Lop Nor marshes. About half the Australian continent – the western half – lacks co-ordinated drainage. Streams that do flow fail to integrate themselves into defined networks. At the final extreme, some parts of the Saharan and Arabian deserts display no signs at all of drainage lines.

11.3a Streamflow characteristics

Desert streams are typically discontinuous in space and ephemeral in time. When streamflow constitutes a response to thunderstorm rainfall, only certain parts of a given network may be affected. Thus, it is possible for flow in a trunk channel to pass by the mounts of dry tributaries, or for a flowing tributary to discharge into a dry trunk.

Runoff from divides into channels, little or not at all impeded by vegetation, is typically swift. In many desert areas, it is made all the swifter because the ground surface is sealed by fine sediment in the silt and clay grades. Infiltration can be completely inhibited. Surfaces of bare bedrock also inhibit infiltration, unless they are much shattered and/or strongly jointed. Both the infiltration rate (volume/time

Fig. 11.1 Flooding on the Georgina River system, in Queensland's channel country (Division of National Mapping, Australia).

ratio) and the infiltration capacity (volume potential) can run very low indeed.

For these reasons, or for some combination of them, low-order desert streams are notoriously flashy. Lag between precipitation and streamflow is sharply reduced below that characteristic of humid climates. When high flows do occur, they can advance downstream in the form of a wave. The storm hydrograph rises almost vertically.

Paradoxically enough, some desert areas are notoriously liable to occasional flooding. Nowhere is this statement more valid than in the Australian Centre. Hurricanes and related storms are capable of escaping from the circulation of the humid tropics, and of bringing heavy rains to the inward-draining streams of Queensland (Fig. 11.1). When this happens, as it did happen in 1949 and 1974, Lake Eyre can be temporarily re-established. The 1974 floods of the interior affected about half the continent.

Very little information exists on the sediment discharge of desert streams. Widespread evidence of stream deposition, both alongside and within channels, shows that sediment load during peak flows is often high. Runoff of sediment, like runoff of surface water, is abundant and rapid, and for the same reasons – along with the fact that rainsplash during a heavy storm can dislodge as much as 25 000 tonnes of sediment per km^2.

11.4 Stream channel characteristics

Desert channels cut into bedrock do not differ importantly from corresponding channels of humid areas. In dominantly sandy or silty areas, however, stream channels in deserts and semi-deserts tend to possess highly distinctive attributes.

In semi-deserts, where streamflow is frequent enough to maintain defined channels, unusual channel patterns develop on gently-sloping spreads of sediment. Whereas the downslope gradient can be of the order of 0.001, the cross-slope gradient can approach zero. Trunk channels can split. A distributary may run for 100 km before it rejoins the parent channel, or before it joins some other trunk (Fig. 8.16). The takeoff stream is called an anabranch. The mechanism of anabranching, little understood as yet, is thought to be analogous to the splitting of channels on deltas and alluvial fans.

Where the bed material is sandy, defined channels have flat floors. A high proportion of fines in the banks will permit these to sustain vertical or near-vertical slopes. Little or no work has been done on the controls of the width/depth ratio of sandbed streams in dry climates. By observation, this ratio can range from about 3:1 to 100:1 or greater.

As dryness intensifies, and as the frequency of streamflow diminishes, the concept of channel-forming discharge becomes increasingly less relevant. On gentle gradients across spreads of sediment, channels become increasingly ill-defined, subdividing into bundles of shallow surface washes. Where channel morphology is controlled by occasional major floods, a single main drainageway consists of a network of minor channels which divide and reunite round elongated sandbars (Fig. 11.2). The resulting reticulate pattern is analogous to the low-flow pattern of braided streams in humid climates.

Where slopes are perceptible but gentle, water may move along paths that are reoccupied with every flood, but where no channel is ever cut.

Except in a single connection, little is known of channel slope in dry climates. There is some evidence that, among low-order channels, profiles tend to be segmented, with slope suddenly decreasing below channel junctions. The exception relates to alternating cut and fill, which has been intensively studied in the U.S. Southwest. The responses are well documented: processes are enigmatic.

If the question were one of simultaneous, or nearly simultaneous, cut and fill throughout a whole region, then it would be possible to appeal to such influences as shift of climate or change in land use. The arguments run somewhat as follows: introduction and intensification of commercial grazing promote increase in the speed of runoff, and therefore tend to promote channel cutting. A climatic shift toward increased humidity, by intensifying the vegetation cover, acts to check the speed, and also the amount, of runoff from divides, and thus tends to promote channel filling. But matters become complicated, when account is taken of effects on sediment supply. Reduction of the plant cover, whether be-

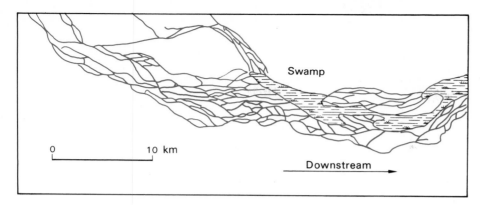

Fig. 11.2   A reticulate channel: diagram closely based on the pattern of Cooper's Creek, inland East Australia. Contrast, with reference to scale, Figs. 8.13, 8.14a, and 8.16.

cause of grazing or because of increased aridity, might well lead to increased sediment supply and thus to channel filling. Intensification of the plant cover, by reducing the sediment supply, may well lead to channel cutting. Furthermore, runoff characteristics can change, without any change in mean values of precipitation. If a given annual rainfall were concentrated in a reduced number of storms, the mean fall per storm would necessarily increase; the storm hydrographs would rise higher than they did formerly. Depending on the influence of sediment supply, the change could readily lead to channel cutting.

Channel changes resulting from changes in land use practice are best documented for the southern Plainlands of the U.S., an area of sub-humid as opposed to semi-arid climate. The wagon trains of pioneer farmers crossed rivers with low width/depth ratios. The spread of cultivation was followed by savage bank erosion. A sudden and very sharp increase in sediment supply caused the rivers to modify their channel shapes, and to develop the very high width/depth ratios appropriate to sandbed streams. Withdrawal of some of the driest land from cultivation, and improved land use practices elsewhere, have subsequently reduced erosion rates, and have permitted the channels to revert to something like their cross-sections of pre-settlement days.

In the dry Southwest, however, a single channel reach can be affected by cut and fill at the same time. Cut in the upstream part combines with fill in the downstream part to reduce channel slope (Fig. 11.3). The change, however, is often minuscule – for

instance, a reduction of slope through the affected reach by only 0.1%. A change of this kind presumably has some physical cause; but that cause so far defies identification. The matter is all the more mysterious, because the type of change in question may affect one trunk channel, but not its next neighbour.

11.4a Stream termini

Streams which fail to escape from deserts terminate in permanent lakes such as the Great Salt Lake of Utah, in ephemeral lakes of considerable depth, such as Lake Amadeus in Western Australia, in salt marshes, or in playas. Commonly associated with all of these is evidence, in the form of shorelines above the level of lakes, marshes, or playas, of a high lake stand at some former time.

Playas are flat expanses at or near the lowest point of a desert basin, or at least at the downstream end of the channel of an ephemeral stream (Fig. 11.4). Many playas in existing climates are subjected to occasional flooding to shallow depth. Playa lakes,

Fig. 11.4   Playa in a minor enclosed desert basin: generalised from actual conditions in inland N.S.W., Australia.

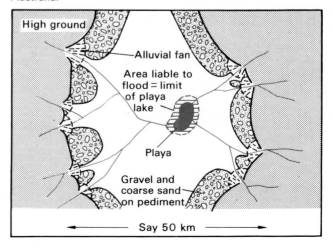

Fig. 11.3   Stepwise change in channel gradient, effected in some ten years or at most in a few tens of years: highly diagrammatic as illustrated, but based on numerous actual cases, particularly in New Mexico.

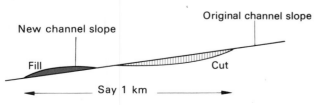

therefore, make up a subset of ephemeral lakes. In the absence of floodwater, the playa surface ranges from clayey to saline. For reasons not completely understood, even next-neighbouring playa lakes can range from fresh, with clay floors when they dry out, to saline, when their floors, on drying, glitter with encrusted salts. The floors also vary from dry to moist, according to the relationship between the ground surface and the water table. Field vehicles drive safely across dry floors, but bog down and even sink in moist floors.

Many playas are completely devoid of visible plants. Some are too saline for anything to grow on them. If scrub vegetation can move in from the margins, it is liable to be killed off by returning drought, or to be drowned in a new flood. But in some cases, what would otherwise be called a playa is vegetated by salt-tolerant canes that root in expanding clay soils.

Playas are subject to deflation (see section 11.6). If the wind blows mainly in a single direction, then the lee side of a playa will be rimmed by a lunette. This is a dune of crescentic plan, built of an admixture of sand blown right across the playa, and grains or aggregates of silt, clay and colloids blown off the playa itself. Lunettes slope more steeply toward the playa, less steeply away from it.

Reduction of a standing lake to the playa state, and disruption of incoming drainage lines, typically leads to the formation of a whole set of ephemeral lakes, and eventually of playas. Ponding, deflation, and the construction of lunettes act irregularly in space, so that individual playas can be strung along the line of an ephemeral channel. A former lake of moderate size may be represented today by, say, a couple of dozen playas ranging up to 5 to 10 km across, and by many tens of tiny playas as little as 200 to 300 m across. With the smallest features of all, it is difficult to separate playas from simple blowouts.

11.4b Desert water chemistry

Information on the chemistry of desert waters comes mainly from studies of enclosed lakes. Generally speaking, desert rivers are more saline than are rivers of humid climates, because of the selective concentration of soluble materials in desert soils. But the highest salt concentrations occur in desert lakes, which receive streams with high solute loads and lose only water by evaporation. The Great Salt Lake records salinities of more than 200 000 parts per million; the value for the Dead Sea is over 225 000 parts per million. Nearly one-quarter of the Dead Sea consists of solutes.

11.5 Desert surfaces

In physical terms, desert surfaces may be broadly classified as rocky, stony, sandy, and silty/clayey. At a rough guess, each sub-class accounts for a quarter of the world's total extent of deserts. Some surfaces, including the floors of former lakes and the flats of existing and former marshes, are highly saline.

Not surprisingly, the generic names applied to sundry types of desert surface are of Arabic origin. One-word equivalents for certain of these names simply do not exist in English. Thus, the word *hammada* (or *hamada*) connotes a more or less horizontal desert surface of bare bedrock. The word *reg* roughly translates as gravel desert, and *serir* as a gravel surface where the stones are closely packed. These two terms, however, are being superseded. For a desert surface littered with abundant rock fragments, whether rolled gravel or not, the Australian aboriginal name gibber is increasingly used. Where the fragments are packed closely enough to cover the surface completely, they form desert armour (Fig. 11.5).

More than one type of material can supply the litter of loose stones. Conglomerate, breaking down under the attack of weathering, provides rounded pebbles ready-made. Tough sandstones, quartzites, and fine-grained metamorphic rocks provide angular fragments such as those in Fig. 11.5. A great deal of gibber in Australia comes from the breakdown of siliceous duricrusts (see Chapter 18, section 18.3).

Sandy deserts are dominated as to form, but not necessarily as to fraction of coverage, by features constructed of wind-laid sand. The leading types of feature will be discussed separately below. The ultimate source of sand supply is probably, for the most part, weathered sandstone; but the main proximate source is alluvial sand in and bordering stream channels. Finally, extensive silty/clayey surfaces occur in the central parts of enclosed desert basins. We are not here considering the floors of playas, which may run for instance to lengths of 12 km and widths of 8 km, but vast spreads of the order of 100 km across. Almost featureless on the surface, and with imperceptible surface gradients, sedimentary desert plains have so far attracted little attention in the field. They tend also to be overlooked in the literature.

11.5a Features built of sand

The Arabic word *erg* connotes a sandy desert surface. Its usage has, however, become somewhat loose. By some writers it is applied to all desert surfaces where the relief is dominated by sandbuilt features; others restrict it to sand seas.

Minor features related to irregularities in the ground surface include sand shadows and sand drifts. Sand shadows accumulate on the lee side of obstacles. Sand drifts extend downwind from the mouths of topographic notches, and from the mouths of valleys that open on to generally lower ground. Both types of feature can be matched in stream channels. Sand drifts are analogous to the sandbars constructed downstream of bridge arches during high flow. They are responses to the reduction of flow velocity as streamlines diverge. Both types, again, can be matched among the forms constructed of fine blowing snow.

Extensive sand sheets (Fig. 11.6) are constructed either of unusually coarse sand or of unusually fine sand. Sand grains of 0.5 mm diameter or greater will fall through a vertical distance of 10 cm in about 0.5 sec. Thus, they move by rolling rather than by the leaping motion called saltation (see below). There is little chance that a surface formed on coarse sand will develop the local irregularities, of 25 cm relief or more, which are critical for the start of dune-building. Sand grains of 0.03 mm diameter and smaller can be carried in suspension by winds of modest velocity. When they fall, they form a powdery sheet too even to exert irregular drag on the moving air above.

Dunes come in more than one type. The list of types varies from author to author, being in part controlled by the characteristics of the particular deserts known. Repetitious types include star dunes, transverse dunes, sand seas, barchans and seifs. Star dunes, as yet very little studied, are massive sand-piles with multiple, radiating and crested arms. Transverse dunes are suspected to be analogous to waves on water, and, like these, to be raised roughly at right angles to the dominant wind. In sand seas, the wave pattern becomes complex, like that of a very choppy sea surface.

Barchans are crescentic in plan. Both in plan and in profile, they are streamlined on the upwind side (Fig. 11.7). Sand moves up the upwind face by rolling, and also by saltation. This latter process is a leaping motion, in which a moving grain follows a ballistic trajectory. Arriving at the crest, the sand simply falls on to the front, or slip, face. Because most new sand lodges at or near the top of this face, the slope angle tends to increase. When the angle reaches 33–34° of arc, negative feedback comes into play. Gravitational pull overcomes the internal friction of the sand, and the slope fails. Failure is characteristically very localised, with a shallow stream of sand coursing down the face, and coming to rest at a surface slope of 30° or so. Descending a slip face 25 m high, the slipped sand would advance the foot of the dune by about 5 m. Because at least part of the lodged sand is supplied from the upwind

Fig. 11.5 Desert armour: rock particles completely protect the silt- and clay-sized particles beneath, inhibiting blowing, washing and rainsplash alike.

Fig. 11.6 Sand sheet built of coarse sand blown mainly from sandbed channel complex, marked by trees about 2 km in the distance. A supplementary source is a dead dunefield still farther off. Distance between crests of waves in the sheet is about 25 cm. One year before the photograph was taken, this was sheep-grazing country that supported saltbush.

Fig. 11.7 Barchans, northern Peru, seen from the air (Ward's Natural Science Establishment, Inc.).

Fig. 11.8   Air view of seifs in the Simpson Desert (Division of National Mapping, Australia).

face of the barchan itself, and because repeated slips will advance the whole dune foot, the barchan migrates in the downwind direction.

Speed of migration is a linear function of the height of the barchan, in the approximate form

$$v = 27 - 0.625H,$$

where v is speed of migration in m/yr, and H is the maximum height of the barchan, in metres. This equation predicts a movement of rather more than 11 m for a barchan 25 m high. That is to say, the whole slip face should fail about twice a year on the average. The equation also suggests that migration rate is zero where H = approximately 43 m, a value similar to that of maximum barchan height as measured in the field.

The asymmetry of barchans in profile is precisely opposite to that of lunettes. Whereas the slip face of a barchan faces downwind, the steeper face of a lunette, like that of a coastal dune, faces upwind.

While the relationship of dune form to wind direction remains incompletely understood, it does appear to be well established that barchans are constructed by sand-driving winds (about 5 m/sec and upwards) which blow predominantly in a single direction. By contrast, sand-driving winds from two directions, oblique to one another, build seifs – long ridges of sand, also called linear or longitudinal

dunes (Fig. 11.8). Seifs can attain heights of 100 m, widths of 350 m, and lengths of 100 km or more. Close to the sand source – such as the reticulate channel of a sandy stream – seif spacing is sub-regular and close (Fig. 11.2). Farther away from the sand source, the ridges converge, decreasing in number but increasing in height. They tend to stabilise at a spacing of 500 to 1000 m between crests. In a given dunefield, seifs can be remarkably parallel and remarkably evenly spaced.

Between the seifs come interdune corridors. The values that have just been listed make it clear that two-thirds or more of the total surface can be underlain by something other than sand; for, whereas sand is concentrated in the seifs, the interdune corridors are typically underlain by silt and clay, the product of the deposition of desert dust.

11.6 Deflation

Removal of sand, silt and clay by wind action is the process of deflation. A smooth dusty surface resists wind erosion, but if the dust is bombarded by saltating sand grains, it will be whipped upwards. Particles in the clay grade are known to travel up to 5000 km outside the deserts from which they come. Particles in the silt grade, disturbed by wheels or hooves, will fall back to the surface in the absence of wind; but in turbulent air, they can be held aloft and carried through considerable distances. Saltating grains return to the surface in distances measurable in centimetres, but continued saltation permits these also to move far. The velocity of a moving sand grain is about one-tenth of the velocity of the wind near the ground. A representative value for annual distance travelled by a single grain is 20 km.

Deflation rates vary greatly, according to the type of surface affected. Typical long-term rates are summarised in Fig. 11.9. On surfaces of bare rock and on desert armour, potential for deflation is controlled by the rate of physical weathering into sand-sized particles, which runs very low. Gibber surfaces where spaces exist among individual stones are more readily affected, although, even for them, the maximum possible lowering by deflation averages only about 1 mm/yr. To reduce the elevation of the ground surface by one metre, 1000 to 10 000 years would be needed.

Farther along the scale, deflation rates continue to increase, as do the rates of lowering into which deflation rates can be translated. It must, however, be understood that the translation assumes zero input into the subsystem. In actuality, inputs must be expected. Active alluvial fans are likely to be growing, so that, even at maximum deflation rates, their surface elevations will continue to increase. Playa

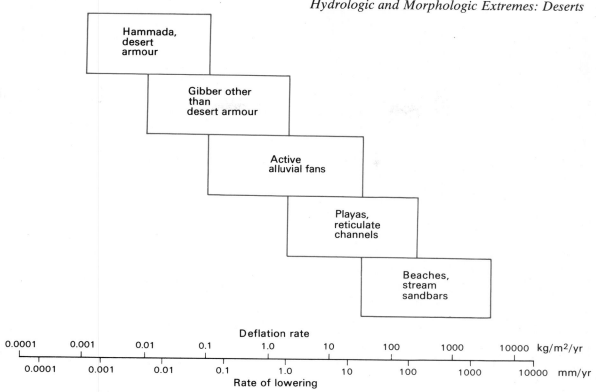

Fig. 11.9 Deflation rates and rates of surface lowering, for sundry surface situations: considerably generalised from the data of I.G. Wilson.

floors and the beds of reticulate stream channels can receive sand from upwind in addition to losing it downwind; furthermore, a reticulate channel receives inputs of sand from upstream. On the available evidence, net deflation of reticulate channels can neither be proved nor disproved, except where the ground alongside a channel is other than sandy. For playas, things seem clearer: the construction of lunettes along the downwind margins indicates net deflation. As the diagram shows, the rate of lowering could be as high as 100 mm/yr, or 100 m in 1000 years. At the low extreme rate of 1 mm/yr, the surface would be reduced by 1 m in 1000 years, and 10 m in 10 000 years. The difference between elevation of the playa floor and the elevation of a former lake beach may only be a rough guide to the former depth of lake water.

High long-term rates for the deflation of beaches and of sand bars in streams can only be sustained if the supply of sand is constantly replenished. No beach, and no bar, could long survive lowering at a mean rate of 1 m/yr. Even higher rates of deflation are observed in the short term on newly exposed sandy soils, where net lowering through as much as 10 m is associated with net deflation rates ranging up to 100 000 kg/m^2/yr.

11.6a Erosion by blowing sand

In the early 1900s, blowing sand was widely assumed to be an effective agent of erosion. It is true that rock fragments locked into a silty or clayey surface can be faceted and polished by windblown sand, and also that the sand grains themselves are polished or frosted. But the density of sandflow falls off rapidly, further than some 5 cm above a sandy surface. Even over a stony surface where grains bounce upward, most flow is concentrated in the lowest 10 cm. As an agent of undercutting or of drilling, blowing sand has little effect. Mushroom-shaped pedestals, and cavities in rock faces, are now ascribed to differential weathering, in the action of which water is important.

One set of erosional features forms an exception – yardangs, the long sinuous ridges or streamlined whalebacks cut in fine-grained materials such as silts. Individual yardangs, comparable in length to some seifs, are aligned on the direction of prevailing wind. Crest heights on the largest of them are as much as 200 m above the floors of intervening troughs.

11.7 Deposition of deflated materials

Desert dust can be washed into dune sand by percolating rainwater. If it is carried beyond the desert margins, and if it accumulates in perceptible sheets, it forms loess, as in NE China. But loess is not associated only with deflation from deserts: it is also supplied by deflation from sandbed channels that carry glacial outwash.

Where a sandy desert reaches the coast, some as yet unknown fraction of the sand will be blown into the sea. But unless it is blown beyond the limit of wave action, or unless waves are dominantly destructive, it will be returned at least to the beach subsystem.

Regional deposition, sufficient to create a sandy desert, may require a timespan of 1 to 2 million years. It also seems likely to require climatic shift. Alluvial sand is deposited along the desert margin during wet climatic phases, and is deflated and redeposited downwind during dry phases. Deflation in the source areas eventually produces surfaces of exposed bedrock, or stony surfaces where little sand remains. The upwind margin of the sand spread will accordingly be eroded. When another wet climatic phase supervenes, broad extents of desert margin will be fixed by vegetation. The sand supply will be renewed, but deflation must await the onset of a further dry phase.

11.8 Deposition of other materials

A well-known general model of the desert setting, derived from the U.S. Southwest, includes a mountain rim – often defined by lines of faulting, where the rim has been strongly elevated – alluvial cones (steep) and alluvial fans (less steep) along the mountain foot, bare pediments next below, and an area of sedimentation in the basin centre. Silty/clayey desert surfaces obviously belong in the basin centre, as also, on the whole, do sandy surfaces and particularly dunefields.

For the world's deserts in general, this model is highly simplistic. It does, however, serve to direct attention to the importance of the deposition of material in the sand grade and upward at and near any existing mountain front. It also serves to indicate that a truly enclosed basin tends to be infilled by the products of the erosion of its margins. Unless deflation – output – can keep pace with alluviation – input – then the long-term prospect is that a given basin centre will become ever more deeply buried.

11.9 Shift of climate: the general problem

The behaviour of pluvial lakes in desert basins was used in Chapter 1 (Fig. 1.1 and text) to illustrate the principle of negative feedback. Pluvial lakes as evidence of former wetter climates, and dead dunes as

evidence of former drier climates, are, as noted earlier, used in Chapter 18 (sections 18.5a, 18.5b: see also section 18.1) and in Chapter 19 (section 19.4b). There is much to suggest that, as in other connections, change of climate has been step-functional. Where a sequence of any detail has been established, it appears that climatic shifts that are morphogenetically significant occur at intervals of the order of 5000 to 10 000 years. In the minimum span of one million years necessary for the formation of an extensive sandy desert, 100 to 200 changes from wetter to drier conditions, or from drier to wetter, may be envisaged. In the much shorter term, a change in the dominant direction of sand-driving winds could transform dune alignment throughout a given desert in as little as 200 years – unless, that is to say, the alignment were fixed, and remained fixed, by vegetation.

The pattern of seifs in the Australian continent, which is known in acceptable detail, forms a swirl round the continental centre. Attempts to relate that pattern to existing wind systems, for instance to an anti-clockwise anticyclonic swirl of surface winds, may be misdirected. Indeed, they would seem inevitably misdirected, unless the wind patterns that shaped large fields of now-dead dunes are assumed to be identical with the wind patterns of the present day.

The types of problem that arise in desert surface chronology, and the kinds of inferences that can be drawn, are well exemplified by the area illustrated in Fig. 11.10, and by neighbouring areas of similar kind. The diagram depicts The Salt Lake, in north-western N.S.W., Australia, and its surroundings. At the present day, the floor of the Salt Lake basin is a playa, surrounded by inactive dunes which are vegetated by tussock grass and low scrub. The level of a

Fig. 11.10 Basic planimetric information for the area of The Salt Lake, N.S.W. Australia.

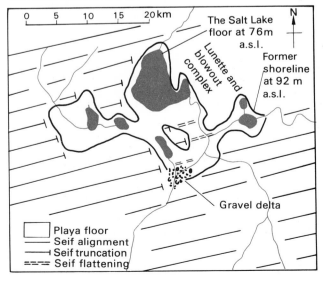

former standing lake is recorded by a shoreline about 16 m higher than the playa floor. This shoreline is defined not only by deposits of evaporite, but also by gravel deltas on the lines of the largest incoming streams. On the western side, the shoreline abruptly truncates the ends of seifs. Near the former delta at the southern end, seifs at low level have been flattened off. We infer, first, that the lake formed amid already-existing dunes, and that the flattening of some dunes resulted from wave action in the rising waters.

Controlled both in height and in extent by negative feedback, the former lake surface stood at or close to a single level, long enough for the shoreline crusts to attain thicknesses of up to 2 m. It seems highly likely, then, that climatic conditions during the creation of the shoreline remained essentially stable. Calculations suggest that precipitation during the life of the lake was about one and one half times that of today – some 330 mm/yr as opposed to the present 220. Increase of stream discharge at high flows is indicated by the delta gravels: existing streams shift nothing larger than coarse sand. But although a standing water body was maintained, and although the stream networks were far better integrated than they are today, drainage lines were not extended. Neither The Salt Lake nor neighbouring similar lakes overflowed their rims of dunefield.

When lake level fell, the fall was sudden. There is no sign of a shoreline intermediate between the 16 m shoreline, with its crusts and deltas, and the bundles of very low shorelines along the immediate margins of the playas. These low shorelines relate to lakes up to 3 m deep – lakes that, because their beach deposits are poorly sorted, seem to have been ephemeral. Whenever the playa floors dried out, deflation from them, and sand blown across the floors, supplied material for the construction of lunettes on the downwind (ENE) side. Sandblowing was vigorous enough to reactivate at least some

seifs, causing them to extend their ends across the line of the 16 m shore. Best desplayed in a neighbouring basin, new seifs of modest length developed on gentle rocky slopes downwind, and also upon the playa floor itself. Since these new seifs are now at least partly fixed by vegetation, we must conclude that they relate to a climate drier than that of today.

The minimum chronological sequence to which this inferential sequence leads is presented in Table 11.1. A radiocarbon age for the lakeshore crust, obtained by analysis of a sample from a next-neighbouring basin, is 14 500 years, but contamination by rainwater may have pushed this apparent age forward. Work in other generally neighbouring areas gives some 15 500 years ago as the likely date of the last high lake stand. Earlier high stands occurred at about 23 500 and 33 000 radiocarbon years ago. Because the high lake stands appear to have been separated by intervals of lunette formation, we seem to be dealing with the alternation of two kinds of steady-state condition, and with step-functional change from one kind to the other, and back again.

Surprisingly, this formerly arid and now marginally sub-arid region is well endowed with datable radiocarbon. Although termites vigorously attack all dead wood they can get their jaws into, the evaporite crusts of the former lakes contain enough carbon for laboratory analysis. In addition, and more importantly, the whole region is littered with aboriginal remains. On the silty slopes immediately west of The Salt Lake playa, these include thousands of stone implements. Throughout the region, they also include cooking sites. Some such sites yield charcoal, the traces of the actual fires. All yield soot, embedded in the sort of low-grade brick of fire-fused silt and clay. Because some cooking sites were buried by the growing lunettes, only to be exposed again by deflation, the spasmodic history of the formation and destruction of lunettes can be worked out.

Table 11.1   Minimum Chronological and Morphological Sequence For The Salt Lake Area, N.S.W.

Climate	Evidence
1. More arid: mean annual precipitation significantly less than 220 mm	Seif dunes that predate the high lake stand: by definition, these were at one time live
2. Less arid: mean annual precipitation about 330 mm; peak streamflows greater than today's; by inference, dunes fixed by vegetation	Lake established; some dune ends truncated, some flattened; climate stable (only one high shoreline); evaporites deposited and deltas formed along high shoreline; streams carried gravel on to deltas
3. More arid: mean annual precipitation significantly less than 220 mm; net direction of sand movement same as in Phase 1	Sandblowing renewed; some old seifs extended; new seifs built on playa floors and elsewhere; lunettes constructed; extension of seifs follows previous lines, new seifs follow parallel lines
4. Slightly less arid; by inference, dunes fixed by vegetation	Very low shorelines constructed on playa margins, probably by short-lived playa lakes
5. Today's climate: mean annual precipitation 220 mm, but very variable with spells of prolonged drought	Dunes largely fixed, although some deflation occurs: sand can blow from dune ends, some root systems of trees become exposed, sand sheets blow out from valley bottoms

Chapter Summary

Deserts occupy about *20% of the world's land* area. They are *located* in *subtropical areas of high pressure*, with major extensions into *regions of rainshadow*. *Hydrological conditions are extreme*; *desert streams* are *discontinuous* and *ephemeral*, with *flashy* regimes, and *high sediment loads* to which the effect of *rainsplash* contributes largely.

On sandy and silty alluvium, *streams of semiarid regions anabranch on gentle slopes*. *Sand-encumbered channels are frequently reticulate*. Some *large channels* in regions of *subhumid climate* have *alternated between widening and narrowing*, in response to changes of land-use practice. Mechanisms of *slope adjustment* within single reaches, in regions of subarid to arid climate, are *largely obscure*.

Stream termini include *standing lakes, ephemeral lakes*, and *playas*. *Playa floors*, clayey to saline, are *subject to deflation*; *lunettes* are built on the downwind sides.

Desert surfaces range from *rocky (hammada)*, through *stony (gibber)*, to *sandy* and *silty/clayey*. *Close-packed stones form desert armour*. *Sand-built features* include the obstacle-related or notch-related minor forms of *sand shadows* and *sand drifts*; *sand*

sheets; and *dunes*. The commonest types of dune are the *barchan, crescentic in plan* and with a slip face on the downwind side, and *seifs*, which are *linear*. Barchans migrate in the downwind direction; seifs extend themselves in the resultant direction of sandblowing by two sets of wind.

Deflation results in the *export of dust* from desert regions. *Sand moves* chiefly in the lowest 10 to 5 cm of air, *by saltation*, and is *contained* within its desert. *Regional deposition can create a sandy desert in 1 to 2 million years*, but the alignment of dunes can be changed in about 200 years.

Materials other than sand accumulate in alluvial cones and fans, and *as silt and clay* in the *lower parts* of desert basins. A basin centre tends to be buried beneath its own sediments.

Detailed study of any given desert region tends to reveal evidence of *shifts of climate* – specifically, of conditions both less arid and more arid than those of the present day. Climatic shift appears to have been *step-functional. Former wetter conditions* are indicated by evidence for *standing lakes. Former drier conditions are indicated by sandblowing* where dunes are now fixed, or more or less fixed, by vegetation. *Increasing aridity* was typically accompanied by *dismemberment of drainage lines*, with growing dunes disrupting stream channels, and with standing lakes being replaced by complex sets of playas.

12 Surface Process-Response Systems: Land Margins

The usually narrow zone that separates the open sea from firm land has fascinated mankind throughout recorded history. In early Roman times it separated two different worlds – so different that unfinished naval battles were broken off at sunset, the crews going ashore to resume the fight next morning.

Except on an atlas map, there is no way to draw a line between land and sea. Waves and tides ensure that, although we usually talk of a shoreline, we are actually dealing with a zone that is inconstant both in space and in time. The rapid response of many shorelines to the processes acting on them has deeply impressed all observers with the complexity of the land-marginal system. Especially is this so wherever the sea gains markedly on the land. One set of human responses has been the construction of numerous systems of classification of land margins.

12.1 Ways of looking at land margins

The margins of the lands can be called coasts, in which case they are only defined in vague terms. There is no way of stating how far inland or seaward the coast extends. Coasts are apt to be classified in terms of origin. A well-known classification includes the following types: drowned, deltaic, volcanic, diastrophic (= produced by crustal movement), wave-erosional, marine-depositional and biogenic (= built by organisms such as corals). But the classification says nothing about what is happening now – whether the sea is gaining on the land, or the land on the sea; whether or not the earth's crust is rising or sinking; and whether or not the edge of the land is being cut back by erosion or built out by deposition.

In contrast to coasts, shorelines can be defined and subdivided in considerable and precise detail, in morphological terms. But they too are liable to be affected by shifts in space – the extension of the land at the expense of the sea, or of the sea at the expense of the land.

12.2 A general classification

Four sets of processes act on the land margins – submergence and emergence, and erosion and deposition. Submergence occurs wherever the relative level of land and sea is raised. Emergence occurs wherever that level is lowered. We need to speak of the relative level, because the cause of change in any given instance may be complex, or in fact unknown. Submergence (alternatively called a positive strand-line movement) could result either from crustal depression or from general rise in ocean level. Emergence (alternatively called a negative strand-line movement) could result either from crustal uplift or from a general fall in ocean level. The general difficulty is that, with ocean level far from stable, no single basis of reference exists.

Erosion and deposition are easier to deal with. If the sea is tearing down beaches and attacking cliffs, its work is obvious. Equally obvious is the seaward extension of coastal deltas and of successive bands of beach.

The general classification (Fig. 12.1) allows for the interplay of the indicated four sets of processes. Where net deposition and emergence combine, the land will gain on the sea. The land will also gain, although less rapidly, where emergence outpaces erosion, or where deposition outpaces submergence. Where erosion and submergence combine, the sea will gain on the land. The sea will also gain, although less rapidly, where erosion outpaces emergence, or where submergence outpaces deposition. Where all processes are perfectly balanced – which is unusual in the real world – the land margin remains stationary in space. To put the matter in other words: land margins can be treated in terms of input/output, or of equilibrium/disequilibrium.

Some process-combinations are more obvious than others. The dominance of emergence over erosion is illustrated by coasts where waves are cutting cliffs, but where postglacial rebound of the

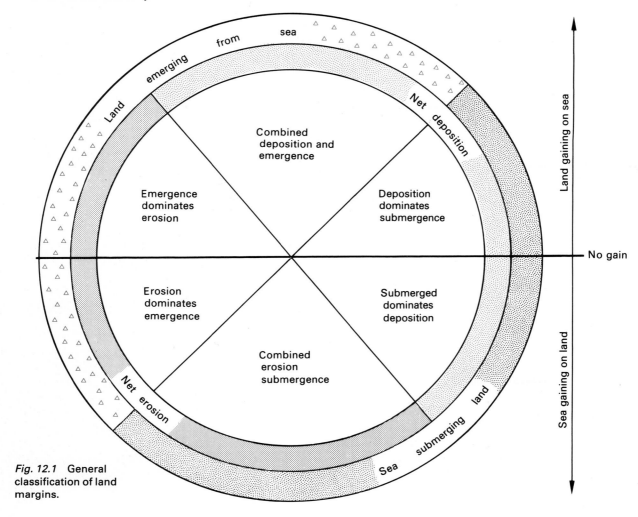

Fig. 12.1 General classification of land margins.

crust ensures a net gain of land over sea (Fig. 12.2). Deposition dominates submergence in the central portion of the Mississippi delta, which is rapidly extending itself seaward, despite the subsidence of the underlying crust beneath the weight of deltaic sediment. Crustal subsidence brings the sea in, displacing the land margin in the inland direction. Deltaic deposition pushes the land margin seaward. In the total balance, the central area of deposition shows a net gain of land over sea.

In the marginal areas of major deltas, including that of the Mississippi, crustal subsidence provoked by sedimentary loading outpaces the deposition of river sediment and the outgrowth of mudflats. Thus, the sea invades land areas (Fig. 12.3). The net outcome is submergence.

Huge extents of the world's coastlines testify to a combination of submergence and erosion. Outside the limits of postglacial crustal rebound, and aside from coasts where crustal uplift is outpacing the general rise of postglacial ocean level, coastal drowning is the rule. Wherever drowning has overtaken high ground, cliffed coasts are usual: waves beat directly on exposed bedrock.

Fig. 12.2 Effect of postglacial isostatic rebound at Little Cumbrae, Bute, Scotland: cliff and wave-cut platform have been lifted out of reach of the sea (Aerofilms).

Fig. 12.3 Subsidence and marine invasion on the margins of the Mississippi Delta.

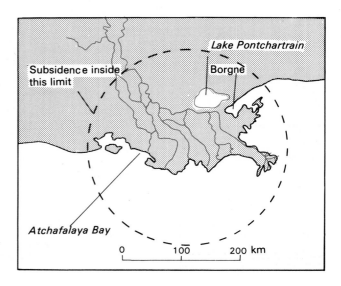

Fig. 12.4   Beach ridges at Dungeness, Channel coast of England: in the area shown, ridges have lengthened themselves from right to left of the view (Aerofilms).

The combination of erosion and emergence, with erosion dominant, is hard to demonstrate: for, by definition, erosion destroys the evidence of former shoreline positions. The combination of deposition and emergence is also hard to demonstrate, but for a different reason. It is thought by some workers to explain the construction of large beach ridge complexes (Fig. 12.4), but there is much disagreement about the changes of ocean level that may be involved.

12.3 Strandline movement: evidence and causes

The most easily recognisable, and the most convincing, evidence of negative strandline movement consists of former shores which now lie above the reach of the waves (Fig. 12.2). These are loosely and inaccurately known as raised beaches – inaccurately, because they have not necessarily been uplifted, and loosely, because they can include not only beach deposits, but also erosional features such as cliffs, stacks, caves, and wave-cut platforms. Positive strandline movement is very widely demonstrated by the drowning of land margins. The sea has invaded what were formerly the lower ends of valleys on land. Underwater exploration is progressively revealing shoreline features beneath the present surface.

Causes of strandline movement are three in number: crustal deformation, worldwide changes in ocean level, and glacial loading and unloading.

Crustal deformation in this context is taken to exclude the effects of glaciation and deglaciation, and to be confined to the movements of folding, faulting, and warping, which include mountain-building. Such movements are called tectonic (from Greek tektonikos, construction). Tectonic strandline movements, then, result from crustal deformation, such as is notably taking place along the western margin of South America. Here the coast is strongly emergent.

Worldwide changes in ocean level are called eustatic, a Greek-derived word that roughly means general balance. The best-studied eustatic movements are those resulting from the storage of water in continental ice caps – negative movements of the strandline – and from its release during deglaciation – positive movements. The amplitude is of the order of 200 or 300 m, the value adopted varying with calculations of the existing volume of glacier ice and with the recognition of drowned shorelines at depth.

There is now considerable evidence that ocean level at glacial maximum was about 200 m below its present mark – low enough, that is, to expose the continental shelves. Because the volume of existing glacier ice is thought to be one-third the volume at glacial maximum, one might at first suppose that complete melting-off would raise ocean level by another 100 m, but allowance for the increase in ocean area reduces the value to some 30 m. In any event, the prospect is horrifying. Quite apart from climatic and social effects on land, the reforming of the world's continental glaciers would dry out the world's commercial seaports. The melting of existing glacial ice would drown the world's coastal cities. One of these things appears sure to happen sooner or later.

Eustatic movements of the strandline caused by the waxing and waning of continental glaciers are called glacio-eustatic. Depression and elevation of the continental crust, in response to loading or unloading, are called isostatic movements (from Greek, = equal balance). Erosion, by reducing the mass of a continental block, should eventually promote isostatic uplift, but such uplift on a continental or subcontinental scale is impossible to separate from movements with other causes. Isostatic depression of the crust beneath large deltas – a response to loading – has already been mentioned. Isostatic depression of the crust beneath ice caps, and rebound on deglaciation, are discussed in Chapter 13. Such movements are styled glacio-isostatic.

12.3a Complications and combinations

Although tectonic, eustatic, and isostatic movements of the strandline can be separated as to cause, they cannot be separated as to effect. Rebound of large parts of North America and the Baltic area after glacial melting has converted considerable extents of sea floor to dry land. That is to say, the total area of ocean surface has been reduced. If this were all, ocean level should have responded by a worldwide rise, and should still be so responding. However, if crustal depression beneath land ice means a displacement of mantle material beneath the affected crust, and if postglacial rebound means a replacement of mantle material, then glacio-isostatic movement of large areas of crust need not necessarily be reflected in significant eustatic changes of the strandline.

Again, it is possible to argue that the flooding of the continental shelves by the glacio-eustatic rise in ocean level, during deglaciation, involves sufficient loading to depress the continental margins, at least where shelves are wide, and thus to exaggerate the general drowning. So far, there is very little to go on in this connection. About changes in the total capacity of the ocean basins we know less still.

Such changes are certainly marked in the geological record. During the last 600 million years, the sea has widely invaded the land margins on three widely-separated occasions. We can only guess that the basic cause was a change in the behaviour of convection currents in the mantle, or perhaps in mantle composition. The hypothesis here is that the weak outer mantle can undergo chemical changes that alter its density. A reduction of mantle density would allow some foundering of the continental masses, and thus permit the oceans to invade their margins. The invasions in question each took 50 or even 100 million years. They were followed by uplift, which exposed as new land areas the sedimentary deposits laid down in the invading seas. Clearly, the time scale for marine invasion and retreat is far greater than that of the few million years at most that measures our present ice age. Nevertheless, a connection exists.

Assume for the sake of argument that in the present ice age there have been four glacials and – including the one in which we are living – four interglacials. Each glacial is marked by a fall of ocean level, during which coastal rivers deepen their valleys, and when delta-building occurs beyond the present shoreline. During interglacials, ocean level recovers. Coastal valleys are flooded, and beaches are formed at the recovery mark. Now, a given interglacial beach, on a coast that seems to be tectonically stable, is lower than the beach formed in the preceding interglacial. The extensive glacio-eustatic rise and fall of ocean level appears to be superimposed on a general, and independent, eustatic fall. This latter fall can scarcely be explained except by a change in the total capacity of the ocean basins. It amounts to some 30 m in probably half a million years.

A general and overriding difficulty is that of demonstrating tectonic stability. In areas that are tectonically unstable, or that have experienced glacio-isostatic movement, old shorelines are tilted. The highest postglacial beach in the Scandinavian area ranges between 0 and 275 m above present sea level. The tilt angle, calculated from a differential uplift of 275 m over a distance of 1200 km, is only 0.015°: but it can be clearly defined by the field evidence. Old shorelines that record eustatic movements only must be horizontal, strictly parallel in section to the present shoreline. Strict parallelism, if it exists anywhere, is hard to prove, because the remains of old shorelines are discontinuous in space, and because correlation in time is often dubious. A small group of workers takes the extreme view that erosional platforms related to former ocean levels can be correlated throughout the world.

12.4 Shorelines and the short-term view

The study of land margins as shorelines concentrates on short-term responses to wave action. It is scarcely necessary to point out that shorelines are highly dynamic systems, often capable of very rapid responses. Changes between successive tides, or during a single storm, are far swifter than those produced by eustatic or isostatic means: and tectonic changes are only rapid when they involve sudden faulting.

Waves are set off by submarine earthquakes, by coastal landslips, and by the calving-off of icebergs. They are also raised by the drag of wind on the ocean surface. Earthquake waves, tsunamis, are so rare as to produce generally insignificant responses, despite their great size and energy. Travelling at velocities measured in hundreds of km/hr, they can arrive as enormous coastal breakers with intervals of more than an hour between successive crests. Waves raised by landslips and newly released icebergs are significant only in deep sheltered inlets. For almost all of the world's shorelines, and for almost all of the time, the waves that count are those generated by the wind. In the light of the importance of wave action on shorelines, it seems ironical that the formation of wind-generated waves is not completely understood.

12.4a Wave geometry

Waves are defined in terms of length, height, period and steepness (Fig. 12.5). The period is the time interval between one crest and another. For large waves generated by vigorous storms in the southern ocean, the period ranges from 12 to 15 seconds.

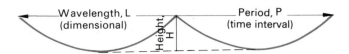

Fig. 12.5   Wave geometry.

Steepness, the ratio H/L of height to wavelength, varies greatly. It is affected by the conditions of wave formation in the generating area, and – most importantly – by the bottom slope up which waves approach the land margin.

12.4b Controls in the generating area

Wind speed, wind duration, and length of fetch combine to control the size and proportions of waves. Their effects can be separated for purposes of analysis, but have to be combined in the consideration of any real wave train.

Up to a certain limit, wave height varies with the length of fetch – the distance through which the generating winds blow over water. The effect of fetch can readily be observed on small inland lakes on a windy day. If strong and persistent winds blow over an ocean surface, then wave height increases in proportion to the square root of the fetch, up to a fetch of about 1000 km (Fig. 12.6). An increase of wind velocity produces a far more rapid effect than an increase of fetch, but to describe its influence we need to assume a very long fetch and a long wind duration to begin with. In these circumstances, wave height varies with the 1.85 power of wind velocity (Fig. 12.6). There is a cutoff point somewhere near 30 m, beyond which waves dissipate as much energy as they receive, with the result that further growth is checked.

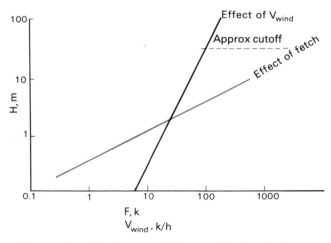

Fig. 12.6   Effect of fetch, and of wind velocity for infinite fetch and duration, upon wave height (vertical scale).

For a given fetch and wind velocity, wave height increases with wind duration, again up to a limit. This limit is fixed in part by the escape of waves from the generating area beyond which decay and dispersion set in.

12.4c Wave groups outside the generating area

The widely held but never tested belief that every seventh wave is larger than the other six has some kind of vague basis in fact. Waves on an open coast rarely belong to a single group. Waves belonging to two groups of different wavelength will alternate between being out of phase, and thus low, and in phase, and thus high. But the seventh wave merely carries a magic number. All kinds of combinations are possible. Waves at sea very frequently combine groups of contrasted wavelength, steepness, and direction of travel.

Waves generated by powerful storms over the southern ocean, in the neighbourhood of New Zealand, can reach the shores of Alaska, some 12 500 km distant. As they travel, different wave groups sort themselves out. Wave velocity and wave period vary with the square root of wavelength (Fig. 12.7). For instance, waves 500 m long travel at 100 km/hr, with a period of 18 sec, while waves 5 m long travel at 10 km/hr, with a period of 1.8 sec. Groups with longer wavelength outrun groups with shorter wavelength in the process called dispersion.

In addition, the leading wave of a group will disappear. Because the rate of disappearance is only about half the rate of onward movement, wave groups of great length are able to propagate themselves for very long distances. Disappearance of the leading wave is part of the decay process: the other part consists in the fact that, when waves leave the

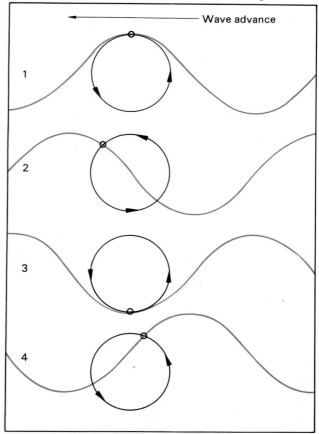

Fig. 12.8 Orbital path of a water particle in wave motion in open water.

generating area, they start to lose height. Numbers of studies suggest that the decay pattern is exponential, with wave height being divided by a constant factor for every addition of distance, but the results of tracking wave groups across the entire Pacific suggest that complications of decay rate may apply to waves of very great length.

12.5 Waves on the shoreline

Shorelines are shaped by waves. Many of them exert much control, in turn, on wave shapes. In order to understand what happens, we need to know how and why incoming waves can break, and to identify the basic shoreline components.

12.5a Breakers

Water in waves moves in circular orbits, the direction of movement at the top of the orbit being in the same direction as that of wave advance. Fig. 12.8 shows the orbital path of an imaginary water particle at the surface. In water, as in many other media, the wave form travels but the substance of the medium does not – except, with ocean waves, for surface water moved by the drag of the wind.

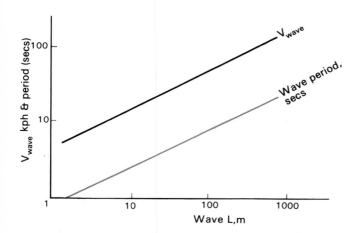

Fig. 12.7 Wave velocity and wave period in relation to wavelength.

The orbital radius falls off sharply with depth, until at the equivalent of about half a wavelength, the orbital motion cuts out entirely. There is thus a limit below which wave action on the sea floor is negligible to zero.

Breakers are waves in which the orbiting pattern collapses. When waves run into shallowing water, they increase their steepness. Water cascades down the wave front in spilling breakers, or simply jets forward in plunging breakers (Fig. 12.9). Breakers form when water depth is reduced to between 1.1 and 1.5 times the value of wave height. The theoretical maximum possible steepness is 0.14, on the face of it an unimpressive fractional value. But, for swell waves with a period of 12 sec – which is by no means exceptional – this translates into a wavelength of 260 m, a height well in excess of 30 m, and a wave velocity of some 20 m/sec. Pressures exerted on a cliff face by waves of this description can run as high as 30 tonnes/m².

Fig. 12.9 Plunging breaker, with wave height about 5 m and wave period about 12 sec: a component of the Southern Ocean Swell, N.S.W. coast, Australia.

12.5b Destructive and constructive waves

Many workers emphasise the contrast between high-energy and low-energy coasts. The former type is subject to the action of powerful breakers, the latter controlled by very gentle offshore slopes and very low waves. It is tempting to suppose that high-energy coasts are liable to shoreline destruction, whereas low-energy coasts experience little if any shoreline erosion, and frequently record dominant deposition. However, the real relationship between wave energy and construction/destruction is greatly complicated by wave steepness. Generally speaking, flat waves are much more likely to be constructive than are steep waves; but contrasted effects can occur on different levels of a given beach.

Californian beaches, which are among the most intensively studied in the world, tend in many cases to vanish during the winter and to reappear in summer. Storm waves generated in the North Pacific during winter arrive as destroyers, tearing away the beach sand and shifting it offshore. Constructive summer waves, shorter and flatter, shepherd the sand back again.

12.5c The beach material budget

Progressive construction means addition to the budget of beach material: progressive destruction means a loss. Beyond these considerations comes the initial supply of sediment to a beach, and its eventual removal.

Beach sediment is supplied partly by direct wave attack on the land, partly by rivers, and partly by the break-up of shell and coral material. The proportion supplied by each of these means varies drastically from one beach to another. For the world as a whole, it seems impossible to determine. Calculations of erosion rates suggest that rivers deliver about twice as much sediment as is supplied by wave action: but against this, much river-borne sediment enters storage in a few major deltas. Only where a beach consists wholly of broken shell or coral can it be proved to be supplied entirely by wave action. Such action can, however, be suspected of being strongly dominant where, on a cape-and-bay coastline, beach sediment is not transported round headlands from one beach to the next.

Sediment is lost to a beach system whenever wind blows sand inland. Very strong destructive waves may move some material so far offshore that it goes into storage on the sea bed: and the geologic record includes much sedimentary rock that originated as shallow inshore deposits. The most dramatic loss of all is that demonstrated for the present day coast of California.

On one view, the real land margin is the outer limit of the continental shelf: this is where the edges of the continental masses descend into the deep oceans by way of the continental slope. Every continental slope that has been investigated in detail is found to be gashed by submarine canyons. Where the slope is wide, the submarine canyons head far out to sea. They can collect, and feed down, loose sediment on the shelf: but where the shelf is narrow to absent, and the continental slope comes close inshore, they can tap the very beaches. Such are the conditions on parts of the California coast Beach sand vanishes down the canyons, to be incorporated in a further stage of the sedimentary cascade.

12.6 Shoreline components

Really detailed study of any shoreline is likely to demand a detailed classification into numerous components. For present purposes it is enough to distinguish the offshore, foreshore, and backshore zones. These are all defined in terms of cross-section (Fig. 12.10).

The offshore zone extends from the breaker zone (a component of the sea surface) to low tide level. The foreshore lies between low tide level and the swash (= uprush) limit of storm waves at high tide. The swash limit is usually marked by a crest of beach material, the berm crest. The backshore zone extends from the berm crest to the base of the cliff, the most seaward sand dune, or the most seaward beach ridge, whichever may be present.

12.6a Activity on and adjacent to the backshore

If the whole beach system is being built out seaward, then it is likely to form a series of beach ridges (Fig. 12.4). But on stationary or submergent shorelines, other than those with cliffs, onshore winds may pile

Fig 12.10 Basic components of the shoreline.

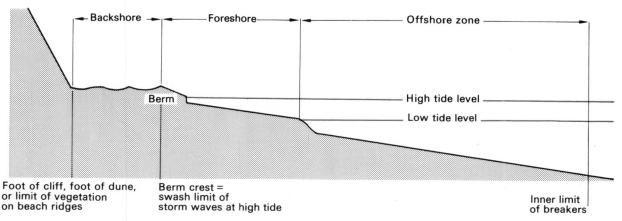

Foot of cliff, foot of dune, or limit of vegetation on beach ridges

Berm crest = swash limit of storm waves at high tide

Inner limit of breakers

up enough sand to form coastal dunes. The source of the sand is the exposed foreshore at low tide. Offshore winds capable of driving sand merely feed it into the sea; but sand raised by onshore winds lodges, to begin with, close to the backshore limit. What happens next depends on local conditions. If the land behind the backshore is low-lying, a prominent sand ridge, the foredune, is apt to form at the backshore limit. Sand carried over the foredune can spread inland, as a sheet varying from smooth to highly irregular. The blowout process leads to the formation of U-dunes (parabolics) which in plan arch boldly away from the shoreline (Fig. 12.11). U-dunes and thick sand sheets alike can advance with sheer fronts (Fig. 12.12).

Dune migration is liable to be highly irregular in time. The feet of too many walkers, or the destructive passage of dune buggies, can provoke a sudden advance. In the longer view, greatly increased storminess during the Middle Ages drove Atlantic beach sand as far as 100 km inland in southwest France, in the district of Les Landes.

Other things being equal, the possibilities of sand-driving are greatest where a wide sandy foreshore is exposed on account of a high tidal range. Such conditions are well exemplified in the Channel Islands. During the same stormy period that moved the sands of Les Landes inland, the western side of the island of Jersey was itself affected. Sand moved inshore in a hummocky sheet, some of it eventually reaching, and ascending, a 200 m high cliff and spreading over the farmland on the plateau above.

12.6b Activity on the foreshore

Waves breaking on the foreshore wash beach material landward in the swash, and tear it back seaward on the backwash. On very many beaches, the dominant waves approach obliquely. In consequence, the swash drives beach material obliquely landward, but the backwash rips it back directly. The combination of swash and backwash promotes a net zigzag transport of material along the beach, that is greatest where the direction of wave approach is about 45° to shoreline, and that can amount to one million tonnes of sand a year through a given section of beach profile.

Beating waves bash beach fragments against one another, rounding sand grains and smoothing off pebbles and boulders. There is a limit of about 0.3 mm of diameter, below which beach sand is not further reduced in shape or size, but boulder and pebble beaches can lose some 5% of their bulk a year. This means that the substance of a boulder or pebble beach can be cycled from coarse rock particles to sand grains in as little as twenty years.

Detailed studies of foreshore activity make much of the calibre of beach material. In general, foreshore slope increases with increase of calibre: pebble beaches are steeper than sand beaches. Mineral constitution can also be important. Wave action on the foreshore can selectively increase the heavy mineral content at the expense of quartz sand. Heavy minerals concentrated on the eastern beaches of Australia include titania, TiO_2, which is widely dredged for use as a whitening agent in paints.

12.6c Activity in the breaker zone

Breakers in the breaker zone can pick up sand from depths as great at 50 m. Lifting bottom sand by turbulent action on the sea bed, they feed it into the longshore current system. Like the swash-and-backwash transport of the foreshore, the longshore current depends on the oblique approach of dominant waves, and thus on a net flow of water in one of the two possible directions parallel to the beach. Again, like the transportation on the foreshore, the longshore current sets in the down-wave direction. At a rough guess, it shifts about twice as much sediment as does the swash/backwash mechanism.

On many shorelines, an important mechanism that feeds beach sediment into the surf zone is the rip, also called the rip current, or the tidal rip. Rips are powerful currents, often very localised, that run downshore. They discharge the surplus of water that is piled into the foreshore zone by incoming breakers. So systematic can they be, that they are well known to endanger bathers on particular parts of individual beaches. In the shoreline systems of water movement, they correspond to the outbursts of polar and arctic/antarctic air that correct the tendency of progressive over-storage (Chapter 14, section 14.7d).

Fig. 12.11 View into the U of a parabolic dune: near ends are vegetated: Myall Lakes, N.S.W., Australia (A.D. Short).

Fig. 12.12 Advancing front of coastal dune, Ninety Mile Beach, Northland, N.Z. (New Zealand High Commissioner).

(a)

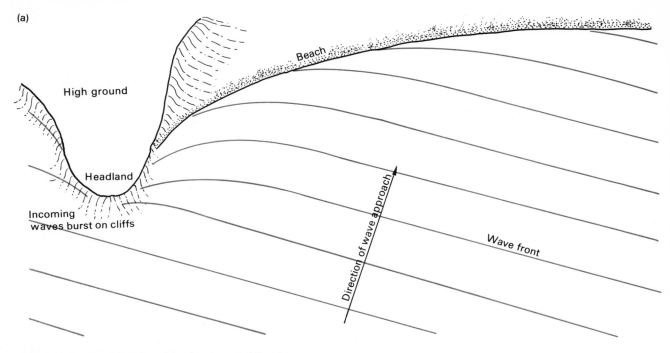

High ground

Beach

Headland

Incoming
waves burst on cliffs

Direction of wave approach

Wave front

Fig. 12.13 Wave refraction: (a) refraction of obliquely
approaching waves against a single headland and a long
beach, when the beach plan becomes a logarithmic spiral;
(b) refraction of directly approaching waves against two
headlands and a circular-arc beach.

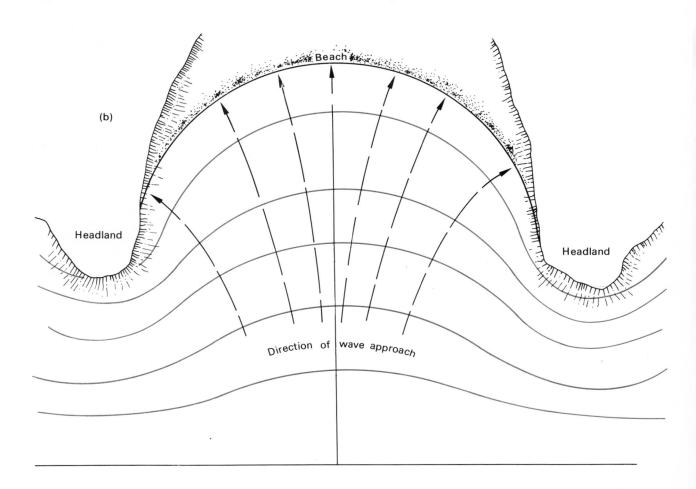

(b)

Beach

Headland

Headland

Direction of wave approach

Fig. 12.14 An abrasion-platform: coast of north Somerset, England. Lines of rock blocks and seaweed roughly indicate outcrops of horizontal limestone beds, that are gently truncated by the platform.

12.6d Wave refraction

Refraction, the bending of wave fronts, happens wherever waves advance obliquely into shallowing water (Fig. 12.13). It can begin seaward of the breaker zone, and can continue as far as the fore-shore. The farther a wave advances across a shallowing bottom, the greater the drag, and thus the closer the spacing between successive crests. Waves refracted round a headland concentrate their attack on the promontory, including the outer flanks. In a bay between two headlands, wave energy is dispersed. Destruction of the headland can be accompanied by beach-building in the bay.

12.7 Features of destruction

Wave attack on the land is concentrated near water level. Undercutting of cohesive rocks produces vertical cliffs. In less cohesive materials, such as thick glacial deposits, the slopes yield by slumping, well before vertical faces are attained. Enormous very steep cliffs, with heights as great as 1000 m, have

been cut on the windward shores of the Hawaiian island chain, and on some very exposed shores of Northwest Europe.

Again in cohesive rocks, and especially where the beach material is coarse and the tidal range is great, the offshore zone can include a wave-cut surface (Fig. 12.14). This is called an abrasion-platform, having been long regarded as shaped by the back-and-forth drag of beach sediment, and by the pounding of sediment-charged waves. Since wavebeat can reduce the beach material itself, there is no reason to suppose that bedrock in the offshore zone is immune to mechanical erosion. The actual processes at work have, however, been so little studied, that the significance – if any – of chemical and biologic action remains essentially unknown.

The abrasion-platforms of western Europe so impressed early geologists, that these workers made exaggerated claims about the power of the sea to plane off extensive areas of the continental margins. In addition, as will shortly be seen, the abrasion-platforms of midlatitude shorelines are replaced by features of different origin on the shorelines of the tropics and subtropics.

Fig. 12.15 Stacks, and an arch with which caves are associated (Hopewell Rocks, New Brunswick: Canadian Government Travel Bureau).

Minor erosional features of shorelines include stacks, caves, and arches (Fig. 12.15). They develop by highly selective wave erosion which concentrates on points and lines of weakness. Sea caves are excavated by alternating compression and sudden decompression of air in cavities such as joint planes. Arches are formed by wave attack on a jutting headland, refracted waves coming in from both sides. Stacks are merely detached bits of a cliff coast, either separated from the outset or detached by the collapse of an arch.

12.8 Features of construction

Refracted waves in a bay between two headlands spaced fairly far apart will form a beach that in plan is often very close to the arc of a circle. Such a beach is apt to be mistaken as to origin. It is easy – too easy – to suppose that the plan and profile of the beach cause incoming waves to break concentrically along it. On this view, the pattern of beach controls the pattern of refracted wave fronts. The opposite is true: the pattern of wave fronts controls the pattern of the beach. Beaches generally are set at right angles to the net direction of waves that advance at 8 m/sec or more.

Where waves are refracted round an offshore islet, lines of beach can link the islet to the mainland. Each link is a tombolo (Fig. 12.16). If refraction and wave energy are strong enough, the potential two

Fig. 12.16 Island linked by tombolo to the mainland: beach activity is greater on the left side than on the right. Mayo coast, Ireland (J.K. St. Joseph).

Fig. 12.17 Intermittently growing spit projecting part-way across an estuary: Barmouth, Wales (J.K. St. Joseph). Three main episodes of growth can be made out.

links can be driven together to form a single link, which will, however, possess a shoreline on both sides.

Successively added beach ridges have already been illustrated, in the context of advancing shorelines. Whatever they demand by way of slight changes of the strandline, or specially favourable conditions of offshore slope and wave form, to promote progressive construction, they illustrate a direct seaward advance of the land. Where, by contrast, beach sediment is dominantly carried along the shoreline, significant construction is also possible in the form of spits. These are projections of the

beach into, and even right across, an inlet (Fig. 12.17). Numbers of spits in western Europe have been observed in great detail, because of their propensity for diverting river mouths and blocking harbour entrances. If longshore transportation of beach sediment is wholly in one direction, spits grow only in that direction; but reversals of direction can cause spits to extend into an estuary from both sides.

Extension of a spit is very often step-functional. During each episode of extension, the newly added extremity is usually curved landward by waves refracted round it. Thus, large spits have complexly hooked ends. It is perfectly possible for the spit to lengthen itself at the same time that the feeding beach is being eroded.

12.8a Barrier islands

On the south Atlantic and Gulf coasts of the U.S., among other areas, occur numerous barrier islands (Fig. 12.18). They act as shorelines, in that ocean waves break against them, but are separated by tidewater from the land proper. Water jets through the inlets of a barrier system on the rising and falling tides, scouring the floor of the inlet channel and constructing fans of beach sediment on either side.

Inshore displacement of a barrier can obviously take place. The sand of which a barrier island is constructed typically rests on lagoon deposits over which it has been driven landward. Erosion of the seaward face and washover of sand during storms combine to promote displacement before destructive waves. The origin of the barriers in the first place is uncertain.

Possible causes include the building of offshore bars by powerful plunging breakers in a gently shelving offshore zone. If the sand supply is generous, such bars can conceivably be built above mean water level, and can be further raised by dune sand. The break-up of very long spits by newly formed inlets is another possible cause. A third is the drowning, by reason of positive strandline movement, of long foredune ridges at former backshore limits. A fourth is the deposition of shoreline sand

on and against minor relief features, once again drowned by a positive movement. More than one process could combine. Very little is known of the rate of landward movement of barrier systems, although some authors are willing to suppose that the existing barriers originated far out on the continental shelf at the time of low last-glacial sea level, and that they have been driven progressively landward as the ocean has risen.

12.8b Tidal marshes

Particularly on the south Atlantic coast of the U.S., extensive swamps occupy wide, shallow, branching inlets. They represent an early stage in the outgrowth of land, under the influence of the deposition of sediment and the successive colonisation by plants. Needless to say, the barrier island systems are important in keeping out ocean waves. In other places, the land grows outward, and vegetation takes over, in the heads of estuaries, in bays that receive no large rivers, and behind spits. The end-product of the processes here at work is the tidal marsh.

Marsh development begins with the deposition of material in the clay and silt grades in the form of mud. According to the inlet concerned, the sediment may come from the land, from surrounding slopes, or from the sea. Between the mean levels of low and high tide, mudflats form. They are colonised slightly below the mean level of high tide by the advance guard of salt-tolerant reeds and rushes, the vegetation of low marsh. It is not possible to describe the sequence in detail without introducing whole strings of botanical names, which will not be used here.

Fig. 12.18 Barrier islands, largely fixed by vegetation, off the shore of the southeast U.S.A. (W.F. Tanner).

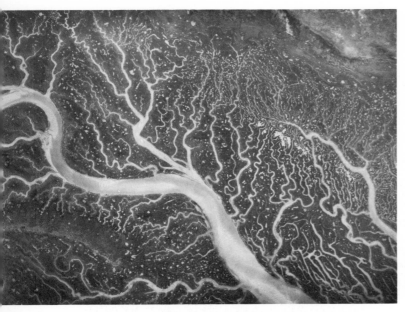

Fig. 12.19   Branching and meandering pattern of tidal creeks: Scolt Head Island Nature Reserve, Norfolk, England (J.K. St. Joseph).

Suffice it to say that low marsh passes – and often quite abruptly – into high marsh, that stands above mean level of high tides, and that is covered by other salt-tolerant plants such as salt grass and tamarisk. It is interlaced by branching networks of tidal creeks, which are characteristically meandering (Fig. 12.19) and obey the number–order and length–order laws derived for drainage networks inland.

12.9 Special features of tropical and subtropical land margins

Because geoscience has developed mainly in middle latitudes, midlatitude components of shorelines are apt to be regarded as normal, and contrasted low-latitude components as exceptional. In the following discussion, the term low-latitude will be used for short, to apply both to the tropics and to the subtropics. Important components of low-latitude land margins are rock platforms, beachrock, mangrove swamps and organic reefs.

12.9a Rock platforms

Many clifflines in low latitudes are bordered by platforms (Fig. 12.20). In contrast to the abrasion-platforms of middle latitudes, the low-latitude features do not typically slope seaward. Although they lie between tidemarks, they are not ground smooth by beach material at high tide: many of them carry no debris at all. They are subject to chemical weathering, and also to attack by organisms and by

crystallising salt. The main control of their nearly horizontal surfaces may be the level of groundwater.

Workers trained on mid-latitude shorelines are inevitably tempted, on first sight, to interpret rock platforms as abrasion-platforms related to a mean sea level slightly higher than the present. There is, in fact, considerable controversy about whether or not the world ocean stood some 3 m above its present level between about 6000 and 3500 years ago. But whatever the fact of this case, the process acting on rock platforms is not that of abrasion by beach material.

12.9b Beachrock

On low-latitude shorelines where groundwater is stored in the backshore zone, beach material can be cemented into rock. The rock becomes exposed by erosion (Fig. 12.21). Although carbonate cement is known to be involved, the specific cementing process is incompletely understood. It probably operates at the top of the groundwater table.

A somewhat remotely related material is coffee rock, so called because its sand grains, stained brown by iron compounds + organic matter, resemble coffee grounds. The relationship to beachrock is that of groundwater control. Coffee rock forms in the subsurface of dune sands, especially where swamps and ponds are present – swamp water in many environments has a strong charge of iron. Within the groundwater table, the sand becomes strongly discoloured and rather weakly cemented. Shoreline erosion can expose coffee rock on the foreshore (Fig. 12.22).

12.9c Mangrove swamps

Mangrove swamps in low latitudes occupy a distinctive ecological niche, corresponding to that of tidal marsh vegetation in higher latitudes. They are the modern counterparts of the coal-swamps of the Carboniferous period of the geologic past. Mangroves are trees that tolerate salt water but that cannot withstand frost. The limits of the swamps are approximately defined by tidal ranges: the swamp

Fig. 12.20   Rock platform at Newport, N.S.W., Australia: bedrock is almost horizontal sandstone, and most large visible slabs are in place.

Fig. 12.21   Beachrock on the shoreline of the Florida Panhandle.

Fig. 12.22   Coffee rock on the shoreline of South Queensland, Australia: part of an old dunefield, now being eroded by waves, appears on the right.

Fig. 12.23 Stilt-rooted mangroves, Hinchinbrook Island, Queensland (R.J. Blacker).

Fig. 12.24 Submarine view of fish amid coral, Bahamas (Bahamas Ministry of Tourism: Roland Rose).

floor is widely exposed at low tide, but is flooded at high tide. For a mangrove swamp to attain any great width, therefore, the foreshore zone needs to slope very gently, and the total delivery of wave energy must be slight. Mangroves support themselves by sending down roots, in the form of props and stilts, into the underlying abundant mud that is necessary for a swamp to develop (12.23). Reduction of sulfates (SO_4 compounds) by bacteria in the mud leads to the release of H_2S and other stinking gases: H_2S, hydrogen sulphide, is the rotten-egg gas.

12.9d Coral reefs

Reef-building corals are limited in their distribution by temperature, salinity, and depth of ocean water. Temperatures below 18°C are lethal, as are those above 36°C: temperatures between 25 and 30°C are the most favourable. Salinities of 27 and 40‰ (parts per thousand) set extreme limits for reef construction, but most living reefs in actuality occur where salinity values are normal for the open low-latitude ocean, in the range from 33 to 37‰. Reef-building corals are closely limited in their vertical range. Below a depth of 30 m they are uncommon; and most growth takes place in depths of less than 20 m. The lower depth limit is fixed by the penetration of light, which is essential to the unicellular algae with which the coral polyps exist symbiotically. The upper limit is approximately that of ordinary low tide, above which the polyps become exposed too long to the air, and are in danger of being heated above 36°C.

Within these limits, reefs are absent from shore-

lines where the water is charged with sediment. They are best developed on and near to the eastern coasts of landmasses, where warm currents run, and in the open ocean.

Their fundamental form is the fringing reef, a shoreline feature. Barrier reefs grow on submergent shorelines, where the upward growth of coral is able to keep pace with the positive movement of the strandline, but where reef construction is displaced away from the shore. The 2000 km-long Great Barrier Reef of the Queensland coast of Australia lies as far as 200 km offshore, rising from a rock foundation 150 m below sea level. Still farther seaward, additional isolated reefs have been constructed on rises on the continental shelf, where the general water depth is as great as 1000 m. Strongly dominant waves and winds can bend an individual reef into the horseshoe or circular shape of the atoll, a curvilinear island or small island chain that encloses a roughly circular lagoon.

Negative feedback affects very many individual reefs. Growth is most vigorous where nutrient supply is greatest – that is, toward the side where the water supply is constantly renewed. But this is the side of wave approach, where wave attack is at its most powerful. In consequence, upward growth is checked not only by exposure at low tide, but also by the destructive mechanical action of ocean waves. The frontal slope of the reef, and also any downwind and downwave lagoon, is freely supplied with broken coral debris. In addition, reefs incorporate the material fixed by calcareous algae. So complex are the contributions that the simple term coral reef is less than satisfactory.

Once formed, a reef provides an elaborate habitat

for marine creatures other than coral, notably fish (Fig. 12.24). Sediment-filled lagoons are liable to be colonised by plants, especially mangroves; and numerous low coral islands support coconut palms.

The history of offshore reefs is in some instances highly complex. It involves not only the general rise of the strandline that accompanied the melting of continental glaciers, but also subsidence of the crustal foundations on which reefs have been built. The depth of coral proved by borings is far in excess of the 20 to 30 m in which reef-builders mainly grow, amounting in some cases to 600 m or more. Each reef system constitutes an individual case: only detailed investigation on the spot can prove the relative effects of general submergence, general slight oscillation of ocean level, and local subsidence of the foundations. Indeed, the matter is not limited to local subsidence: local elevation is also possible. Subsidence, if rapid enough, drowns a reef in a sea too deep for continued growth. Elevation, if rapid enough to outpace a general rise of ocean level, raises a reef above the range of tides and kills it there, as it has done, for example, in parts of Florida and New Guinea.

Chapter Summary

Land margins are acted on by *submergence* and *emergence*, and by *erosion* and *deposition*. *Submergence and emergence* can result from *general changes in ocean level*, from *crustal movement*, or from *some combination*. A *general classification depends on net advance or retreat* of the land margin, with a stationary state unusual in nature.

Change in the relative level of land and sea, from whatever cause, is called *strandline movement* – positive for a relative rise, negative for a relative fall. Strandline movements may be *tectonic*, resulting from *crustal deformation*; *eustatic*, resulting from a *change in the capacity of ocean basins or* from the *waxing and waning of* continental *glaciers*; or *isostatic*, resulting from the *loading or offloading of ice*. The terms glacio-eustatic and glacio-isostatic are used where appropriate.

Glacio-eustatic movements range between about *200 m below present ocean level* to about *30 m above*. They are *superimposed on a eustatic fall* of unknown origin, that may be related to the vertical movement of continental masses, such as has taken place several times in the geologic past.

Wave action is vital to the understanding of shorelines. Waves can be set off by submarine earthquakes, by coastal landslips, and by the calving of icebergs, but the *most important factor is the drag of wind on the ocean surface*. *Wave geometry is defined in terms of length, height, period and steepness.*

In the generating area, wave size and proportion is controlled by *wind speed*, wind *duration* and length of *fetch*. Up to about 1000 km, wave *height increases in proportion to the square root of* the *fetch* length, if wind duration is long enough to raise the maximum possible waves. Wave *height increases rapidly with* increase of *wind velocity* – with the 1.85 power of

wind velocity, again if wind duration is sufficient, and if fetch is long enough. There is a cutoff at about 30 m, beyond which further growth is checked. The influence of wind duration is partly set by the escape of waves from the generating area.

Wave velocity and wave period vary with the square root of wavelength. Therefore, long waves outrun short waves: wave groups travelling out from a generating area sort themselves out, in the *dispersion* process. Leading waves of a group dissipate, and wave height declines with increasing distance of travel. Thus, as they advance across the ocean, waves *decay*.

Water in waves moves in circular orbits, which decrease rapidly in radius with increasing depth. *In shallowing water the orbits become distorted*, and *water falls out* of advancing waves as plunging or spilling *breakers*, which form where water depth is from 1.1 to 1.5 times the wave height. Waves beating on the shoreline can be either *constructive or destructive*, depending on their own geometry and on the beach slope or beach position.

Beach sediment is supplied by direct *wave attack* on the rocks of the land, by *rivers*, and by the *destruction of shells and coral*. The relative contributions are impossible to determine. Sediment is *removed* from beaches *where it is blown inland, carried far out to sea* in heavy storms, or *fed into submarine canyons*.

Shoreline components include the backshore, the foreshore and the offshore. Some backshores include a *foredune*: sand can be driven far inland beyond the backshore limit. Activity on the *foreshore* is *dominated by swash and backwash*. In the *offshore zone* (roughly equivalent to the breaker, or surf, zone), there is often a *longshore current* promoted by obliquely-approaching waves. It probably shifts twice as much sediment as does the swash/backwash mechanism. On some shorelines, *rips feed back accumulated water* through the breaker zone.

Waves are refracted as they move into shallowing depths. Refraction ensures a concentrated attack on headlands.

Wave attack on bedrock *produces cliffs by undercutting*, but the cliff slope is liable to control by rock strength. In the offshore zone, *wavebeat can develop an abrasion-platform*. *Minor erosional features*, some of them dependent on wave refraction, include *caves, arches and stacks*. *Minor depositional features include tombolos, beach ridges, and spits. Barrier islands are dubious as to origin*, although obviously depositional.

Tidal marshes belong with advancing shorelines where mud progressively accumulates. They *eventually develop meandering tidal channels. On tropical and subtropical shorelines*, they are *supplanted by mangrove swamps*.

Tropical and subtropical shorelines display many *special features*. The *abrasion-platforms* of middle latitudes are *replaced by* more or less horizontal *rock platforms*, that are *weathered by chemical and biologic processes. Groundwater cements* beach sediments into lime-rich *beachrock*; swamp conditions in sand sheets produces *coffee rock*. Both rock types can be exposed by shoreline erosion. *Reef-building corals* are strictly *limited* in their distribution by *water temperature*. Where they can grow, they construct *fringing reefs, barrier reefs* or *atolls, according to* the local *tectonic setting*. Where they occur, they provide complex habitats for submarine life.

13 Surface Morphological Systems: Land Ice

The central theme of this chapter will be the structural, erosional, and depositional forms produced by ice on the land: hence the treatment under the head of morphological systems. Glaciers themselves are best regarded as process-response systems. It was, however, morphological evidence that first proved the former extension of glaciers far beyond their existing limits – and, indeed, the former presence of ice where, as in the British Isles, not even permanent snow caps exist today.

13.1 Brevity of glaciation on the geologic time scale

An ice age, or glaciation, is both rare and brief in terms of earth history. There are indications, admittedly very rough, that glaciations recur at average intervals of about 250 to 300 million years, a fact which has encouraged some scientists to search for periodic astronomical causes. The most attractive cause, however, the period of rotation of our galaxy, does not fit: it only measures about 200 000 000 years. Whatever the actual cause, it is quite obvious that glacial conditions are unusual. On the assumption – once again, very rough – that a given glaciation lasts about 5 000 000 years, then the total duration of glacial conditions amounts to only about 2% of geological time. For the other 98%, the world's land surfaces are everywhere shaped by agents other than land ice.

It follows that, when glaciation does set in, land ice modifies pre-existing landscapes. It can do so in spectacular fashion.

13.1a Modification of valley heads

The least possible effect of glaciation in mountains is to create small glaciers in low-order valleys. As mean temperatures fall and a permanent snowcap is established on the highest ground, the greatest depths of snow accumulate in the depressions of valley heads. Under its own weight, the snow expels its contained air and turns to ice. One can readily simulate the process by tightly squeezing a snowball.

A small glacier in a valley head is called a cirque glacier, after the erosional form that it produces. The French name *cirque* is identical with the Latin word *circus*, meaning an amphitheatre. Cirques are the amphitheatre-shaped hollows at the heads of glaciated valleys (Fig. 13.1). Each cirque typically measures from 0.5 to 1.0 km across and replaces more than one low-order stream valley.

Glacier mechanics are still incompletely investigated. For this reason, they are to some extent controversial. Nowhere is this more so than in respect of cirque glaciers. The net effects on the other hand are vividly apparent. Not only can cirque glaciers excavate amphitheatre-like hollows: they can destroy the original divides between adjacent low-order and pre-glacial streams; they can produce steep headwalls and sidewalls from which bedrock has been split or torn off; and they can over-deepen their central beds, so that their melting on deglaciation reveals a rock-cut lip and an enclosed shallow lake, a tarn. The best guesses at process involve a combination of frost-shattering with rotational slipping. In rotational slipping, a cirque glacier is imagined to act somewhat like a slump block in a landslide. Discussions of frost-shattering concentrate on what happens in the bergschrund, the deep crack that, close to the headwall, separates the ice of the cirque glacier from the adjacent bedrock.

Where two headwalls or sidewalls intersect, they do so in a sharp ridge called an arête (Fig. 13.1). Such a ridge typically exhibits the effects of frost action that is not infrequently still in progress. The deep biting of three – or sometimes more – cirques into a peak produces the pyramid-like horn shape (Fig. 13.1) which is a widespread result of intense mountain glaciation.

13.1b Modification of trunk valleys

A representative temperate (= non-polar) glacier will move at a rate from 2–15 m/yr at its base, and at 15–75 m/yr at the surface. Take the value of 15 m/yr as a rough average for the whole. Now compare this

Fig. 13.1 Modification by ice of fluvial topography: (a) summit, divide and valley head converted respectively to horn, arête and cirque (Devils Paw Mountain, British Columbia: Canadian Government Office of Tourism) (b) glacier trough (Nant Ffrancon, North Wales: J.K. St. Joseph).

with an equally rough average of 1.5 m/sec for streamflow. The ratio of cross-sectional area between a glacier channel and a stream channel with equivalent discharge is reckoned in the millions – in this case, some 3 000 000. Simply because it moves slowly, a valley glacier needs a large channel.

The calculation runs as follows: discharge, $q = wdv$, where w is channel width, d is mean channel depth and v is flow velocity. The product wd, width times depth, is the cross-sectional area of the channel, which can also be written as A. Now, since discharge is the same for the glacier and its meltwater stream, cross-sectional area is controlled entirely by velocity. More precisely, the difference in cross-sectional area between the glacier and its meltwater stream is controlled by the difference in velocity. If we use the subscripts s and g for stream and glacier, then

$$q = v_s A_s = v_g A_g$$

and the ratio of cross-sectional area between glacier and stream, A_g/A_s, is the same as the ratio v_s/v_g between stream velocity and glacier velocity. As

already seen, the ratio of velocity, and therefore the ratio of cross-sectional area, can easily exceed one million.

We observe the effects of the velocity difference wherever a stream of meltwater issues from a glacier snout (Fig. 13.2). We also observe it in the colossal proportions of former glacier channels. These, often loosely termed glaciated valleys, are best called glacier troughs (Fig. 13.1).

Fig. 13.2 Meltwater stream issuing from ice tunnel: Franz Josef Glacier, South Island, New Zealand.

Glacier troughs are cut in mountain country where the snowline descends below the outlets of head valleys, and where cirque glaciers merge into trunk (or valley) glaciers. The most obvious effect to be revealed when the glaciers melt away is the shaping of glacier channels into U-profiles. Considerable erosion is involved in the carving of a glacier trough, especially in the destruction of mountain spurs. A valley glacier cannot turn as readily as a stream can: moreover, it seems to display no tendency to swing from side to side.

Mountainous areas can be assumed to be based on rocks that are generally resistant. The major lines of weakness are faults and shatter-belts, that are often straight in plan. Picked out by the preglacial erosion of rivers, these lines become even more strongly emphasised in the landscape by the work of ice.

Where mountains run down to the sea, and where former glaciers extended beyond the present coastline, the lower ends of the troughs are flooded by the postglacial sea. The flooded valleys are fiords (Fig. 13.3). Fiord coastlines are found in British Columbia and the Alaska Panhandle, in southern Chile, in South Island, New Zealand, in Labrador and Greenland, and in western Scotland and Norway. Particularly in the North Atlantic examples, the drowning is only partly due to the postglacial rise of sea level: indeed, in some areas, the land has risen more than has the sea. Glacier tongues can erode well below the level of the sea. Where they represent the

valley-guided basal flow of ice caps, as they did in the North Atlantic area, it is easily possible for erosion to continue where the trough floor is 5000 m or more below the sea surface.

With the huge dimensions of major glacial troughs goes a striking disparity of size between trunk troughs and tributaries. During the peak of glaciation, low-order tributary glaciers join high-order trunk glaciers at the same surface level. But imagine the circumstance where a tributary is only one-quarter as wide as the trunk: it is likely to be also about one-quarter as deep. Therefore, when the glaciers melt away, the bottom of the tributary trough comes in well above the floor of the trunk trough. The tributary remains as a hanging trough (Fig. 13.4a).

If a long valley glacier, fed by abundant snow on high ground, extends far below the snowline, then it can actually shear off the ends of tributary stream valleys. These, characteristically V-shaped in contrast to the U of tributary glacier troughs, are hanging valleys (Fig. 13.4b).

Trunk valleys undergo much modification in profile. Not merely are they deepened: their slopes become highly irregular. For all its great erosive power, a valley glacier can seem extraordinarily

Fig. 13.4a Three hanging troughs: the main valley is occupied by the Athabasca Glacier (Jasper National Park, Alberta, Canada: Canadian Government Travel Bureau). Hanging troughs also appear in Figs. 4.5 and 13.3.

Fig. 13.4b Hanging valley, V-shaped in section: between the geologist and the valley intervenes part of a deep glacier trough (J.C. Knox).

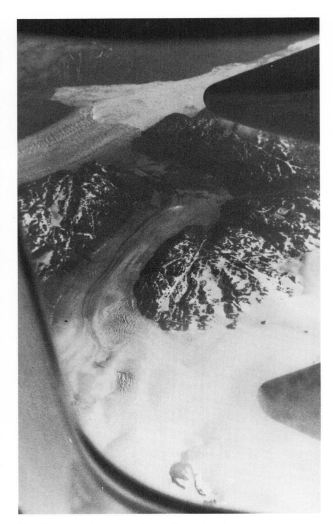

Fig. 13.5   A rock bar at the head of Loch Coruisk, Scotland (J.K. St. Joseph).

Fig. 13.6   A ribbon lake in Waterton Lakes National Park (Canadian Government Office of Tourism). Compare also Loch Coruisk, Fig. 13.5.

Fig. 13.7a   Outspilling glacier, Greenland.

sensitive to differences of rock resistance. Especially resistant outcrops will be left, after deglaciation, as rock knobs, or even as rock bars extending right across the floor of a trough (Fig. 13.5). Hollows excavated below the general line of the profile become, after deglaciation, the sites of long, narrow, deep lakes, the widely familiar ribbon lakes of glaciated highland (Fig. 13.6).

It seems likely that the lake troughs, many of which extend as far as 300 m below the water surface, represent responses to a whole range of processes, not all of which may have operated in a particular case. Some writers have speculated that former deep weathering (see Chapter 18, section 18.3) may have prepared the way for irregular erosion, although no such consideration can apply where all the remaining bedrock, including that of the trough sides, is sound. Others rely on compression of the glacier, as where a trough cuts through a large resistant outcrop: reduced erosion at the sides must be compensated for by increased erosion of the glacier bed. Additional possibilities exist. None of them affects the fact that ribbon lakes are strongly characteristic of glaciated highland. Whether or not their water level is raised by a morainic barrier, they do involve deep erosion of the bedrock floor.

13.1c Modification of mountain divides

Because it is so deep in comparison with water in an equivalent stream, glacier ice may spill out of valleys or topographic basins, across divides that would firmly contain streams. Among the most striking examples of spilling are those at the margins of the Greenland ice cap, where glaciers stream outward between protruding mountains (Fig. 13.7a). Compression along the margins of an outspilling glacier is compensated by increased erosion at the base: what is lost in width must be made up in depth. In such circumstances, the trough through the mountains can be cut very deeply indeed. Along the axial divide of the Scandinavian Peninsula, and also in the Northwest Highlands of Scotland, glacier channels have been cut clean through the pre-glacial divides: what were passes in pre-glacial times are now troughs at least, and troughs containing deep lake-basins at the extreme (Fig. 13.7b). All are called glacial breaches.

On a somewhat smaller scale, ice accumulating in

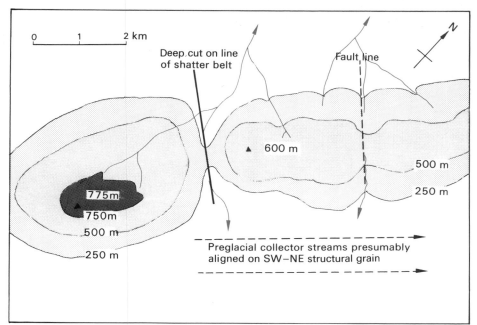

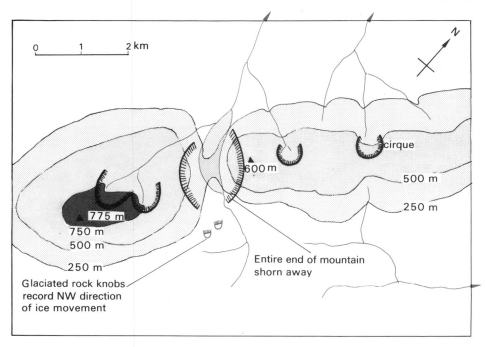

Fig. 13.7b Breaching of a preglacial divide on the line of a pass: upper, preglacial topography; lower, postglacial topography. Based on the area of Altan Lough and Mt Errigal, Donegal, Ireland.

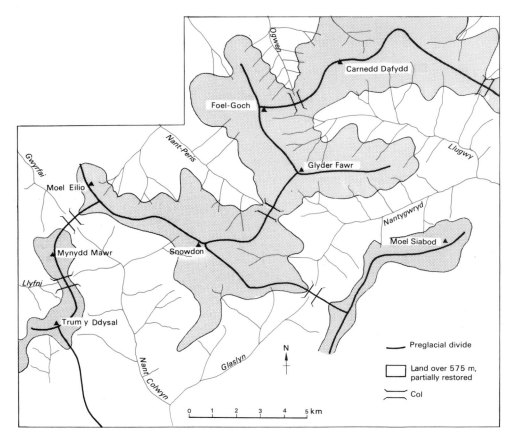

Fig. 13.8 Conversion from dominantly centripetal to dominantly centrifugal drainage by means of glacial breaching: Snowdon area, North Wales.

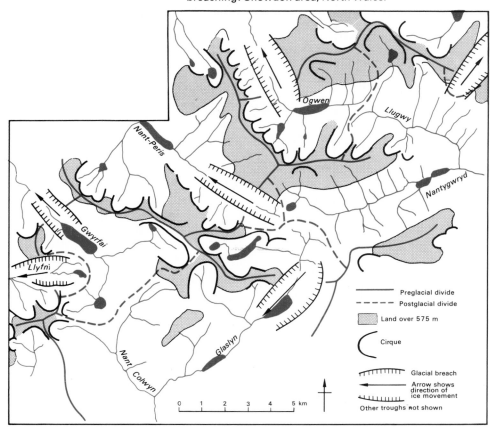

a widely opened upper valley may be forced to spill out through the surrounding passes, especially where its downvalley path is obstructed by abundant ice coming from a distance. In this way, a preglacial pattern of inward-radial drainage, collected by a single outgoing trunk stream, can be converted to a postglacial pattern of mainly outgoing radial drainage, as in the Snowdon district of North Wales and in the Glengesh Plateau of Northwest Ireland (Fig. 13.8). Breaches seem anomalous in the fluvial landscape. Many of them carry streams right through what should be a continuous line of high ground based on resistant bedrock. The direction of ice movement through a glacial breach is commonly recorded by rock knobs, scoured on the up-ice side and plucked on the down-ice side.

13.2 Step-functional advance

Valley glaciers advance when the climate cools off, and melt back when it warms up. The relationships between climate and glacier behaviour can, however, be complex. In addition to changes in temperature, changes in precipitation can also be involved; and different glaciers can record very different response times. Thus, although the rather coarse record of the past resolves itself into more or less synchronous advances or recessions, glaciers of today need by no means behave in phase with one another. For valley glaciers, the response time may range from about five to fifty years.

Glacier advance, when it does occur, can be abrupt, in the form of a surge. This is a wave that moves down-valley faster than the mean velocity of the ice: a rate of about four times mean ice velocity seems fairly common. At the actual snout, the surge may mean a quite sudden extension of the glacier, perhaps by as much as 100 m/day.

Glacier surges have much in common with queues of traffic on the open road. Individual vehicles escape from the queue in front, but others add themselves on behind. The form persists, although the components are interchanged. In traffic, however, the average speed per vehicle is greater than the speed of the queue. With glaciers, the surge speed is greater than the velocity of ice, except at the very extremity.

13.3 Mass balance and glacier equilibrium

A glacier can be treated in the same way as any other part of the hydrologic cascade: it can be measured for input–output relationships. The mass balance is the ice budget. In a rather simplified fashion, we can

assume that snowfall above the snowline is total input. Total output consists of moisture evaporated below the snowline, and of discharged meltwater. The latter is equal to the discharge of outflowing streams less the amount of rain that falls on the glacier. For glaciers that reach the sea, icebergs must be added on the output side.

Quite apart from surges, it is rare for inputs and outputs to be precisely matched. But valley glaciers especially tend to strike a balance, under the control of negative feedback.

Recall that such glaciers descend below the snowline. The larger the glacier, and the faster the rate of movement, the lower it will descend. But descent below the snowline means entry to levels of higher and higher temperatures, where melting becomes progressively severe. Therefore, the position of the snout of a given glacier at a given time roughly represents a balance between rate of ice supply and rate of melting. An increase in supply would extend the glacier, but only to a new point of complete melting. A decrease in supply would cause glacier recession, but recession would be checked at a new point where a reduced rate of melting matches the reduced rate of supply.

13.4 Glacier termini

The ends of glaciers are usually called snouts for valley glaciers and fronts (or ice fronts) for continental glaciers. Glacier snouts in particular tend to be stationary, because progressive thinning of the ice below the snowline finally produces a depth too small for the ice to move. Fig. 13.9 shows what happens to the Saskatchewan Glacier. Shortly below the snowline the mean depth is about 450 m, and the

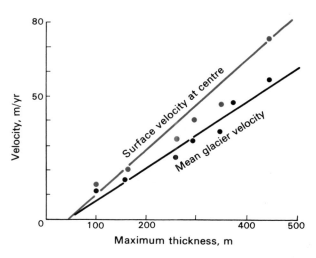

Fig. 13.9 Velocity/thickness ratio for the Saskatchewan Glacier: down-trough direction is to the left.

Fig. 13.10 Thrusting at glacier snout: a large block of rock is emerging along a thrust-plane in the dirty, rigid, thin ice (Franz Josef Glacier, South Island, N.Z.).

mean velocity 55–60 m/yr. As melting progressively reduces glacier depth, mean velocity falls off, becoming zero for a glacier depth of about 35 m. A typical snout includes a large fraction of rock material, that is of course not subject to melting; and pressure from the moving ice immediately up-valley can cause marked thrusting (Fig. 13.10).

The unsorted debris that accumulates at the snout is a terminal moraine, also known as an end moraine. Moraines are presumably formed at each position of the snout during spasmodic glacier advances, but are removed when the ice moves still farther forward. A distinctly receding glacier will stagnate and eventually melt for part of its length up-valley of the end moraine, forming a new moraine at the new limit of movement. Alternatively, an episode of melting back may be followed by a partial re-advance, with a new moraine formed at the new limit of advance. Such a sequence, roughly corresponding to a repetition of two steps back and one step forward, has been followed by the Austerdalen Glacier in Norway (Fig. 13.11). Nine episodes of recession and partial re-advance in 187 years gives an average of about 21 years per episode. The mean net retreat distance per episode is about 220 m, producing a mean retreat rate of about 10 m/yr.

Moraines left by receding glaciers can often be matched from one valley to the next, indicating very strongly that net recession is typically step-functional.

13.4a Moraines on and in glaciers

Numerous observers have been highly impressed by moraines on valley glaciers – and, where information is available, by morainic material inside glaciers. The rock debris of moraines on glaciers is concentrated in highly distinctive strips. It is supplied mainly or entirely by frost action on the exposed valley walls. Falling onto the margin of the glacier, it forms thick bands of lateral moraine (= moraine along the side). Where two valley glaciers unite, the lateral moraines inside the confluence combine in a medial moraine (= moraine along the middle of the glacier). Very large trunk glaciers with many tributaries carry large numbers of medial moraines, always one less than the number of individual glaciers in the system. Their chief importance is probably to demonstrate that, unlike most rivers, glaciers fail to mix their tributary contributions with the trunk flow.

13.5 Piedmont glaciers

Very large glaciers descending from mountains on to the surrounding plains spread out in lobes (Fig. 13.12). They can construct substantial lobate moraines. In this sense they are intermediate between valley glaciers and ice sheets, the margins of which are also typically lobate.

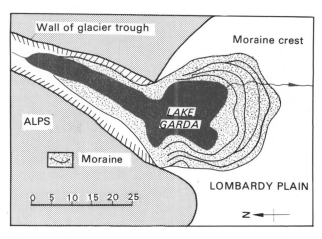

Fig. 13.12 Lobate moraine complex on the Lombardy Piedmont, Italy: the morainic dam raises the water level by 160 m, but a maximum depth for Lake Garda of 645 m indicates erosion of as much as 485 m into bedrock. The most likely process of deep erosion is repeated surging of the glacier over a wet floor.

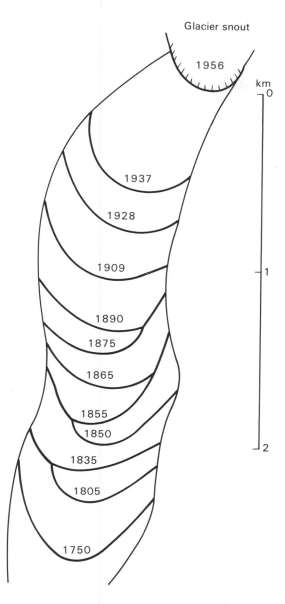

Fig. 13.11 Sequential moraines, considerably restored in places, for 200 years of net recession of the Austerdalen Glacier, Norway: freely adapted from the data of C. Embleton and C.A.M. King.

13.6 Continental glaciers

During the glacial maxima of the present ice age, continental glaciers repeatedly extended themselves far beyond the existing glacial limits. In so doing, they vastly modified the landscapes which they overspread and which bordered them. Their effects are conveniently discussed under the subheads of central areas, near-marginal areas, marginal areas and extra-marginal areas.

The central areas are dominated by erosion, although depositional forms can be constructed during glacier retreat. At the extreme, the bedrock can

be very largely stripped of its preglacial waste mantle, appearing after deglaciation in a generally subdued landscape, with much variation of detail which includes highly disorganised drainage, multitudinous and mainly shallow lakes, and equally multitudinous bogs (peaty) and swamps (very wet) (Fig. 13.13).

Two puzzles appear here: the nature of the preglacial surface, and the origin of some deep lake basins. The best known of the surfaces eroded by continental ice are those of the Canadian Northland and of the lands – especially in Finland – that surround the Baltic Sea. Both surfaces have been developed on the cratons of antique continental cores. Their preglacial form cannot be recon-

Fig. 13.13 Effect of glacial scour on the basement rocks of part of the Laurentian Shield: area shown is about 75 km² (original photo supplied by the Surveys and Mapping Branch, Department of Energy, Mines and Resources, Canada).

structed with certainty; but it may well have been that of a subdued terrain, where cratonic geology and landscape history coincided only by accident with the development of maximal continental glaciers. The huge deep lakes of North America, that range in a swinging arc from the Great Bear Lake to Lake Ontario, may be another matter entirely. Undoubtedly occupied by ice, at the latest not many thousands of years ago, these lakes lie athwart or very close to the edge of the North American craton. The relationship of their geological position to glacial erosion remains obscure. On the one hand, the Great Lakes especially are associated with huge morainic lobes formed by tongues of ice that moved through the lake basins: on the other hand, the lake basins are well inside the extreme ice limit. One thing, however, is sure: much of the material stripped by ice from the craton has either lodged in the surface deposits of the Middle West, or is stored in the southern trench of the Mississippi River, or has already reached the Mississippi Delta.

13.6a Deposition in near-marginal areas

Inside a given end moraine, stagnation and eventual melting of the ice border will produce till – a type of deposit that ranges from sand and gravel, more or less washed, to rock fragments ranging widely in size and set in a clay matrix. Much of the till is simply lowered onto the ground as the ice melts away. Initially it forms an irregular surface on which postglacial drainage has to start all over again.

A spread of till may be variegated by eskers, winding ridges of sand and gravel that originate as the coarse bedload of meltwater streams on, in, or beneath the decaying ice. Eskers can themselves be variegated by deltas at the position of each temporary ice margin. But the most impressive forms are those of drumlins, small streamlined hills shaped in glacial deposits, which typically occur in large groups known as swarms (Fig. 13.14).

The word drumlin is originally Gaelic, standing simply for little hill; but it has been taken over for the shape here in question. Drumlins are asymmetrical: they are widest and highest toward the up-ice end. Streamlined like aerofoils, they record the direction of ice movement. Internally they range from sand and gravel to extremely clay-rich till. They probably result from the glacial overrunning of glacial deposits, such as sand and gravel discharged by an advancing ice lobe, or clayey till deposited during an earlier advance: their streamlined form, which may result partly from erosion, and their occurrence in large groups, recall the random deformation of bed material observed in braided stream channels.

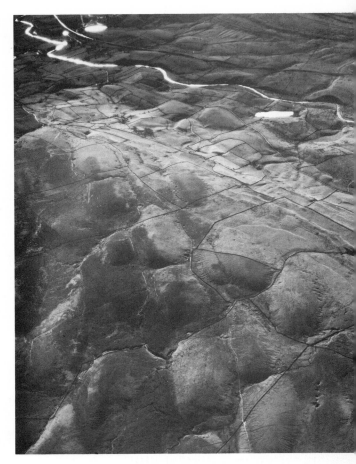

Fig. 13.14 Drumlins in Ribblesdale, Yorkshire, England: the moulding ice moved into the distance (J.K. St. Joseph).

13.6b Deposition in marginal and extra-marginal areas

In addition to concentric lobate moraines (Fig. 13.15), depositional forms in marginal areas include the already-mentioned deltas where meltwater streams burst through the ice margin, and belts and spreads of glacial outwash.

Outwash is rock debris swilled away from the ice front by meltwater streams. Strictly speaking, it includes the sediments of eskers and of esker deltas. Most of it, however, if it does not eventually reach the sea, is deposited as bands of sediment in valleys – valley trains – or as extensive spreads on low ground – outwash plains, or outwash aprons (Fig. 13.16).

The most intensively studied outwash plains are those of Iceland where, however, the sudden outbursts of meltwater may be related to the supply of volcanic heat from the subsurface. If continental glaciers in non-volcanic (better, non-thermal) regions are also liable to sudden outbursts, which can be described as glacier collapse, then we are certainly dealing with a mode of step-functional behaviour.

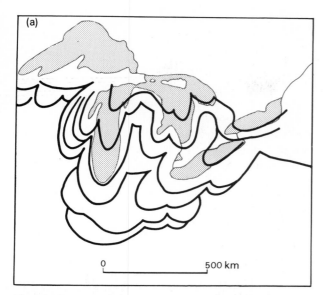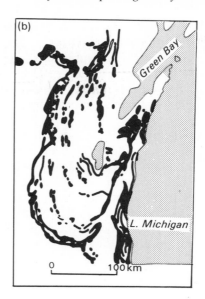

Fig. 13.15 Lobate moraines in the U.S. interior, in part of the Lake Michigan area: (a) general pattern, much simplified; (b) detailed pattern for the latest stands of the Green Bay sub-lobe.

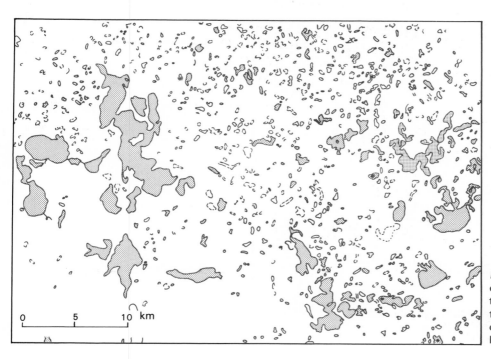

Fig. 13.16 Part of a pitted outwash plain in Minnesota: the map actually understates the number of kettles, since closed depressions without lakes or swamps are not shown.

13.6c Pitted outwash plains

When a continental glacier melts back, the stagnating marginal belt can become disrupted into isolated blocks which are buried in outwash sand and gravel. Mean annual temperatures close to the freezing mark combine with the high input of heat energy required for melting to ensure that the blocks survive long enough to be buried. When melting does occur, the cover of outwash collapses, forming the roughly circular depressions known as kettles (Fig. 13.16). A plain of outwash material pocked by kettles is a pitted outwash plain. Depending on the local position of the water table, kettles will be dry, will contain bogs, or will be occupied by lakes.

13.6d Modifications of drainage

Postglacial drainage makes an almost completely new start on the scoured landscapes revealed by the melting of the central ice. There must be some kind of general slope, which will partly control the drainage direction; and there are likely to be numerous short channels cut by debris-laden streams that flowed under the ice during the retreat stages. There are also many small topographic depressions where water collects and into which it runs. But, as a whole, postglacial drainage is highly disorganised. It cannot be analysed in terms of repeatedly branching networks.

Thick deposits, whether of till or outwash within the outermost moraine, or of outwash sheets beyond that moraine, obliterate the preglacial surface. Here again, postglacial drainage makes an almost completely new start, except that some stream trunks are likely to occupy the spillways cut by glacial meltwater. Notably in the central U.S. Midwest, the chief components of the postglacial drainage system differ strongly from the corresponding components of the preglacial system. Large preglacial streams flowed northward across the postglacial ENE–WSW line of the Ohio River, a principal collecting stream of today which in preglacial times did not exist. Generally similar effects are known, wherever thick spreads of glacial deposits have been explored by drilling.

Drainage on a newly-exposed till sheet, such as that on a scoured expanse of bedrock, is typically deranged to begin with. How long it takes for drainage to become integrated on scoured bedrock is not known; but a maximum interval can be established for till sheets. In parts of the U.S. Midwest, the last major ice advance failed to cover all the territory that was glaciated in the previous advance. Drainage lines on the till of this latter advance have become organised into branching networks in not more than 70 000 years. Drainage on the till of the last major advance of all is still deranged after some 7000 years of action (Fig. 13.17). We see that 7000

years is not enough, but that 70 000 years is more than enough, to destroy the many irregularities of the depositional surface, to fill or drain lakes, to obliterate swamps, and to replace deranged drainage by orderly branching systems.

Some of the most striking effects upon drainage are those produced at and close to the former ice-fronts. A continental glacier advancing across country is certain to dam approaching drainage. With a thickness approaching 1000 m quite close to the front, an ice sheet can override relief which would confine and guide stream networks. Wherever the glacier moves against the slope of the ground, surface drainage will be ponded, diverted, or – as is usual – both. Ponding is especially pronounced where the crust is depressed beneath the weight of the ice: depression extends beyond the ice margin, especially during retreat stages if the response of the crust to unloading lags behind the melting-off of the ice.

An incredibly complex sequence of ponding and diversion resulted from the alternating advance and recession of ice lobes, during the last glacial, in the Great Lakes area of North America. At the extreme, ice occupied and extended beyond all the lake basins. The only outlets led to the Mississippi. During retreat stages, large proglacial (= ice-marginal) lakes came into being, interconnected by systems of spillways. At one stage of marked glacier recession, the chief drainage outlet followed the valley of the Ottawa River. Recovery of the crust from its load of ice later established the existing line of the St. Lawrence River.

On the European mainland, step-functionally receding ice of the Scandinavian cap permitted waters brought by generally north-flowing rivers to discharge westward along the ice margin, and to cut wide sandy channels which are today followed by rivers, roads, railroads and canals (Fig. 13.18). One of the best studied cases, however, is that of Glacier Lake (= proglacial lake) Harrison in the English Midlands (Fig. 13.19). As is normal in all well-

Fig. 13.17 Drainage patterns on the most recent till sheet and on an older sheet: some swamps on the most recent till drain in more than one direction, and much potential drainage line is occupied by swamp, whereas swamps have vanished from the surface of the older till, and the drainage nets there are completely organised into dendritic systems.

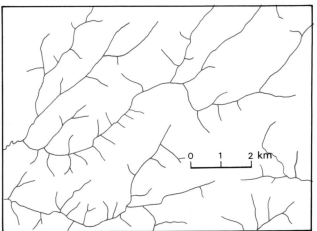

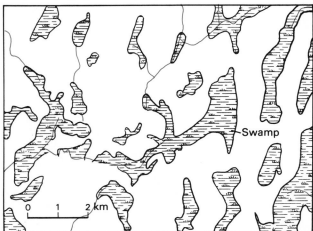

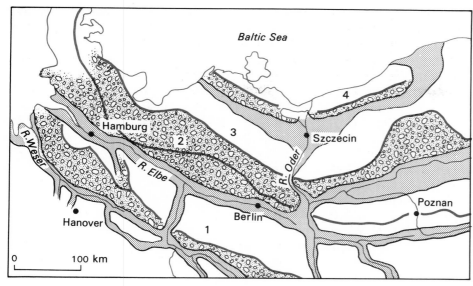

Fig. 13.18 Ice-marginal drainage ways on part of the North European Plain (somewhat simplified).

Ice front

Proglacial drainage-way

Outwash

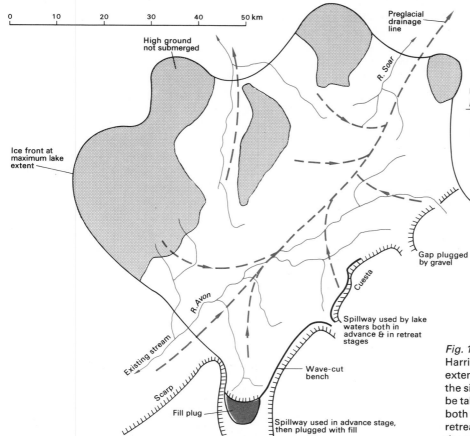

Fig. 13.19 Glacier Lake Harrison at its maximum extent of about 1900 km²: the situation illustrated can be taken to approximate both the advance and the retreat situations, but see the accompanying text.

investigated cases, the Lake Harrison sequence is complex in detail. In summary, it involves the advance of ice across a stream-cut landscape where the main collector stream ran northeastwards. A lake was dammed between the ice front and the high ground, both during glacier advance and during retreat. At the maximum, the ice reached the high ground, overriding it in the northeast and plugging one of three gaps with outwash gravel. In the southwest, a minor ice lobe entered a second gap, partially plugging it with till. In the centre, the third gap acted as a principal outlet for the lake water: the wave-cut bench of the lake shore runs into the gap. Thick deposition of till and lake sediments provided an entirely new surface for post-glacial drainage to work on. The main collector stream grew headward from the southwest, where the ice melted off first: and the chief drainage direction today is toward the southwest, precisely the opposite of that existing before the ice came in.

13.6e Catastrophic drainage changes

In the Pacific Northwest of the U.S.A., most of the Clark Fork valley of the Columbia River system was repeatedly dammed during the last glacial. Dammings during earlier glacials can be guessed at, but have not yet been demonstrated. The resulting lake, Glacier Lake Missoula in western Montana, attained an area of some 7500 km². The area is not particularly impressive, but the volume was very great: against the ice dam, the water stood 600 m deep. Failure was sudden. Initial discharge through the outlet ran at least as high as 10 million m³/sec. Striking across northern Idaho, the floodwaters spread across 20 000 km² of basalt plateau country in southeast Washington, before being collected by the lower Snake and the middle Columbia Rivers. Bedforms were developed on a gigantic scale – individual channels, waterfalls, braids, potholes, ripples, and eroded hills of loess (Fig. 13.20).

The processes at work were partly similar to those operating on the beds of streams generally; but the widespread evidence of shattering and plucking of basaltic bedrock, especially in the western part of the affected area, testifies to the operation of enormous vortices which have been compared to tornadoes in the atmosphere. In addition, the already rough nature of the bed of the flood promoted the formation and sudden collapse of air bubbles, with incredible hammer-blow effects on the underlying rocks. For this effect to occur, flow velocities had to reach or exceed 30 m/sec where the water depth was 100 m or more. Such depths and velocities were attained in places, for flow across some parts of the former low divides approached 3 million m³/sec.

Investigations of the resulting topography have emphasised erosion rather than deposition, although depositional features are by no means absent. Particular stress has been laid on the cutting of channels, that on a normal scale are large and numerous, but that in the total context amount only to minor components in the general assemblage. For this reason, the affected region has been named the Chanelled Scablands.

13.7 Ice as rock

At the present day, some 15 million km² of the earth's surface is covered by land ice. During glacial maxima, the figure was about 45 million km². About two-thirds of the cover has melted off, between maximum conditions and the present day. The total volume has also been reduced to about one-third of what it formerly was.

The locking up of seawater in the form of continental glaciers amounts in effect to the loading of the glaciated areas by additional rock. Two effects follow: the ocean surface falls, and the loaded areas respond by crustal depression. Converse effects follow deglaciation: water is restored to the oceans, which rise in level; and crustal depression is replaced by elevation.

The ice-controlled rise and fall of ocean level has been discussed in Chapter 12. It remains to treat the effects of glacial loading and unloading here. An ice thickness of 3500 m, which is about the maximum possible for an unconfined ice sheet, is equal in weight to about 1250 m of continental crust. The existing ice caps of Antarctica and Greenland do in fact depress the crust beneath them by the stated amount, some 1250 m. When an ice cap melts off, the change in loading is followed by rebound, compensatory displacement taking place in the outer mantle.

Relative response times are highly significant. Most of the water melted out of a waning ice cap rapidly finds its way to the oceans, which respond immediately by a rise in level. On the other hand, a continental glacier takes centuries to disappear entirely. Melting has most severely affected the North American and Scandinavian areas, where existing glaciers are equivalent to only one-hundredth part of the ice caps of glacial maxima. Most of the last melting took place between 10 000 and 5 000 years ago.

The same two areas are the most strongly affected by crustal rebound, which is still in progress and may require another 10 000 to 15 000 years to complete. Rebound has carried shorelines in the northern Baltic Sea and around the southern end of Hudson's Bay some 300 m above existing sea level. Because

this level is 200 m above the low stand of glacial maximum, total rebound so far ranges up to 500 m. Allowance for the probable loss of the earliest part of the record suggests 600 m as a more likely value. Now, when rebound sets in, it works very rapidly at first. As little as 1000 years may suffice to accomplish half the recovery measured for 10 000 years. Very rapid initial rebound has allowed the rise of the crust to outpace the rise in ocean level.

13.8 Ice in the ground

Where mean annual temperature is 0°C, patches of soil will fail to thaw completely in the warmer season. Where the mean descends into the range of −1 to −4°C, ground ice forms and persists. The frozen ground, and its contained ice, are called permafrost. The relevant climate is called periglacial.

The depth to which permanent freezing extends is fixed by the interplay of chilling from above and warming from below. The chilling involves loss of heat to cold overlying air. The warming is effected by the escape of heat from the earth's interior, which establishes a temperature gradient of some 1°C for 30 to 60 m of depth. Somewhere in the subsurface the two effects balance out: the ground at greater depths is unfrozen. Generalisation is very difficult,

Fig. 13.20 Gigantic headcuts in the basalts of the Snake–Columbia Plateau, resulting from the outburst floods of Glacier Lake Missoula: flow direction was toward the observer (John S. Shelton).

partly because some permafrost, especially in Siberia, has been left over from times of colder climate than that of today. But some of the available data suggest that for every 3°C below zero, permafrost will penetrate an additional 100 m. Record depths are reported from northern Siberia – no less than 1600 m.

As a foundation material, permafrost could hardly be worse. It offers special problems, and severe problems, all of its own. If the surface layer melts, as it widely does during the high-sun season, it loses all bearing strength. During the winter, the freezing and expansion of ground ice causes vigorous heaving. Subsiding houses break apart. Melted roadbeds lose all their strength. Winter freezing can distort bridges into jacknife shape.

In respect of the surface morphology of the land, a minor fraction of ground ice produces a major fraction of results. It is the approximate 10% contained in ice wedges. Most ground ice simply occupies pore spaces, or consists of films, lenses, or layers that grow more or less horizontally in the subsurface. Wedges strike straight down. They are

Fig. 13.21 Polygonal ground: (a) stone polygons, about 2 m across; (b) low-centre polygons, up to 7 m across. Both sites are in the Northwest Territories of Canada (S.C. Zoltai).

Fig. 13.22 (below)   A 45-m high pingo in the Northwest Territories (S.C. Zoltai).

responsible for the patterns of polygonal ground (Fig. 13.21).

Severe dry climates where mean annual temperatures run at −7°C or below, and which include weather spells with temperatures of −15°C or below, will cause silty material to crack. Water fills the cracks in the high-sun season, but freezes during the winter. Repeated cracking, filling, and freezing creates vertical ice wedges which can penetrate as deeply as 10 m. If the cracked waste mantle contains stones, these will become concentrated along the lines of cracking, as the interior surface of a polygon experiences frost-heave; and many fragments will be turned on edge. On slopes, the stone nets are stretched into stone lines.

Pingos (an Eskimo name) are ice-cored hills (Fig. 13.22). They typically occur on the silty floors of former or present-day lakes. They probably originate in the bursting of sub-surface water, under pressure, through the permafrost layer. Effectively speaking, a thick ice lens is locally injected. In many cases, it bulges strongly enough to break the centre of its sedimentary cover. Although pingo ice amounts only to an estimated 0.1% of all ice in the permafrost, it has understandably attracted much attention.

Polygonal ground can be matched, at least to

Fig. 13.23   Stone-banked terraces in the Northwest Territories (S.C. Zoltai).

some extent, elsewhere: morphologically speaking, it compares closely with the patterns produced on the beds of dried lakes and reservoirs. Pingos, on the other hand, are unique to the permafrost belt.

Although seasonal thawing of the surface layer – the active layer – increases the difficulties of control on all surfaces, it is particularly troublesome on slopes. Here, the shrink and swell of the freeze/thaw alternation is combined with downslope movement of the waste mantle during the thaw season. The active layer undergoes a slow flowing movement, made possible by the lubrication of abundant soil moisture. A net result is a kind of treacly displacement, that produces festoons in the affected material. At the surface, lobate turf-banked or stone-banked terraces come into being (Fig. 13.23).

In a sense, the stone lines already noticed as drawn-out stone polygons represent the channelled surface flow of the products of weathering by frost. They constitute one end-member set of a rather vague but impressive series of morphological features. The next set is that of blockstreams which range from narrow trains in old channels to broad

Fig. 13.24   A form of blockstream: sarsens sludged into the Valley of the Stones, Wiltshire, England (J.K. St. Joseph). Local bedrock is Chalk, a limestone, but the sarsens are portions of a siliceous crust formed at the level of the plateau top. For another example of periglacial sludging, see Fig. 18.1.

Fig. 13.25   Snout of a rock glacier: although it superficially resembles an end moraine, it is in reality the end of a completely mobile rock-ice system (J.C. Knox).

spreads across entire valley bottoms (Fig. 13.24). The agency of movement ranges from seasonally contracting and expanding ice in streams of blocks, to melting and thawing ice in a mud/rock mixture. At the extreme end of the series come rock glaciers, bulky combinations of ice and rock fragments, that move down-valley much as ice glaciers do (Fig. 13.25).

13.9 Loess

Wherever the climate is very cold, chemical weathering is held to a minimum. Weathering products range from blocks of frost-riven rock, through sand grains to silt particles. Chemical reduction to clays is very slow, although clays of preglacial or interglacial origin can, of course, occur. Loose surficial material in the silt grade is closely, but by no means exclusively, associated with near-glacial conditions and climates. As a surface deposit it is known as loess (pronounced lerss).

The name was actually first applied to the thick windlaid sediments of northeast China, that have been blown by the winter monsoon winds out of the Gobi Desert. But the best studied loess deposits are those of the southern U.S.S.R., central Europe, and the U.S. Midwest. These all originated as glacial outwash. They were whipped up by the storm winds that crossed the outspill channels of glacier meltwater. Winds cannot lift rock particles. They can lift clay particles, and can carry these for very great distances. They can lift silt particles, but can rarely carry them very far. In the silt grade, of very small quartz fragments and of aggregates composed of individual clay particles, the sediment load of the wind settles out rapidly.

It seems to be typical that the recession of a continental glacier is followed by highly variable climate, with stormy dryness setting in early (Chapter 18, parts of section 18.1a). Strong dry winds collect silt from outwash plains and the deposits of meltwater spillways, dumping it as loess sheets on the neighbouring terrain downwind, as soon as wind velocity drops enough to let the dust settle. Parts of the U.S. Midwest and of the Danube Basin in Europe are overspread by loess that, derived from glacial outwash, ranges down from 3 m in thickness, according to the downwind distance from the supplying valleys.

Chapter Summary

Glaciers are process-response systems, but *former glaciation* is *demonstrated by morphological evidence.*

Glaciation is short-lived on the geologic time scale, accounting for perhaps 2% of geologic time. The cause of ice ages is unknown. The attractive hypothesis of *galactic rotation does not fit the geologic record.* But, whatever the cause, glaciation is so brief that *land ice modifies pre-existing landscapes.*

When glaciation sets in, the *heads of river valleys are converted into cirques. Some divides* between adjacent cirques *are converted to* the sharp ridges of *arêtes.* Cirque-bitten *peaks become horns.*

Trunk valleys, occupied by glaciers with low velocities compared to stream velocities, *develop very large cross-sectional channel areas. Stream channels are replaced by glacier troughs. Tributary valleys come in as* troughs, left after glacier melting as *hanging troughs, or as* truncated stream valleys, left as *hanging valleys.*

Glacier troughs drowned by the postglacial sea *form fiords. Trough floors above sea level are* typically *diversified* by rock knobs, rock bars, and the basins of ribbon lakes.

Glaciers can punch through pre-glacial divides in glacial breaches, greatly modifying the pattern of postglacial drainage.

Glacial advance and recession is typically step-functional. Advance often involves forward *surges; recession* often involves *stagnation* of the ice margin. *Temporary balance* is recorded by *end moraines.*

For *continental glaciers,* the *central area is mainly erosional.* The *near-marginal* area is *mainly depositional,* involving *till sheets* which *may be deformed into drumlins. Deposition in marginal and near-marginal areas* includes the production of *lobate moraines, deltas,* and spreads of *outwash.* Plains of outwash can be pitted by *kettles.*

Glaciers tend to derange drainage. They make *glacial breaches* in highlands, *blanket lowlands* with glacial and glacier-associated deposits, and *dam and divert streams* along their margins. Every closely-studied instance proves highly complex. The *most catastrophic case* is probably that of the sudden release of ice-dammed water across the *Chanelled Scablands* of the U.S. Pacific Northwest.

Ice incorporated in continental glaciers *represents water abstracted from the world ocean.* It also represents an *added continental load.* The *overloaded crust sinks. When the ice caps melt, the crust rebounds,* but with some lag. Rebound after the last episode of melting is *especially* pronounced *in North America and Scandinavia,* where melting-off has been greatest. The total rebound, probably *already* by *500 m,* is still in progress.

Ice in the ground forms and persists where the soil and subsoil fail to thaw. *Frozen ground is permafrost: the climate is periglacial.*

As an engineering prospect, permafrost is formidable. As a control of surface morphology, it depends primarily on the formation and development of *ice wedges,* which *are mainly responsible for patterned ground.* Patterns include those of *stone polygons* (that stretch into stone lines on slopes); the ice-cored hills of *pingos; turf-banked terraces; blockstreams;* and rock glaciers.

The *silt spreads of loess* are *typical* of, *but not exclusive* to, regions of present or former periglacial climate. In *northern China,* loess deposits are carried by the monsoon winds *out of the interior deserts. Elsewhere,* however, they are typically derived from glacial outwash, being composed of silt-size particles whipped up by storm winds during *an early stage of glacier recession.*

14 The Solar Energy Cascade

All life on earth depends on energy radiated by the sun. Solar energy activates ocean currents, controls the climates in which we live, drives our weather systems, and governs plant growth. But solar energy in about the same amount that strikes Earth also falls on our nearest-neighbouring planets Venus and Mars. These are neither inhabited nor habitable. The difference from Earth results from the special properties of our atmosphere. This chapter will deal with the chief general consequences of the effect of solar energy on the earth–atmosphere system.

The lowest imaginable temperature is absolute zero, $-273.7°C$. At this mark, no object could radiate energy. At all higher temperatures, energy is given off at a rate dependent on temperature. The radiated energy travels in sine waves, with wavelength also controlled by temperature: the higher the temperature, the shorter the wavelength. Also, the shorter the wavelength, the higher the frequency.

14.1 Scales of measurement

Frequency is measured in cycles per second. Solar radiation arriving at the outer limit of the earth's atmosphere moves in waves ranging from about 1 cycle/sec to more than 10^{22} cycles/sec (Fig. 14.1). Wavelength can be measured in metres, although other units are normally employed. A common measure is the micron, $1/1000$ millimetre. As Fig. 14.1 shows, solar radiation at the outer limit of the atmosphere ranges in wavelength from about 10^{14} to about 10^{-8} microns. Amount of radiation is measured on a scale based on the calorie. The calorie – familiar in relation to food intake, but often only vaguely understood – is the amount of energy needed to raise the temperature of one gram of water through $1°C$. Energy sufficient to raise the temperature of the water through $1°C$ is our basic unit of radiation. It is the langley (ly). We allow for the time element by writing of ly/sec, ly/min, ly/hr, and so on. With a surface temperature of $6000°C$, the sun gives off energy at the rate of 56×10^{26} ly/minute – 5600 million million million million units per minute. But by the time that the sun's radiation has reached the outer limit of the earth's atmos-

phere, it has been much diluted. It amounts only to an average of 2 ly/min.

At this rate, it would take water fifty minutes to go from zero to boiling point. The rise, however, would not stop there. In less than another hour, the temperature would be high enough to roast people. Since the solar system has been in existence for 4500 million years, and since people live on earth without being roasted, there must obviously be some kind of negative feedback at work. It there were not, the earth's atmosphere would run at flame heat, if indeed it existed at all.

14.2 Cascade components

Radiation at the outer limit of the atmosphere ranges down on the scale of wavelength, and up on the scale of frequency, from radio waves, through microwaves, infra-red rays, visible light, ultra-violet rays, and x-rays, to gamma rays. There is some overlap between one waveband and another. Since, however, we are dealing with an energy cascade, we need not examine the components in detail. It is enough to notice that occasional bursts of solar emission in the radio waveband upset radiocommunication on earth; that radiation in the range from

Fig. 14.1 Solar radiation at the outer limit of the atmosphere: range of visible light at the earth's surface also indicated.

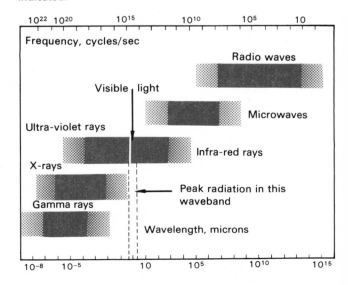

microwaves through gamma rays can be dangerous or even lethal to animal life, including human life; and that solar radiation is strongly concentrated in the waveband between 2 and 0.2 microns, which includes the band of visible light.

Life forms that can see, notably including ourselves, have evolved sight organs which respond to what we recognise as visible light. The very narrow waveband of visible light can easily be broken down into sub-components – red; red + yellow = orange; yellow; yellow + blue = green; blue; blue + violet = indigo; violet. The breakdown is readily effected by glass prisms, which bend the incoming rays of light apart, and by water droplets which, by similar bending, produce rainbows. Red, yellow, blue and violet are the only true colours.

Space photography shows that the earth is a blue planet. It looks blue from outside space, just as the sky looks blue from the earth's surface. The incoming solar radiation is selectively scattered in the range that we recognise as that of blue light.

14.3 Radiation bars and radiation windows

The earth's blueness well illustrates the selective filtering effect of the earth's atmosphere on solar radiation. An equally selective effect applies to outward radiation from the earth. Radiation in certain wavebands will pass through the atmosphere with little or no obstruction. In other wavebands, radiation is checked. It passes easily through radiation windows, but is reflected, scattered, or absorbed at radiation bars.

14.3a Bars to incoming radiation

Very little is yet known about what actually happens in the earth's outermost atmosphere. It is scarcely possible to define the atmosphere's outer limit, although it does seem likely that the earth's atmosphere merges with the atmosphere of the sun, at a height very roughly of the order of 80 000 km above the earth's surface. Beyond some 2000 km, the atoms of atmospheric gases are broken down by solar radiation into negatively-charged particles (electrons) and positively-charged particles (protons). The effect of gravity is negligible: the movement and concentration of the particles is governed by the solar wind and the earth's magnetic field. The particle-equivalent of hydrogen dominates above 2500 km, and that of helium between 2500 and 2000 km (Fig. 14.2).

Fig. 14.2 The vertical structure of the atmosphere: on the right, total structure; on the left, structure of the lowest 100 km.

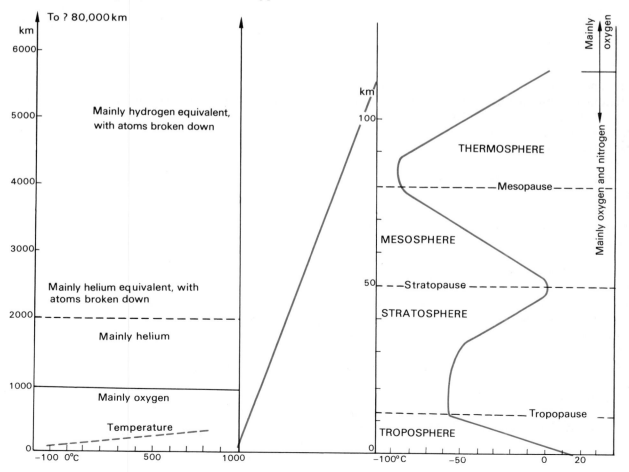

Between 2000 and 1000 km the atmosphere consists mainly of helium gas, between 1000 and 115 km mainly of oxygen gas. Below 115 km occurs the nitrogen-oxygen mixture of our familiar low-level air. Overlapping this mark, chiefly in the range from 80 to 300 km aloft, ultra-violet rays from the sun combine with space radiation (cosmic rays) to ionise atoms or molecules. They dislodge one or more electrons from the atom or molecule system, causing the surviving particle to be electrically charged – that is, to become an ion. Concentrations of ions respond to the discharge of unusual amounts of energy from solar flares by producing the northern lights (aurora borealis) and the southern lights (aurora australis), those spectacular displays in the night sky which can appear as far as 35° of latitude from the poles. Particles ejected from the solar flares energise and displace particles in the outer atmosphere, that otherwise would remain trapped in radiation belts. Of more practical importance, the concentrations of ions between 300 and 80 km aloft are great reflectors of radio waves. By bouncing these waves back and forth to the earth's surface, they make possible radiocommunication round the earth's curve.

We cannot help but wonder if radiocommunication from somewhere in space is not being blocked out by the concentrations of ions.

The absorption of ultraviolet and cosmic rays makes the outer atmosphere hot. At least, it would do so if the outer atmosphere had any significant density. The hypothetical temperature at 300 km aloft is some 1000°C. Below this level we have reasonably firm information, including a mark of about −90°C at 80–85 km aloft (Fig. 14.2). Beyond this height, and in respect of temperature, the atmospheric shell is called the thermosphere. Therm, as in thermometer, thermos flask, and thermal underwear, stands for heat (Greek *thermē*). The absorption effects produced in the thermosphere decay in intensity with the approach to earth.

Below the thermosphere, between 80 and 50 km aloft, lies the mesosphere (Greek *mesos* = middle). Very little seems to be known about the mesosphere, perhaps because it is difficult to investigate with the aid of artificial satellites, and because the outer atmosphere and outside space attract more ready attention than does this particular atmospheric shell. Since temperature in the mesosphere varies considerably with height, from 0° at the bottom to −90°C at the top, we infer that the atmosphere here can absorb radiation, but that it is heated primarily by radiation from below. Even at 50 km, we are still in one sense discussing the outer atmosphere. Pressure is only one-thousandth of its value at sea level, and meteoric dust from space combines with ice to form the noctilucent (= night-glowing) clouds of high latitudes.

The downward transition from the thermosphere to the mesosphere is the mesopause. It constitutes a temperature low (Fig. 14.2). The transition at the base of the mesosphere is a temperature high, called the stratopause. Beneath it lies the stratosphere. Extending down from 50 to a somewhat variable base, the stratosphere (Latin *stratum* = layer) acquired its name in the early years of atmospheric sounding, when only the lowest portion had been explored. Temperatures between the base and about 20 km up are more or less constant. We now know, however, that temperature increases upward through the stratosphere, from about −55° to 0°C, indicating heating from above. The main process of heating is absorption of ultra-violet rays by ozone, triple-linked atoms in O_3 molecules, as opposed to the double-linked atoms in oxygen molecules, O_2. Nearly all atmospheric ozone is concentrated in the stratosphere. The possibility of disrupting the ozone layer, and so of destroying a major radiation bar, has caused aerosol sprays to fall into disfavour, and has led – at least for the time being – to the abandonment of U.S. plans to manufacture supersonic aircraft, which operate in the lower stratosphere.

At the base of the stratosphere comes the tropopause, which at a height from 5 to 16 km separates off the troposphere (Greek *tropē* = turning, changing, altering). Temperature in the troposphere decreases very rapidly with height, from about 20° to −55°C, showing that heating is applied mainly from below. At the same time, the troposphere includes noteworthy bars to incoming radiation, in the form of ozone, water vapour, clouds, dust, and even the molecules of atmospheric gas. It is within the troposphere that most of our weather is generated. The fundamental reason for this is that the troposphere contains almost the whole of atmospheric water vapour.

14.3b Bars to outgoing radiation

Whereas incoming radiation is strongly concentrated between the wavelengths of 0.2 and 2 microns, and within this range inside the visible band close to 1 micron, outgoing radiation from the earth spreads itself mainly through wavelengths from 5 to 25 microns in the infra-red waveband, with a concentration between 10 and 15 microns. Now, the atmosphere, and especially the troposphere, is only slightly transparent to radiation from earth. There are two minor windows, through which reflected and radiated energy can escape, but the main effect is that earth radiation is absorbed by the atmosphere, and that a vigorous energy exchange takes place between the lower atmosphere and the earth's surface.

14.4 The energy balances

Discussions of the energy balances of the atmosphere and of the earth's surface commonly deal with the disposal of incoming radiation, as this is received at the outer limit of the atmosphere. In reality, we still know very little about the disposition of energy at very high atmospheric levels, apart from what has been stated. At low levels, we do have firm records of drastic temperature changes, some spread over millions of years and others over tens of thousands. Instrumental records can detect changes of 1°C in fifty years for some midlatitude areas. But so far, all the changes have proved reversible. There is no evidence that, in the very long term, the earth's atmosphere is becoming progressively hotter or progressively colder. The same applies in the very short term. We can assume, therefore, whatever complications we build in, that incoming energy is balanced by energy going out. Nothing less could be expected. An increase in energy input, by exciting the earth–atmosphere system, should result in an increased output. Negative feedback again comes into operation in the establishment of an equilibrium situation.

The energy balances relevant to our weather and climate are struck in the lowermost atmosphere. Three-quarters of the whole atmosphere by mass is contained in the troposphere. Indeed, half by mass lies within 5.5 km of the earth's surface. It does not really matter whether we consider the disposition of energy at low atmospheric levels to begin at the stratopause or at the tropopause (Fig. 14.3).

Less than half the incoming radiation reaches the earth's surface. Only a quarter comes in directly. Another fifth arrives after being scattered. It is scattered radiation which can cause sunburn when the sky is overspread by thin cloud. More than half

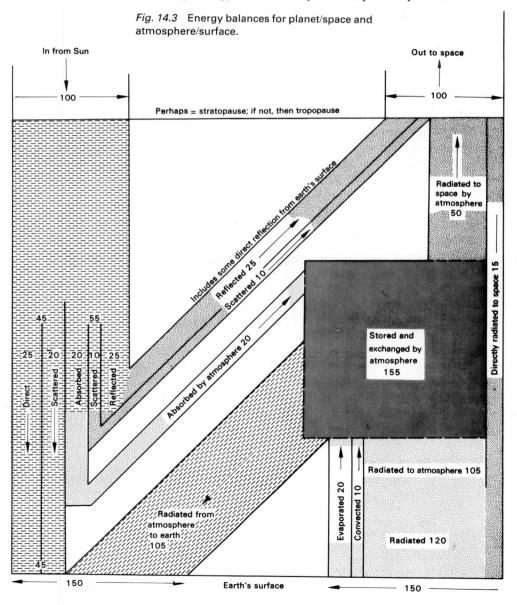

Fig. 14.3 Energy balances for planet/space and atmosphere/surface.

the incoming radiation is absorbed in the atmosphere, or is reflected or scattered back into space.

Taking the total incoming radiation as 100 (= 100%), we see that 45 units fall on the earth's surface, 20 are absorbed by the atmosphere, and 35 are returned to space. Add to the returned 35 the 50 radiated by the atmosphere (mainly by clouds) and the 15 escaping through a radiation window, and output equals input: 35 + 50 + 15 = 100.

Depending on where we start, the next piece of calculation may seem like an arithmetical trick, but actually is not. The atmosphere radiates back to earth 105 units, making the total input to the surface 150 units – 105 from the atmosphere, 45 direct or scattered from space. The output from the surface must also amount to 150 units. It is composed of the 15 escaping units already listed, 105 units absorbed by the atmosphere, 10 units carried into the atmosphere by the convection process, and 20 units transferred by evaporation. The lower atmosphere acts as a system of storage and exchange, with more energy being exchanged between atmosphere and earth than comes in from space. The fairly free entry of incoming radiation, mainly in the waveband of visible light, and the trapping of outgoing radiation in the infra-red waveband, is called the greenhouse effect. To a considerable extent, the glass of a greenhouse does what the lower atmosphere does – lets in light but keeps in heat, and so keeps up the interior temperature by means of heat storage.

14.5 Controls on radiation input at the earth's surface

In the broad view, the screening-out of solar radiation depends on the effective thickness of the atmosphere. This thickness increases as the sun's rays become increasingly oblique to the surface. In middle latitudes, the noonday sun is far less powerful in winter than in summer, because the winter rays traverse the atmosphere very obliquely. Similarly, the morning and afternoon rays are less powerful

than the noonday rays. Low suns are typically red, because energy in the wavebands that include blue light has been blocked out.

The effectiveness of energy received at the earth's surface depends not only on filtering by the atmosphere, but also on how widely the incoming energy is spread. Rays striking the surface at a low angle have not only come through a great thickness of atmosphere: their energy is also spread over a large area.

Take the simple case where the noonday sun is overhead at the equator (Fig. 14.4). The effective thickness of the atmosphere is least at the equator and greatest at the poles. The angle at which solar radiation strikes the earth's surface is 90° at the equator and 0° at the poles.

Fig. 14.5 is drawn on the assumption that all radiation bars that influence weather at low levels are contained below a height of 50 km, in the stratosphere and the troposphere. It also assumes a certain relationship between effective atmospheric thickness and the influence of radiation bars. But whatever detailed modifications could be inserted, the general effect is beyond dispute. The effective thickness of the atmospheric filter reduces incoming radiation by half, at about the polar circles. The added effect of the widening spread of radiation, as latitude increases, means an additional reduction by a factor of one-half, at the latitude of 45°. The combined effect is to reduce incoming radiation per unit area of surface at 40° of latitude to half the equatorial value, and to zero at the poles.

14.5a Variation in reflectivity

The discussion so far has taken no account of the various types of surface on which incoming radiation falls. In the real world, there are wide variations in the percentage of radiation that is reflected. This percentage is called the albedo. Fig. 14.6 illustrates

Fig. 14.4 Spreading of rays at the surface as obliquity of incidence increases.

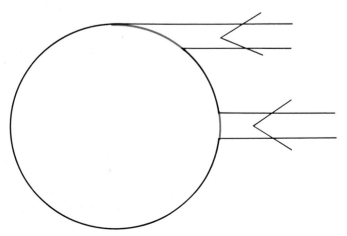

A bundle of rays falling on this area vertically would spread over this area at an angle of 1°; intensity is reduced to less than 2%

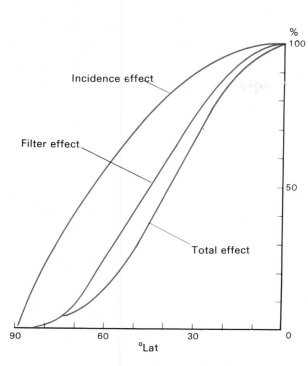

Fig. 14.5 Incidence effect, filter effect and total effect, on receipt of radiation at the surface.

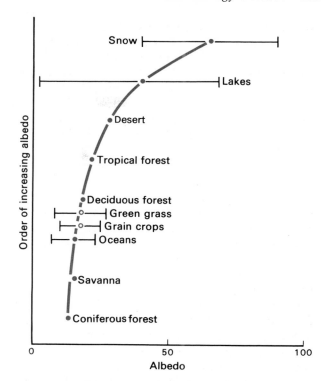

Fig. 14.6 Variation in albedo for various kinds of surface: 0 = no reflection, 100 = total reflection; bars show ranges.

the variation in albedo for various kinds of surface. In the range from coniferous forest to tropical forest, the albedo runs at quite modest levels. Even for tropical forest, it corresponds to less than 10 of the 45 units that arrive in either direct or scattered form. The fairly high albedo of deserts results from the reflectivity of bare rock and dry sand. Lakes are important as reflectors on a regional scale only when they cover a large fraction of the total surface, as they do in the Canadian north. The most powerful widespread reflector is a cover of snow, for which the albedo value can run as high as ninety. At the extreme, less than 10% of incoming radiation is used in the melting process. This fact has prompted suggestions of decreasing the albedo of the north polar ice, as for instance by sprinkling it with coal dust, and so causing the ice to absorb enough energy for melting to come about.

14.6 Heat exchange across latitude

For the lowermost atmosphere as a whole, incoming radiation exceeds outgoing radiation below the 40th parallels of latitude. There is an excess heat supply. Between the 40th parallels and the poles there is a radiation deficit. But the tropics and subtropics do not become progressively warmer, any more than areas in middle and high latitudes become progres-

sively colder. Some kind of energy exchange takes place. Heat is transferred from low to high latitudes by winds (about two-thirds) and by ocean currents (about one-third).

14.6a Deflection of heat transfer

If the earth did not rotate on its axis, the transfer of heat from equatorial to polar regions would be extremely simple. It would resemble the flow of water through a pipe connecting two containers with contrasting water levels – a flow from high to low. But heat transfer on a rotating sphere works differently. The winds and ocean currents that transport heat are deflected by the earth's rotation. The deflecting effect is called the Coriolis (pronounced Coreeolee) effect, after the French mathematician who first defined its theory in full. In practice, we speak of the Coriolis force, although strictly speaking no force is involved; but it simplifies calculations to assume that a deflecting force is actually at work. The Coriolis force, then, deflects winds and ocean currents to the right in the northern hemisphere, and to the left in the southern hemisphere.

The earth spins toward the east. The equatorial radius is some 6000 km, the equatorial circumference 37 700 km. A point on the equator is rotating at 1570 km/hr. At 45° latitude, the radius of the parallel is down to 4240 km, the circumference down to 26 660 km, and the rotational speed down to 1110 km/hr. Now fire a rocket due north from the equator. For the sake of convenience, take its average ground speed as 9425 km/hr, so that it will take exactly an hour to reach a target on the 45th parallel. Its path will have an eastward component, the difference between the 1570 km through which the launching site has rotated and the 1110 km travelled by the target (Fig. 14.7). If the rocket had been fired from the 45th parallel to the equator, the deflection would have been toward the west.

14.7 Major systems of pressure and winds

The Coriolis force enables systems of high and low pressure to exist. If winds were not deflected, air could flow straight in and prevent a low from developing, or straight out, draining a potential high. Deflection tends to make air blow round pressure systems.

Fig. 14.8, drawn for a northern-hemisphere situation, shows how. The wind is raised by the difference between high and low pressure. The pressure difference sets up the pressure gradient force, which drives the air from high pressure and toward low pressure. The deflecting force, acting at right angles to the direction of flow, turns the wind from its straight-line path from maximum to minimum pressure (Fig. 14.8a). As the wind direction changes, so does the direction of the deflecting force. An equilibrium state is only reached when the wind is blowing at right angles to the pressure gradient (Fig. 14.8b). This part of the diagram represents conditions above ground, at levels high enough for the influence of surface friction to be negligible. In reality, surface friction makes itself quite strongly felt at low levels. The outcome is shown in Fig. 14.8c. Friction acts in the opposite direction to that of airflow. When the combined effect of friction and of the Coriolis force is equal to, and diametrically opposed to, the pressure gradient force, an equilibrium state exists. The wind direction becomes the resultant of the pressure gradient force and the Coriolis force, and the wind blows obliquely across the direction of the pressure gradient. Over land, winds at low levels traverse the pressure gradient at an angle of about 30°. Over the ocean, where friction is much less than over land, the angle is about 10°. The flow of air at low levels into a system of low pressure is compensated by lifting within the system and by outflow aloft. Conversely, the low-level outflow of air from a high is compensated by subsidence within the system.

14.7a Tropical and subtropical systems

The equatorial belt is characterised by generally low pressure. The strong excess of heating causes the air to expand, to rise, and to flow outward at high levels in the upper troposphere. Air is drawn in at low levels. The inflowing winds are the Trades, which like all winds are named for the direction they come from, not for the direction in which they blow. The inflow of the northern hemisphere constitutes the northeast Trades, that of the southern hemisphere the southeast Trades.

The equatorial belt is rainy (Fig. 14.9). Rain is

Fig. 14.7 The Coriolis effect, illustrated for deflection to the right in the northern hemisphere.

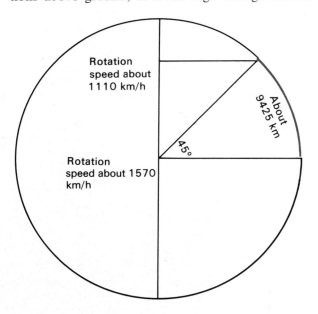

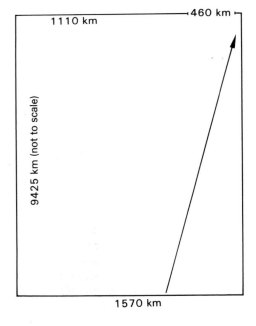

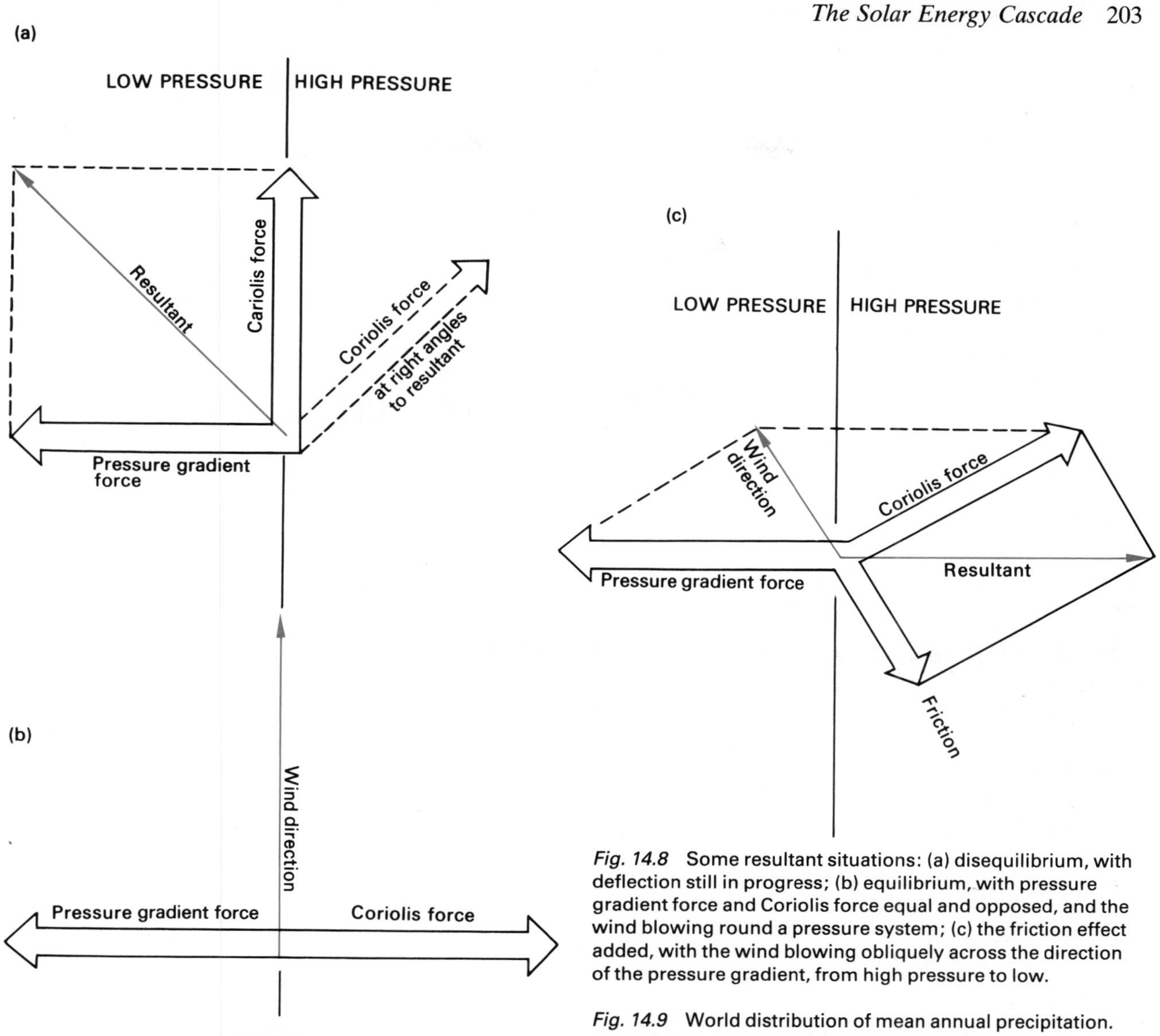

(a)

LOW PRESSURE | HIGH PRESSURE

Coriolis force

Resultant

Coriolis force at right angles to resultant

Pressure gradient force

(b)

Wind direction

Pressure gradient force | Coriolis force

(c)

LOW PRESSURE | HIGH PRESSURE

Wind direction

Coriolis force

Pressure gradient force | Resultant

Friction

Fig. 14.8 Some resultant situations: (a) disequilibrium, with deflection still in progress; (b) equilibrium, with pressure gradient force and Coriolis force equal and opposed, and the wind blowing round a pressure system; (c) the friction effect added, with the wind blowing obliquely across the direction of the pressure gradient, from high pressure to low.

Fig. 14.9 World distribution of mean annual precipitation.

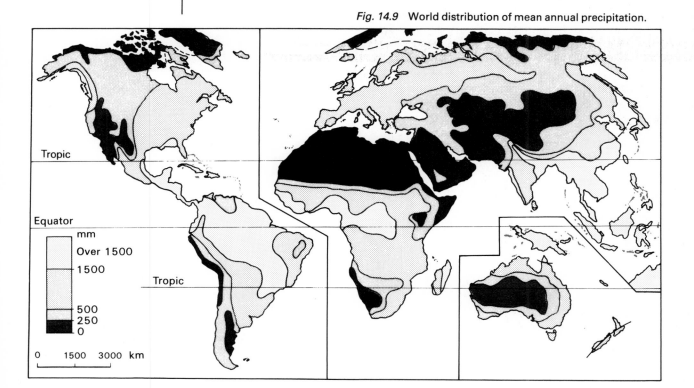

Tropic

Equator

mm
Over 1500
1500
Tropic
500
250
0

0   1500   3000 km

supplied by thunderstorms, by rather weak individual systems of low pressure, and by activity along the intertropical convergence. This is a narrow belt, usually located close to the equator, where winds from the two trade systems can come together. The strong probability is that the northern air will be less warm, and therefore more dense, than the southern air, or that the opposite condition will hold. As a result, lifting of air and rainfall is frequent along the intertropical convergence.

Within the trade-wind belt, rainfall on the western sides of continents is low to negligible, because the winds blow offshore and because offshore waters are usually cool (Chapter 17). On the eastern sides, where winds blow onshore and the offshore waters are typically warm, the amount of rainfall is much greater. The contrast is best seen in the southern continents. It is true that very low rainfall characterises the Pacific coastland on the U.S./Mexican border, and also the Atlantic coastland at the western extremity of the Sahara; but the irregularities of land-and-sea distribution confine rainy trade wind climates to the Caribbean Islands in the Atlantic, and replace the trade winds over the lands of eastern Asia by monsoonal systems.

Being generated by heat, the equatorial low-pressure system is recognised as a thermal low. Straddling the tropics come two belts of high pressure known as the subtropical highs. Propagated downward from above, these systems are dynamic highs. They feed the trade winds on their equatorward sides, sustain westerlies on their poleward sides, and are primarily responsible for the aridity of low-latitude deserts.

In common with pressure systems of middle and high latitudes, the subtropical highs vary in intensity and location from one part of the year to another.

Fig. 14.10 Cause of high-sun and low-sun seasons in areas outside the tropics.

The tilt of the earth's spin axis against the plane of orbit ensures that all parts of the world between the tropics and the poles experience a high-sun season and a low-sun season (Fig. 14.10). Because a land surface heats and cools more quickly than does an ocean surface, the major pressure belts are considerably affected by land-and-sea distribution.

In the high-sun season of the southern hemisphere, the subtropical highs are displaced somewhat to the south. They retract from the heated lands to the oceans, while systems of low pressure appear overland on the map of averages (Fig. 14.11). In the southern-hemisphere low-sun season, the subtropical highs link up as a continuous band, although the highest pressures of all occur in cells located over the oceans.

In the high-sun season of the northern hemisphere, a belt of subtropical high pressure reduces itself to two major cells, located over the Atlantic and Pacific oceans, while a rather weak low appears on the average map in the southwest U.S. and northwest Mexico, and a vast and intense low forms over Asia. In the northern low-sun season the high-pressure cells join in a continuous band, but the greatest single cell is located over eastern Asia, too far north to be regarded as subtropical.

14.7b Monsoonal systems

Because it is so large, and because much of it is located in middle latitudes, the Asian landmass is able to disrupt the subtropical and mid-latitude systems of pressure and wind. Systems of low pressure formed in the high-sun season draw air in. Systems of low pressure formed in the low-sun season feed air out. The seasonal winds are the monsoons. The climates which they dominate are monsoon climates. Low-sun seasons are typically dry, high-sun seasons typically rainy.

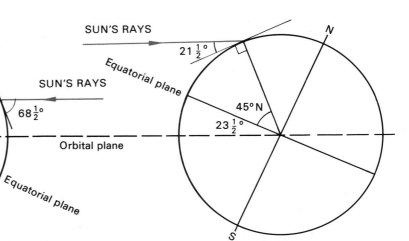

NORTHERN SUMMER   NORTHERN WINTER

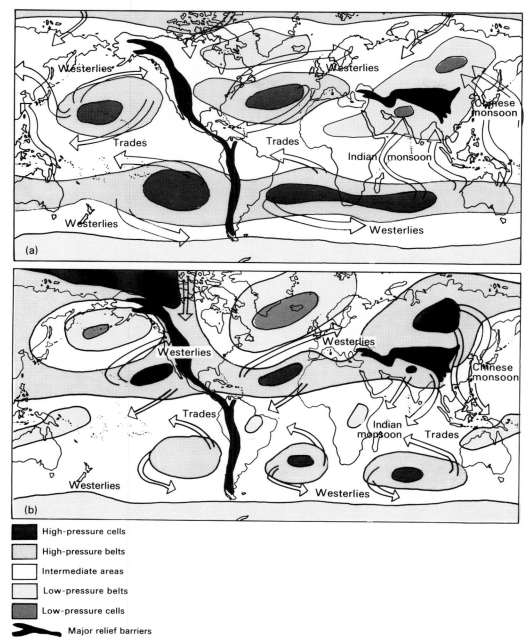

High-pressure cells

High-pressure belts

Intermediate areas

Low-pressure belts

Low-pressure cells

Major relief barriers

Fig. 14.11 Highly diagrammatic array of major cells and belts of high and low pressure, with associated wind systems (a) northern mid-summer; (b) northern mid-winter. Major relief barriers also shown.

The Indian monsoonal system is separated from the monsoonal system of eastern Asia by the relief barrier of the Himalayas and the plateaus of Tibet. This barrier reaches 6 km above sea level, effectively dividing the lower tropospheric circulation of India from that of the remainder of the continent. In the low-sun season, the northeastern part of the Indian subcontinent is occupied by a cell of high pressure, part of the subtropical band. The very dry subsiding air in this cell is responsible for the climatic state of the Rajasthan desert. The outflow at low levels is part of the trade wind circulation.

In the high-sun season, intense heating replaces high pressure by low. Although the Rajasthan desert experiences no rainy season, the rest of the subcontinent does. The flow of the northeast Trades is reversed into that of the southwest monsoon; and the shift of circulation is enough to bring the intertropical convergence over the subcontinent. The relief barrier to the north prevents inflow from that side, regardless of pressure conditions over eastern Asia.

The main Asiatic monsoon is driven by the winter high of Siberia and by the thermal low of the high-sun season. The monsoonal circulation extends across the equator. The summer inflow originates in the high-pressure cell generated over Australia in that continent's low-sun season, while the winter outflow drains to low pressure over northern Australia and the adjacent low-latitude ocean, or to the seasonal low of the northern Pacific (Fig. 14.11).

At least one-third of the world's people live in the rainy climates of the Asiatic monsoons. In India and China, the primary support of human existence is intensive subsistence agriculture, which depends completely on the monsoon rains. Its most distinctive – but by no means exclusive – form is wet-rice cultivation.

14.7c Midlatitude systems

Antarctica excepted, there is very little land in the southern hemisphere beyond 40° of latitude. This part of the discussion will therefore be limited to what happens in the northern hemisphere.

The midlatitude areas are often called the westerlies belts, because large parts of them record, on the average, a low-level flow of air from west to east. This low is sustained in part by the subtropical highs, and in part by temporary pressure systems developed within the westerlies belt itself. Such temporary systems tend strongly to move westward, in accordance with the total flow (Chapter 15). Apart from the Asiatic monsoonal system, the average January map for the northern hemisphere shows two well-defined systems of low pressure over the northern oceans. The average July map shows weak development of low pressure over the northern lands, the Asiatic monsoon again excepted. More than anywhere else, however, these averages represent reduction to order by brute force. For immediate purposes it is enough to notice that the average low-level flow of air in the westerlies belt contains a poleward component. This is how heat is mainly transferred from lower latitudes towards higher. Ocean currents act in the same directions as do the winds. Both over the northern Pacific and over the northern Atlantic, and also in the adjacent coastlands, air temperatures at low level are very high for the latitude.

In the middle latitudes occurs the third great concentration of precipitation on the globe (Fig. 14.9). In North America and South America, westerlies blowing from the Pacific encounter relief barriers. Precipitation falls on the coastland and on the adjacent mountains. The water vapour that supplies rain and snow to the North American continent east of the Cordillera comes very largely from the Atlantic and the Gulf of Mexico. Europe opposes no continuous belt of high mountains to the westerlies. Penetrating far eastward across the Eurasiatic Plain, the dominant winds keep annual precipitation as high as 200 mm/yr at distances of 7500 km from the Atlantic. It is distance, not elevation, that

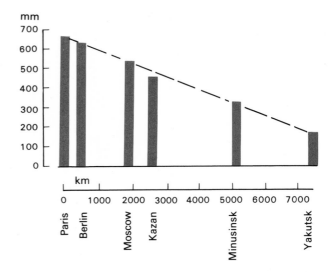

Fig. 14.12 Midlatitude decrease of mean annual precipitation across Eurasia, with increasing distance from the Atlantic: the amount/distance relationship illustrated is almost exactly linear.

controls the eastward diminution (Fig. 14.12).

The main expanse of dense population in North America coincides roughly with the main concentration of precipitation east of the Cordillera. The main concentration of population in Europe (including the western U.S.S.R.) coincides quite closely with the main concentration of precipitation on low ground. It is of course immediately clear that the climate–population linkage here is something very different from the intensive subsistence agriculture of the monsoon lands. At the same time, the intensive commercial farming of Europe and North America has been developed precisely in those parts of the westerlies belt that receive adequate precipitation.

14.7d High-latitude systems

At the simplest, low-level conditions in high latitudes can be looked on as involving thermal highs. The energy deficit of the earth's surface means a chilling of the lowermost atmosphere, so that the air contracts and becomes denser. At the simplest again, thermal highs mean outblowing winds at the surface, that will be northeasterly in the northern hemisphere and southeasterly in the southern. At the same time, the net poleward transportation of air in the westerlies belts implies an accumulation in the polar regions. The accumulation results in low-level outbursts toward the equator. One outburst in July 1975 destroyed Brazil's coffee crop. Another in early 1977 brought snow, for the first time on record, to the Bahamas. In October 1979 heavy snow caused great damage to apple orchards in the Shenandoah

Valley. Some of the most spectacular effects, however, are produced when the cold air at low levels is overridden by warmer and moister air aloft; precipitation can then arrive as freezing rain which puts a thick coating of ice on every exposed surface (Fig. 14.13).

14.8 Circulation aloft

The analysis of major systems of pressure and circulation began, inevitably, with observations at low levels. The foregoing account treats the low-level situation. When we look at what happens in the upper troposphere and the lower stratosphere, a highly contrasted picture emerges. The major circulation is not two-dimensional, as shown by maps of surface pressure and low-level winds, but three-dimensional. It includes vertical circulation in addition to horizontal circulation.

No completely satisfactory picture has yet been drawn. Fig. 14.14 is one possible interpretation. In the vertical view, the fronts are major boundaries between pairs of cellular or turbulent systems. The net poleward transport of air in mid-latitudes operates both at low levels and above ground. Equatorward of the polar front, the vertical circulation is mainly cellular, with the trades and the westerlies constituting the low-level return currents. Poleward of the polar front, upward and downward turbulent movement is prominent.

Fig. 14.13 Shorting of electricity cables during an ice storm in early March 1976 (*Capital Times*, Madison, Wisconsin).

Fig. 14.14 A sectional model for the northern-hemisphere circulation of the troposphere.

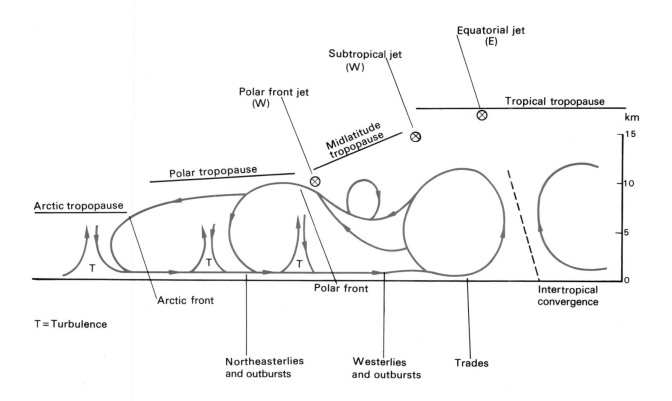

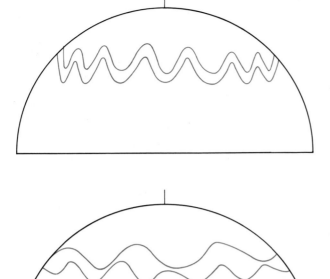

Fig. 14.15 In-phase and out-of-phase situations for the northern-hemisphere polar front jet: highly diagrammatic.

14.8a Jet streams

The tropopause in cross-section is both uneven and broken. At some times and in some places, the boundary between troposphere and stratosphere disappears. Near where this happens – and for reasons incompletely understood – wind systems of extremely high velocity are generated near the top of the troposphere (Fig. 14.14). These are the jet streams, which attain velocities of 400 km/hr or more, equal to the greatest speeds ever recorded at ground level. There are three jet stream systems. The equatorial jet blows from the east across southern Asia, during the season when the southwest monsoon is blowing in at low levels. Its nature is still quite obscure. The subtropical jet is associated with the break between the mid-latitude and tropical heights of the tropopause, the polar jet with the slighter break between the mid-latitude and polar heights. Whether or not there is an arctic or antarctic jet is not yet known. There may not be room for such jets in the general circulation.

The subtropical and polar jets consist of stream families, rather than of single streams. All jets are given to meandering, just as very many single-channelled rivers are (Chapter 8). When the wave trains of a family of jet streams are out of phase, their total oscillation is damped down. But when the waves are in phase with one another, the total

oscillation is amplified (Fig. 14.15). In-phase behaviour means that cold air will be transported in some areas toward the equator, while warm air will be transported toward the poles in other areas. The influence of the jet streams aloft is transmitted downward. The dominant flow direction – westerly when oscillations damp one another out, and highly variable from northerly to southerly when the oscillations reinforce one another – greatly affects weather at low levels. Also, the meander curves of the mid-latitude jets, and especially of the polar jets, strongly influence the generation of the temporary systems of low or high pressure in the midlatitude belt of westerly airflow at low levels (Chapter 16).

Chapter Summary

A common *measure of wavelength of solar radiation is the micron*, 1/1000 mm. *Radiation* striking the earth –atmosphere system *is strongly concentrated between 0.2 and 2 microns*, a waveband which *includes visible light*. The *unit of radiation is the langley* – one langley is the amount of energy needed to raise the temperature of water through 1°C. *At the outer limit of the earth's atmosphere*, solar *radiation has been diluted to 2 ly/min.*

Above about 2000 km, the atoms of *atmospheric gases are dissociated into particles*. The *nitrogen –oxygen mixture* (with other minor contents) of our own familiar air *extends up to about 115 km.*

Ionised layers in the range 80–300 km *reflect radio waves*, making possible radiocommunication round the earth's curve; they also *produce the auroras* under the influence of particles ejected from solar flares.

The *outer atmosphere absorbs ultra-violet and cosmic rays*, which heat it: hence the name *thermosphere* for the air above about 80 km. *Between 80 and 50 km lies the mesosphere*, heated mainly from below.

The *stratosphere*, between 50 km and a variable level averaging about 11 km, *contains nearly all atmospheric ozone*, which by *absorbing ultra-violet rays* ensures that the *stratosphere is heated mainly from above*. The *troposphere*, the atmospheric shell next to the earth's surface, *contains nearly all* the atmosphere's *water vapour and cloud*. Most of our *weather is generated in the troposphere.*

Outgoing radiation from the earth is *concentrated between 10 and 15 microns*, in the infra-red waveband. The *troposphere especially* is only *slightly transparent* to *radiation from earth*.

In the very long term and also in the very short term, *incoming energy is balanced by outgoing energy*. The *atmosphere* acts as an *energy filter* and *energy reflector*, so that *only about half the incoming energy reaches the earth's surface* directly. However, the atmosphere *also* acts as *an energy store and a heat exchanger*. Thus, if incoming radiation from the sun is given the value of 100 units, 155 units are stored and exchanged by the atmosphere, 50 being radiated out to space, and 105 back to earth.

The amount of *radiation received at the earth's*

surface depends on the *length of travel through the atmosphere* by the sun's rays, and on the *angle* at which the rays strike the surface. Both effects are *functions of the height of the noonday sun*; the angle of incoming rays is of course also affected by the time of day.

Absorption of incoming radiation can be greatly *affected by albedo*, the percentage reflectivity. A *snow cover can have a very high albedo*.

Radiation excess in low latitudes and radiation *deficit in high latitudes imply horizontal transfer*. This is effected by winds and ocean currents.

Winds and ocean currents are deflected by the so-called *Coriolis force*, to the right in the northern hemisphere and to the left in the southern. The Coriolis force (more properly styled an effect), *causes winds to circulate round* areas of *high or low pressure*, and thus enables these to persist. *At very low levels, friction permits inflow* to low pressure *or outflow* from high pressure.

The *contrasts of* incoming *radiation and* the *Coriolis effect* combine to *produce the global circulation*. This includes the *equatorial belt of thermal low*, which is rainy, with *much rain produced along the intertropical convergence*. *Trade winds* blow into the belt of equatorial low pressure from the *subtropical belt* (often, a belt of cells) *of dynamic high pressure*. The *intensity* of the subtropical highs *varies with the seasons*.

Monsoonal systems disrupt the general circulation. They draw in oceanic air in the high-sun season, and deliver continental air in the low-sun season. The *Indian and Asiatic monsoonal systems are divided by a relief barrier*.

In middle latitudes, dominant airflow at low levels is west to east. The westerlies are fed by outflow from the subtropical highs, and by temporary pressure-systems generated in the westerlies belts themselves.

The world's major concentrations of people occur either in the monsoon lands of Asia or in the rainy parts of the westerlies belts. In the one, food is provided by intensive subsistence agriculture; in the other, by intensive commercial farming.

Airflow at low levels in high latitudes includes outbursts of cold air that can extend across the tropics, sometimes with disastrous results. These outbursts are the return flows of air driven toward the poles in the westerlies belts, at whatever level. *No completely satisfactory picture has yet been drawn for the three-dimensional circulation of air in middle and high latitudes*.

The *boundary between* the *troposphere and* the overlying *stratosphere is the tropopause*. It is *often broken. Near the breaks*, winds of very high velocity develop. These are the *jet streams. In middle latitudes*, the jet streams come in *families*. They also *tend to meander*. In-phase meander amplifies the oscillation. Out-of-phase meandering supresses it. Thus, *weather at low levels is powerfully affected* by the dominance of west–east or of north–south – south –north flow.

15 Atmosphere Process-Response Systems: Large Scale

Large-scale process-response systems of the atmosphere include the general circulation discussed in the previous chapter. The present chapter will deal with the structured sets of attributes that we recognise as types of climate.

The list of attributes that could usefully be examined is dauntingly long. The most important items are temperature, precipitation, incoming radiation, absorption and reflection of the radiation, winds, sunshine, cloudiness and evapotranspiration. For each item, daily, monthly, seasonal and annual variation could be examined. Merely to keep the operation manageable, we are obliged to treat each climatic subsystem as a cascade, and to concentrate on a few of its main attributes in discussing process-response.

15.1 Different kinds of climate

Since the earth's atmosphere is a single fluid system, it might at first seem surprising that there is not a steady decline in temperature, away from the equator and toward the poles. It might also seem surprising that quite abrupt changes occur in mean annual precipitation, as for instance away from a rainy coastland toward a parched interior desert. But in the real world, quite sharp step-functional change of climatic type is at least as common as gentle gradation.

There are two reasons why this should be so. High ground can form sharp climatic barriers. Mountain ranges, drying out the winds, can separate a rainy climate on the seaward side from an arid climate to landward. High ground, however, serves only to complicate the general picture. The air of the troposphere contains boundaries of its own.

These boundaries, aptly named fronts (Chapter 16) separate one kind of air from another. Instant mixing simply does not occur. Dry cold air on one side of a front just does not fuse straightaway with wet warm air on the other side. Although the situation in the atmosphere cannot be directly seen by eye, it can be simulated in a liquid medium. Try gently squeezing a jet of ink into a vessel of still water. The ink-stained water remains distinct from its clear surroundings for a considerable time.

15.1a Preferred frontal positions

Fronts in the lower air tend to occupy and re-occupy certain positions. Their average annual or seasonal positions can be mapped. On one side of an average position, one type of air dominates. The region on the other side is dominated by air of a contrasted kind. Wherever this happens, there is liable to be a sharp change in wild vegetation across the preferred line.

The circumstance has been best studied for North America east of the Cordillera. In the interior of the continent there is very little ground more than 500 m above sea level, all the way from the Arctic Ocean to the Gulf of Mexico. Nevertheless, the natural vegetation undergoes changes that, in the small-scale view, are distinctly sharp.

The northern tundra – dominated by lichens, mosses, and low shrubs – gives way southward to the boreal (= northern) forest, with its needleleaf evergreens, larches, birches and willows. The boundary between tundra and boreal forest closely matches the average summer position of the arctic front (Fig. 15.1). The southern limit of the boreal forest closely matches the average winter position of the arctic front. Next southward comes mixed hardwood forest, where oak and maple are increasingly prominent. This belt lies north of the average position of the arctic front in late winter.

To the south and southeast lie the southern woodlands, where internal distribution of hardwoods and needleleaf evergreens is largely a matter of soil type, These woodlands lie southeast of the average winter extent of Pacific air, which forms part of the system of westerlies. Between the average winter limit of Pacific air on the southeast, and the average winter position of the arctic front on the north, there comes a great wedge of natural grassland.

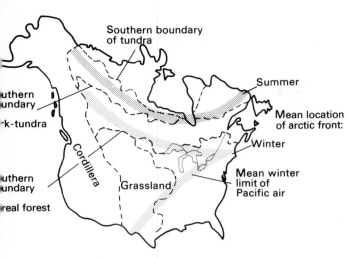

Southern boundary of tundra

Summer

Mean location of arctic front:

Winter

uthern undary

k-tundra

Cordillera

Mean winter limit of Pacific air

uthern undary

Grassland

real forest

Fig. 15.1 Some preferred frontal positions and major vegetational boundaries, for North America east of the Cordillera (selectively adapted, and considerably simplified, from the work of R.A. Bryson, W.R. Barreis, and W.M. Wendland).

15.1b Air-mass dominance

The frontal positions mapped in Fig. 15.1 separate one kind of dominant air mass on the one side from another kind on the other side.

An air mass is a large body of air, which varies little in physical properties throughout its extent. Most air masses form (= acquire their characteristics) at low levels in the troposphere. The developing air mass stagnates or circulates over a particular part of the earth's surface, long enough to reach equilibrium with that surface. Equilibrium mainly concerns temperature and humidity. A given air mass will be cold or warm, dry or humid, throughout its extent, according to the character of the underlying surface. A low-level air mass will be perhaps 3000 km across and 5000 m deep, and will reach equilibrium with the underlying surface in a week or ten days.

The area where an air mass forms is its source region. Certain source regions produce air masses again and again. When winds blow outward from the source regions, they tend to follow particular paths. They export dryness, humidity, heat, or cold, as the case may be. Repeated formation of air masses in particular source regions, and repeated export of particular weather conditions, permit certain types of air to dominate various parts of the earth's surface.

The simplest examples of low-level air masses involve the winter systems of high pressure over the northern continents, and the winter systems of low pressure over the oceans in high latitudes. The high-pressure systems discharge dry, frigid air toward the equator. The low-pressure systems dis-charge moist air, that can be warm for the season, on to the exposed coastlands in the belt of wester-lies.

The subtropical high-pressure systems are domin-ated by air that sinks from aloft. Conditions at ground level in the subtropical deserts can be re-garded as reaching equilibrium with the subsiding air.

15.1c Air masses and the general circulation

The primary classification of air masses identifies the type of source region. This can be either continental, c, or maritime, m. Continental air masses form over land or over a frozen ocean. They are typically dry. Maritime air masses form over the open ocean. They are typically humid.

The secondary classification locates an air mass within the general circulation. Arctic and antarctic air masses, A and AA, form poleward of the arctic and antarctic fronts. Polar air masses, P, form between the polar and the arctic/antarctic fronts. Tropical air, T, is produced mainly by the subtropical high-pressure systems. The word tropical is used quite loosely in this connection, being applied to any air mass that forms between the polar front and the equatorial belt of low pressure. Equatorial air, E, forms inside that belt.

Where it seems useful to indicate that the origin of a given air mass involves subsidence, the letter S can be added. Similarly, if an airstream is part of a monsoonal circulation, the letter M can be applied. Airstreams are often warmer or colder than the surfaces over which they flow. The letters W and K (German *kalt* – cold) express the relationships in question.

To give examples: the winter systems of high pressure over the northern continents are sources of cPK and cAK air (continental polar cold, continen-tal arctic cold). The winter system of low pressure over the northern oceans promote the interplay of mTW air (maritime tropical warm) and mP air (maritime polar). What the mP air is like when it strikes land depends on where it has been. If it has looped widely toward the equator before once more turning poleward, it will arrive as mPW (maritime polar warm), whereas a direct trajectory will allow it to remain mPK (maritime polar cold). The subtro-pical deserts are dominated by cTS air (continental tropical, subsiding), which blows out on the pole-ward sides as cTW airstreams (continental tropical, warm). The regions of equatorial rainforest are dominated throughout the year by E air. The mon-soon systems of eastern Asia suck in $mEWM$ air in summer (maritime equatorial, warm, monsoonal), although a long passage can convert this to $mTWM$.

Table 15.1 A Classification of World Climates

CLIMATIC TYPES		DOMINANT AIR MASS TYPES		DOMINANT WINDS AND WIND CHARACTERISTICS	
		summer	winter	summer	winter
LOW LATITUDE	EQUATORIAL	E		weak, variable	
	TROPICAL	E	cT	weak, variable	trades
TRADE WIND	HUMID	mT, cT, mP	mT, mP	trades	trades to westerlies
	DRY		cT	trades to weak and variable	
SUBTROPICAL	WEST COAST	cT	mT	weak, variable	westerlies
	EAST COAST	mT	cP to mT	monsoonal inflow	westerlies
MIDLATITUDE	WEST COAST	mT	mP, cP	westerlies	
	TRANSITIONAL	mT, mP, cT	mP, cP	westerlies	westerlies to weak and variable
	WARM CONTINENTAL		cT	often weak and variable	
	COOL CONTINENTAL	mP, cP	cP, cA	westerlies	weak, variable
	EAST COAST	mP, mT	mP, cP	westerlies	
HIGH LATITUDE	SUB-POLAR	mP, cP, cA		westerlies to weak and variable	
	POLAR	cA, cP, mP		highly variable both in direction and strength	
MOUNTAIN		not applicable			

These monsoon systems pour out continental air in winter. For the Indian monsoon, it is *cTKM* air. For northeast Asia, it is *cPKM* air.

15.2 Climatic systems boundaries

If the rest of the world could be treated as eastern North America is treated in Fig. 15.1, the task of fixing the boundary between one climatic system and another would be far more satisfactory than, in practice, it actually is. As yet, however, either the necessary information does not exist, or it has not been duly analysed. We are thrown back mainly on actual records of temperature and precipitation.

Many schemes of climatic classification have been proposed. Anyone is as free to construct one of these schemes as to devise a new map projection. Two schemes have in fact proved highly influential, those of W. Köppen and of C. W. Thornthwaite. The latter based his classification on levels of potential evapotranspiration (compare Chapter 5, section 5.7 and Fig. 5.13). Köppen worked back from major vegetational boundaries to the temperature/precipitation relationships which seemed to control those boundaries. Another approach however is possible, to construct a genetic classification which depends primarily on the place of a particular climatic type in the general circulation, and to relate climate, insofar as it may be possible, to kinds of air-mass dominance. The genetic classification, pioneered by H. Flohn and transferred to the basis of a world map by E. Neef, underlies the pattern of distribution shown in Fig. 15.2 and the listing of climatic types in Table 15.1.

15.2a Qualifications and elaboration

While based primarily on the progression through the circulation belts from equator to poles, the classification also takes into account the effect of land-and-sea distributions, which are particularly important in the range through tropical to midlatitude climates. In the absence of detailed information and analysis, such as has been mentioned, for types of air-mass dominance, the listings in Table 15.1 are to some extent inferential, although certainly well founded in most instances. Both because the classification is genetic, and because air-mass incidence cannot yet be analysed proportionately for the whole world, the central characteristics of the climatic types are more important here than are the numerical characteristics of the boundaries between one type and another. Individual stations (Tables 15.1, 15.2) have been selected to illustrate central characteristics.

Mountain climates, not being located with reference to the general circulation, stand apart from the rest.

CLOUD COVER summer winter	LEADING CHARACTERISTICS OR PRECIPITATION REGIME	ILLUSTRATIVE STATION
abundant	year round; single or double maximum	Singapore
abundant scanty	strong concentration in high-sun season	Kandi
moderately abundant	year round: single or double maximum	Sydney
rare	irregular; summer maximum in mean regime	Alice Springs
scanty abundant	strong concentration in winter	Rome
abundant, esp. in summer	year round; variable summer concentration	Macon
abundant	year round; winter maximum possible	Nantes
abundant moderately abundant	year round; summer maximum usual	Berlin
scanty	irregular; summer maximum in mean regime	Astrakhan
moderately abundant	strong summer concentration	Moscow
abundant	year round; early winter maximum possible	Eastport
variably dense to thin	summer concentration usual	Trout Lake
variable, often thin	summer concentration usual	Coppermine
		Säntis

Table 15.2   Mean monthly and annual temperature and precipitation for selected stations: temperatures are actual values, precipitation values are mainly rounded

		J	F	M	A	M	J	J	A	S	O	N	D	Yr
Singapore	°C	25.8	25.7	26.5	26.9	27.2	26.9	27.2	26.8	27.1	26.9	26.3	26.0	26.6
	mm	105	45	105	80	95	95	55	65	135	160	145	95	1200
Kandi	°C	25.0	28.0	30.8	31.7	30.9	28.0	26.4	25.4	26.2	27.9	26.9	25.1	27.7
	mm	0	1	5	30	70	145	190	335	215	55	5	0	1051
Sydney	°C	21.9	22.0	21.0	18.5	15.4	12.9	12.2	13.7	15.4	17.1	19.0	20.8	17.5
	mm	170	175	190	180	200	215	100	115	85	155	105	170	1860
Alice Springs	°C	27.9	26.5	24.3	19.5	14.6	11.2	12.4	15.4	18.3	22.1	25.0	26.7	20.3
	mm	20	90	30	5	15	3	10	3	3	25	25	57	286
Rome	°C	6.3	7.7	10.7	14.0	18.5	23.2	26.0	25.0	22.6	16.9	12.1	8.4	16.0
	mm	65	75	35	30	45	20	8	12	50	90	130	80	630
Macon	°C	10.3	10.6	13.9	18.9	22.9	26.9	28.0	26.9	22.7	16.7	12.9	10.9	18.5
	mm	95	100	165	100	100	80	150	105	75	65	70	125	1230
Nantes	°C	4.1	5.7	8.5	11.5	14.0	17.1	19.3	18.6	16.2	12.6	8.5	5.1	11.9
	mm	62	45	45	40	50	40	38	60	62	65	70	75	652
Berlin	°C	−1.8	0.0	4.0	9.0	14.2	17.0	18.9	18.1	14.8	8.8	4.7	0.6	9.0
	mm	50	35	30	55	50	60	90	70	55	55	50	40	640
Astrakhan	°C	−7.0	−7.0	−0.3	10.0	17.8	23.0	25.5	24.0	17.2	9.7	2.0	−4.0	9.2
	mm	9	12	12	10	22	17	17	13	18	20	13	12	175
Moscow	°C	−10.0	−8.2	−4.8	3.5	10.1	15.6	18.5	16.1	11.1	4.2	−2.7	−7.8	3.8
	mm	25	25	35	30	50	75	80	90	55	45	35	35	580
Eastport	°C	−4.7	−5.0	−0.5	3.7	8.3	12.3	15.0	14.5	13.0	9.0	3.8	−3.1	5.5
	mm	70	60	70	55	55	70	70	75	80	60	85	70	820
Trout Lake	°C	−23.1	−21.7	−13.9	−5.1	3.9	11.2	16.0	14.4	9.2	1.4	−8.9	−19.6	−3.0
	mm	30	30	20	30	50	80	110	105	80	57	57	35	684
Coppermine	°C	−27.3	−29.1	−24.0	−17.1	−5.1	3.5	8.9	8.0	2.5	−5.1	−14.0	−25.6	−10.0
	mm	20	10	20	15	30	20	40	40	35	35	20	12	277
Säntis	°C	−8.8	−8.5	−5.8	−2.9	0.7	3.8	5.9	6.0	3.8	0.0	−4.8	−7.7	−1.5
	mm	215	200	180	175	170	245	320	295	230	170	275	190	2665
Rangoon	°C	25.8	26.9	28.5	30.3	29.1	27.5	27.0	27.0	27.4	27.6	27.5	25.7	27.5
	mm	1	5	5	75	310	545	605	515	450	220	80	25	2836
Cherrapunji	°C	12.0	13.0	16.5	18.0	19.4	20.0	20.0	20.0	20.4	19.0	16.0	13.0	17.3
	mm	30	45	200	800	1700	2980	2400	1820	1180	350	70	2	11577
Verkhoyansk	°C	−47.2	−42.9	−29.0	−12.9	2.8	12.8	15.1	9.0	−1.4	−12.6	−35.7	−45.2	−15.6
	mm	5	4	4	4	8	30	35	20	10	10	9	7	146
Vostok	°C	−34.0	−44.0	−59.0	−67.0	−66.0	−67.0	−67.0	−64.0	−65.0	−58.0	−47.0	−33.0	−55.9

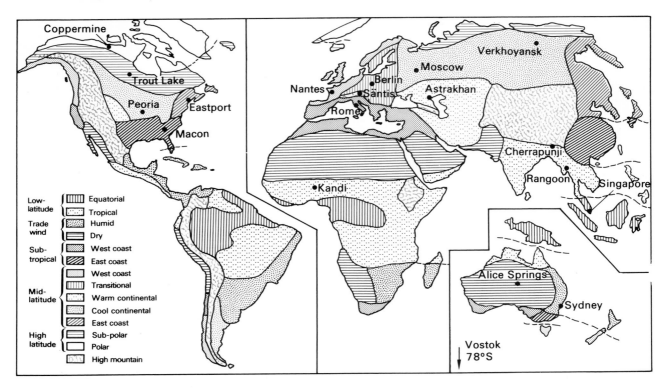

Fig. 15.2 Simplified distribution of climatic types, with locations of selected stations.

As shown, the degree of contrast between the high-sun season and the low-sun season varies much from one climatic regime to another. Both for very low and for very high latitudes, the terms summer and winter are misnomers. Lowland areas in low latitudes have no winters: those in very high latitudes have no summers. The terms are employed here merely for the sake of convenience. Certain cutoffs of mean monthly temperature have been retained from the Köppen classification, by way of indicating the occurrence or non-occurrence of a winter season, a hot summer, or a severe winter: but these cutoffs should be regarded as descriptive rather than classificatory.

In the graphs for individual stations, the lines for mean monthly actual temperature T, and mean monthly precipitation, P, are enclosed within bands of uncertainty. These bands, labelled only in Fig. 15.3, show the expectable variation within a twenty-five year period. The variation has been established by means of the analysis of magnitude and frequency (see Chapter 6, section 6.6), which has also provided the mean values used here. In most cases the monthly means are sensibly identical to the long-term arithmetic means, but mean monthly precipitation is mostly stated to the nearest 5 mm (Table 15.1). The graphical analysis of magnitude and frequency does not justify greater precision where values are large.

The information supplied for individual stations includes the soil-moisture balance, as computed from a comparison of mean precipitation with potential evapotranspiration.

Four stations listed at the end of Table 15.2 and located in Fig. 15.2 do not figure in the list of illustrative stations in Table 15.1. Their characteristics will be used to exemplify certain extreme developments.

15.2b The major subdivisions

The winterless climates of low latitudes occur within the range of E air, which is dominant throughout the year in equatorial climates proper, and which alternates seasonally with mainly continental air in the tropical variety of low-latitude climate. This part of the classification well illustrates the difficulty of selecting climatic names: the extent of trade-wind climates reaches well inside tropical latitudes. Moreover, the influence of the trade winds themselves declines markedly from the coasts inland, to be largely replaced in continental interiors by the subsiding air of seasonal to permanent highs.

Just as tropical climates are transitional between low-latitude and trade wind climates, so subtropical climates are transitional between tropical and midlatitude climates. Their intermediate character is best illustrated in west-coast situations, where the high pressures which elsewhere feeds the trade winds dominate in summer, but where the westerlies come in by winter.

Midlatitude climates constitute a large, extensive, and highly variable subset. It is to these that the term temperate is often applied. Their continental variants are however temperate only in the sense of being less hot in summer, and in some areas less arid, than climates of lower latitudes, and in being to some extent less cold in winter than climates of high latitudes. As is usual when large landmasses come into question, it is impossible to make a general statement about climatic characteristics and distributions without entering an immediate qualification. The world record low winter temperatures for low-level stations are recorded for climate of the cool continental type (see discussion of Verkhoyansk, below). Subclassification of the midlatitude climates is possible with reference to warmth and duration of summer and to the coldness and duration of winter, but the chief genetic differences among them relate to the relative dominance of the westerlies. These can be blocked out of continental interiors by high-pressure systems, mainly but by no means exclusively in winter. In addition, a poleward deterioration of climate is well displayed in North America, and a combined poleward and eastward deterioration in Eurasia. Westerlies are strongly dominant and often strong, both in west-coast and in east-coast situations, where they average out the rotary circulations and eastward drift of travelling lows.

Sub-polar climates are dominated by the westerlies, but are liable to incursions of colder air, especially in winter. In a sense, these again are transitional. Fig. 15.1 indicates that they occur poleward of the mean position of the Arctic front, both in summer and in winter, but the fluctuations of temporary frontal positions permit the incursion of westerlies from time to time. Predictably, rain-bearing winds are most easily able to enter in the high-sun season. Polar climates represent a further deterioration, with incursions of the westerlies decreasingly frequent, and with the seasonal concentration of precipitation largely suppressed in consequence.

Mountain climates are modifications of the climates of surrounding lowland areas. On any regional basis, they form special cases. Possible modifications include suppression of the seasonal regime of mean temperature, and either partial suppression or powerful exaggeration of the seasonal regime of precipitation. Selected cases both of mountain climates and of the gross modification of climate by land-and-sea distribution will be discussed below.

15.3 Climates of low latitudes

Singapore and Kandi (Figs. 15.3, 15.4) illustrate some of the principal manifestations of low-latitude climate. The noonday sun is never far from the zenith, and twice a year is directly overhead; its

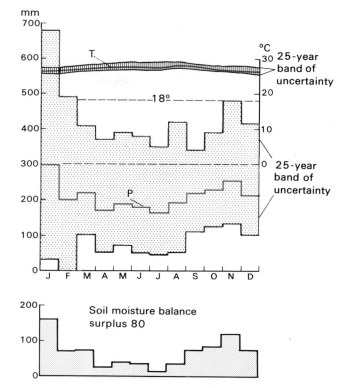

Fig. 15.3 Graphs for equatorial climate: Singapore.

minimum elevation at Singapore is 65°, and at Kandi 55½°. Nor is there marked variation in length of daylight between one part of the year and another. With temperature range small – although two maxima are possible, as at Kandi – and with no winter season, as indicated by the cutoff of 18°C for the coolest month, a very great deal depends on the precipitation regime. A double maximum is again possible here; but where a marked seasonal regime occurs, its significance lies basically in the incidence of a dry season, if any.

15.3a Equatorial climates

Dominated by *E* air, and with copious year-round rainfall, these climates can sustain rainforest. It is not common for any month to average much less than 60 mm. Rain comes from the numerous weak lows and the frequent thunderstorms associated with the intertropical convergence.

At the selected station, Singapore (Fig. 15.3), variation in mean monthly temperature is almost negligible, and is totally irrelevant to plant growth. The band of temperature uncertainty is very narrow. In the graph of rainfall, there is no real suggestion of seasonality. On the other hand, the rainfall expectancy is far more variable than anyone living in the middle latitudes, and cherishing the stereotype of constant rain and humidity, might be tempted to

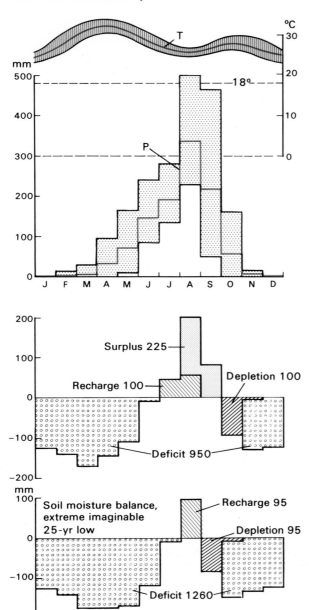

Fig. 15.4 Graphs for tropical climate: Kandi.

time, at Kandi it is strongly seasonal. Even in the worst twenty-five year case, there is still a marked peak in August. Seasonal concentration of rainfall means that soil/moisture relationships come strongly into play.

In the average climatic year, Kandi experiences only four months when precipitation exceeds potential evapotranspiration, and only two months when there is a soil moisture surplus. Streams in the region are seasonal only. Groundwater tables, and levels of well water, are subject to powerful seasonal fluctuations. The considerable net excess of soil moisture deficiency over surplus requires drastic adaptations on the part of plants. Growth is typically rapid in the wet season, with its dominant E air and thunder rains, but ceases for annuals in the dry season, when cT air blows in from the subtropical highs. To relate vegetation type to climate is perhaps as difficult in this case as in any, on account of the long-term impact of man-made fires; but, without human intervention, this climatic type would probably sustain dominant wooded savanna – tall tropical grass with interspersed drought-resistant trees.

Intermittent drought is a perpetual menace. The broad belt of latitude that straddles the 10° parallel of latitude in Africa, and that includes Kandi, was ravaged by persistent droughts during the later 1960s and the earlier 1970s. As shown by the band of uncertainty, the rainy season can be very variable. The worst imaginable twenty-five year situation involves too little rainfall to produce any surplus of soil moisture at all.

15.4 Trade wind climates

Next poleward from the regions of low-latitude climates come those of trade wind climates, abutting generally on regions of tropical climate. They experience real winters, but also hot summers, with mean monthly temperatures rising above 22°C (Figs. 15.5, 15.6; Sydney and Alice Springs). It is convenient to distinguish under this heading between humid climates and dry climates, even though the separation is not sharp everywhere. Humid trade wind climates are those with dominant on-coast winds, as widely developed in the Caribbean and on the eastern sides of South America, Africa, and Australia, whereas dry trade-wind climates include those of the tropical-subtropical deserts. On lowland coasts with onshore winds, mean annual precipitation can fall off very rapidly in the inland direction, for instance from 1500 mm or more a year to less than 500 mm, within a distance of 75 km. In reality, many trade-wind coastlands are backed by high ground; local to regional rainshadow produces marked effects.

guess. Six months of the year will probably record 50 mm or less in a twenty-five year span. At the other extreme, six months in a period of equal length are likely to record more than 400 mm. In the average year, no month records a deficit of soil moisture.

15.3b Tropical climates

Rainfall at Kandi is strongly concentrated in the middle part of the climatic year. Whereas variation in monthly rainfall at Singapore is irregular through

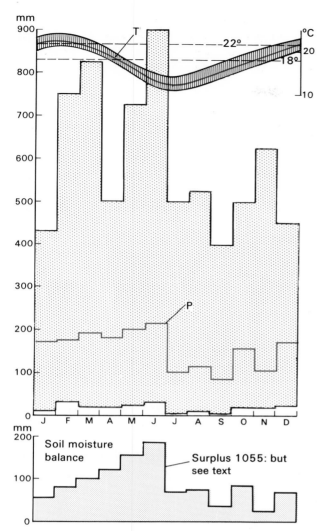

Fig. 15.5   Graphs for humid trade-wind climate: Sydney.

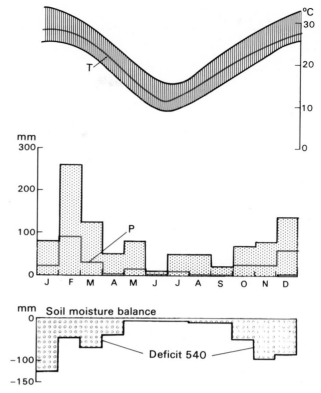

Fig. 15.6   Graphs for dry trade-wind climate: Alice Springs.

15.4a Trade wind climates: humid

Despite the occurrence of a perceptible winter season, Sydney rarely experiences frost and very rarely indeed records snow. When really cold weather comes, it results from an unusually powerful and far-travelling outburst of polar air, such as caused Jamaicans to shiver at the Christmas of 1971, or destroyed the Brazilian coffee crop in the southern-hemisphere winter of 1975. Sydney recorded only two snowfalls in a run of about 100 years.

Precipitation is year-round. Depending on location, it may tend to display a peak in midsummer (equatorward locations) or in the winter (poleward locations). It is typically copious, but also typically erratic. While the band of uncertainty for temperature at Sydney is narrow, that for precipitation is extremely wide – perhaps a surprising fact for dwellers in midlatitudes, who may tend to think of the

trade-wind systems as thoroughly dependable. Although the average record for Sydney indicates a soil-moisture surplus throughout the year, drought is in actuality a persistently recurrent risk.

The humid trade-wind climates are forest climates, with rain-forests on exposed coastlands differing from equatorial rain-forests only in their plant constitution and in their comparative poverty in climbers and trailers; but, as humidity and precipitation decline, and/or as the chance of drought increases, the forests become composed of trees capable of withstanding both drought and fire. The fire danger of course increases with the liability to incursions of cT air, which in turn depends to a considerable degree on the extent of continent next to the west.

15.4b Trade wind climates: dry

In these climates, the horizontal flow of the trade winds is largely replaced by the subsidence of air in temporary, seasonal, and permanent cells of high pressure. Alice Springs (Fig. 15.6) provides an apt contrast to Sydney. Its range of temperature is the greater, as befits a far inland station. At some time in a twenty-five year interval, mean temperatures of less than 18°C can be expected through a six-month span – although of course, the probability of a continuous run of six months with temperatures this low is small. The band of uncertainty for temperature is wider than at Sydney.

Nearly the entire year, in a twenty-five year period, is liable to zero precipitation; and entire years with no rainfall do in fact occur. At the other extreme, February falls can exceed 250 mm, while three months in the year are liable to falls of 100 mm or more. Channel systems in the region are not adapted to contain falls of this magnitude; in consequence, although this is a desert climate, severe overland flooding occurs from time to time. In the average regime, all months record a deficiency of soil moisture.

Although the concentration of average precipitation occurs in summer, when incursions of maritime air are easier than at other times, average values say very little about the likelihood of events in a real individual year. But even on the average assessment, the deficit of soil moisture is equivalent to about half a metre of rain a year – a sobering thought, surely, for those who would capitalise on the high input of solar energy, and the chemically undepleted character of the soils, to bring deserts under cultivation. Meantime, despite much extensive ranching, desert climates retain widespread and fascinating adaptions to aridity and to undependable rainfall, both in their vegetation and in their wildlife. Vegetation is discontinuous: there is too little moisture to sustain a complete cover. Plants and animals alike employ mechanisms evolved to counter the effects of very low rainfall and a highly irregular supply.

15.5 Subtropical climates

West-coast and east-coast climates are affected by the westerlies; but both the onset of the zonal winds and the regime of precipitation contrasts strongly between eastern and western sides of continents. In both types, summers are hot, with mean monthly temperatures in the mid-year rising well above 22°C, but winters are more severe than in the trade wind climates, with mean temperatures characteristically falling to 10°C or below.

15.5a Subtropical west-coast climates

There is nothing in low latitudes to compare with the climates of subtropical west coasts. Whereas low-latitude precipitation is dominantly convectional, and concentrated (if at all) in the hot season, that of subtropical west coasts is dominantly cyclonic, and concentrated in winter. Indeed, there is nothing to match the winter concentration of precipitation anywhere in the world. In direct opposition to the

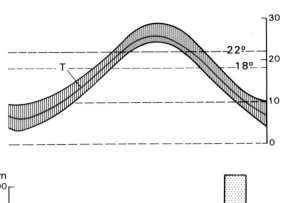

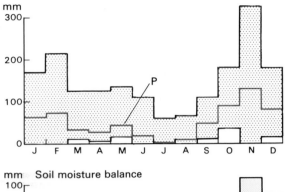

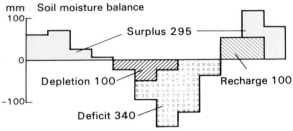

Fig. 15.7 Graphs for subtropical west-coast climate: Rome.

circulation regime of low-latitude tropical climate, the subtropical west coasts are affected in summer by a poleward extension of the subtropical highs, while in winter they come under the influence of the equatorward shift of the westerlies.

Precipitation in total is moderate to somewhat scanty. Variability from year to year can be considerable. In the twenty-five year span at Rome (Fig. 15.7), about six of the twelve months are liable, at some time or other, to experience no precipitation at all. In the average regime there is a distinct soil moisture surplus in winter, and a roughly matching deficit in summer. Streamflow therefore tends to be highly seasonal.

The fact that subtropical west coast climate is most extensively developed in the Mediterranean lands of Europe, North Africa, and the Near East means that certain complications, special to this region, figure prominently in descriptive accounts. The Mediterranean is subject to invasion, especially in late autumn and early spring, by very vigorous Atlantic lows that are switched southward by the blocking action of temporary high pressure systems. As a result, the moderate precipitation totals of the average regimes

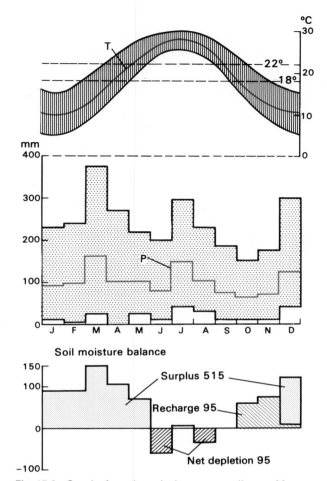

Fig. 15.8 Graphs for subtropical east-coast climate: Macon.

tion is ill-marked. Totals run high, especially near the oceans on the east. There is a considerable surplus of soil moisture; and the annual regime can include, on average, no spell of actual deficiency.

Although there is a considerable difference in temperature between the hottest and coldest parts of the year, a winter resting season for plants may be slight to absent. On the whole, wild vegetation runs to evergreen forest – in eastern Asia, evergreen broadleaf, but in the U.S.A. to a combination of evergreen broadleaf and needleleaf.

It is scarcely possible to discuss the place of this climatic type in the general circulation without specific reference to the actual major cases. Precipitation in the U.S.A. region is supplied largely from the Atlantic and the Gulf of Mexico, but the transporting winds represent a combination of outflow from high-pressure cells offshore, inflow into travelling lows of the westerlies belt, and regional summer indraught provoked by the differential heating of land and sea. Such differential heating is exacerbated for Asia, where the relevant climatic region can be included in the subset of monsoonal climates (see below). Nevertheless, even here, the dramatic seasonal contrast of precipitation regime that characterises monsoonal climates as a group is modified toward a year-round occurrence of rainfall.

15.6 Midlatitude climates

As a first approximation, these may be taken as the climates dominated by the zonal westerlies throughout the year. Particularly in respect of the Northern Hemisphere, however, a qualification must at once be entered: with increasing eastward depth into Asia, the effect of continentality becomes increasingly powerful. This effect is expressed in a progressive decrease of moisture with increasing distance from the principal source, the North Atlantic Ocean; by a gross exaggeration of seasonal regime of temperature; and by the increased incidence of polar and arctic air, particularly in winter, with the attendant weakening of the zonal circulation at low levels.

15.6a Midlatitude west-coast climates

Poleward from the regions of subtropical west-coast climate, no season escapes the flow of the westerlies. Summer temperatures fall off: the season can no longer be called hot (Fig. 15.9: Nantes). Precipitation comes throughout the year, with no great variation through the typical twenty-five year span, and with only a very subdued peaking in the earlier part of winter. Soil moisture surplus for the year as a whole is considerable in the average regime; any

can include, and smooth out, occasional very heavy and flood-provoking storms.

As in all climates with strong seasonal contrasts of precipitation, wild vegetation in regions of subtropical west coast climate is highly specialised – in this instance, to withstand summer drought. The natural cover ranges from evergreen hardwood forest in rainier parts, through scrub forest and to scrub proper. Outside the tall forests, aromatic plants are numerous and common.

15.5b Subtropical east-coast climates

Climates in this bracket are extensively developed only in the northern hemisphere, where extensive land occurs in the appropriate latitudes. The band of uncertainty for temperature (Fig. 15.8: Macon) is broader than in the subtropical west-coast climates, in accordance with the incursion at times of continental air. Precipitation falls the year round; dominantly convectional rain in summer and cyclonic rain in winter may produce two distinct maxima, but at numerous stations the seasonal regime of precipita-

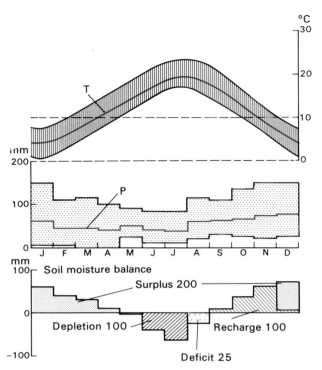

Fig. 15.9 Graphs for midlatitude west-coast climate: Nantes

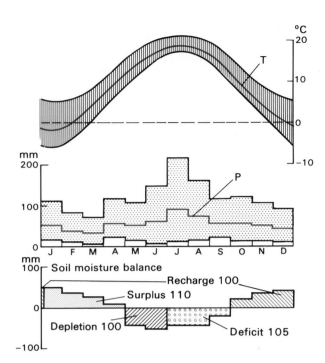

Fig. 15.10 Graphs for midlatitude transitional climate: Berlin.

actual deficit in the late summer is likely to be small.

Most extensively developed in western Europe, midlatitude west-coast climates are the most highly documented, and the most intensively studied, of all. Not surprisingly, in view of the density of settlement in much of the area affected, they have had their local variations worked out in great detail. Thus, to refer only to the British area, they are seen to vary from really winterless (and also possibly summerless) in the southwest, to distinctly more extreme, but also much drier, in the sheltered southeast, and to effectively sub-polar on the highest ground of the north. The range and character of wild vegetation is also known in detail. On the whole, midlatitude west-coast climates are forest climates, the forest ranging from broadleaf deciduous to needleleaf evergreen according to the rigour of the temperature regime, with passages into heath or highly-drained land and to peat where the climate is persistently damp and cool and the soil waterlogged.

The graphs for Nantes repay careful reading. As shown, all months but September can expect to record precipitation of no more than 25 mm, and in parts of the year as low as zero, in a twenty-five year span. Monthly temperatures are far more variable upward than downward. Thus, though one agricultural risk is the year without a summer – say, with the warmest month recording *maximum* temperatures of 12° to 15°C – another risk, which always seems to come as a surprise in this climate, is drought.

15.6b Midlatitude transitional climates

In North America, the relief and climatic barrier of the Cordillera separates the climates of the western coastal areas from those of the interior. No corresponding barrier exists in Eurasia. Whereas in North America there is a mountain belt between the region of midlatitude west-coast climate and the region of midlatitude cool continental climate, maritime influences in the form of mild winters and year-round precipitation under the dominance of the westerlies give way gradually across the Eurasian Plain to the continental influences that produce really severe winters and a pronounced summer concentration of the precipitation. It proves convenient to distinguish a transitional climatic regime, illustrated here by the graphs for Berlin (Fig. 15.10). Precipitation is still year-round, but with a strong suggestion of a convectional summer maximum. The total is closely similar to that of Nantes, with indeed a somewhat reduced risk of occasional drought; but against this, the combination of temperature and precipitation regimes indicates a real deficit of soil moisture in high and late summer.

As shown, mean monthly winter temperatures at Berlin fall below zero Centigrade. In contrast to conditions at Nantes, a seasonal cover of lying snow can be expected. Variability of winter temperature is greater at Berlin than at Nantes, in accordance with the considerable disturbances to the flow of the

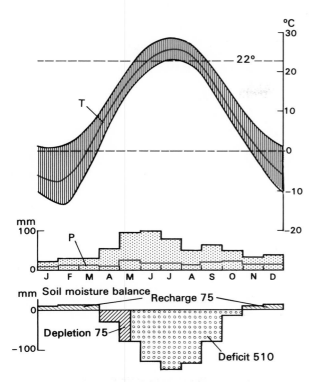

Fig. 15.11 Graphs for midlatitude warm continental climate: Astrakhan.

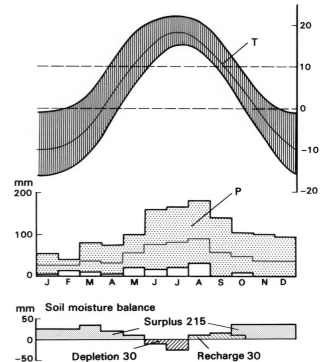

Fig. 15.12 Graphs for midlatitude cool continental climate: Moscow.

westerlies that are imposed by blocking highs. Although most frequent in winter, such highs can develop at any time of the year, serving in many instances to reduce the incidence of cloud.

15.6c Midlatitude warm continental and cool continental climates

These climates are respectively warm and cool, chiefly in contrast to one another, although the mean temperature regime for Astrakhan includes a hot summer season, while that for Moscow does not (Figs. 15.11, 15.12). Although the mean annual temperature for the former station is very close to that of Berlin, while that of the latter is distinctly below, the really striking contrast with most westerly stations is in the length and severity of winters; mean temperature runs below zero for three and a half months at Astrakhan and for five months at Moscow.

Both stations, and the climatic regions to which they belong, lie within the circulation belt of the westerlies, in the sense of lying between the cells of subtropical high pressure and the frontal zones of the far north. Both regions however are powerfully subjected to the blocking action of highs in winter, when in any event there is little surface heat to promote precipitation. The light winter precipitation

falls as snow. Summer rainfall is convectional, producing a marked seasonal concentration; this is best displayed for Astrakhan by the band of uncertainty than by the graph of seasonal regime, but is pronounced at some other stations in the same climatic region. Summer concentration of precipitation at Moscow succeeds, in respect of the mean regime, in inhibiting a soil moisture deficiency – a clear sign of the effectiveness of the westerlies, in combination with rather low mean temperatures – whereas a very marked deficit characterises Astrakhan.

The effect of the direction from which moisture is supplied is well illustrated by the contrast in total precipitation between Moscow and Peoria. This latter station, located in the North American region of cool continental climate, will be used below to illustrate the poleward degradation of temperature. It is worth observing at this juncture, however, that whereas the continental climates of Asia constitute an eastward progression from the Atlantic coastlands of Europe and the transitional region of the eastern European Plain, the region of cool continental climate in North America constitutes a poleward degradation from the regime of the subtropical east coast. With some three months averaging below-freezing temperatures, Peoria has a real, indeed a marked, winter; with a maximum of precipitation in late spring and early summer, it displays clear continental characteristics. On the other hand, no month averages below 50 mm of precipitation, while

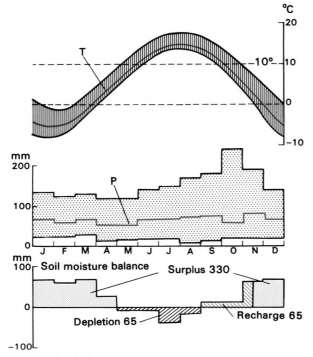

Fig. 15.13 Graphs for midlatitude east-coast climate: Eastport, Maine.

the annual total of 985 mm testifies to the effectiveness in moisture supply of the Gulf and the Atlantic, even 1500 km inland.

Wild vegetation in the regions of transitional and midlatitude continental climate is powerfully influenced by effective moisture supply in the warmer areas and by plant-effective temperatures in the cooler. In the more genial parts, deciduous broadleaf forest is typical. This gives way progressively, as winter temperatures decline, to needleleaf evergreen forest, and that in turn to needleleaf deciduous forest. In the areas of warm continental climate in Asia, wooded steppe passes into steppe grassland, which in turn merges into scrub, and eventually into desert where local to regional rain-shadow comes into operation.

15.6d Midlatitude east-coast climate

As on midlatitude west coasts, precipitation falls throughout the year. Totals are typically high. At Eastport, Maine, there is very little in the way of a seasonal precipitation regime (Fig. 15.13), although the pattern of the band of uncertainty shows that the heaviest monthly totals of all are likely to come in the autumn and early winter. In the average regime, there is no soil moisture deficit.

These various moisture characteristics depend on the east-coast situation. The net westerly airflow is effected by the average of circulation round travell-

ing lows, many of them large and intense, with their energy gradients in winter greatly increased by the contrast between the cold continent inland and the less cold ocean offshore.

The east-coast situation is also responsible for the temperature contrast with west-coast climates. At Eastport, winters can be savage. Nearly eight months a year average below 10°C; about three and a half months average below zero. Here are the effects of continental winter chill in middle to high-middle latitudes. The oceanic effect, on the other hand, is displayed in the contrast with Moscow in precipitation total and precipitation regime.

15.7 Sub-polar and polar climates

The region of sub-polar climates mapped for North America (Fig. 15.2) is more or less axially traversed by the mean position of the Arctic front in summer (Fig. 15.1). One might guess, perhaps, that the climatic regime in question is, on the average, dominated by arctic air for about six months in the year. Some doubt probably comes in here, as a result of the synthesising of information from more than one author, and also on the count of referring vegetation to climate, or vice versa. If sub-polar climate is to be matched with particular types of

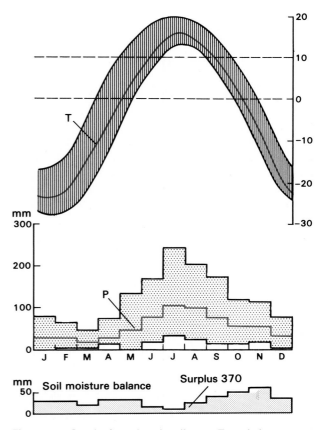

Fig. 15.14 Graphs for sub-polar climate: Trout Lake.

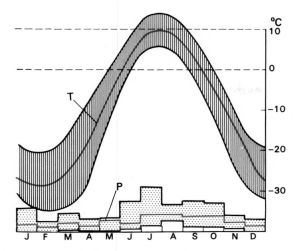

Fig. 15.15 Graphs for polar climate: Coppermine.

tation, surplus soil moisture nourishes lakes and bogs in every small hollow. To some minds, the polar climates are the climates of high-latitude deserts; and, as in low-latitude deserts, vegetation is typically discontinuous.

Coppermine in the average regime experiences no month with an average temperature as high as 10°C. Its high-sun season would prove cold in the winters of the more genial midlatitude climates. Only three and a half months average above 0°C. Expectable variation of mean temperature is considerable, but, in the extreme twenty-five year case, only a four-month span would see mean temperature rising above 10°C, while, in the corresponding worst case, nine months would average below freezing.

15.8 Mountain climates

The records for Säntis (Fig. 15.16) should be compared to those for Nantes, Rome and Berlin. The most obvious differences are those – needless to say – in temperature, and also in precipitation amount. With an average fall of more than 2500 mm annually, Säntis compares in respect of precipitation with such rainy stations as Singapore, Macon, and Sydney, receiving in fact about twice as much as any of these, and about four times as much as Rome,

vegetation, then these must apparently be taken to include park-tundra and continuous tundra, and to exclude both softwood boreal forests to the south, and the highly discontinuous tundra of the polar belt.

The precipitation total for Trout Lake (Table 15.2, Fig. 15.14) is surprisingly high in view of the generally low temperatures. There can be little local heat energy to promote falls. The seasonal regime, however, provides a clue. The winter minimum indicates that during this season the westerlies rarely enter. The summer maximum corresponds to the seasonal poleward shift of the westerlies belt, and to intense storms which bring large amounts of rain and snow. With every storm goes heavy cloud. Predictably, there is no soil moisture deficit in the average regime; soils remain sodden throughout the year. Although for about three months the average temperature runs above 10°C, so that summers are roughly comparable to those at Eastport, although distinctly shorter, more than six months a year average below freezing.

At Coppermine (Fig. 15.15), which on average lies within the Arctic front the year round, precipitation (almost wholly snow) is low but year-round. There is a hint of a subdued peak in the latter part of the high-sun season. No graph of soil moisture balance is presented for this station, because a change in the relationship between temperature and moisture demand, corresponding to a change in plant structure, makes the equations for evapotranspiration unreliable here. For what it is worth, one may say that a deficit is highly improbable at any time. This is especially so because the subsoil is permanently frozen – down to depths of 1600 m in Siberia – and because soil moisture is made available only by the thawing of a superficial layer which may be no more than 5 cm thick. Despite the low preci pi-

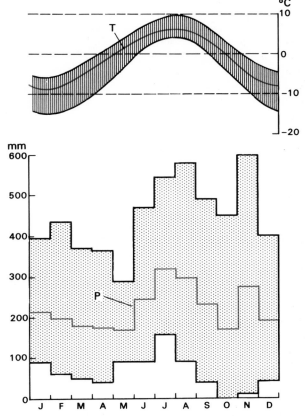

Fig. 15.16 Graphs for an example of mountain climate: Säntis.

Nantes, or Berlin. The effect here involved is that of relief. Air rising up the Alpine flanks expands, cools, condenses and precipitates its moisture. Seasonal regimes of precipitation at mountain stations can be highly variable from one station to another. Säntis combines a major summer maximum with a brief secondary maximum in early winter. The second of these is due to the seasonal incursion of the westerlies before the full onset of winter cold. Low temperatures and abundant precipitation combine to ensure a constant surplus of soil moisture.

Height reduces temperature to levels unthinkable on adjacent lowlands, with about five months averaging below freezing. The seasonal spread of temperatures, however, remains modest. In addition, there is one physiologically and commercially important effect that does not show in the average values. Great heights puts the Säntis station well above much of the lower troposphere. Sunshine is frequent, little filtered, and powerful. The unwary can combine frostbite and sunburn.

15.9 Special considerations

Because we are dealing with the real world, with its real and irregular distribution of continents, oceans, and major mountain areas, any climatic classification, no matter how based, must relate to a greater or lesser extent to the major pattern of the earth's relief. Some of the considerations involved have already been examined. In what now follows, particular attention will be paid to the climatic influence of the Asiatic landmass on its southeastern parts, to the interchangeability of eastward and northward distance, to ocean currents, and to the extreme combination of latitudinal and altitudinal effects. In all this, we shall be concerned either with extreme climatic developments, or with a strong tendency toward them.

15.9a Monsoonal climates

In Southeast Asia, in the broad sense that includes the Indian subcontinent on the one side and China on the other, the general circulation of the trade winds, the westerlies, and the intervening semipermanent high-pressure cells is overriden by seasonal wind systems. These in the high-sun season blow landward, into thermal lows. In the low-sun season, they blow out from the land from thermal highs. The massive relief barrier of the Himalayas and Tibet separates the monsoonal system of the Indian subcontinent from that of eastern Asia.

At Rangoon (Fig. 15.17) the temperature regime is not at all significantly different, in relation to plant growth, from the regimes of the low-latitude climates illustrated by Singapore and Kandi. But not only does Rangoon receive more than twice as much precipitation as Singapore; it also records a pronounced dry and an enormously wet season.

The wet season falls in mid-year, when the continental indraught is operating. Rain of up to 900 mm/month in the average twenty-five year span is equivalent to a sustained fall of 30 mm/day. During the dry season, more than four months record an average deficit of soil moisture. River levels fall drastically and plant growth ceases. In the commonest developments of monsoon climate in low latitudes, evergreen rainforest is replaced by deciduous forest with a marked resting season.

In a way, the climate of Rangoon can be regarded as a gross exaggeration of the climate of Kandi – grossly exaggerated, that is to say, in the wetness of the wet season. But just as at Kandi, the risk of a shortfall of moisture is grave. In the worst imaginable twenty-five year case, not much more than a three-month span has a soil moisture surplus. In the longer term, the monsoon rains are capable, throughout vast regions, of failing altogether in some years. Their pattern of failure is random. Since monsoon rains are the climatic basis of the rice cultures of eastern Asia, random failure means the sudden, widespread, unpredictable, and disastrous onset of famine.

Cherrapunji (Fig. 15.18: temperature and mean rainfall only) displays a still further exaggeration of climate. Its mean temperatures are brought down, and its precipitation greatly increased, by its situation on the flanks of the Himalayas, in an area where the monsoon inflow is forced to converge laterally and to ascend. Precipitation of a quantity and intensity unimaginable in midlatitudes is the consequence. The average August records no less than 3000 mm, equivalent to 100 mm/day; but since the rain is not spread evenly, either through the days of the month or through the hours of the day, individual falls of 50 mm/hour are not uncommon. At the extreme, twelve-month totals exceed 25 000 mm, and totals for single months have been known to surpass 9000 mm.

Even so, Cherrapunji is not, as it was long held to be, the wettest place – or, at least, the wettest recording station – in the world. Similar effects both on temperature and on rainfall in a low-latitude situation are produced in parts of the Hawaiian group, which is completely exposed to the north-east trades in mid-Pacific. The station on Mt. Waialeale on the island of Kauai lies at about 22°N latitude and at about 1550 m above sea level. Its mean annual temperature is accordingly depressed to some 15°C, similar to the value for Cherrapunji. Its mean annual rainfall, 12 350 mm, is the greatest known.

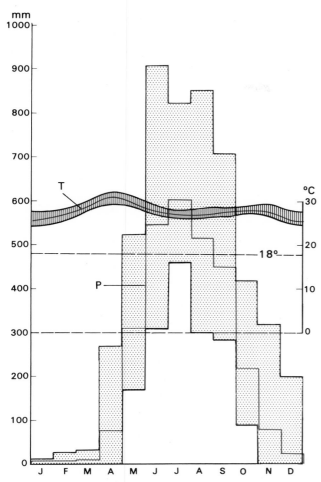

Fig. 15.17 Graphs for strong monsoonal development of tropical climate: Rangoon.

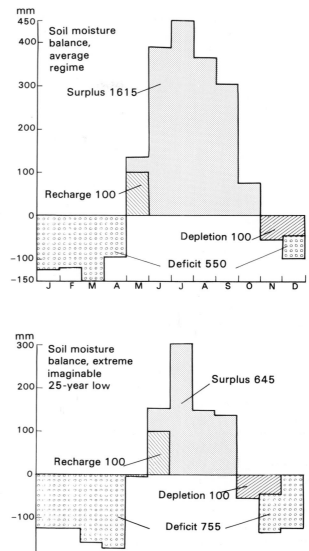

15.9b Distance inland and distance poleward

Attention has already been drawn to the eastward degradation of winter climate across the northern extent of Eurasia, and also, although possibly with less strong emphasis, to the matching poleward degradation in the eastern part of North America. The underlying cause, already hinted at, is the contrast in the distribution of major relief. This contrast ensures that, as between the two land-masses, the latitudinal and longitudinal effects of location are largely interchangeable.

The thermal progression across northern Eurasia, as far as the eastern extremity of cool-continental climate, is largely a matter of increasingly severe winters and of the associated broadening of annual temperature range. Verkhoyansk in northeastern Siberia (Fig. 15.19) provides a frequently quoted case. Its mean annual temperature is well below that even of Coppermine. On the other hand, it does

have some kind of summer season, with about a two-month span averaging above 10°C. There is, in the annual regime, a summer deficit of soil moisture, despite a maximum in the regime of precipitation, wherein the annual total is only about 150 mm. All these are characteristics of a deep inland situation. In the winter season, Verkhoyansk lies within a principal source region of *cPK* air. Outward radiation through clear winter skies has generated temperatures approaching −70°C.

The eastward progression of temperature regime across Eurasia is illustrated in Fig. 15.20, the northward progression in the North American continent in Fig. 15.21 (see Fig. 15.2 for locations). Both sets include a powerful depression of mean temperature and an increase in annual range, the one with distance eastward and the other with distance northward. Verhoyansk averages −15.6°, Coppermine −10.0°: the two stations are at almost identical latitudes.

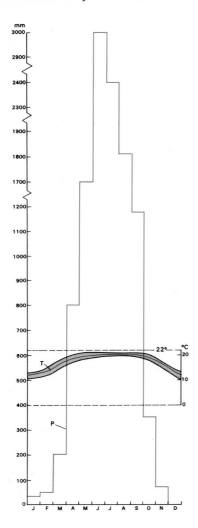

Fig. 15.18 Graphs for the combination of strong monsoonal development with the effect of height: Cherrapunji.

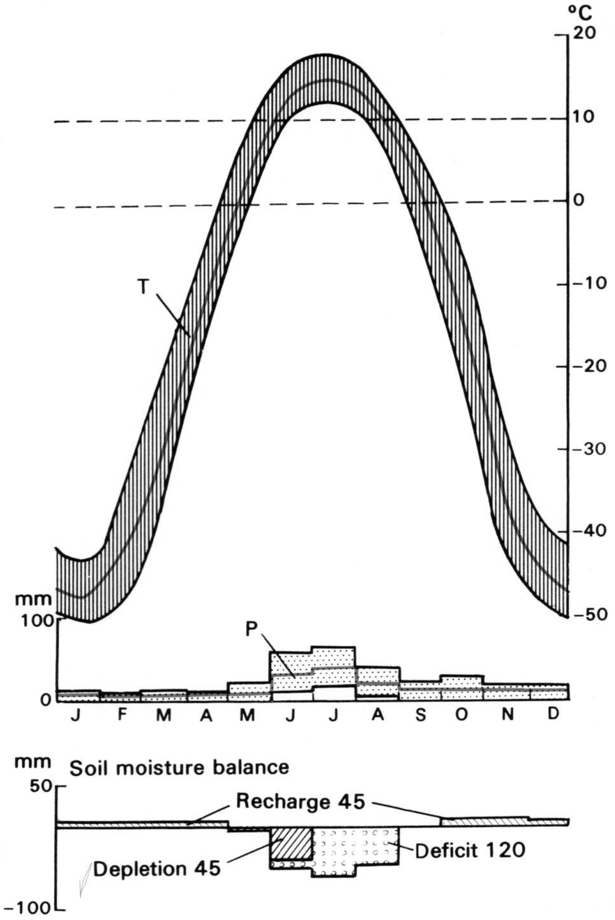

Fig. 15.19 Graphs for the extreme development of cool continental climate: Verkhoyansk.

15.9c Combined effect of latitude and height

The Antarctic station of Vostok (Table 15.2) somewhat beyond 78°S latitude and 3400 m above sea level, records a marked seasonal variation of temperature, with a range of 34°, comparable to the ranges recorded for Trout Lake and Coppermine. Here is, in the main, the influence of distance from the equator. Height depresses the actual values, which indeed correspond to a lapse rate roughly similar to that of the free air in middle latitudes.

15.9d Climatic effects of ocean currents

It is not possible to describe climates in full, and above all the midlatitude climates, without making reference to ocean currents. Some reference has in

fact been made in the previous chapter. The central discussion of ocean currents will appear in chapter Seventeen. At this point, it will suffice to look at certain major influences.

Equatorward-moving currents, on the western sides of landmasses and in the tropics and middle latitudes, inject unduly low temperatures into the local climatic regimes. They thus tend to stabilise the air of such winds as may blow onshore, especially in summer, and in consequence to reinforce summer dryness where this is included in the climate. On the other hand, the most extensive region of summer-dry climate is the European Mediterranean, which is mainly out of the influence of currents in the open Atlantic. The drying effects of cool currents are exerted particularly along the western desert coasts of South America and southern Africa.

Farther poleward, the currents off western coasts flow away from the equator. They transport heat.

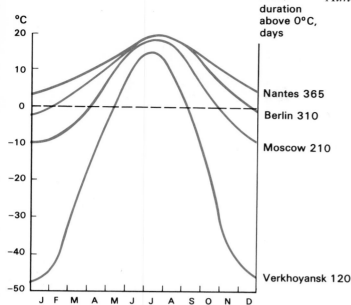

Fig. 15.20 Comparative temperature regimes, for eastward progression across Eurasia.

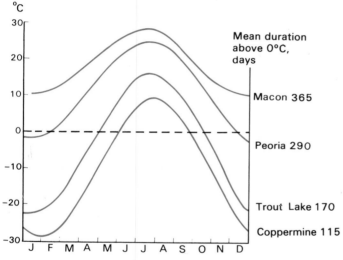

Fig. 15.21 Comparative temperature regimes, for northward progression across North America.

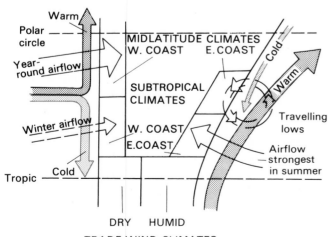

Fig. 15.22 Abstract model of the climatic relationships of ocean currents and wind systems, for coastal climates in the subtropics and middle latitudes.

They tend to destabilise the air of onshore westerlies, by heating it from below. They are responsible for considerable poleward extensions of midlatitude west-coast climate (Fig. 15.22).

On the eastern sides of landmasses, midlatitude ocean currents flow mainly away from the equator. They carry part of the heat involved in the development of subtropical east-coast climates, and also affect the areas of midlatitude east-coast climates. For this latter class, however, an additional influence sets in – cold currents moving equatorward close inshore. Until these disappear below the surface, they contribute signally to the unpleasant character of the east-coast winters.

Chapter Summary

Studies in North America have shown that *climatic boundaries can be defined by preferred frontal positions. Fronts are the boundaries between air masses. Air masses are large bodies* of air with *distinctive physical characteristics.* They *originate in source regions,* where they reach equilibrium with the underlying surface. Air masses are *classified* according to *type of source* region as either *continental* or *maritime;* according to *position in the general circulation* as *equatorial, tropical, polar,* and *arctic or antarctic;* and as *cold or warm* in relation to the underlying surface.

Although preferred frontal positions cannot be defined for the entire *world, climates classified by location in the general circulation can be referred to patterns of air-mass dominance. Low-latitude climates are dominated,* either seasonally or throughout the year, *by equatorial air. Trade-wind climates are dominated by tropical air,* either maritime for humid climates or continental for dry climates. *Location with respect to continental landmasses is highly important in subtropical and midlatitude climates,* wherein west-coast and east-coast types are distinguished. To varying extents, all climates in these classes are *affected by the westerlies. Sub-polar and polar climates come under the influence of air masses originating poleward of the westerlies belt.*

Selected stations illustrate the central characteristics of individual regimes. The specific information will not be repeated in this summary. It can be said, however, that *climate can be assessed and compared in terms of temperature, including range and variability, precipitation similarly considered, and the soil moisture balance.*

Aside from the climates located with references to the general circulation *come mountain climates. Special cases* include climates where *seasonal regimes are greatly exaggerated* by the effects of location, latitude, height, or some combination of these. A particular set of *special influences* is that of *ocean currents,* which bears heavily on the climates of the subtropics and of coastal areas in midlatitudes.

16 Atmosphere Process-Response Systems: Medium and Small Scale

In order to understand the atmospheric systems treated in this chapter, it is necessary to understand the ups and downs of air in the troposphere. Movements are controlled by the interplay of pressure, temperature, and moisture content.

Pressure is controlled primarily by the earth's gravity. The thicker and denser a column of air, the greater is its pressure on the surface. Pressure is measured in millibars, each one-thousandth of a bar, the fundamental unit. One bar = 1000 millibars (mb) is a convenient unit in the metric system. It is also close to the mean surface pressure of 1013 mb. For practical purposes, we can take surface pressure of 1020 mb and upwards as high, and pressure of less than 1000 mb as distinctly low.

Temperature affects pressure by causing air to expand or contract. Low temperature at the surface means contraction, increased density and increased pressure. High temperature produces opposite effects – expansion, decreased density and decreased pressure.

Moisture content is highly influential in the absorption, storage and release of heat energy. The energy needed to vaporise water comes largely from the vaporising air. When the vapour reconverts to the liquid form, the energy used in vaporisation is released, warming the air involved. The heat here in question is latent (= hidden, concealed) heat. Between vaporisation and condensation, it remains stored in the water vapour.

16.1 The tropospheric lapse rate

A lapse rate is the rate at which temperature changes as height increases. Unless otherwise specified, temperature is assumed to decrease with increase of height in the troposphere, as is generally indicated in Fig. 14.2. A mean temperature of about 20°C at sea level, and of −55° at the top of the troposphere, at an average height of some 11 km, means a lapse rate of about 6.5°C/km.

The tropospheric lapse rate reflects the heating of the troposphere mainly from below, and the influence of air pressure on temperature. As has been seen, only a minor fraction of total energy is supplied by the absorption of incoming solar rays. Most is supplied by radiation from the earth's surface, with supplements carried up by convection and by evaporated moisture. A further minor supplement consists of energy carried up by turbulence. The observed general lapse rate corresponds to the expansion of air as pressure decreases, upward from the earth's surface. At the top of the troposphere, air pressure is of the order of 200 mb, only one-fifth of its mean value at sea level. Energy used in distending the atmospheric gases is not available to keep up the temperature.

Air in the troposphere that is forced to rise will expand and cool down. Air that is forced to descend will contract and warm up. The cooling/heating effects of expansion and compression are readily demonstrated with the aid of an ordinary bicycle. Open a valve on an inflated tyre. The escaping air, expanding, strikes cool on the finger. Plug the outlet of the tyre pump (a firmly-pressed finger will serve) and pump repeatedly. The contained air gets hot.

16.2 Observed lapse rates

Above a point on the earth's surface, and at a given time, the lapse rate in the troposphere is likely to differ from the mean. The lapse rate that actually exists is the environmental lapse rate, ELR. For the sake of illustration, assume that the environmental lapse rate is in fact identical to the mean rate. But rising air will not cool at this rate. Its internal lapse rate will depend on whether or not condensation takes place. Without condensation, the rate of cooling is the dry adiabatic lapse rate, $DALR$. The term adiabatic means without exchange of heat between the rising column and its surroundings. Very little exchange happens in reality, because air is a very poor conductor. The dry adiabatic lapse rate is 9.8°C/km.

If condensation does take place, then the released latent heat reduces the rate of cooling. The new lapse rate is the saturated adiabatic lapse rate, $SALR$. It depends considerably on temperature. Because warm air can hold more moisture than cool air can, the heat release from condensing warm air is greater than that from condensing cool air. The $SALR$ ranges from as low as 4°C/km to as high as 9°C/km. For the sake of simplicity, a value of 5.5°C/km will be assumed in the immediately following discussion.

Fig. 16.1 presents the purely hypothetical case of air rising through a height of 5 km and cooling all the time at the $DALR$. With a starting temperature of

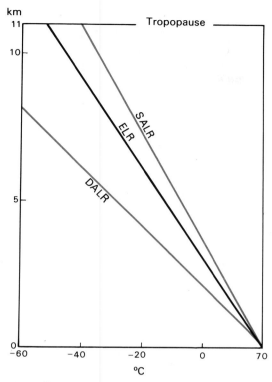

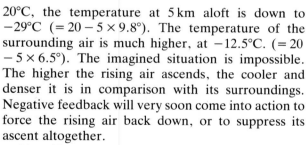

Fig. 16.1 Effect of environmental lapse rate: dry air will not ascend because *DALR* > *ELR*, but condensing air will accelerate its ascent because ELR > SALR.

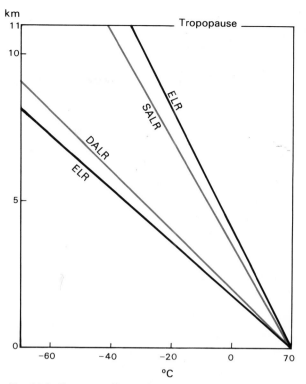

Fig. 16.2 Extreme effects of environmental lapse rate: if *ELR* < *SALR*, no air can ascend; if *ELR* > *DALR*, all air will accelerate its ascent.

20°C, the temperature at 5 km aloft is down to −29°C (= 20 − 5 × 9.8°). The temperature of the surrounding air is much higher, at −12.5°C. (= 20 − 5 × 6.5°). The imagined situation is impossible. The higher the rising air ascends, the cooler and denser it is in comparison with its surroundings. Negative feedback will very soon come into action to force the rising air back down, or to suppress its ascent altogether.

Also illustrated in Fig. 16.1 is the case of a rising column that condenses moisture all the way up. The higher it goes, the higher its temperature, and thus the lower its pressure, than that of the rising air. Positive feedback speeds the ascent. The situation is one of marked disequilibrium. At the assumed 11 km level of the troposphere, the rising air has cooled to −40.5°C, but the *ELR* has brought the surrounding air down to −51.5°C. The rising air keeps going. It will not, however, fly off the planet. After another 2 km, the zero lapse rate of the lower stratosphere will cause it to be surrounded by air warmer than itself. Negative feedback takes over, and the ascent is halted. It is clear nonetheless that some rising columns, such as thunderheads, that in the tropics can reach 20 km, dent the stratosphere. So do nuclear blasts. The cap of the mushroom cloud spreads itself precisely where the surrounding temperatures equal those of the rising column.

16.2a Stability and instability

Now look at the results of variations in the *ELR*. Fig. 16.2 illustrates the two extreme cases, one where the *ELR* is less than the *SALR* (4°C/km is assumed), and the other where the *ELR* is greater than the *DALR* (11°C/km is assumed). For the very low *ELR*, all rising air, condensing or not, will be surrounded by air warmer and therefore less dense than itself. The contrast increases with increase in height. Negative feedback suppresses all upward movements. The condition is one of total equilibrium, called absolute stability. While such a condition is not particularly common in nature, it does serve to show how a lessening of the *ELR* tends to inhibit lifting.

For the very high *ELR*, all rising air, even if not condensing, will be warmer and less dense than the surrounding air. Positive feedback will accelerate all upward movements. The disequilibrium condition, that of absolute instability, cannot occur in nature except temporarily and on a local scale, as when warm air is overrun by cold. But the imaginary general example shows how an increase in the *ELR* tends to promote instability. It goes part of the way towards explaining the development of thunderstorms over heated land.

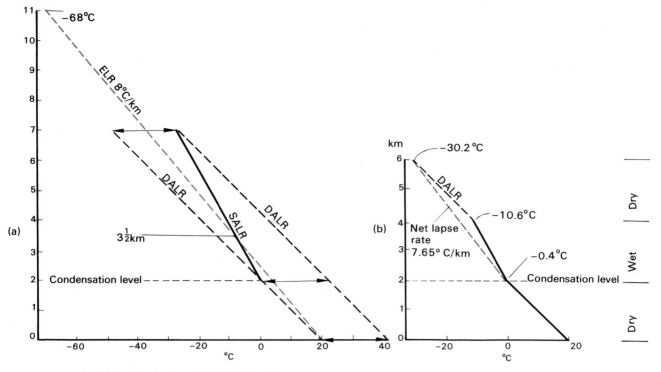

Fig. 16.3 Conditional and potential instability: (a) conditional; (b) potential. For explanation, see text.

16.2b Conditional instability

When the *ELR* is less than the *DALR* but more than the *SALR*, as in Fig. 16.1, the situation is one of conditional instability. Rising air can become warmer than the surrounding air, if it is forced far enough above the condensation level.

A steepening of the *ELR* increases the likelihood that instability can be induced. Fig. 16.3a assumes an *ELR* of 8°C/km. Once more, we start with a temperature at the surface of 20°C. At first the rising air cools at the *DALR*, but the rate changes to the *SALR* at 2 km aloft, where condensation begins. If the air continues to rise to some 3½ km, it will be no cooler than the surrounding air. Further rise will make it warmer than the surrounding air and its ascent will be speeded up by positive feedback. The release of conditional instability is the main driving force of thunderstorms over heated land.

16.2c Potential instability

Instability can also be brought on by the ascent of air along a front, or over a relief barrier; but in these instances, the environmental lapse rate is not particularly relevant. We are dealing not with the relation of an ascending column of air to its surroundings, but with the internal lapse rate of the whole airstream. Fig. 16.3b shows what happens if – as is quite common – the rising air is moist in the lower part and dry in the upper. We are assuming an airstream 4 km deep with the lower 2 km moist and the upper 2 km dry. It is forced to rise through a height of 2 km; and the condensation level, as before, is 2 km above the ground. The lower 2 km of air is raised to the height of 2 to 4 km. Condensing out moisture, it cools at the *SALR*. The upper 2 km is raised to the height from 4 to 6 km. Failing to condense out moisture, it cools at the *DALR*. The net lapse rate of the airstream is 7.65°C/km, distinctly greater than the *SALR*, which as before is taken as 5.5°C/km. Whatever the actual value of the

Table 16.1   Effect on air temperature of flow over a mountain barrier

Height (km)	Temperature of surrounding air °C at ELR of 8°C/km	Temperature of rising air with condensation level at 2 km	Temperature of descending air with crest of rise at 7 km
7	−36.0	−27.1	
6	−28.0	−21.6	−17.3
5	−20.0	−16.1	− 7.5
4	−12.0	−10.6	2.3
3	− 4.0	− 5.1	12.1
2	4.0	0.4	21.9
1	12.0	10.2	31.7
0	20.0	20.0	41.5

SALR, the net lapse rate for the airstream must be greater. Thus, the moist air in the lower part can continue to rise. The whole thickness will be overturned. Here is the state of potential instability. The potential for overturning exists, provided that the airstream is forced above the condensation level.

16.2d Temperature change during descent

Descending air warms at the *DALR*. On the lee side of a relief barrier that releases conditional or potential instability, the air will be warmer, level for level, than on the windward side. The temperature difference at the height corresponding to condensation level on the windward side, and at all lower elevations, will be equal to the difference between the temperature actually attained at the top of the rise, and the temperature computed from the *DALR* (Fig. 16.3). The top of the rise is here taken as 7 km. Both Fig. 16.3 and Table 16.1 show that the descending air can become very much warmer than the ascending air. It is certain also to be dry, having shed much of its moisture during the rise, in addition to increasing its temperature and vaporising capacity during descent.

16.2e Allowing for variations in the *SALR*

Because the *SALR* is actually variable, graphs of the kind represented by figs. 16.1–16.3 would be cumbersome to use. Instead, weather forecasters and the planners of air flights use aerograms. A simplified aerogram is presented in Fig. 16.4. The bases of

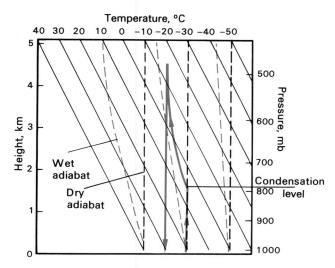

Fig. 16.4 A simplified aerogram: heavy arrowed line shows temperature change for air rising above the condensation level, and then descending.

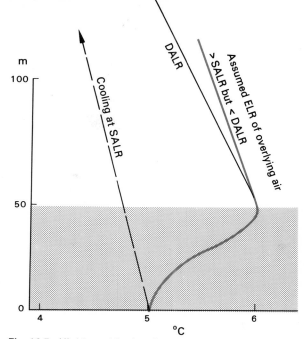

Fig. 16.5 Highly stable situation: inversion next the ground.

reference here are height and temperature, but in reality pressure is normally substituted for height: pressure can be measured when height cannot, as by devices for the sounding of the lower atmosphere.

The graph is so constructed that rates of cooling at the *DALR* plot as straight vertical lines. Height is plotted on a plain vertical scale. Sea-level pressure is assumed to be 1013 mb, the mean value, above which pressure at higher levels is marked on a logarithmic scale.

The graphs for cooling at the *DALR* are called dry adiabats. Graphs for cooling at the *SALR* are saturated (or wet) adiabats. Because the SALR varies with temperature, the saturated adiabats are curved in the diagram. The more complicated aerograms used in standard practice permit the condensation level to be determined. Temperature change with height is determined by following the dry adiabat from the starting temperature up to the condensation level, and then working along the curve of the appropriate saturated adiabat. The figure shows the effect of upward movement of air at the *DALR* to about 1.5 km, of subsequent cooling at the *SALR* to about 4.5 km, and then the effect of descent, with heating at the *DALR*, back to a pressure of 1000 mb.

16.3 Inversions

At some times and in some places, the decrease of temperature with increasing height is changed to a temperature increase with height. The lapse rate is said to be inverted. The weather situation is an inversion (Fig. 16.5). If you, the reader, live in an industrial area, then some inversions can endanger your health, or even your life.

16.3a Inversions next to the ground

Inversions at very low levels are caused by the chilling of air next to the ground. Chilling may result from radiational cooling of the ground surface during the low-sun season, under clear night skies and in calm air. Alternatively, it can affect air that is very gently drifting across a cold surface – too gently for turbulence to carry the chill upward. An inversion means very great stability. If we imagine air to rise, and to cool at the minimum possible rate, the *SALR*, a very long ascent would still be necessary for the rising air to escape the influence of negative feedback (Fig. 16.5). In actuality, no rise will occur until the ground surface has been heated up, or until the inversion is dispelled by winds.

Inversions at low levels are responsible for trapping the notorious smogs of large industrial cities. They prevent pollutants from escaping. If these pollutants include sulphur dioxide, SO_2, a particularly unpleasant event can occur. If the air above the smog layer is clear, then the sun's rays will cause the sulphur dioxide to oxidise into sulphur trioxide, SO_3. This combines with the water droplets of fog to produce H_2SO_4, sulphuric acid

$$H_2O + SO_3 \rightarrow H_2SO_4.$$

Although the acid is very diluted, it is still harmful to breathe. It was one of the substances contained in the London smog of 1952, which accelerated the deaths of 10 000 people.

16.3b Cold-air drainage

Many mountain valleys generate their own winds – up-valley by day, down-valley by night – in response to the daily regime of heating and cooling, and the resultant changes in air temperature and air pressure. Plateau country, including that of quite low relief, is affected chiefly by cold-air drainage. The still, clear long nights which promote low-level inversions cause the plateau surface to become overlain by cold dense air. Sooner or later, this will spill off. It drains either into valleys cut in the plateau or off the side.

The down-draining air moves in pulses. On the local scale, it becomes trapped in valleys, producing very low temperatures in the valley bottoms, and inverting the lapse rate. Field areas where the author has suffered the chill of cold-air drainage include the Lea Valley on the northern side of the London metropolis, the valley of the Hawkesbury River in New South Wales, Australia, and the dissected plateau country of Southwestern Wisconsin. All three areas contain commercial apple orchards. These are located on the plateaus, and are concentrated on the margins, where cooling air has the best probability of draining away.

Pulses of cold-air drainage on the local scale have durations measurable in minutes, or even seconds. On the regional scale, a pulse may last through a whole day, or even through several days. The best-known regional pulses are those that generate the mistral and the bora. The mistral is a wind that blows down the Rhone valley in southern France. The chilled air which generates the mistral is cooled over the Central Plateau in the low-sun season. In the same season, chilled air above the plateau country of Yugoslavia is apt to spill down toward the Adriatic sea in the winds called bora, which are so strong that barriers have been built along the roads down the plateau edge, to prevent cars from being swept away.

16.3c Inversions aloft

The zero lapse rate of the lower stratosphere acts like an inversion, preventing the rising air of thunderstorms from escaping altogether from the lower air. A more obvious case of inversion aloft is that which frequently accompanies high pressure. The subsiding air in a system of high pressure warms up as it descends; but its descent and warming are apt to be checked at the upper limit of turbulence – say from 500 to 750 m above ground. Above this mark, the lapse rate is controlled by the compression of the descending air. At lower levels, the control is exerted from the ground upwards. If the ground surface is cooler than the temperature projected for the air aloft, then an inversion above ground is likely to form.

Fig. 16.6 illustrates what happens. To keep things as simple as possible, a temperature of 10°C is assumed for the air next to the ground, with an *ELR*

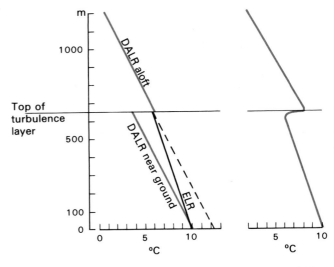

Fig. 16.6 Highly stable situation aloft: inversion at top of the turbulence layer.

Fig. 16.7 Cirrus, moving toward the observer's right.

for the lower air of 6.15°C. This reduces the temperature at the top of the turbulence layer, taken as 650 m thick, by a neat 4 degrees, to 6°C. Now, if the air above the turbulence layer could go on sinking, it would warm up (at the *DALR*) to more than 12°C by the time it reached the ground. Conversely, if air at ground level could rise to the top of the turbulence layer – as some of it surely will – its temperature will fall off, again at the *DALR*, to less than 4°C. The discontinuity between the two *DALR* graphs becomes, in the real world, an inversion in the actual lapse rate. Such an inversion often traps a layer of cloud aloft, just as fog is trapped by an inversion next to the ground.

16.4 Lapse rates and types of cloud

Although we cannot measure lapse rates without instruments sent up by rocket or balloon, we can easily interpret them, whenever clouds appear in the sky. There are only three basic types of cloud – stratus, which is layered; cirrus, which is fibrous; and cumulus, which is heaped.

16.4a Stratus

Layer clouds and their relationships to inversions have already been treated. A fog is merely stratus next to the ground. It is of course possible for more than one stratus layer to form, if more than one inversion occurs in the tropospheric lapse rate. Stratus is not photogenic: no illustration is offered here. Precipitation from stratus, if it occurs at all, consists of fine drizzle, slight powdery snow, or fog droplets.

16.4b Cirrus

Fibrous clouds take their name from the Latin word for curl – that is, a curl of hair. They appear as streamers, often with hooked ends, well above the ground, in the general range from 6 to 12 km aloft (Fig. 16.7). They form high in the troposphere. Consisting of sparse ice crystals, they often presage the approach of the warm front in a travelling low (see below). They contribute little or nothing to precipitation that reaches the ground.

16.4c Cumulus

In Latin *cumulus* = a heap. Heap clouds can be, but are not necessarily, great bringers of rain, snow, or hail.

Fair-weather cumulus forms during calm days in the high-sun season, especially in the summers of middle latitudes. Heating of the ground surface increases the lapse rate, and causes columns of air to rise. Each column supports its own cloud (Fig. 16.8). The cloud bases come where the rising columns, expanding and cooling, start to condense out their water vapour. They neatly define the condensation level. In between the rising columns, descending columns are at work. They vaporise moisture. Their effect is often demonstrated by the dissolution of the cloud margins: descending shreds of cloud simply vanish.

Where fair-weather cumulus forms, there is not enough moisture in the air to promote continued uprise. When the moisture supply gives out, so does the supply of latent heat. Fair-weather cumulus does not become deep enough to promote rainfall.

Where the moisture supply is abundant, condensation will continue as an air column goes on rising. The release of latent heat, reducing the lapse rate, makes the whole column unstable – far warmer than the surrounding air, so that positive feedback continues. The resulting cloud form is the thunderhead, cumulonimbus (Fig. 16.9), that discharges copious rain, snow or hail. It also promotes lightning.

Cumulonimbus forms in moist air that is strongly heated from below, or along the cold fronts of temporary lows, or where moist air is forced to rise up the flank of a relief barrier. From bases 1 to 3 km above ground, the clouds can reach very high levels. Given sufficient moisture to promote continued condensation, there is nothing to halt the convective ascent until the cloud reaches the top of the troposphere – indeed, many vigorous thunderheads push up the stratospheric boundary.

What actually goes on inside a cumulonimbus is not wholly certain. Direct observation is difficult: instrumented aircraft, flown in to make recordings, may be ripped to pieces by the violently turbulent air. Supercooling is certainly important. Supercooled water vapour remains a gas, well below the temperature at which the rising air is saturated and would in other circumstances condense its moisture content. Supercooled droplets remain liquid, well below the apparent freezing point. In ways not wholly understood, the supercooling process contributes both to the formation of hailstones and to the separation of electrical charges within a thundercloud. About the upper third of the cloud, above the temperature level of $-20°C$, becomes positively charged, while the remainder carries a mixed or a negative charge. In contrast to the lower part of the cloud, the earth is positively charged. What we think we see as cloud-to-ground strokes of lightning are really double, beginning as negative downward strokes but also including positive, upward, returning strokes. Flashes also occur within and between clouds. Thunder is the noise of the exploding air that surrounds the flash.

16.4d Hybrid forms of cloud

When a cloud sheet is broken up by instability within a shallow layer, it becomes patterned, displaying waves or arrays of individual cloud cells. It is inhibited from deep development by overlying stable air. The hybrid clouds are classified according to height – stratocumulus up to about 2.5 km, altocumulus (alto = high) between about 2.5 and 5.5 km, and cirrocumulus above 5.5 km. Like other cloud types, these hybrid forms provide ready information about the lapse rate.

16.5 Temporary weather systems

Short-lived weather systems involve the temporary development of high or low pressure, or the temporary development of stable or unstable lapse rates, or some pressure/lapse rate combination. Generally speaking, stable lapse rates are associated with high pressure, unstable lapse rates with low pressure. The small-scale systems to be discussed are, in order of increasing violence, land and sea breezes, dust devils, thunderstorms and tornadoes. Medium-scale systems include the travelling highs and lows of the westerlies belts and tropical depressions which include hurricanes.

16.5a Small-scale systems

Localised heating, whether from the ground alone or by means of the release of latent heat during condensation, drives the small-scale systems. Upward convection of air is essential to them all.

Land and sea breezes can only be produced in stable air that, aside from the breezes themselves, is calm or only very gently drifting. They are generally associated with conditions of high atmospheric pressure, which provide the essential clear skies. Incoming radiation by day heats up the ground surface, causing a shallow overlying layer of air to expand. There is a horizontal pressure gradient from sea to land. The pressure differential drives the sea breeze. By night the ground surface cools down, and the shallow overlying layer of air contracts. The

Fig. 16.8 (above)   Fair-weather cumulus.

Fig. 16.9 (below)   Cumulonimbus: the cloud is producing thunder and lightning and is discharging hail.

horizontal pressure gradient is reversed and the land breeze rises. Net compensating movements aloft may be disguised in the general flow of air.

Although land and sea breezes develop well into middle latitudes – including the coasts of the Great Lakes – they are of chief climatic importance in low latitudes. The land breezes, often blowing at about 2 m/sec, have a great deal in common with cold-air drainage and frequently move in pulses. The sea breezes, blowing at something like 5 m/sec, are stronger and more evenly sustained. They are most powerful in the afternoon and early evening. Capable of reducing air temperatures by 10°C, they can have great climatic significance in coastal belts where they are frequent. Distance of inland penetration varies much, although 10 to 50 km is not unusual. The record distances of penetration are probably those in Western Australia, where the sea breeze sometimes makes its influence felt at Kalgoorlie, 600 km from the ocean.

Dust devils are desert whirlwinds. Whirlwinds occur quite widely outside deserts, but do not become visible except when they carry up leaves or rubbish, whereas many desert surfaces freely supply dust. The whirlwinds form in calm air under a strong sun. Calmness prevents the air close to the ground from being mixed, so that the lapse rate near the

surface increases (Fig. 16.10). The very lowermost air becomes unstable. Small cells burst upward, spinning as they go. Their ascent is soon checked, when the rising air is reduced by expansion to the temperature of the air that surrounds it.

Individual thunderstorms have lives of one or two hours. Nothing more need be said other than what was said about cumulonimbus clouds, except to emphasise that thunderstorms are temporary weather systems associated with high lapse rates in moist air.

Tornadoes in turn are associated with thunderstorms – usually squall-line thunderstorms. The squall line is a long narrow belt of gusty thunderstorms developed along, or ahead of, the cold front in a travelling low (see below). That is to say, the squall line can be located ahead of the position of the cold front at ground level. It seems possible, at least in some instances, that ground friction can so retard the cold front that the cold air coming from behind can override the warm air ahead (Fig. 16.11). If this should happen, the effective lapse rate ahead of the line of cold front at the surface becomes very great. Explosive convection takes place. Alternatively, the squall line may be maintained by the return downdraughts of thunderstorms along the cold front. In either case, thunderstorms along the squall line are likely to be severe.

Tornadoes are funnel clouds and their associated winds (Fig. 16.12). They have been best studied in the interior plainlands of the U.S.A., where they occur mainly in spring and early summer. Like thunderstorms, tornadoes are in some respects impossible to monitor – their winds can blow at up to 100 m/sec. Their mechanics remain incompletely understood. What is known is, that a funnel develops in an individual thundercloud, that the winds revolve round the centre of the funnel, and that

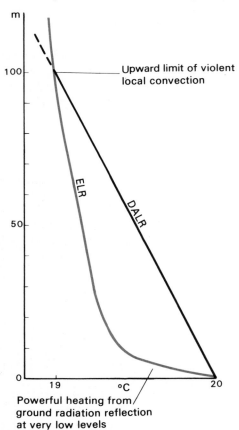

Fig. 16.10 Lapse rate array for dust devils and other small whirlwinds.

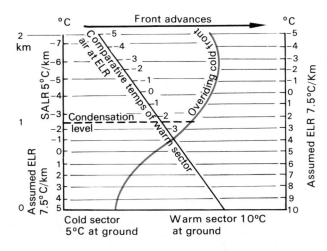

Fig. 16.11 Conditions for explosive convection set by overriding cold air.

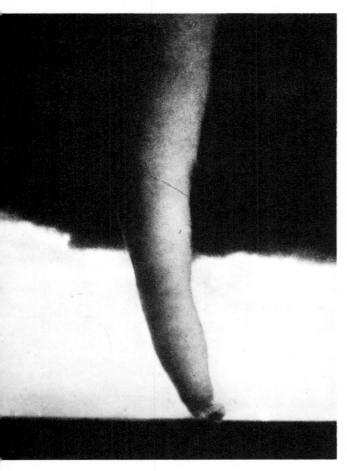

Fig. 16.12   Funnel cloud of a tornado on the southern U.S. Plains (Paul Popper).

Fig. 16.13   Tornado damage at Murphysboro, Illinois (Paul Popper).

pressure in the centre is very low. It seems probable that the development of tornadoes is favoured by a combination of a dry westerly jet at high level with a moist southerly jet at low level. The subsiding air aloft holds the underlying moist air down to below about 2000 m. In combination, the two airstreams produce an extraordinarily unstable lapse rate. A possibility is being explored that heat energy additional to that released by condensation is brought in by a dry hot jet shortly in advance of the cold front.

In contrast to uncertainties about mechanisms, the effects of tornadoes are grimly certain. When a funnel touches down, savage destruction results both from the high winds and from the sudden reduction of air pressure (Fig. 16.13). Closed buildings in the ground path of a tornado will explode.

Over land, a touching-down tornado sucks in much dust and debris. Over the sea it can only suck in water. Here is the origin of waterspouts.

16.5b Tropical lows

Whereas temporary lows in the westerlies belt come largely in one sort, those of the belt of high rainfall of the equatorial and low-latitude monsoonal lands are distinctly varied. To a large extent, they cannot be discussed in the terms used in relation to mid-latitude systems – partly because in very low latitudes the Coriolis effect is weak, or even absent.

Hurricanes excepted, tropical lows are better classified as disturbances than as low-pressure systems proper. They promote instability and precipitation in situations of converging airflow from which lifting

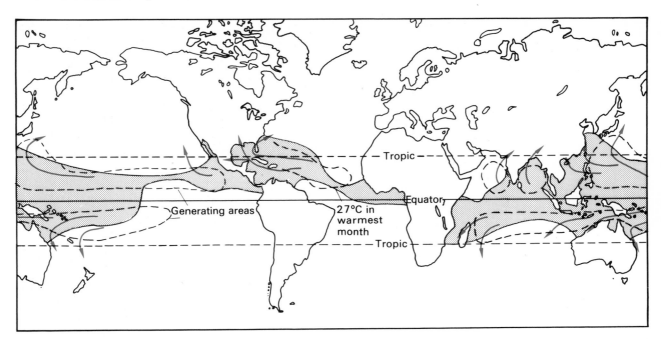

Fig. 16.14 Generating areas for hurricanes: generalised main trackways also shown.

must result. Condensation and the release of latent heat in the lifted air triggers off convection. The list includes disturbances along the intertropical front, which, however, is discontinuous and by no means explains all low-latitude rain; wave disturbances in the trade winds aloft; and low-level disturbances in onshore monsoon winds, wherein the boundary between disturbances and identifiable low-pressure systems becomes blurred.

Hurricanes are tropical cyclones – storm systems with revolving winds and revolving clouds. In east Asia they are called typhoons (Chinese *tai fung* = big wind). Averaging 650 km across, hurricanes generate violent winds which often surpass 50 m/sec. They develop in low latitudes where the temperature of the sea surface runs at 25°C or above (Fig. 16.14). Their main heat supply is the underlying sea, although the release of latent heat by the condensation process contributes to their highly unstable lapse rates. If 100 to 200 columns of cumulonimbus come into being, then the total assemblage can combine into a single and very vigorous low. The spinning storm (Fig. 16.15) will then move, often erratically, through the trade wind belt, and quite often escape into the belt of westerlies.

As with tornadoes, hurricanes have been best studied for the U.S.A. – in this case, the southeastern states. The thrust of the shoulder of Brazil into the southern hemisphere ensures an anomalously high rate of delivery of warm equatorial water into the northern-hemisphere oceanic circulation. The surface waters of the Caribbean Sea and of the Gulf of Mexico are very warm for their latitudes. Hurri-

canes escaping from the tropical circulation and intruding into the westerlies belt can cause extreme damage, to some extent from the high winds but chiefly from flooding. Whereas tornado damage is local, hurricane damage is regional.

16.5c Travelling lows of the westerlies belts

The development of temporary pressure systems in the westerlies belts is affected by temperature conditions at the earth's surface and also by the northerly or southerly flow induced by waves aloft (Fig. 14.15). However, the average lows for January over the northern Atlantic and Pacific, while associated with ocean surfaces that are warm for their latitudes, merely represent a summation of numbers of individual travelling lows. The main control over the development of travelling highs and lows is the interplay of convergence and divergence in the middle and upper troposphere.

Convergence aloft means subsidence, accompanied by high pressure and diverging flow at low levels. Divergence aloft means ascent, accompanied by low pressure and converging flow at low levels. Belts of convergence and divergence are systematically associated with the paths of upper-air waves. Fig. 16.16, which is drawn for the northern hemisphere, shows convergence on the equatorward track and divergence on the poleward track. On the southern to southeastern side of the diverging flow, fronts appear on the map of weather at the surface. The formation and strengthening of fronts is called frontogenesis.

Like the average January systems of low pressure

Fig. 16.15 Air view of Hurricane Gladys, 1968, from Apollo 7 Spacecraft: about 240 km SW of Tampa, Florida, looking toward Cuba (NASA photograph). Winds in the hurricane are reaching speeds up to 150 km/hr.

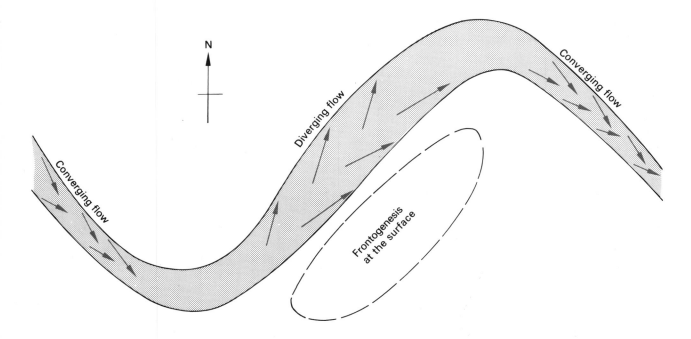

Fig. 16.16 Relationship of frontogenesis at the surface to jetstream flow aloft.

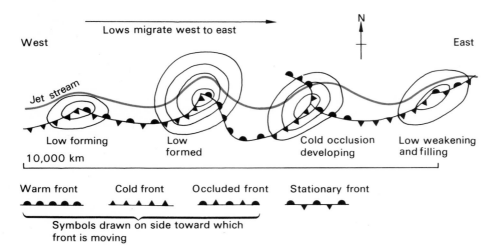

West   Lows migrate west to east   East

Low forming
10,000 km
Low formed
Cold occlusion developing
Low weakening and filling

Warm front   Cold front   Occluded front   Stationary front

Symbols drawn on side toward which front is moving

Diagram can be read either as showing a family of lows at an instant of time, or as typifying the life history of an individual low

Fig. 16.17 Family of lows: alternatively, life history of a single low, in relation to the jetstream track.

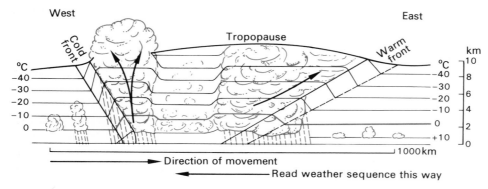

West   East
Cold front   Tropopause   Warm front

°C
−40
−30
−20
−10
0

°C   km
−40   10
−30   8
−20   6
−10   4
0   2
+10   0
1000 km

Direction of movement
Read weather sequence this way

Fig. 16.18 Weather array and weather sequence in a travelling low of the midlatitude Westerlies belt.

mentioned above, the fronts illustrated for the general circulation in Fig. 14.14 are the result of repeated individual events. At the same time, they are real. Frontogenesis in the westerlies belts is a continuing process. Convergence at low levels produces a narrow boundary band, between relatively warm (and usually relatively wet) air on the equatorward side and relatively cold (and usually relatively dry) air on the poleward side. The boundary band is the front. It is not easily dissipated by mixing.

The full details of frontogenesis are complex. In the simplest view, the front in an individual low makes its appearance on maps of surface as more or less straight, but rapidly becomes kinked. A warm front and a cold front separate out. They are so named for the temperature changes which they bring, as the whole low travels eastward. A typical speed of travel for shallow lows is from 20 to 50 km/hr, according to season. The movement of such lows is controlled mainly by what is happening aloft: their paths tend to resemble the path of the jetstream (Fig. 16.17). Deep lows can move much more slowly, and can seem to be only loosely affected by what is happening aloft.

While the upper-air waves tend to move downwind, certain bends display preferred positions. One of these is over the North American Cordillera,

somewhere near the U.S.–Canadian border, where the jetstream turns back toward the equator. The subsequent bend back to the northeast lies somewhere in the southern part of the interior plainlands. Depending on just where this bend is located at a given time, winter lows developed in association with divergence aloft can move rapidly northeastward through the Midwest, causing heavy snowfall, or can be displaced eastward, bringing severe weather to the Atlantic coastland.

The sequence of weather produced by a passing low is identical with the sequence along a section through the system. Fig. 16.18 presents a summary picture, from which detailed variation of cloud type is mostly omitted, and where lapse rates are somewhat generalised. The polar air in advance of the warm front is flowing away from the observer. In it, patchy cloud forms. Some 10 to 30 hours in advance of the passage of the warm front on the ground, the approach of the system is heralded by strings of cirrus near the top of the troposphere. As the warm front continues to advance, the associated cloud becomes thicker and denser, and the cloud base descends. Precipitation breaks out well ahead of the warm front on the ground, first as drizzle or light snow, then as a heavy downpour. Total duration runs typically from 10 to 25 hours.

The passage of the warm front on the ground is marked by a falling-off of precipitation, or by cessation altogether. But the weather in the warm sector (= between the warm and the cold front) is typically muggy, and the sky is usually murky. There can be a continuous cover of low cloud. The time needed for the passage of the warm sector depends on the angle between the warm and cold fronts, and the position in the system: near the centre precipitation can be continuous.

Where the warm and cold fronts are clearly separated, temperatures rise as the warm front passes and tropical air flows in.

As the cold front approaches, deep dense clouds appear. These are often cumulonimbus, formed in moist tropical air which is violently upthrust by cold polar air coming in behind. Precipitation occurs ahead of the cold front on the ground, during the passage, and also behind the front. It includes intense rain or sharp heavy snow from thunderstorms, and hailstorms, according to season, locality, and what is going on aloft. The precipitation belt is, however, narrower than that associated with the warm front – say, 5 to 15 hours in duration.

Temperature falls, sometimes very sharply indeed, as a cold front passes. Some of the most noteworthy cold fronts are those that traverse southeastern Australia; they can bring temperatures down by 10°C in half an hour. Behind the cold front, polar air takes over again for the time being. It may generate convectional showers, being heated and moistened by the ground surface that was previously overspread by warm tropical air and that has been heavily rained on. Such showers, dying away as the hours pass, are the clearing-up showers of Britain.

If conditions in the middle and upper troposphere are inconducive to extensive lifting, then no deep development of cloud is possible. Such conditions are promoted by the inflow of warm air aloft, with a resulting decrease in lapse rates. Precipitation associated with the fronts is then much reduced, and the low-pressure system can eventually fill.

The cold front in a particular low, being vigorously pushed forward by the cold air behind, may overtake the more gently drifting warm front. When this happens, the air in the warm sector is lifted off the ground. It is said to be occluded (= shut off). The single front at low levels is an occlusion, shown on the weather map by a combination of the symbols for warm and cold fronts (Fig. 16.17). An occlusion is either cold or warm, depending on the temperature change when the front passes at ground level. Cold occlusions are common in the winter months, both in North America east of the Cordillera and in northwest Europe. If the fronts shown in Fig. 16.18 closed up, they would produce a cold occlusion, because the air behind the cold front is cooler than

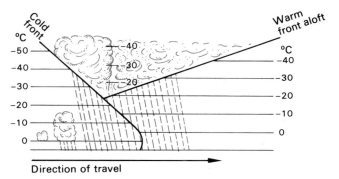

Fig. 16.19 Weather array and weather sequence for a cold occlusion.

the air ahead of the warm front (Fig. 16.19). With an occlusion, the weather sequence of a fully-formed low is compressed into a single band of precipitation, which for a cold occlusion falls largely behind the front at ground level.

16.5d Temporary highs of the westerlies belts

In a way, many travelling highs in the mid-latitude westerlies can be looked on as merely the counterparts of travelling lows. If pressure is low in some areas, it must by definition be high elsewhere. Numbers of cells of high pressure do in fact appear, drifting eastward at speeds like those of travelling lows. If their skies are clear, they promote warm sunny weather in summer and frost and low-level inversions in winter. Inversions aloft produce covers of stratus.

Both for weather and for climate, the most significant highs of the westerlies belts are those which persist for days or weeks – and, on rare occasions, months – at a time. These are blocking highs. Forming when extreme development of upper-air waves has reduced the westerlies system to an array of cells, these highs can stagnate in particular places. They not only control the weather of the areas which they cover, but also divert approaching lows. In early 1947, blocking highs over Scandinavia caused Britain to be overspread by bitterly cold air from the east. When the highs finally shifted, rapid snowmelt caused unprecedented flooding on some eastern English rivers. In November 1966, blocking highs over Scandinavia and adjacent areas of the North Atlantic diverted all approaching lows to southerly tracks. Severe flooding occurred in parts of the Mediterranean area – notably in Italy, where water and mud up to 7 m deep brought much destruction to the art treasures of Florence. In 1975 and 1976, the blocking pattern was such that large parts of Britain and neighbouring parts of the European mainland suffered prolonged drought. If the summer blocking pattern causes lows to stagnate over Britain, then a year without a summer can result. In 1954 the maximum July temperature in southern England was 13°C. The fact that temperatures of 13°C were registered in the following November did nothing to retrieve the year's lost harvest.

Chapter Summary

Pressure of the atmosphere *is controlled primarily by gravity* but is *also affected by temperature. Moisture content influences* the *storage and exchange of heat.*

Reduction of pressure with height means corresponding *reduction of temperature.* The *rate* of temperature reduction is the *lapse rate,* which *for the troposphere as a whole is about 6.5°C/km. Rising dry air cools at 9.8°C/km,* the *dry adiabatic lapse rate, DALR. Rising air* that is *condensing out moisture is warmed by latent heat.* It *cools at the saturated adiabatic lapse rate, SALR,* which according to temperature *ranges from 4°C/km to 90°C/km.*

What happens to a rising column *depends on the environmental lapse* rate, *ELR,* and on whether or not the column is condensing. A *very low ELR* inhibits ascent and *promotes stability.* Conversely, a *very high ELR* promotes ascent and thus *instability.*

When the ELR is between the DALR (less than this) *and the SALR* (greater than this) the *situation* is one *of conditional instability.* If the air rises far enough to become warmer than the surrounding air, by cooling first at the *DALR* but afterwards at the *SALR,* instability will be released.

If the *upper part of an airstream is dry* while the *lower part is moist, ascent* (e.g. over a relief barrier) *can steepen the net lapse rate* and *promote overturning.* The *situation* is one of *potential instability.*

Descending air heats up at the DALR. Hence the common contrasts between the two sides of relief barriers in middle latitudes – *exposed sides rainy and cool, lee sides dry and warm.*

Inversions are *reversals of the lapse rate.* For some distance, *temperature increases with height. Inversions next to the ground* hold down *fog* and trap the pollutants of *smog. Some* are produced simply by *radiational cooling, others* by *cold-air drainage.*

Inversions aloft include the *zero lapse rate of the low stratosphere,* and *those produced by subsidence in a high* as far down as the top of the turbulence layer.

Cloud types, their form and development, are *closely associated with lapse rates. Stratus belongs with inversions* in the troposphere, *including next to the ground. Cirrus forms* well above the ground, *at freezing levels. Fair-weather cumulus results from condensation in rising columns,* but not enough moisture is present to promote long ascents with cooling at the

SALR. Where the moisture supply is abundant, prolonged ascent and cooling at the *SALR* causes *cumulonimbus,* including *thunderheads,* to form. A deep cumulonimbus can dent the base of the stratosphere. *Supercooling* in a thunderhead *contributes,* in ways not fully understood, *to the formation of hailstones. Lightning* results from a *differentiation of electrical charges.*

Hybrid forms of cloud are *classified according to height –* in descending order, *cirrocumulus, altocumulus,* and *stratocumulus.* The patterns result from *instability within a shallow layer.*

Land and sea breezes result from the pressure gradients produced by *differential heating of land and sea surfaces. Sea breezes* blow especially in the *afternoon and early evening, land breezes* in the *later night hours and the early morning.* They can be climatically important, chiefly in low latitudes.

Dust devils are *desert whirlwinds* resulting from the *intense heating* of a shallow layer of air *near the ground,* when the *ELR* is otherwise stable.

Thunderstorms have been discussed. *Tornadoes* are usually associated with *squall-line thunderstorms.* They are *visible as funnel clouds.* They involve *explosive convection* and *very high windspeeds.*

Tropical lows are *distinctly varied.* Apart from hurricanes, most are best regarded as *disturbances. Hurricanes* are *revolving tropical storms,* with *violent winds* and *heavy rain.* Their energy comes largely from the underlying seas over which they form.

Frontal lows form in the *westerlies belts.* They appear on the *weather map* as *kinks in the polar front. Exaggeration of the kink* separates out the *warm front* and the *cold front.* Characteristically, lows on the map are *associated with diverging flow in the upper-air westerlies.*

The *sequence of weather* produced by a passing low *depends on* the observer's *position.* For a fully-formed low, where the warm sector will pass over the observer, the *approach* is *heralded by cirrus. Lowering and thickening cloud* eventually *delivers precipitation,* which begins *well ahead of the warm front* on the ground. As the *warm front passes,* the *wind shifts* and the *temperature rises.* In the *warm sector, muggy weather* and *murky skies* are typical. *Dense clouds,* often *cumulonimbus,* are associated with the *cold front,* which can bring heavy thunderstorms with rain, snow or hail. *Behind* the cold front, *temperatures drop.*

The *cold front can overtake the warm front,* causing the air in the warm sector to be lifted off the ground. The weather sequence is then compressed. The *combined front at ground level* is an *occlusion.*

Some *highs* in the westerlies belts merely drift with the general circulation. Those *of greatest interest* in respect of *weather spells* are *blocking highs.* These come into being when *great exaggeration of the upper-air waves* has reduced the westerlies to a *system of cells.* Blocking highs mean *dry conditions for the areas they cover.* They *divert approaching lows.* For Europe, lows will be directed northeasterly or southeasterly, *according to the blocking pattern.*

17 Ocean Process-Response Systems

The ocean systems respond to inputs and outputs of radiation, water and solutes. They also respond to the rotations and gravitational pulls of the earth-moon-sun system. Their response to the friction of wind-raising waves was discussed in Chapter 12. This chapter deals with the vertical movements of tides, with associated tidal currents, with seawater chemistry, with the interacting effects of temperature and salinity, and with the major surface currents of the oceans.

17.1 Tides

Atmospheric and earth tides are not discussed in this book. Their environmental effect on life on earth is negligible to zero.

Ocean tides are oscillations of sea level that, for the most part, occur once daily (diurnal tides) or twice daily (semi-diurnal tides). There can be complications. Some coasts experience mixed tides – semi-diurnal for part of the month, diurnal for the remainder. In some sea basins, but mainly in large inlets, the tides include harmonic oscillations of short period, for example ¼ day, ⅛ day, and so on. But the semi-diurnal tides are basic to the whole discussion.

17.1a Tide-generating forces

Tides are raised by the relative motions and gravitational pulls of the earth, moon and sun.

For the sake of simplicity, begin by considering only the moon and the earth. Although we usually speak of the moon as revolving round the earth, in actuality the two bodies form a binary (= dual) system which rotates round the centre of gravity of the system (Fig. 17.1). Because the mass of the earth is far greater than the mass of the moon, this common centre of gravity lies beneath the earth's surface. As the moon rotates round the centre in one direction, the earth may be thought of as backing off in the other.

The joint movement sets up what is often called the centrifugal force of rotation. As with the Coriolis effect, no real force in involved, although it proves convenient to assume a real force is working out some of the equations for the generation of tides. An alternative term is centrifugal acceleration.

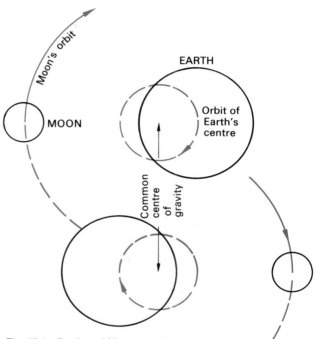

Fig. 17.1 Earth and Moon as a binary system.

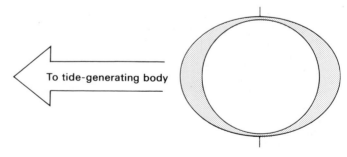

Fig. 17.2 Tidal bulges on opposite sides of the Earth.

Whereas centrifugal acceleration is constant at all points on the earth's surface, the gravitational pull of the earth–moon system is not. The differences between the two are the tide-generating forces. In combination, the relative masses of the earth and moon, and their relative distances from the common centre of gravity, ensure that there is an excess pull toward the moon on the moonward side of the earth, and an excess pull away from the moon on the opposite side (Fig. 17.2).

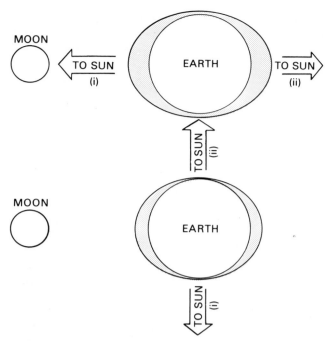

Fig. 17.3 Combination and opposition of lunar and solar tide-generating forces.

Similar relationships apply to the earth–sun system. But because gravity depends not only on the masses of two bodies but also on the square of the distance between them, the solar tide-generating force is less than half the corresponding lunar force. When, with respect to the earth, the sun and moon are more or less lined up, tidal ranges are greatest. Spring tides occur. They arrive twice a month – when the sun and moon are roughly lined up on the same side of the earth, or when they are roughly lined up on opposite sides. At these times, the solar tide-generating force is added to the lunar force. When, with respect to the earth, sun and moon are roughly at right angles, the low ranges of neap tides are recorded. The solar tide-generating force is subtracted from the lunar force (Fig. 17.3).

17.1b Tidal ranges

Ranges for spring tides in the open ocean, and also on many coasts, run at 2 m or less. Some large inlets, including certain river mouths, experience ranges of 10 m or more. The world's record is held by the Bay of Fundy, where a standing wave, rocking backward and forward in the Bay, strongly reinforces the normal tide, producing ranges up to 17 m. In the open ocean, as will be discussed under the next head, tidal ranges tend to increase from mid-ocean toward the margins of the land.

As seen in Chapter 12, tidal range can be significant in the shaping of land margins, especially with regard to abrasion-platforms and estuaries. The morphological effects of tides are, naturally enough, generally greatest where range is greatest. High ranges in estuarine settings command attention in another context also – that of the use of tides to generate electric power. Some working installations in fact exist. In comparison with major hydro and thermal plants, however, they rank only as pilot projects. Their principle is simple enough. Barrages built across inlets with large tidal ranges include turbines with blades of variable pitch – that is, turbines capable of working both on the incoming and on the outgoing tide. Because generation comes to a halt during high tide and during low tide, tidal power appears unlikely to become more than a supplement to other forms. At the same time, it is likely to be considerably exploited in the future.

17.1c Tidal waves

The term tidal wave here refers to the moving crest and trough of a single tide. It emphatically does not stand for waves generated by earthquakes.

Fig. 17.4 is a quite highly generalised diagram of the progress of a tidal wave: the diagram is based on what actually happens in the North Atlantic. As shown, the wave crest rotates round a central point – the amphidromic point (Greek *amphi*, around, and *dromos*, course). At the amphidromic point itself, the amplitude of the tidal wave is zero, and the velocity of travel is zero. With increasing distance from the amphidromic point, amplitude and velocity both increase. Because the moon takes 24 hr 50 min of earth time to complete one orbit, as observed from earth, the crests of semi-diurnal tide waves come at intervals of 12 hr 25 min. If the tides are

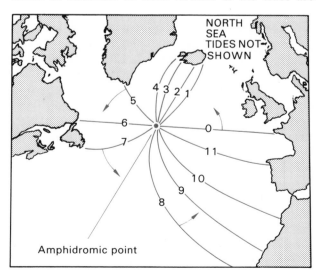

Fig. 17.4 Rotation of the tidal wave in the North Atlantic: various complications, including those of marginal seas, omitted.

diurnal, then the spacing between crests equals the length of one apparent moon orbit.

Rotary circulations of tidal waves result from the combined effects of tide-generating forces, the earth's rotation, and the Coriolis effect. Consequences in the open ocean are generally simple, even though the array of ocean tides includes amphidromic points. Intricate complications can typify basins that are partly enclosed – such, for instance, as the real North Sea, where tidal regimes are affected by the shape of the basin in plan, and by variations in depth. Some tidal subsystems work as if their amphidromic points were on land.

17.1d Tidal currents

Particularly in shallow waters and along irregular coastlines, the rise and fall of the tides can be accompanied by powerful currents. There must of course be some net transport of water landward on the flow (rising tide) and seaward on the ebb (falling tide). Landward and seaward flow also takes place in inlets. But, in addition, the tidal movements can generate currents in the sea. Such currents are often complexly related to the rise and fall of tides. They do not all flow forward ahead of the rising tidal crest, or rearward behind the falling crest. On a highly broken coastline, such as that of the Channel Island Group in the English Channel, they can attain velocities of 4 m/sec, or about 14.5 km/hr, and prove highly dangerous to small craft.

17.1e Tidal bores

In some inlets which narrow landward, the incoming tide arrives as a steep-fronted wave, the tidal bore. While sideways compression of the tide front has some effect in some cases, bores in general should be regarded as resonance effects, comparable to the effect of a bathroom echo on a singing voice. Positive feedback causes the wave to rise high enough that the front becomes very steep, even if it does not actually break. On the English river Trent, the bore disappears in certain reaches, only to re-form farther upstream, demonstrating clearly that something more than mere compression is involved.

17.2 Seawater chemistry

At first glance, seawater chemistry may seem entirely unrelated to tidal movements. There exists, however, a relationship of a kind, in that contrasts in chemistry contribute to vertical motions in the ocean. But, whereas with tides we are dealing with oscillations of the surface, with differences in water chemistry we deal with the movement of individual bodies of water, sometimes through very great depths.

Every bather in the sea knows how the water tastes – salty. The dissolved materials that produce the salt taste are dominated by sodium chloride, NaCl, which in solid form is common salt. Salt crystals prepared from seawater have a somewhat complex taste, because of a 15% admixture of substances other than sodium chloride.

Rivers deliver solutes (= solutions, dissolved materials) to the oceans. Average river water has a salinity (= salt content) of about 0.13‰ (parts per thousand), whereas the average salinity of seawater is about 35‰, more than 250 times as great. In the early days of earth science it was assumed that the oceans are storage systems that become progressively saltier as rivers continue to deliver solutes. In actuality, storage in the oceans accounts only for about 1% of the materials cycled through the sedimentary cascade. At present rates of delivery, the total salt content of the oceans could be supplied in as little as 12 000 000 years. The oceans have existed for far longer than this. In addition to input mechanisms, output mechanisms are also strongly at work.

17.2a Contrast between chemistries

Fig. 17.5 uses slightly rounded values to illustrate the contrast between the respective chemistries of river water and seawater. The solute load of average river water consists of more than one-third bicarbonate ion, HCO_3. Calcium, rather more than one-fifth of the whole, is also prominent. Average river water can be described as bicarbonate solution, in which calcium bicarbonate, $Ca(HCO_3)_2$, dominates. There is also a significant silica component, SiO_2. Average seawater, on the other hand, is a sodium chloride solution, where calcium is down to 1% of the solute total, and where dissolved silica has vanished entirely.

Part of the contrast results from the selective extraction of calcium carbonate and silica by marine organisms, which use these materials in the construction of their hard parts. Shellfish, corals and fish provide obvious examples of users of carbonates. But large quantities, both of carbonate and of silica, are taken up by the tiny plants and animals of the plankton. Plankton is the collective name for life forms that drift or float, some of them rising and falling through shallow depths as the intensity of light varies through the day. Many members of the plankton are so small that individuals can only be seen under a microscope. In total, however, they

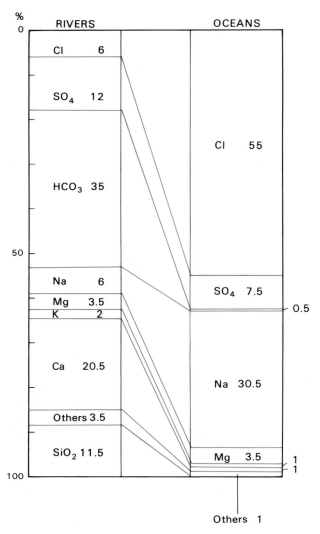

Fig. 17.5 Contrasted chemistries of river water and ocean water.

250 000 tonnes/km². In the Pacific alone, they form at the rate of 10 million tonnes/year. The expression so-called is used here, because the nodules typically include fifteen or more metal compounds, nearly half of them compounds of metals other than manganese.

Land-contained ores of aluminium, cobalt, copper, lead, manganese, nickel and zinc promise to become exhausted within 50–100 years. The manganese nodules of the Pacific alone contain enough reserves of these elements to last from 1000 to 500 000 years, according to species.

17.2b Variations in seawater chemistry through time

A set of fixed ratios among the various salts in sea water by no means denies the possibility of variation in salinity through time. In any event, those fixed ratios are unlikely to have applied to the original ocean.

What the world ocean was like before 3500 million years ago – if indeed an ocean existed at all – is obscure. If there was an ocean, it was probably strongly acid. The fossil record, beginning at least 3000 million years ago with bacteria and algae, indicates a cooling of the earth's surface below 100°C. Permanent water bodies could exist, because they would no longer boil off as gas.

Up to 2000 or 1500 million years ago, the atmospheric and oceanic environments remained oxygen-poor. So much is shown by the chemical nature of iron-bearing sediments. Ever since, life forms have developed in an oxygen-rich atmosphere, and in a correspondingly oxygen-rich sea. But the ocean environment of life has probably been affected by changes of salinity through time.

As discussed in Chapter 4 (section 4.3e), salt can be removed from the ocean in large quantities, given a suitable combination of climate and geologic setting. Now, many marine invertebrates (= animals without backbones) have blood plasma that strongly resembles ocean water. For many species, quite small changes in water composition, either in total salinity or in chemical proportions, can be lethal. We draw the general conclusion that many forms of marine life are adapted to an unchanging salinity, and to an unchanging proportional distribution of dissolved salts.

During the Permian period of geologic time, 280 to 225 million years ago, large-scale removal of oceanic salt took place. Huge salt deposits dating from the Permian period supply the basic evidence. It has been suggested that the massive extinction of marine life forms during the Permian period was largely a result of the partial de-salting of the oceans.

amount to a great deal. When they die, their soft parts rot. Their hard parts – often highly ornamented spheres – rain down upon the ocean floor, contributing to bottom sediment.

Even if large amounts of carbonates and silica were not cycled through organisms to the ocean bed, it seems highly probable that seawater chemistry would not change. The ratios among different solutes represents a set of chemical equilibria. Weathering of carbonate rocks on the world's lands, for example is balanced by deposition of carbonates in the oceans. Input of solutes that tends to unbalance the chemical equilibria is offset by output, either through the intermediary of life forms, or by direct precipitation.

More than 100 million km² of ocean floor is covered by red clay, formed by chemical reaction in the ocean waters. Chemical precipitation also forms the so-called maganese nodules. These concretions, although averaging only 4 cm in diameter, can total

Looking further back in time, we consider our own blood plasma. The blood plasma of human beings closely resembles seawater diluted with three times its own volume of fresh water. Modern fish, including marine fish, possess blood plasma very like ours. The circulatory systems of fish are separated from seawater by the gills, which let oxygen in but keep salts out. It can be argued that the ancestral form common both to fish and ourselves evolved in an ocean where the proportions of solutes resembled those of today, but where the total concentration was low. If this is so, then we are looking back only 500 million years. The first animals with backbones, fish, appeared in the Ordovician period of geologic time, 500 to 430 million years ago. The first amphibians, legged and backboned animals capable of moving overland, appeared as late as the Devonian period, 405 to 340 million years ago. Two choices are open. Either gills were evolved in an ocean far less salty than the oceans of today, or backboned animals in general have reduced the levels of salt in their blood plasma, at a rate of reduction common to all.

17.3 Spatial variation of salinity in surface waters

Belts of higher than average salinity straddle the tropics in both hemispheres (Fig. 17.6). They correspond, although rather roughly, to the subtropical belts of high atmospheric pressure, where rainfall at sea is low, and where little water flows off the lands. An intervening belt of lower salinity reflects the heavy rain of equatorial areas. The most noteworthy dilution of surface water in low latitudes is effected off southeast Asia, where copious river discharge reinforces the effect of equatorial and monsoon rainfall. Poleward of the subtropics, salinities decline into high latitudes, partly because of reduced evaporation, but also because of precipitation in the westerlies belts and the melting of polar ice.

Surface salinity in the open ocean does not vary a great deal, ranging mainly between 34‰ and 37‰. The value rises above 40‰ in parts of the Red Sea, where evaporation is strong and runoff from the land negligible. Values as low as 5‰ are recorded for the inner arms of the Baltic Sea, where evaporation is low and runoff high. Melting pack ice (= frozen seawater) contains very little salt, and therefore dilutes the surface water; but it is less effective than melting icebergs which, originating as land ice, can attain very great thicknesses. Surface salinity runs below 34‰ in a broad band of ocean encircling Antarctica, in the Arctic Ocean, and off Newfoundland.

17.4 Salinity, temperature, density and pressure

Density is mass per unit volume – for instance, grams per cubic centimetre, the measure used here. On account of its higher charge of solutes, seawater is denser than river water. But because of the contraction or expansion caused by changes of temperature, density must be referred to some point on the temperature scale.

Fig. 17.7a shows the effect of varying salinity on water at 0°C. As salinity increases, so does density, approaching a value of 1.03 at a salinity of 35‰. This diagram is slightly generalised: as will be noticed shortly, pure water (salinity = 0‰) reaches its greatest density at +4°C, not at 0°C, but the scale of the diagram is too small for the difference to appear.

Fig 17.7b shows the effect of varying temperature on water with a salinity of 35‰. Density decreases as temperature increases.

For the open oceans, the density differences resulting from contrasts in temperature range more widely than those resulting from contrasts in salinity, but the very low salinities of highly diluted waters involve far lower densities than those associated with high temperatures.

It will be obvious that salinity and temperature often act in opposite ways as controls over density. Whereas in the equatorial belt high temperatures and abundant rainfall combine to promote low densities, the low rainfall of the subtropics tends to produce high salinities and therefore high densities, while high temperatures tend to reduce density.

Where rivers in high latitudes reduce surface salinities by dilution, and thus decrease densities, low temperatures act to increase densities.

Density is the main control over the sinking of surface water, and over the uprise of water from a depth. For our purposes, we can assume that pressure increases at a uniform rate as depth increases (Fig. 17.7c). Because sinking or rising water and the surrounding water will be affected to the same

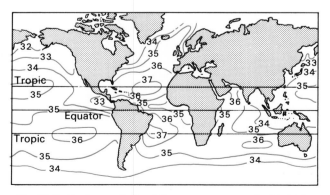

Fig. 17.6 Surface salinity of the open oceans: considerably generalised, and with no allowance for seasonal variation.

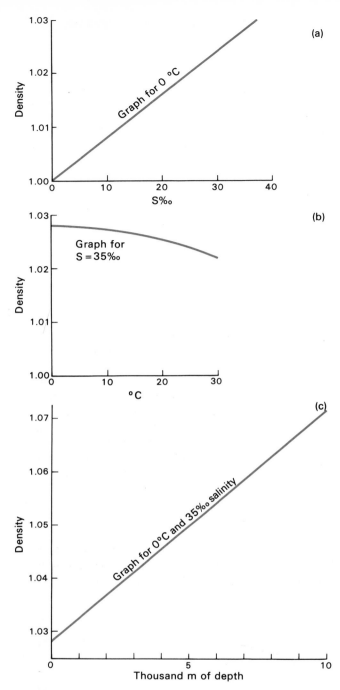

Fig. 17.7 Relationships of density to salinity, temperature and depth as a surrogate for pressure.

extent by change in pressure, sinking or ascent are governed by the temperature/salinity relationship. Because salinity contrasts at depth are mainly slight, temperature differences are the chief causes there of differences in density.

17.4a The thermocline

In some ways, the temperature structure of the ocean can be compared to that of the lower atmosphere. A surface layer, say 100 m thick, records variable but generally high temperatures. Liable to

mixing by waves and surface currents, it is the counterpart of the atmospheric turbulence layer (Fig. 17.8). Next beneath comes a layer about 1000 m thick, the thermocline. Here temperatures fall off very rapidly with depth, just as they decrease with increasing height in the troposphere. Beneath the thermocline, in the deep water, temperatures are little above freezing point. They decrease very slowly with increasing depth.

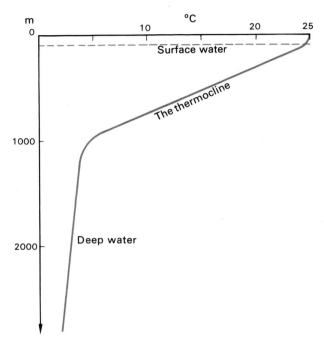

Fig. 17.8 The thermocline.

17.4b Origin of deep water

Nothing in the behaviour of the atmosphere matches the formation of oceanic deep water. Formation takes place in the polar seas when pack ice is produced during the winter season.

Pure water is densest at 4°C. Density decreases beyond that point. It decreases still further when ice expands as it freezes. Freezing point for pure water is 0°C. As salinity increases, freezing point decreases: water with a salinity of 35‰ will not freeze until its temperature is down to −2°C. An increase in salinity rapidly brings down the temperature of maximum density (Fig. 17.9).

Up to a salinity of about 25‰, temperature of maximum density runs above freezing point. Water cooled below the temperature of maximum density remains at the surface, and freezes over when freezing point is reached. For seawater, freezing means the release back into the ocean of most of the salt content. If enough salt is released, the surface waters become dense enough to sink. More important is what happens at salinities greater than 25‰. The water becomes denser as it becomes colder. It sinks to great depths. Subsiding cold water along the margins of the Antarctic continent, where surface salinities run between 33 and 34‰, is the main supplier of the slowly-moving chilled masses of the ocean deeps.

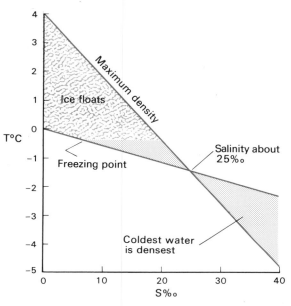

Fig. 17.9 Relationship to salinity of temperature of maximum density.

Fig. 17.10 Surface water masses and convergences: highly generalised.

17.5 Water masses

Just as in the atmosphere, so in the ocean – large homogeneous masses can be identified. But whereas air masses can form in a matter of days, water masses are permanent features of the oceans in existing climatic conditions. Each water mass has its distinctive temperature/salinity value.

A complete account, or even a total summary, of the distribution of water masses is not yet possible. Distribution relates not only to surface waters, but to waters throughout the entire ocean depths. A great deal remains to be learned about subsurface waters. Information available so far shows that a typical vertical section is more complicated than the abstract version of Fig. 17.8, which makes little allowance for variations in salinity at depth.

The distribution of surface water masses is mapped in Fig. 17.10. Boundaries consist of convergences, where temperature/salinity contrasts cause one or other mass to sink at its edge, and divergences, where separating surface flow causes subsurface water to rise. But subsidence also occurs in regions of high surface salinity, and along the margins of freezing polar seas. From a human viewpoint, the most important regions of the ascent of subsurface waters are those on the western margins of continents in tropical and subtropical latitudes.

17.5a Upwelling

On the western sides of Africa and the Americas, in the tropics and the subtropics, surface currents move toward the equator (see below). Because of the

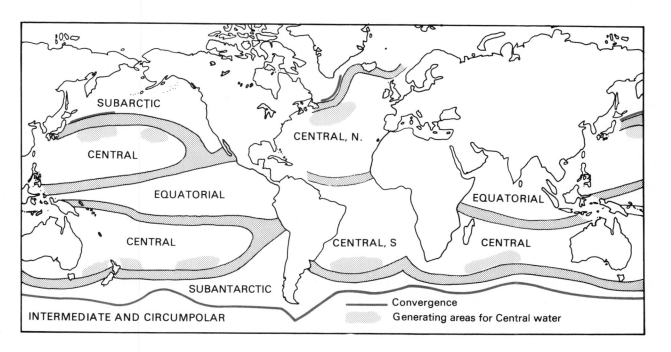

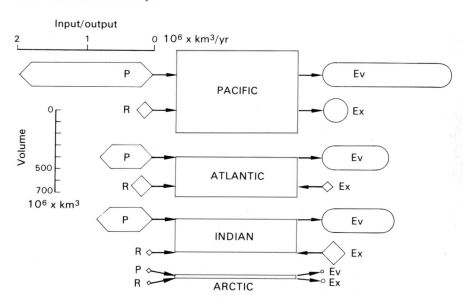

Input/output

Fig. 17.11 Water balances of the oceans: P = precipitation, R = runoff, Ev = evaporation, Ex = exchange. Ocean extents are drawn proportional to volume; inputs and outputs are also drawn proportional to volume, but to a different scale.

Coriolis effect, the currents swing away from the land. Close inshore, the surface water is replaced by water that wells up from rather shallow depths, not usually exceeding 300 m. But the effect can be profound. The rising water brings up nutrients which sustain huge populations of fish – for instance, the anchovies of the inshore waters of Peru. On the negative side, the corresponding effect in the northern hemisphere keeps Californian inshore waters cool, greatly abbreviating the season of ocean bathing.

17.5b Exchange among oceans

Surface currents and the movements of deep water combine to influence exchange between one ocean and another. Fig. 17.11 illustrates estimated total exchange. The symbols for the oceans are proportional in area to ocean volume. Symbols for inputs and outputs are proportional in width to the amounts supplied or abstracted. As shown, the Arctic Ocean is largely self-contained. The deep water generated round the edges of Antarctica moves very slowly. The Indian and Atlantic Oceans profit and the Pacific loses, by net exchange of water. In all cases, input by precipitation is far greater than input by runoff, or than input by exchange where this occurs. For the Pacific, output by evaporation is far greater than output by exchange. The most important exchanges are those affecting the surface waters within ocean basins – for example, between the South and North Atlantic.

17.5c Exchange between the open oceans and enclosed seas

Contrasts in temperature/salinity relationships between the open ocean and a landlocked sea can readily promote a two-way exchange of water. At the simplest, the exchange is effected on two levels.

The Mediterranean Sea, with a deficit of runoff against evaporation, tends to become progressively saltier. Below the level of the sill at the Straits of Gibraltar, about 300 m below the surface, salinity rises above 35‰. The saline water at depth spills out over the sill, flowing for very great distances at intermediate depth into the Atlantic. The Atlantic surface water enters the Mediterranean in a compensating flow (Fig. 17.12a).

Connected by a narrow and shallow strait with the Mediterranean, the Black Sea records an excess of runoff over evaporation. Its surface salinity is low. But saline Mediterranean water is able to spill in because of density differences. The surface outflow from the Black Sea to the Mediterranean is partly compensated by an inflow near the bottom (Fig. 17.12b). Becoming progressively more saline and deficient in oxygen, the lower layers of the Black Sea can support no marine life. They have died.

Along the southern coast of Norway, valleys come down to the sea in the form of deep fiords – glacier troughs drowned by ocean waters. A typical fiord possesses a threshold – a sill of decreased depth, that is probably an end-moraine. Water running off the land, cold but fresh, floats above the warmer, but saline and denser, Atlantic seawater that crosses the sill (Fig. 17.12c). The trapped Atlantic waters are important in winter fisheries.

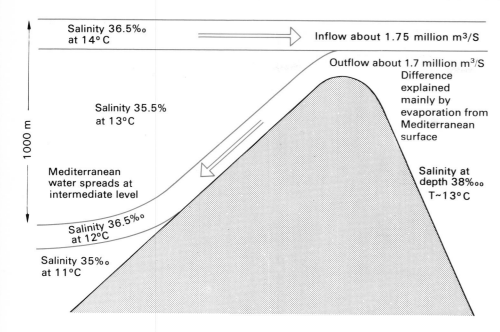

Salinity 36.5‰ at 14°C

Inflow about 1.75 million m³/S

Outflow about 1.7 million m³/S
Difference explained mainly by evaporation from Mediterranean surface

Salinity 35.5% at 13°C

Salinity at depth 38‰ T~13°C

Mediterranean water spreads at intermediate level

Salinity 36.5‰ at 12°C

Salinity 35‰ at 11°C

1000 m

Fig. 17.12a Exchanges of water between the Atlantic and the Mediterranean.

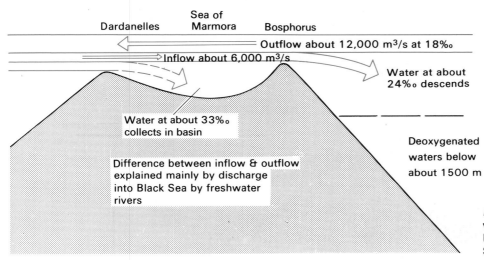

Dardanelles

Sea of Marmora

Bosphorus

Outflow about 12,000 m³/s at 18‰

Inflow about 6,000 m³/s

Water at about 24‰ descends

Water at about 33‰ collects in basin

Difference between inflow & outflow explained mainly by discharge into Black Sea by freshwater rivers

Deoxygenated waters below about 1500 m

Fig. 17.12b Exchanges of water between the Mediterranean and the Black Sea.

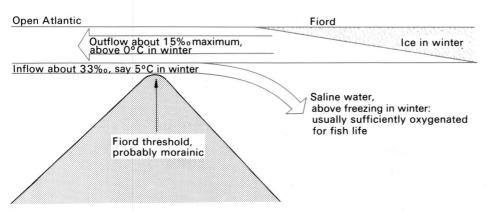

Open Atlantic

Fiord

Outflow about 15‰ maximum, above 0°C in winter

Ice in winter

Inflow about 33‰, say 5°C in winter

Saline water, above freezing in winter: usually sufficiently oxygenated for fish life

Fiord threshold, probably morainic

Fig. 17.12c Exchanges of water between the open Atlantic and a Norwegian fiord.

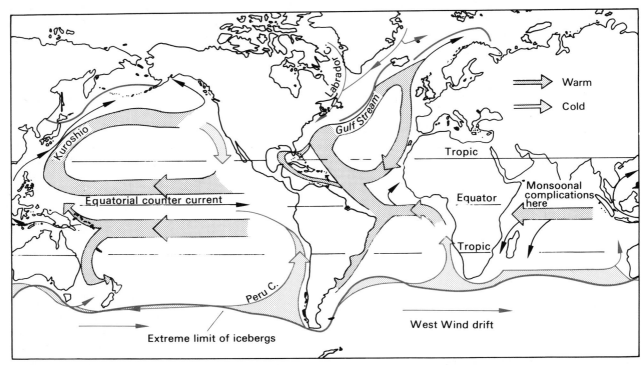

Fig. 17.13 Major ocean current systems: considerably simplified. Compare Fig. 14.11.

17.6 Ocean currents

Major systems of surface currents suggest an immediate comparison with the major systems of atmospheric pressure cells and their associated winds (Fig. 17.13). There is a well-defined current flowing from west to east in the Southern Ocean, in the belt of westerly winds. In contrast to winds, currents are named for the direction in which they flow; thus, the Antarctic Circumpolar Current (also called the West Wind Drift) is an easterly current. Transporting some 175 m³/sec, it is the greatest of all individual currents. Complications result from the effects of monsoon winds in the northern part of the Indian Ocean, but the southern part contains a well-defined oval cell – a gyre (Greek *guros*, ring). Other gyres are present in the North and South Atlantic and the North and South Pacific. Westerly currents in the trade-wind belts and easterly currents in middle latitudes are connected by return flows, equatorward on the eastern sides of ocean basins and poleward on the western sides. A major northeasterly branch separates itself from the North Atlantic gyres.

The Circumpolar current of the southern hemisphere extends to unusually great depths – as much as 3000 to 5000 m. The northward-flowing limbs of the North Pacific and North Atlantic gyres, on the western sides of these two oceans, attain depths of 400 and 800 m, respectively. Elsewhere, surface cur-rents are usually about 100 m deep. Because their greatest velocities appear at or very close to the surface, one might suspect that they are driven entirely by the dominant winds. In actuality, the matter is far more complicated than this.

17.6a The drive of surface currents

Take the North Pacific gyre. Make a first assumption that its southern limb is driven by the frictional drag of the trade winds. Then, because of the Coriolis effect, the surface water should move obliquely to the right of the wind direction – as indeed it does. In middle latitudes, in the westerlies belt, the surface current flows toward the east – obliquely to the right of the mean path of the westerly winds, which have a poleward component of direction. Return flows connecting the westerly current in low latitudes with the easterly current in middle latitudes have already been specified.

But if the low-latitude westerly current is a response to the drag of the trade winds, the movement of water below the surface is a response to the drag of water at the surface. Below the surface, the direction of movement should be north of west. Increasing deflection with increasing depth, down to about 100 m, tends to establish a net transport of water at right angles to the wind direction. Since we are discussing the northeast trade winds, the net transport of water should be toward the northwest into the central area of the gyre.

Similarly, the net transport in the westerlies belt should be toward the southeast, once again toward the central area of the gyre. But accumulation of water in the central area would mean a surface slope from the centre toward the margins, and thus a tendency for water to flow outward and to be deflected by the Coriolis effect into a path round the centre.

This line of reasoning can be looked on either as a two-stage argument, or as a chicken-and-egg proposition. Moreover, nothing has yet been said about the influence of large-scale differences in salinity. Such differences should tend to encourage flow from areas of high salinity toward areas of low salinity, except that the Coriolis effect comes into play again. No rigorous explanation of the drive of surface currents can be offered in purely verbal terms, except for the general statement that a given major current represents a state of dynamic equilibrium among gravity, pressure differences, friction and the Coriolis effect.

The Equatorial Countercurrent at the surface is associated with a belt of divergence in which the sea surface is elevated. The combined outcome of surface slope and the Coriolis effect is a surface current flowing toward the east. The equatorial belt of low pressure occupies an average position somewhat to the north of the geographical equator; so does the Equatorial Countercurrent.

As information continues to come in, the current systems of the equatorial belt promise to prove as varied as the low-pressure disturbances of the equatorial atmosphere. An Equatorial Undercurrent has been identified, both in the Pacific and the Atlantic. Reaching velocities of up to 1.5 m/sec at depths of 50 to 150 m, it rarely reaches the surface, but is instead accompanied by a weak westerly flow at the surface, very close to the equator.

17.6b The Gulf Stream system

Because the shoulder of Brazil projects into the Atlantic south of the equator, it diverts part of the southern-hemisphere equatorial current into the northern hemisphere. Excess water and excess warmth are delivered to the Caribbean Sea and to the Atlantic off the West Indies. A convenient starting-point for discussion is the Florida Straits, through which the Florida current, flowing from the west, delivers 25 to 30 million m³/sec, about a hundred times the flow of the lower Amazon at bankfull stage. The Florida Current is driven partly by an eastward slope of the ocean surface.

Past the eastern exit from the Florida Straits, the flow turns northeastward, being joined by waters of the western Atlantic. The combined current, the Gulf Stream, is some 100 km wide and 800 km deep, with a transverse slope of 1 m/km and a surface velocity of 2.5 m/sec. Its discharge averages about 70 m³/sec.

The Gulf Stream keeps close to land to begin with, but beyond Cape Hatteras, about 35°N, the swing of the North Atlantic gyration takes it out to sea. On the oceanward side, parts of the current break away in meanders and detached eddies. On the side toward the land, the Gulf Stream is sharply separated from colder but less saline water by a front, comparable to the fronts that separate air masses in the lower atmosphere. Between the front and the coast lie the waters of a branch of the Labrador Current. These eventually sink, vanishing from the surface circulation.

Especially off the northeastern U.S. and Newfoundland, the Labrador Current is a highly significant supplier of the nutrients on which fish feed. Furthermore, the shallow waters of the Grand Banks, where glaciers have dumped large quantities of moraine on the continental shelf, are highly suitable for trawling grounds. Against this, warm air drifting off the Gulf Stream, and becoming chilled over the Labrador Current, produces frequent fogs. The hazards that fogs pose to navigation are made worse by icebergs. Some 10 000 bergs are estimated to calve off the Greenland glaciers every year. Fed into the Labrador Current by the Greenland Currents, many of these go far enough south to enter transatlantic shipping lanes. The world's imagination seems to have been permanently caught by the Titanic disaster of 1912: the so-called unsinkable ship, on its maiden voyage, collided with an iceberg and was sunk. More than 1500 passengers and crew went down with the ship.

As soon as it begins to diverge from the coast, the Gulf Stream tends to subdivide into separate threads. Subdivision is especially pronounced in the northeastern Atlantic, where the stream becomes known as the North Atlantic Current, and where wind drag increases. Part of the North Atlantic Current contributes to the return flow of the surface gyre, but parts run northeastward, fringing northwestern Europe. Onshore westerlies, blowing off the warm waters of the North Atlantic Current, make the climates of northwest Europe anomalously warm for the latitudes concerned. On the shores of Norway at the Arctic Circle, mean January temperatures lie between −5° and 0°C, whereas the corresponding value for the eastern margin of the Canadian Archipelago is −25°C.

It seems possible that the strength of the North Atlantic Current varies in a pulsatory fashion. If so, the current should exert pulses of climatic influence on northwestern Europe. However, detailed information is as yet so scanty, and so many other variables are involved in year-to-year climate, that the matter remains rather speculative.

Eventually the northeast-flowing waters subside beneath the much colder but far less saline waters of the Arctic Ocean. The inflow to the Arctic might be distinctly greater than it is, were not some Atlantic water barred out by rises of the ocean floor – notably the sill that connects Scotland to Greenland by way of the Faroe Islands and Iceland. A suggestion comparable to that of melting the Arctic ice by sprinkling it with coal dust is that of blasting out part of the sill, in order to admit enough warm Atlantic water to result in general melting. A parallel suggestion has been made for the Bering Strait between North America and Asia.

17.6c The Japan Current System

What has been said about the Gulf Stream System permits the account of the Japan Current System to be abbreviated. The powerful western limb of the North Pacific gyre is the Kuroshio, about 80 km wide, 400 m deep, and with a delivery of some 45 m³/sec. There is no great branch to correspond to the North Atlantic Current; but the Labrador Current of the Atlantic does have a counterpart, fed by Arctic waters through the Bering Strait, in the Pacific. The dominant winds and the configuration of the eastern coastline of the ocean ensure that the North Pacific current splits into two branches. Flowing towards the south, the return limb of the gyre moves water into lower and lower latitudes, constituting the cold California Current. The northerly Alaska Current – actually only one component of a complex circulation in the northeast Pacific – carries warm waters close inshore, greatly ameliorating the winters of the Alaskan shores.

17.6d Southerly and northerly currents of the southern hemisphere

Southerly (warm) currents affect the eastern coastlands of Australia, southern Africa and much of South America. The trade winds tend to keep the warm water close inshore and also to transport warmth to the lands. Both in volume and strength, these currents are liable to vary somewhat from year to year. On the eastern coast of Australia, where sea bathing and surfing have attained cult status, an increase in the flow of the East Australia current results in a greatly increased shark danger off the beaches of Sydney.

Cold northerly currents flow off the coasts of Africa and South America. The northerly West Australia Current is unusual in being warm. As noticed above, the equatorward return flows of the current gyres are typically cold; and, where they are strong, are sufficiently diverted by the Coriolis effect to involve upwelling in the close offshore waters. But because the trade winds blow away from the land – incidentally reinforcing the Coriolis effect – the cold return flows have only modest effects on the climates of the adjacent coastal areas. If air is drawn in, it is drawn in to a heated land and is unlikely to bring rain: its moisture-holding capacity increases as its temperature rises. Desert comes down to the sea in southwest Africa and in tropical South America, just as it does in the Northern Hemisphere at the western edge of the Sahara and in Lower California.

By far the greatest of the northward return flows in the southern hemisphere is the Peru Current. It includes large amounts of water diverted from the Antarctic Circumpolar Current by the long southward projection of the South American continent. If air moves onshore from the Peru Current to the Peruvian desert, it comes in an airstream so stable that fog, rather than rain-bearing cloud, forms as the air rises over the foothills of the Andes.

Exceptions occur when the northern part of the Peru Current is overspread by a skin of equatorial water, which though saline is so warm that it floats on the surface. Then, onshore winds bring in moist air that has been strongly heated from below. Instability released by the ascent toward background of mountains causes the formation of cumulonimbus cloud which releases heavy downpours. Nothing in the design of the regional irrigation farming, or the type of rural housing, provides for these. Storm rains rip out irrigation channels or choke them with flood-borne debris. Flat roofs with parapets, and with no or inadequate outlets for water, may collapse. The sun-dried bricks of house walls melt away in the rain. At sea, the upwelling cold water of the inshore strip is covered by the warm equatorial veneer. Disaster is liable to strike the inshore schools of fish, the fishing industry, and also the multitudinous flocks of fish-eating birds.

On account of their timing – when events of this kind take place, they come in the Christmas season – the happenings are collectively known as *el niño*, the Christ-child. There can be little doubt that we have to do with some kind of pulsatory rhythm, broadly akin to that suggested for the Gulf Stream. *El niño* arrives at intervals of about nine years. For some periods, the nine-year spacing has in fact proved remarkably regular.

17.6e The long view

The best prospects for investigation of the short-term influence of surface currents on climate, and especially of short-term changes in that influence, are offered by the Gulf Stream system, *el niño* and the Japan Current system. But there seems to be no reason why additional current systems should not be studied in the same way and with the same objectives. If pulses can be identified, then variations in the transport of heat energy can be calculated. Climatic forecasting could conceivably, by this means, become a great deal more precise than it is today.

Chapter Summary

Ocean *tides are raised* by the *gravitational pulls* of earth, moon and sun. *For the lunar* (moon-generated) *tide*, there is an *excess of pull toward* the moon, *on the side* of the earth *that faces the moon*, and an *excess pull away* from the moon *on the opposite side*. *Similar considerations* apply to the *solar* (= sun-generated) *tide*. When *moon and sun pull together*, *spring tides* occur. *When they pull at right angles*, *neap tides* occur.

The *greatest tidal ranges* are recorded *in* certain *inlets*. Some inlets have been harnessed for the production of hydro power. Some inlets experience *tidal bores*.

Tidal waves in an ocean or sea basin *rotate round an amphidromic point*. The tidal range *increases with distance* from the amphidromic point.

Average river water has a *salinity of 0.13‰, average sea water* one of *35‰*. Average *river water* is a *bicarbonate solution*, average *sea water* a *sodium chloride solution*. Observed *seawater chemistry indicates* a set of *chemical equilibria*. Materials *precipitated out* include *red clay* and *manganese nodules*, the latter a *probable source of metallic ores*.

Seawater chemistry tends to remain *constant through time*, in respect of the *relative proportions of solutes*, but massive *de-salting during the Permian* period may have been responsible for massive *extinctions of marine life*.

Surface *salinity in the open ocean does not vary much*. Unusually *high values* are recorded for the desert-bordered *Red Sea*, and unusually *low values* for the *Baltic Sea*, where run-off from the land far exceeds evaporation.

Whether surface water remains at the surface, or sinks down, depends on density. *Density increases as salinity increases, and as temperature decreases*. The *total effect* is to produce a *sharp* downward *decline in temperature between* about *100 and 1000 m* down: this is the *thermocline*. *Below* the thermocline lies *deep water*, with temperatures *little above freezing point*.

Density complications occur *close to the freezing point*. *Surface water can be made more saline by freezing*; very little salt is taken up by the ice. *Above* a salinity of *25‰, increasing cold means increasing density*, regardless of freezing. The cooling water sinks. Subsiding cold water around the margins of Antarctica is the main supply for the chilled masses of the ocean deeps.

Like the low troposphere, the *ocean surface* can be *classified into* masses – in this case, *water masses*. Their *boundaries are convergences* (zones of sinking) *and divergences* (zones of upwelling).

Exchange of water among oceans is modest in comparison with the inputs of runoff and the outputs of evaporation. *Exchange between the open ocean and enclosed seas* involves *complex effects of differential density* through depth.

Major surface current systems include prominent *gyres*. The *flow directions resolve* the influences of *wind drag*, *density* differences, *friction* and the *Coriolis effect*. Complications at depth include progressive divergence of flow direction from surface wind direction, and systems of countercurrents.

The *Gulf Stream system* is a *major climatic modifier* for the North Atlantic and its eastern borderlands. The *Labrador Current contributes nutrients to* the *fisheries* of the Grand Banks, but *also carries icebergs* into the transatlantic shipping lanes. The *Japan Current* system in the North Pacific *corresponds to the Gulf Stream* system in the North Atlantic. In *midlatitudes in the southern hemisphere, warm currents* affect the *eastern sides of landmasses*, and *cold currents* (except in Australia) the *western*. The *Peru Current* is a most noteworthy cold current, *important in fisheries*. Its climatic influence is cut off from time to time by the spread of warm equatorial waters, in the *el niño* event.

In the long term, continued *study of variations in ocean currents* promises to *assist climatic forecasting*.

18 Atmospheric Step Functions I

In dealing with environmental systems, it is impossible to follow the advice given to Alice in Wonderland – namely, to start at the beginning, go on until you reach the end, and then stop. The operations and interconnections of environmental systems are not linear. If they were, it would be possible to start with a set of axioms (= established principles, maxims, self-evident truths) and to work through to the end by means of logical reasoning. The analysis of systems would resemble exercises in Euclidean geometry.

In practice, it proves extremely difficult to discuss one system without reference to other systems. Possible exceptions are the solar energy cascade at the outer limits of the earth's atmosphere, and gravitational pull. But considered solely for their own sake, these are matters of astrophysics and geophysics, not of environmental components, attributes and behaviour.

Moreover, there is no best fixed order of major environmental topics, or of minor topics within a major topic. Even though the weathering of some original granitic crust supplies a convenient starting-point for the tracing of the rock material cascades, it does not really matter where one enters any given cascade. In pursuing a circle, such as a circle of the recycling of rock material, one must by definition come back to the starting-point. Although the distant history of the atmosphere and the ocean can be provisionally reconstructed, what matters in the present is the nature of working systems as they are observed today.

Climatic systems can be examined for their own sake, but can scarcely be discussed without reference to the exchange of energy between the atmosphere on the one hand and land and sea surfaces on the other. Climate, depending ultimately on the solar energy cascade, defines the atmospheric environment, including that of land areas. But land areas are themselves great modifiers of climate, especially of climate near the ground. Climate in the lowest 1 m of air, below the level of instrument readings in a meteorological screen, can differ strongly from the climate of the lowermost troposphere as a whole. Still greater contrasts apply to the lowest 2 cm of air,

wherein seedlings begin their life above ground. This layer can develop climates – microclimates – all of its own.

At certain junctures, then, it becomes useful to step back a little, and to see where we have got to so far in the discussion of interrelated and interacting systems. The present is one such juncture. Previous chapters have dealt with some leading aspects of geoscience, under the general heads of the recycling of rock materials and of water, the shaping of the land-surfaces and of land margins, with the dynamic behaviour of the oceans, and with the behaviour of the lower atmosphere as expressed in weather and climate. It is now time to examine the interaction of climate and landscape in relation to the effect of climatic change upon landscape type and landscape features.

18.1 The concept of morphogenesis

The term morphogenesis derives from the Greek *morphē*, shape, and the compound *-gen*, become. According to the concept of morphogenesis, particular climates control the distinctive evolution of particular landscapes. In some writings morphogenesis is called climatic geomorphology: the two names mean the same thing. The basic idea is that climatic input constitutes the set of processes. Landscape type is the response. Climate is taken to impress a distinctive imprint upon the landscape as a whole.

Somewhat more modestly, it can be claimed that even if entire landscapes do not differ from one another according to climate, certain types of physical features develop only in particular types of climate. Thus, if such features occur outside the limits of the diagnosed climate, they are diagnostic of climatic shift.

The idea of morphogenesis, whether applied to entire landscapes or to particular features, is closely allied to the principles of the analysis of the sedimentary record. Workers on sedimentary deposits pay close attention, not only to the environments of deposition, but also to the environments of the source areas. Sediments delivered to the sea by rivers, or laid down in inland basins, reflect the weathering processes that operated in the areas of sediment supply. Former climatic environments are important throughout. Desert sands can be produced only in regions of arid climate. Evaporites only form where brines are progressively concentrated by evaporation. Glacial materials dating from 250 million years ago, and now contained in regions of tradewind climate, testify to major changes of the depositional environment. So do rocks formed from marine sediments which now occur far inland and at great heights above existing sea level.

The proposition of drastic environmental change is by no means new. However, as will shortly be seen, it raises many difficulties when applied to the landscape as opposed to sediments. Nevertheless, if any part of the concept of morphogenesis is accepted, then a two-stage approach to landscape analysis becomes possible. Either the wholesale alteration of morphogenetic process-response can be traced, or features diagnostic of a particular climate can be relied on.

Implicit in all this is step-functional change. If one set of processes acts on a given area long enough to produce a distinctive landscape or distinctive features, and then another set of processes takes over, acting long enough to produce landscapes or individual features of its own, then the switch from one set of controls to another is assumed to be abrupt. So far, nobody has claimed to identify intermediate forms, such as could be imagined to result from a smooth and gradual transition.

18.1a Some extreme and obvious cases

Glaciated landscapes in areas not now subject to glacial climate, and no longer capable of sustaining valley glaciers or ice sheets, supply one set of extreme and obvious cases. As discussed in Chapter 13, erosion and deposition by ice produce highly distinctive erosional and depositional forms. The fact of glacial morphogenesis is beyond dispute. Where the landforms of glaciation occur in areas not now controlled by glacial climates, a shift of climate and of morphogenetic process has taken place. It is recorded in the relict (= leftover) response of landscape attributes and landscape components.

Similarly, features produced by periglaciation can survive in areas where the processes of frozen grounds are no longer active. Patterned ground defined by former ice wedges, stone nets and blockstreams are widely known from midlatitude regions (Fig. 18.1). The ice wedges have long since melted, the stone nets are no longer developing, and the blockstreams are stagnant. Here again, one type of climate has taken over from another. There has been

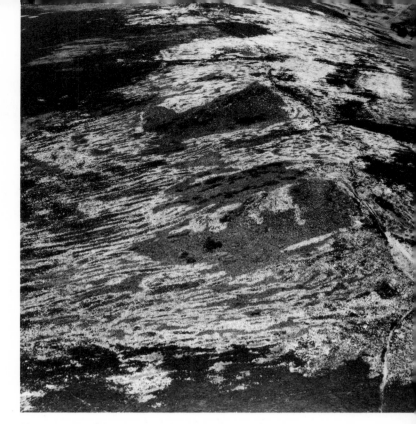

Fig. 18.1 The Stiperstones: stagnant streams of quartz blocks, originally shed by the upstanding outcrops on the hilltop (Salop, Welsh Border: J.K. St. Joseph). Compare Fig. 13.24.

a change in the dominant processes of landscape development. But whereas it is easy to recognise a landscape of former glacial erosion or of former glacial deposition in its entirety, landscapes of former periglaciation cannot be identified on sight. They demand the recognition of diagnostic types of feature which may be of minor importance, or even altogether invisible, in the present landscape. Former ice wedges, and frost-cracks generally, appear only in quarry sections and on air photographs. Relict stone nets and blockstreams are often partly overgrown, and in any case contribute little to the total scene.

18.1b Desert margins: some less obvious cases

Whereas the climatic limits of glaciation and periglaciation are quite narrowly defined, the limits of desert climates are not. It is true that limits for semi-arid and arid climates can be selected, as they have been in Chapter 15: but desert margins are apt to fluctuate quite widely in the medium term, say a term from ten to fifty years. Dry conditions take over, beyond what is supposed to be the desert boundary. Alternatively, regions that average only

Fig. 18.2   A dead dune of the seif type, largely fixed by trees: northwestern N.S.W., Australia.

200 mm of rain a year can from time to time be subjected to extensive flooding.

Where rainfall in low latitudes is scanty, it is also likely to be highly variable. Low-latitude regions with scanty and variable rainfall provide instrumental records of weather and climate which are typically short, if they exist at all; and recording stations are spaced far apart. It thus becomes difficult or impossible to measure by direct means the kind of variability that can reasonably be expected in the medium term. For the long term, only indirect evidence is to be had.

In looking at certain dry regions as they exist today, we observe extensive systems of dead dunes. Such is the case in vast areas of inland Australia. It is obvious that the dunes were once live, consisting of sand which moved whenever the wind blew strongly enough. Today, although much dust and some sand will blow during a windstorm, the dunes are mainly inert, supporting and being fixed by vegetation (Fig. 18.2). The pointed and simplest inference is that the present climate is less dry than some former climate wherein the dunes were live. But we simply do not know, for the long term, how normal it is for episodes of dune development to alternate with episodes of dune stagnation; and we know very little about the respective durations of the contrasting episodes.

Again, it is usual to find evidence of former permanent lakes, well inside the margins of existing deserts. The term permanent in this context does not mean everlasting: but the lakes in question remained filled with water for long periods at a time. Somewhere in the past, the climate was wetter than it is now. The former lakes have been succeeded by dry saline flats. Once again, we simply do not know how normal it is for flooding to occur in existing climates. The huge Lake Eyre basin in the Australian inland has been flooded twice since 1949. While on both occasions the floodwaters were soon vaporised, the events in question show how difficult the use of indirect evidence for variation in weather and in climate can be.

18.1c The humid tropics: some still less obvious cases

Like glacial, periglacial, and desert regions, the humid tropics represent a climatic extreme – in this case, the combination of extreme heat with extreme wetness. We can expect chemical weathering to go deeper, and to penetrate faster, in the humid tropics than elsewhere, and to promote unusually high rates of erosion. But there is as yet no proof that humid –tropical landscapes, with their rainforests replaced by midlatitude forest, would look significantly different in general from the landscapes of middle latitudes. One type of physical feature is somewhat special to the humid tropics: a second may be.

The first of these is the non-cyclic nickpoint, the break of slope in a stream profile that is unrelated to a wave of downcutting. In the natural state, rivers in the humid tropics transport little or no bedload, and little or no suspended load in the sand grade. Lacking abrasive sediment, the rivers are incapable of drilling potholes into the channel bed, where this is cut in bedrock, and are extremely slow in removing outcrops that produce rapids or falls (Fig. 18.3). In consequence, the profiles of tropical rivers are far more liberally broken by nickpoints than would be normal for mid-latitudes.

The second possibly special type of feature is the sharp and very steep-sided ridge cut in silica-poor volcanic rocks (Fig. 18.4). Features of this kind occur widely in the Pacific, for instance in the Hawaiian Islands and in Tahiti. They are spatially associated with frequent thunder-rain and with high temperatures the year round. Nothing comparable seems to have developed on the flow basalts of middle latitudes. But whether or not they are really special, or merely the results of weathering and stripping that are rapid merely because they are intense, remains to be determined.

Fig. 18.3 Detail of rapids at low-flow stage, on a small stream cutting through granite in South Brazil: no bedload, no potholes, no sign of scour.

Fig. 18.4 Maunaluaui Ridge, Hawaiian Islands: cut in basalt in humid-tropical climate (P.J. Gersmehl).

18.2 Appeals to particular landforms

Because non-cyclic nickpoints on midlatitude rivers can be explained by variations of channel efficiency, acting on existing rivers in existing climates, they are irrelevant to the question of morphogenetic change. Regionally-developed forms that are prominent in many landscapes, and that have been used by some writers as evidence of climatic shift, include pediments and inselbergs.

18.2a Pediments

These landforms, as stated in Chapter 9, are rock-cut surfaces which tend very strongly to possess semi-logarithmic profiles. They develop upward from some controlling level such as that of a drainage channel or a basin floor. Their extension involves the destruction of the higher ground beyond.

Pediments were first recognised in regions of dry climate, where they have been mainly studied. An idea has developed that they are, in fact, diagnostic of dry climates. If this were so, then pediments outside the existing limits of dry climates would be diagnostic of climatic shift. In actuality, pediments can be shown to be developing in midlatitudes at the

present day (Fig. 18.5). Moreover, workers on periglacial landscapes recognize cryopediments – pediments formed under the influence of frost action. We must conclude, therefore that pediments are not diagnostic of dry climates, and cannot be used as evidence of process change.

18.2b Inselbergs

The word inselberg (taken from the German) means island-mountain: that is, a steep-sided and usually large residual hill or mountain. The particular form here in question is the sugarloaf hill (Fig. 18.6), first

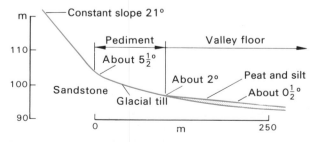

Fig. 18.5 A pediment currently developing in cool-temperate mid-latitude climate: Salop, Welsh Border (from a field survey).

Fig. 18.6 Inselbergs of the Mount Olga group, Central Australia (Australian Information Service). Bedrock is conglomerate; summits rise as high as 500 m above the surrounding plain.

described in detail for regions of tropical climate in Africa. As for pediments, an idea has developed that sugarloaf inselbergs are in some way diagnostic of climate, although there is some doubt about their alleged climatic association. Those who consider inselbergs to be typical of particular climates usually fix on tropical climates of some kind or other.

Now, inselbergs are known to occur and to be still developing in a wide range of existing climates. They acquire their characteristic shapes, very steep sides and rounded tops, by sloughing off sheets of rock that range in thickness from a few centimetres to

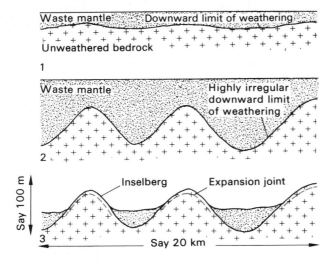

Fig. 18.7 The hypothesis of the origin of inselbergs in the deep-weathering process.

Fig. 18.8 Continuing development of an inselberg in arid climate: Ayers Rock, Central Australia (S.A. Schumm). Bedrock locally is coarse sandstone, elsewhere conglomerate; lifting sheet, about 1 m thick, allows sky to be seen beyond. Compare Figs. 4.7 and 18.6.

several metres. Many of them are composed of rather coarsely crystalline igneous rock which froze deep underground at very high pressures. Erosion, stripping of the cover, has permitted the development of concentric expansion-joints: the outermost shell lifts away as it expands in response to the relief of pressure, and is eventually shed.

In order to claim inselbergs as climatically diagnostic features, it is necessary to suppose that their evolution begins with deep weathering in humid –tropical conditions. The necessary argument runs as follows: deep weathering attacks the weaker rocks around the potential inselberg, preparing rock waste for eventual stripping. When stripping occurs, the inselberg is left as an upstanding feature, subjected thereafter to the expansion and shedding of rock shells (Fig. 18.7).

It is true that very many inselbergs occur in regions of former tropical deep weathering (see next section). But the fact that they are still evolving, even in regions of truly arid climate (Fig. 18.8), suggests that former deep weathering is not relevant to their origin. All that seems to be needed is a rock mass capable of shedding sheets, when internal pressure is released by differential erosion.

18.3 Appeals to the products of weathering

Here we come to more certain ground. The processes and products of weathering in the humid tropics differ greatly from those of humid mid-latitudes. The processes differ in speed and effectiveness; the products differ chemically, mainly because of the selective removal and redistribution which occur in humid–tropical climates.

Chemical weathering in the humid tropics is rather loosely called deep weathering. The expression seems to carry one overt meaning and one loud overtone. The meaning is literal – chemical weathering can penetrate very deeply in the humid tropics, say to 100 m or more: however, it is simply not possible to say how deep weathering has to go before it can be defined as really deep. The overtone is that weathering in humid–tropical conditions is very thorough. Rock materials that would be stable, fairly stable or partly stable in the humid mid-latitudes are mobilised in the humid tropics. They are either carried down into the groundwater system, to be eventually delivered to rivers or to be re-precipitated at depth; or they may be re-precipitated in the subsoil.

Only three chemical substances concern us here: an oxide of aluminium, Al_2O_3; an oxide of iron, Fe_2O_3; and silica, SiO_2. The compounds Al_2O_3 and Fe_2O_3 are commonly called the sesquioxides (Latin *sesqui*, $1\frac{1}{2} = 2:3 =$ two metal atoms to three oxygen atoms). They are chemically stable. They accumulate selectively as other products of weathering are dissolved or washed out. If a soil or subsoil layer highly enriched in sesquioxides is made to dry out, it sets hard like brick (Fig. 18.9). The end product is a form of duricrust (simply = hard crust).

Drying out can result from general downcutting of drainage lines and the lowering of the groundwater table. It can also result from deforestation, such as is in progress today in the Amazon Basin: interruption of the recycling of soil materials through the forest body disturbs the equilibrium of the soil system. But a change of climate sufficient to lower the groundwater table, and thus to cause the topsoil and subsoil to dry out, can work in the same way. This it has certainly done in the past.

18.3a Deep-weathering profiles

Disregarding the question of how deep weathering has to go before it is actually classified as deep, we can recognise two variants of deep-weathering profiles. As already indicated, these can extend to depths of 100 m or more, contrasting strongly with the weathering profiles of middle latitudes, which commonly run between 0.5 and 2 m in depth. In one variant, the profile is strongly differentiated into horizons (Fig. 18.10). The pallid zone, in the lower part, has been reduced (= de-oxygenated). In many sections it is dominated by kaolinitic clay, the familiar china clay of pottery manufacture. It is inferred to record a former, permanent, groundwater table. The upper part of the profile, now the duricrust, originated as a spongy concentration of sesquioxides, close beneath the ground surface. The interme-

Fig. 18.9 Detail of the texture of a ferralitic crust exposed in a cliff section, N.S.W., Australia: clays have been removed from cavities by wave action. The visible crust is about 2 m thick, but the original thickness was greater; soil formation in the existing climate is destroying the crust from the top downwards.

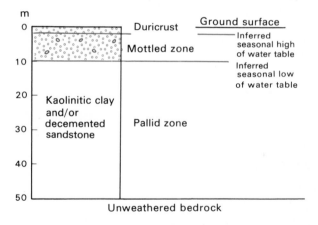

Fig. 18.10 A strongly differentiated deep-weathering profile.

diate horizon, the mottled zone, was formerly within the range of the seasonal rise and fall of the water table. In response to the wet season, the groundwater table rose. Reduction of oxygen caused the formation of clay minerals. During the dry season the level of groundwater fell. Sesquioxides were precipitated. Hence the blotchy combination of metallic (reddish-purple) and pale (clayey) colours.

In the second variant, the surface concentration of sesquioxides is succeeded downward by a thick horizon of clays, blotched and stained throughout by sesquioxides, and appearing mauve, purple and red.

Fig. 18.11 Silica-rich crust, analysing at more than 99% silica, in northwestern N.S.W., Australia: coloured bands on the surveyor's picket are 1 ft (25 cm) long.

Such an horizon seems to develop especially well on silica-poor igneous rocks.

Where deep-weathering profiles, or some parts of them, are particularly rich in Fe_2O_3, they can be sources of commercial iron ore. Far more important, though, is richness in Al_2O_3, which can mean sources of commercial bauxite, the chief ore of aluminium. To keep the record straight, it is necessary to say here that the bauxites of Jamaica, the leading present supplier of the western world, originated with deep weathering, but have been naturally transported to their present positions. They no longer constitute parts of deep-weathering profiles in place.

18.3b Deep-weathering profiles and morphogenetic change

Relict deep-weathering profiles are known far beyond the present limits of humid–tropical climates. It seems certain that they developed in former climates of humid low-latitude types. They contrast strongly with the soils developing in middle latitudes today. The provisional inference – which will presently be confirmed – is that a major morphogenetic shift has taken place.

18.3c Silica-rich crusts

Whereas the selective concentration of Al_2O_3 and Fe_2O_3 is easy to understand in terms of the selective removal of other soil minerals, very high concentra-

tions of silica, SiO_2, pose problems. Nevertheless, they are widespread. Moreover, they typically occur in the tops of weathering-profiles, where pallid and mottled zones are present below. It appears possible that, in some cases, enrichment with silica is independent of deep weathering, in the sense that it occurred later; but elsewhere, the potential crust was a layer of sand or fine gravel in the subsurface, that has been powerfully cemented by mobilised and redeposited silica. One likely means of mobilisation is the uptake of silica by plants: SiO_2 amounts to as much as 20% of the solid substance of bamboo. The grasses whose leaves can cut your fingers are also high in silica.

Silica-rich crusts provide some of the best evidence of the climatic environment of deposition. They are found, mainly in remnant form, round the margins of the lower Mississippi Basin in North America and round the lower Rhine basin in Europe, with an extension round the ancient basin of the English river Thames. In Australia, they are very widespread in huge expanses of the inland regions (Fig. 18.11). The European occurrences include imprints of the leaves, stems and roots of tropical plants – palms, bananas, laurels and plantains. Perhaps more importantly, they also include the fossil burrows of tropical spiders. It follows that the material that is now rigid and highly resistant rock was originally soft enough for plants to root in and for spiders to burrow through.

18.3d The former vegetation cover

The analysis of the vegetation cover, under which soil material that is now duricrust accumulated, and beneath which deep-weathering profiles developed, does not stop with the identification of individual plant species. Wherever complete plant assemblages have been related to former deep weathering, the former climate can be reconstructed. The most detailed reconstruction of all belongs to the London Clay formation of southeast England, from which fossil fruit has been gathered in great quantity. The reconstructed forest of from sixty-five to fifty-five million years ago is of a type now confined to the humid–tropical lands of India and Malaysia. Elsewhere, analysis of leaf shapes permits former climate to be determined. The answers so far have always ranged from humid–tropical to humid–subtropical.

18.3e Timing and implications

Relict deep-weathering profiles and duricrusts extend well into middle latitudes – say, at least as far as 45° of latitude from the equator. Those that have

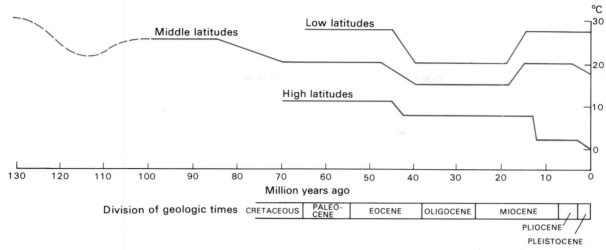

Fig. 18.12 Step-functional change of ocean surface temperature.

been dated go well back into the Tertiary bracket of geologic time, say between 15 and 65 million years ago; and some may well be older still. The changes in question have been far too rapid to be accounted for by continental drift, which merely confuses the general picture, if indeed it changes the picture at all.

We are dealing here with the difference between a tropospheric system adapted to glacial/interglacial conditions, such as those in which we ourselves have evolved and now exist, and a system that includes ice-free polar regions.

Information about low-level air temperature and near-surface sea temperature in the last 100 million years is still scanty. Points in a series are often generalised into curves. But where something approaching a continuous record is available, step functions appear. Fig. 18.12 (admittedly quite highly generalised) presents an outline sequence for the last seventy million years. From seventy to forty-five million years ago, the surface temperatures of low-latitude seas resembled those of the present time. They were considerably lower between forty and twenty million years ago. The temperature graph for rather high middle latitudes broadly reflects that for low latitudes, except for a modest reduction of temperature during the last five million years. The graph for very high latitudes is very different. It shows an abrupt cooling after forty-five million years ago, and a drastic and very abrupt cooling at about fifteen million years ago. It is the differential behaviour during the last fifteen million years which has produced the existing temperature contrasts between high and low latitudes.

Some writers maintain that the cooling at about forty-five million years ago reflects the first establishment of an Antarctic ice sheet; others place this event at about fifteen million years ago, and still

others as late as three and a half million years ago. Obviously, more remains to be learned. Meantime, we must conclude that the general atmospheric circulation of non-glacial times was very different from the glacial (strictly speaking, interglacial) circulation drawn in section in Fig. 14.14.

The switch from midlatitude climates capable of promoting deep weathering and duricrusting to climates such as those of today was sudden. We are talking about a duration of tropical-equivalent climate of from ten to twenty-five million years – or even more – at a time, with the switch needing only from two and a half to five million years. On the geologic time scale, the switch is certainly abrupt enough to be classed as a step function. The net temperature difference from existing midlatitude climates is of the order of from 12 to 15°C, as great as any shift in the known record.

18.4 Climatic shifts of glacial and periglacial episodes

A midlatitude change from existing climates to glacial or periglacial climates would mean a decrease in mean annual temperature of some 5°C. In higher latitudes the difference may amount to 9°C.

Two kinds of step function are here in question. One is the climatic deterioration which made widespread glaciation possible in the first place – the switch from geological normal climates with ice-free poles to glacial climates with extensive ice sheets. The other is the switch from glacial to interglacial conditions, and back again, during a single glaciation.

We know little about the duration of glacials and interglacials except for their most recent history. The total duration of the present glaciation is highly uncertain, as also is the number of glacial episodes that it has contained. The longer the total duration, the larger the number of glacials inferred. A converse statement also holds: the greater the number of glacials thought to be in the record, the longer the inferred total duration.

It is possible to attempt a rough calculation of the fraction of time consumed by switches from glacial to interglacial conditions and back again. There is evidence that one such switch takes about 5000 years. Conversion from the last full-glacial conditions to climates very like those of the present is well documented, especially that about the rise of sea-level as the ice sheets melted. Most of the rise took place in a 5000-year interval. Analyses of minute fossils in ocean cores indicates that the transition from warm (= interglacial) to cold (= glacial) conditions also averages about 5000 years.

A period of 5000 years seems long enough for the build-up of an ice cap. Although the central parts, once established, would probably receive very little precipitation, increase in the energy gradient may well increase storminess and snowfall in marginal areas. An intensified polar front and increase in the travel speed of lows could probably increase precipitation over present values, at 50°N, by some 10%. Ice accumulating at the equivalent of 600 mm of rain a year could reach a thickness of 3000 m in the 5000-year interval indicated.

On this basis, we can calculate that the switches required only from 4 to 8% of the total duration of glaciation during the Pleistocene. To take a specific case: 5000 years is very short in comparison with the 50 000 years duration of the interglacial before the last glacial. Thus, with glacial/interglacial alternations, as with the shifts of the boundaries of low-latitude climates, the changes were abrupt. They were step functions. They resulted from the replacement of one equilibrium state by another.

18.5 Step-functional changes in surface runoff

While changes in climate involve changes both in temperature and in precipitation, it is convenient in several contexts to concentrate on one or the other. The extension of humid-tropical climates into middle latitudes was, in all probability, accompanied by an increase in precipitation in addition to an increase in temperature. The switch from interglacial to glacial climates must have involved some disturbance of the low-level circulation in the westerlies belts, with accompanying change in the precipitation regime. But some of the changes now to be discussed were far more rapid than the flips from interglacial climates and back again. Equally, the temporary equilibrium states lasted for much shorter times than did glacial maxima or glacial minima.

Former increased aridity is inferred from systems of desert dunes that are now dead. Former increased humidity is inferred from former standing lakes in dry areas, and from former large river channels in middle latitudes. In all cases, the landscape changes can be explained by a reduction of precipitation by about one-half in the case of former drier conditions, or by an increase in precipitation by a factor of about two in the case of former wetter conditions.

18.5a Dead dunes

Despite what has been said in section 18.1b about the problems of interpreting the records of desert margins, certain large-scale and widespread effects testify surely to climatic change. For areas of hot dry climate, present or former, temperature is far less important than is precipitation.

Extensive dunefields, no longer subjected to sand-drying, with their dunes fixed and supporting vegetation, occur far outside the existing limits of active dunes. The extreme case is that of the near-arid Kalahari region and the adjacent arid Namib of southwestern Africa. In some former spell of severe dryness, late enough for its products to be recognisable today, linear dunes from the Kalahari–Namib reached from 1500 to 2000 km to the mouth of the Congo (Fig. 18.13). The Congo drainage was so short of runoff that the river could not maintain its course through the barriers of sand. Probably at the same time, the southern boundary of the Sahara shifted equatorward through some 500 km. Dead dunes today extend as far as that past the existing

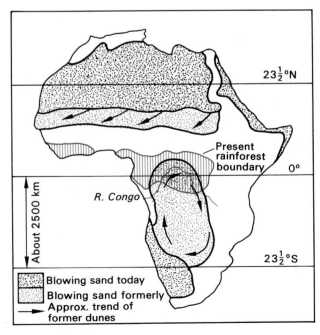

Fig. 18.13 Former active dunefields in Africa.

Fig. 18.14 Shore bench of former Lake Bonneville, cut into the front of the Wasatch Mountains near Salt Lake City, Utah.

limit of sand-driving. Dead dunes are enormously extensive in inland Australia. On a much smaller scale, the Sand Hills of South Dakota record some former episode of increased dryness and of dune activity. A decrease of precipitation to about one-half present values would suffice to explain the observed effects.

18.5b Pluvial lakes

The term pluvial implies a former increased precipitation (Latin *pluvia* = rain). It is not altogether desirable, because it begs the whole question of cause: but we are stuck with it. Pluvial lakes are former lakes which existed as standing water bodies where no lakes or only reduced remnants exist today. They are extremely common in large areas of the world's deserts. We recognise them by their bottom sediments and by their shorelines. The latter are frequently encrusted by precipitated brine salts – evaporites. Bottom sediments also frequently include evaporites, as in the basin that includes the Great Salt Lake.

The best-known pluvial lakes are those of the Great Basin – Lake Lahontan in the west, Lake Bonneville in the east. The largest known is Lake Dieri, the ancestor of Lake Eyre in central Australia, which at maximum extent covered 160 000 km². The ancestor of Lake Tchad, in dry tropical Africa, was also colossal. All of these have left shorelines, in places with well-defined wave-cut platforms (Fig. 18.14). A lake stand recorded in a line of beach indicates the state of dynamic equilibrium discussed in Chapter 1 (Fig. 1.1 and text) and in Chapter 5 (section 5.5a) where output (evaporation) matches input (runoff plus direct precipitation) and the behaviour of the system is controlled by the extent of the lake surface.

It is not possible to relate pluvial lakes in general to glacial maxima, in the sense that it is not possible to prove an exact coincidence in time. Indeed, the dates of the latest arrays of pluvial lakes may eventually be shown to differ according to latitude. Former lakes in very low latitudes existed at different times from those in middle and subtropical latitudes (see Chapter 19, section 19.4b). Where dates are known, the last high lake stands of the subtropics and middle latitudes appear to have come later than the last glacial maximum – say, roughly in the range between 15 000 and 11 000 years ago. When allowance is made for somewhat reduced temperatures, the standing lakes can be explained by precipitation not greater than twice the values of the present day.

The initial filling of the lake basins with water is another matter entirely. Calculations using rates of input of solutes and rates of output of evaporated water are fearsomely imprecise, because there is no simple relationship among salt concentration, precipitation, runoff and evaporation. The only statement possible at the present time is that the infilling of a given basin to the recorded maximum level may have required hundreds of years, or perhaps even thousands. Episodes of drying, on the other hand, were by comparison brief.

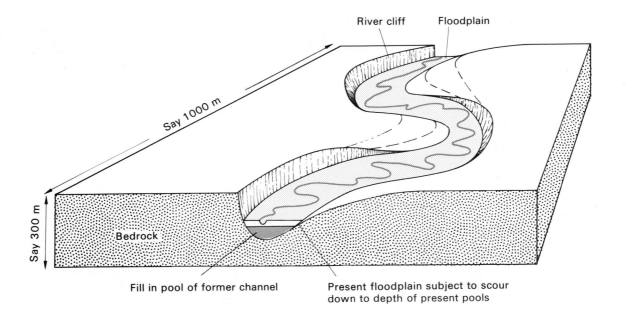

Fig. 18.15 Valley meanders and present-day alluvial meanders.

18.5c Large former river channels

Very many midlatitude rivers possess two sets of meanders, the alluvial (= floodplain) meanders of the valley floors, and the sinuous windings of the valley itself, the latter being cut into bedrock (Fig. 18.15). Streams of this type have long been called underfit, on the assumption that the valley meanders were cut by streams much larger than those of today. Early discussions of the combination of valley meanders with stream meanders were highly unsatisfactory. They assumed, but did not define, a relation between dimensions of meanders and size of stream. Furthermore, size of stream was never defined. As we have subsequently learned, meandering channels are dominantly shaped by discharge at the most probable annual flood (Chapter Six, section 6.6a). If the valley bends are true meanders, as their shapes suggest, then they were formed by mean annual flood discharges far greater than those of today: the numerical relation between meander wavelength and channel-forming discharge has been determined (Chapter Eight, section 8.6a).

However, the relation between meander wavelength and channel-forming discharge is indirect, working through the discharge/width and wavelength/width ratios. If the valley bends are authentic former meanders, then they ought to be associated with large former channels (see, for the wavelength/width ratio, Chapter Eight, section 8.2b). Such former channels in fact exist. They display pool-and-riffle sequences like those of the floodplain meanders of today; and their wavelength/width ratios resemble those of existing streams (Fig. 18.16).

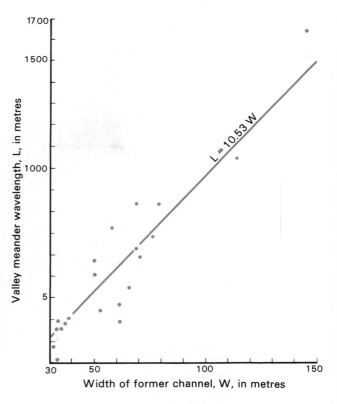

Fig. 18.16 Wavelength:width ratio for valley meanders and former channels: based mainly on components of the Avon and Thames drainage systems, England. Compare Fig. 8.9.

So far, then, the implication of the word underfit – namely, that channel-forming discharge has been greatly reduced – holds good. The question now arising is, where has the water gone to?

Among the various ideas that have been put forward, only two have been seriously and thoroughly considered. One is that stream shrinkage results primarily from river capture: the shrunken (= existing underfit) streams have lost territory, and therefore volume of discharge, to competing neighbours. The other is that the primary cause is change of surface run-off, which in turn implies a change of climate.

Since underfit streams flowing in opposite directions occur on the two sides of very many stream divides, change of surface runoff must be accepted as the primary cause of stream shrinkage. At channel-forming discharge, the former streams that cut the valley meanders may have carried twenty to fifty times as much water as do existing streams. But the contrast is precipitation that was much less than the contrast in channel-forming discharge. An increase in precipitation to 1.5 to 2 times existing values would be enough to bring the former large rivers back.

The last major reduction in channel-forming discharge was certainly step-functional. So was the onset of the former high discharge. One of the very best documented cases is that of Black Earth Creek, Wisconsin (Fig. 18.17). The sequence begins with the discharge of meltwater from an ice front, the outermost front of the last glacial maximum. This dates to somewhere in the range from 14 000 to 12 500 years ago. A braided stream of meltwater carried coarse outwash gravel away from the ice front and down the valley. When the marginal ice became stagnant and eventually melted, water from the new front could escape in another direction. Flow down the valley ceased. The local climate

became dry and windy. Dust whipped up from the main meltwater channels fell on the countryside as loess, blanketing the valley bottom and spreading over the outermost moraine.

An abrupt change to a rainy climate formed a large stream which cut large meanders in the loess and down into the outwash. Then, a change to climate like that of today produced the existing stream. All this sequence has to be compressed into not more than 4 000 years, and possibly into as little as 1 500 years. The range depends on the date accepted for the beginning of glacier decay, from 14 000 to 12 500 years ago, and the beginning of the formation of the existing floodplain, from 10 000 to 11 000 years ago.

A great deal obviously happened in the short interval available. But the record that has been summarised may in fact have been more complicated still. As will be described in Chapter 19, midlatitude stream discharge was increased for a time in the general range from 6 000 to 4 000 years ago, although not to the values computed for the range from about 11 000 to 10 000 years ago.

18.5d Different kinds of underfit stream

Streams with the two sets of meanders, one of the existing channel and one of the valley, are called manifest underfits. They can be recognised on sight – in the field, on maps and on air photographs. Their wide distribution in middle latitudes suggests widespread climatic change. But by no means all midlatitude streams are manifest underfits. Some of them are contained in meandering valleys, but lack stream meanders. Such streams have as yet been little studied. Where they have been studied, they are found to possess pool-and-riffle sequences on the existing channels, with pools and riffles alike spaced

Fig. 18.17 Development sequence for Black Earth Creek, Wisconsin.

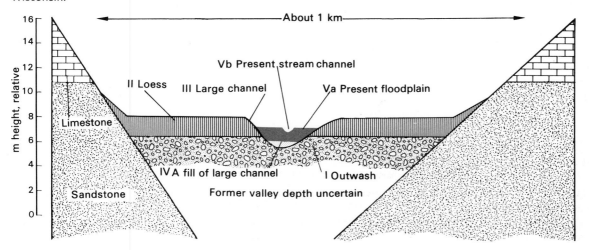

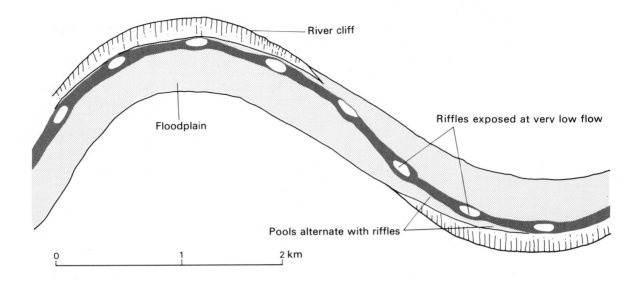

from about 5 to 6 bedwidths (Fig. 18.18). Apart from being forcibly curved round the valley bends that contain them, these streams behave as if their channels were straight – as in some reaches they are. They are called Osage-type underfits, after the Osage River in Missouri. Their channels have been reduced in size by change of runoff, but they have failed to develop meanders on the existing channels. They add to the evidence for general stream shrinkage.

18.5e Confirmation of the step function

For the U.S. Midwest, firm confirmation of the last episode of very high stream discharge comes from the analysis of fossil pollen. The pollen record can be translated by computer into surplus or deficiency of surface moisture, which in turn can be related to dominance by different kinds of air mass. The brief episode of very high discharge on Black Earth Creek corresponds to a sharp blip of excess surface moisture in the translated pollen sequence.

18.5f Earlier episodes of high discharge

The large meanders of Black Earth Creek represent only the latest in what has probably been a long but discontinuous series. Valley meanders cut deeply into bedrock obviously require far more than the few hundred years that produced the large bends of Black Earth Creek in the unresistant spread of loess, or similar bends on the drained floors of ice-dammed lakes. How far back the origin of some valley bends may go is conjectural.

Where anything like firm information is to be had,

Fig. 18.18   Array of pools and riffles in an Osage-type underfit.

it is clear that downcutting into bedrock started well back in the present spell of glaciation. Moreover, at least some of the initially downcutting streams were sensibly straight to begin with. The valley meanders have developed during downcutting, which has typically been spasmodic. The episodes of high channel-forming discharge were separated by episodes of low discharge. In some winding valleys, the delivery of abundant rock-waste during periglacial intervals caused the channels to braid.

Although few sequences can be accurately dated in years, they clearly record step-functional change in surface runoff and channel size, and in some cases also in channel habit. Also, they show that the channel-forming discharges that shaped the valley bends were closely comparable from one episode of high discharge to another. The increase of precipitation to 1.5 or 2.0 times the present seems to have constituted a cutoff point.

18.6 Possible step-functional change in the humid tropics

Two theoretical views exist about what happens to the extent of humid–tropical climate during glacial maxima. It is known that the midlatitude circulation is displaced toward the equator. The poleward margins of the subtropical deserts are invaded by seasonally moist climates, usually by climates of the summer-dry type developed on subtropical West coasts: that is, winter rains fall beyond their equatorward limits of today. One can reason theoretically that the equatorward limits of the deserts should also be displaced, again toward the equator, so that dry

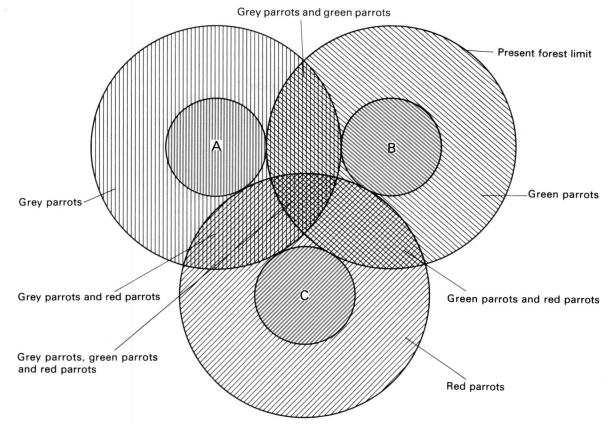

Grey parrots and green parrots

Present forest limit

A

B

Green parrots

Grey parrots

Grey parrots and red parrots

Green parrots and red parrots

C

Grey parrots, green parrots and red parrots

Red parrots

Fig. 18.19 Highly simplified model of speciation in the Amazon rain forest, under the influence of intermittent forest disruption.

climates of the trade-wind belt would invade existing regions of tropical climate and would simultaneously encroach on the existing limits of equatorial climate. This amounts to saying that, from the midlatitudes to the equator, climatic belts would be shifted equatorward and would at the same time be compressed.

The rival view is that the increase in energy gradient between the mid-latitudes and the equator would cause the equatorial rain belt to expand. If so, then at times of glacial maxima, equatorial climate should invade the borders of existing regions of tropical climate.

Types of evidence available in other regions are unavailable in the humid tropics, except that mountain glaciers on very high peaks are known to have been formerly more extensive than they are today. But the Amazon basin provides evidence of another kind – that of speciation and radiation, especially of birds.

Animals evolve more rapidly than plants do. Where their radiation is controlled by the spread of a plant assemblage, such as that of tropical humid forest, they cannot spread faster than the forest does. But if the forest cover is reduced by climatic change into discontinuous areas, then each area may evolve its own species and sub-species of animal life.

If the climate changes back again, and permits a continuous forest cover to grow once more, then the areas of new forest will provide ecological niches for the newly-evolved animal types. These will radiate out from their areas of speciation.

Precisely this kind of thing seems to have happened in the Amazon basin. The results are diagrammatically illustrated in Fig. 18.19. The core areas A, B and C represent the blocks of surviving forest during periods of climatic stress. For the sake of illustration, they are here imagined to evolve related forms of parrot, one grey, one green, and one red. When continuous forest returns, each kind of parrot moves into the newly available areas. Where more than one kind partially occupies a given ecological niche, the parrot population is more varied than if a single kind had evolved in a single continuous forest throughout the region.

Although the presentation of Fig. 18.19 is grossly over-simplified in comparison with actual distribution of the animal populations of the Amazon basin, the real-world situation – great variety of fauna generally, plus concentrations of some forms in particular areas – is readily explicable by the kind of succession indicated. If the explanation is correct, then the Amazon rainforest has indeed been formerly subject to large-scale reduction in total area, and to disruption into separate blocks.

It is impossible to imagine that the reduction and disruption can have been caused by anything other than a reduction of rainfall. That is, the region appears to have been partially invaded by climate too dry to support continuous forest. A reduction of mean annual precipitation by a factor of from one-half to one-quarter would certainly suffice to dismember the blanket cover of rainforest. As yet, it is only possible to suspect that the invasion – or, much more probably, invasions – of less rainy climate coincided with the major extensions of ice in higher latitudes.

Chapter Summary

The concept of *morphogenesis* holds that *physical landscapes are process-response systems* and that *either total landscapes or individual* prominent *features* are *diagnostic* of the controlling *climate*.

Obvious cases include formerly *glaciated* or *periglacial landscapes*, where the climate is no longer glacial or periglacial. *Less obvious cases* include *desert margins*, because we know little about the frequency of episodes of sand-driving in fields of mainly dead dunes, or of flooding in desert lake basins.

Still less obvious cases include landscapes of the tropics. There, *non-cyclic nickpoints are common, but cannot* be used to *distinguish former* humid–tropical *conditions from existing* different *conditions*. *Pediments cannot be* regarded as *diagnostic of dry climates, nor inselbergs* as diagnostic *of humid-tropical climates*.

Deep weathering is more reliable. It *results in the concentration* at and near the soil surface *of sesquioxides*, Al_2O_3 and Fe_2O_3, *or of silica*, SiO_2. The *initial concentrations are soft*. They *set hard when they dry out*. Drying out can result from *deforestation, lowering of the water table*, or *climatic change*. These causes can overlap.

Two main kinds of deep-weathering profile are known. *One is strongly differentiated* into horizons, the kaolinitic *pallid zone* in the lower part, the *mottled zone* with mixed clays and sesquioxides next above, and the *indurated zone, or duricrust*, at the top. The *differentiation* is *thought to reflect* seasonal positions, and development through time, of the *water table*. The *other kind* consists of a *surface concentration of sesquioxides* and a *highly discoloured clayey horizon* beneath.

Silica-rich crusts pose problems of origin, but record some of the firmest evidence of climatic change, in the form of *plant impressions, plant remains* and the *burrows of tropical spiders*. *Reconstructed former plant covers* for deep-weathering profiles generally *indicate the former extension from humid–tropical to humid–subtropical climate* well into existing middle latitudes. The extension *cannot be explained by continental drift*. The *changes*, on the geological time-scale, *were step-functional*. They included the initiation of the present glacial period. *Geologically speaking, ice-free poles are normal*.

The *switch from glacial to interglacial conditions* is *also abrupt*, requiring only from 4 to 8% of glacial (Ice Age) time.

Step-functional change of surface runoff has included the *activation of dunefields*, far beyond existing desert borders; the establishment of *standing lakes*, well inside existing desert borders; and midlatitude changes that produced the *ancestors of streams that are now underfit*. The *last major increase* in midlatitude *streamflow* occurred about *10 000 to 11 000 years ago*. The *changes of surface runoff* can be explained by *reduction of precipitation to* about *half present values, or increase to* up to *twice present values*.

Step-functional change of climate in the humid tropics is strongly suggested by the analysis of *speciation and radiation* in the faunas of *the Amazon Basin*. There, the tropical rainforest appears to have been broken down into discontinuous blocks, probably by decrease in rainfall, at some times in the past.

19 Atmospheric Step Functions II

This chapter will deal with the climatic shifts of the last 10 000 years or so. The first half of this period saw the recovery of the ocean surface from very low late-glacial levels to the level of today. The period as a whole recorded most of the fundamental accomplishments of human kind.

Speech in the modern sense, with however limited a vocabulary, can safely be assumed to have developed long before 10 000 years ago. All existing groups that have cultures typical of, say 50 000 to 100 000 years ago, possess elaborate languages. Stone tools date back many hundreds of thousand of years. Spears, bows and arrows, and domesticated dogs are also recorded quite early in prehistory. But the last 10 000 years saw the beginnings of cereal cultivation, the domestication of cattle, sheep and goats, the development of a whole range of handicrafts – including the making of pottery – the use of spoked wheels and the smelting of metals. Most important of all, the same period brought the establishment of villages and cities, irrigation agriculture and the invention of writing.

Our innovating ancestors lived through climatic shifts which, even if less far-ranging than the full change from glacial to interglacial conditions, or the sharp end-glacial onset of very high surface runoff, were nevertheless noticeable. On one view, indeed, it was precisely climatic shift toward increasing dryness that fostered the domestication of animals, and the turn toward cultivated crops, in the flood-liable valleys of the Middle East. A case can be made out that climatic shift, affecting the flow of the Nile, also affected the fortunes of some of the dynasties of ancient Egypt. When a fairly full and reliable historical record begins, as for the countries bordering the northeastern Atlantic, the social effect of climatic shift can be abundantly documented.

19.1 The nature of the evidence

One main line of evidence is supplied by fossil vegetation, including fossil pollen. Vegetation change, except that due simply to plant migration, implies climatic change. In at least some instances it implies very rapid change – for instance, two hundred years between the melting back of an ice front and the invasion of boreal forest. Variations in surface moisture supply and/or in temperature are recorded in the scouring or filling of stream channels, in the rise and fall of lake levels, and in the advance or recession of mountain glaciers. Tree-ring analysis reveals sequences of favourable and unfavourable years for growth; it has been applied mainly in dry regions, where moisture supply is the chief control.

Archeology (the study of antiquities, mainly prehistoric) also supplies information. In common with pollen analysis, it appeals to radiocarbon dating, on which tree-ring analysis provides a partial check. Archaeology overlaps to some extent with the written record of history. Instrumental records – mainly those of temperature – rarely go back more than a hundred years. Nevertheless, some such records prove capable of revealing step-functional behaviour of climate.

When all the types of evidence are taken together, it becomes possible not only to define a sequence, but also to look for, and sometimes to suggest, a cause.

19.2 Fossil pollen

Pollen is extraordinarily resistant to decay. The one exception to this statement is that it decays very rapidly in a calcareous environment that is subject to drying. For the most part this exception is irrelevant, for pollen studies mainly concern pollen preserved in acid peat.

Each genus of pollinating plant produces pollen of a distinctive kind (Fig. 19.1). As a peat bog grows, it traps and preserves the pollen that falls on it. Inevitably there are complications. In the long term, some pollen types do decay faster than others. Some plants discharge their pollen into the wind – notably the pine, which produces large quantities that may travel downwind for 400 km or more. Other plants –

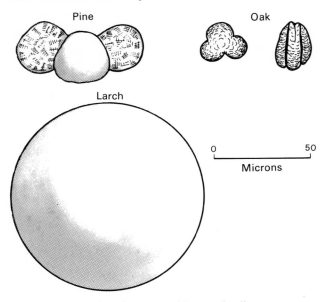

Pine

Oak

Larch

0   50
Microns

Fig. 19.1 Examples of contrasted forms of pollen.

19.2a A basic succession: the Scandinavian record

Pollen analysis was first developed in the Baltic area, where numerous bogs formed after the decay of the last ice sheet. Comparison of the individual successions for different sites, undertaken well before the method of radiocarbon dating had been developed, defined the sequence presented in Table 19.1.

This table is based on findings for northwestern Europe, especially for Scandinavia. The original version was simpler than that presented here; and it only included the names of climatic phases and the descriptions of vegetation. The dates shown depend on radiocarbon analysis (see next section and Chapter 24). The list of probably dominant air masses is partly conjectural, partly based on reliable inference.

There is good reason to think that the timings and directions of climatic shift shown in Table 19.1 were common to all midlatitude regions of humid climate in the northern hemisphere. Where ice was still present after 16 000 B.P. (= Before the Present), as in large areas of the North American interior, the early part of the sequence is missing. In addition, the climates that produced the very high stream discharges of late-glacial times appear to have shifted poleward as the ice sheets melted back. But where glacial complications are lacking, the Scandinavian sequence provides a widely applicable standard. With an average duration of about 1500 years, the

for instance, holly, ivy, mistletoe – are not wind-pollinators. Their pollen falls close at hand. But for all the complications, many records show clearly that vegetation cover has changed through time. Some genera become more abundant during a given interval, others less abundant (Fig. 19.2). It thus becomes possible to reconstruct the vegetation cover of the past.

All the earliest pollen studies concentrated on tree pollen (arboreal pollen, AP). Only when intensive research was conducted into the causes of hay fever did the additional analysis of pollen other than tree pollen (non-arboreal pollen, NAP) become standard. Now analysis deals not only with former forests but with the total vegetation cover, whether forest trees were dominant or not.

Fig. 19.2 A simplified pollen diagram: adapted from the data of A.M. Davis for a bog in the U.S. Midwest. A full diagram would include counts for twenty of more types of pollen.

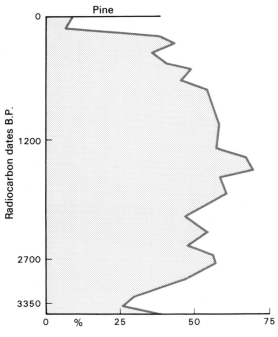

Pine

Radiocarbon dates B.P.

1200

2700

3350

0   %   25   50   75

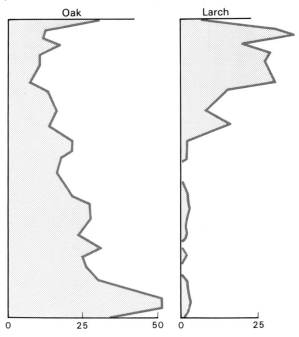

Oak

Larch

0   25   50 0   25

Table 19.1 The Scandinavian Succession

Years B.P.	Climatic phase	Vegetation	Possible dominant air-mass types
– 0 – 1000	Sub-Atlantic	Further reduction in oak forest; beech and spruce extended; renewed growth of bogs	mP, mT
	Sub-boreal	Some reduction in oak forest; pine increases; beech, hornbeam, and spruce enter	mP, mT, cP
– 5000	Atlantic	Mixed oak forest; vigorous bog growth on exposed sites	mP, mT
	Early Atlantic (transitional)		
	Boreal	Pine and birch dominant but some mixed oak forest	cT, cP, mP
	Pre-boreal	Birch and pine forest	cP, some cT
– 10 000	Younger Dryas	Tundra & c	cA, cP
	Allerød	Birch and pine trees widespread	cP, some cT
	Older Dryas	Tundra & c	cA, cP
	Bølling	Some stands of tall trees	cP, some cT
– 15 000	Oldest Dryas	Tundra, plus dwarf willows and birches	cA, cP
– 17 000	glacial ice present		cA

climatic phases are far briefer than the events considered in Chapter 18.

The sequence begins with a halting recovery from full-glacial climates. The three phases of Dryas climate are named for a widespread genus of tundra plant. During these phases, the vegetation cover consisted of sedges, grasses, tundra shrubs, and dwarf forms of trees which could not reach full development in the savage climate of the time. Identical forms occur along the tundra borders of today. In the Bølling phase, temperatures rose high enough to permit many trees to grow above dwarf status, but tundra returned in the Younger Dryas phase. Temperatures in the Allerød phase averaged only 1 to 2°C below those of the present, but a sharp drop, to about 5°C below present levels, brought the tundra back once more in the Younger Dryas phase. Conditions improved through the Pre-boreal and Boreal phases, enough to promote the wide extension of birch-pine forest, and then to bring in mixed

oak woods; but increasing dryness, indicating increased continentality of climate, is recorded by lowered water tables, the drying-out of some shallow lakes, and the scarcity or absence of peat bogs.

With the change from the Boreal to the Atlantic part of the sequence comes a marked change in climate. Mean temperatures ran 1 to 2°C higher than at present. The birch-pine forest receded poleward, being replaced by mixed oak forest. At the same time, precipitation increased – more than enough to offset the rise in temperature. Peat grew vigorously, both in the closed shallow basins of wetlands, and in the blanket bogs of cool rainy uplands.

An abrupt change separates the Atlantic phase from the following Sub-boreal. Where the succession is recorded in the peat bog sequence, a sharp discontinuity occurs. The Sub-boreal record, opening with reduced temperatures, is complicated by the invasion of new kinds of trees; but, in the background, pine forest extends itself at the expense of oak. The strongly maritime climate of the Atlantic phase was replaced by more continental conditions, with increasingly warm summers but also with more severe winters than before. Subsequently the Sub-atlantic phase has brought another increase in wetness – that is, a renewal of increasedly maritime climate – and a reduction of temperatures, at least in summer. The total increase in surface moisture has once again fostered the growth of peat bogs at ill-drained sites.

19.3 Radiocarbon dating

Life forms take in carbon as atmospheric carbon dioxide, CO_2. A tiny fraction of the carbon is radioactive carbon, known as radiocarbon, ^{14}C. When an organism (for example a tree) dies, the intake of carbon ceases: but the decay of radio-carbon goes on. The extent of decay is used in dating (see Chapter 24).

Radiocarbon has a half-life of about 6000 years. Start with an amount of 1 gram. Then, in 6000 years, half a gram will have reconverted to nitrogen, and half will remain as radiocarbon. In another 6000 years, making 12 000 years in all, another one-quarter gram will have reconverted to nitrogen, and one quarter gram will remain: and so on. By comparison with the lifetime of most organisms, 6000 years is very long. It may therefore be assumed that radioactive decay begins when the organism dies. The proportion of radiocarbon to ordinary carbon in a given sample is used to calculate the radiocarbon age.

Uncertainties result from the assumption that the rate of cosmic ray bombardment, and thus the rate of production (and absorption) of radiocarbon, is constant. In reality there are variations. Some of these have been identified with the aid of the growth rings of the bristlecone pine, the oldest known kind of living tree: some specimens have already lived more than 4000 years. But despite the uncertainties, radiocarbon dating provides a time scale some 30 000 years long, with a possible extension back to 70 000 years. It is especially useful for the last 10 000 to 15 000 years, the interval here under discussion. Its supreme advantage is that it enables simultaneous events in different parts of the world to be related to one another.

19.4 Changes in the supply of surface moisture

Simply because the supply of surface moisture is controlled by the precipitation/temperature relation-

ship, not all the evidence available under this head is easy to interpret. The following discussion will be limited to reasonably clear cases.

19.4a Scour and fill of stream channels

The fills of the large former stream channels, that were last swept clean as much as 10 000 or 11 000 years ago, have not everywhere remained undisturbed. Partial re-clearance is on record, both in southern England and in the U.S. Midwest, for the Atlantic climatic phase. We have already noticed that this was a time of increased warmth but also of increased wetness. The channel-forming discharge of rivers increased, not enough to scour the large channels clean, but enough to remove significant parts of the fill. Renewed infilling took place during the Boreal phase. Modest re-excavation, followed once again by infilling, is known on some streams for the Sub-Atlantic phase.

19.4b Lake levels

The pluvial lakes of the Great Basin fluctuated in level in sympathy with the climatic changes of middle latitudes, It is not, however, entirely certain if high lake stands coincided precisely with glacial advances. The last very high mark was reached about 10 500 B.P. During the Atlantic phase, these lakes fell to very low levels, many drying out altogether. The responses here were due partly to change in temperature, for the Great Basin is not readily accessible to rain-bearing lows such as those that affected the U.S. Midwest and western Europe.

Low-latitude lakes have been studied mainly in tropical Africa. The record there completely fails to accord with the standard midlatitude sequence of climatic shift during the last 10 000 years. Three main episodes of high lake stands are known during the period 9500 to 2500 B.P. As with pluvial lakes in general, the changes involved seem to have been abrupt. The total effect was to make tropical Africa moister for most of the time than it is at the present day.

19.4c Mountain glaciers

As integrators of the temperature/humidity relationship, mountain glaciers are highly sensitive. An increase of snowfall causes glaciers to advance; a decrease leads in time to a recession. A decrease of temperature brings the snowline down and permits extension of the ice tongues. An increase of temperature causes melting back. What weight is given to change in precipitation, as against a change in temperature, remains to a large extent a matter of choice.

Mountain glaciers in the northern hemisphere advanced briefly at a time that can possibly be referred to the Younger Dryas climatic phase, when tundra climate returned to the southern part of the Scandinavian Peninsula. A major halt in the melting of the North American ice sheet, probably including some regrowth, dates from about 6500 B.P. Rising temperatures during the Early Atlantic and Atlantic phases appear to have prevented any general glacier advance. Regrowth occurred in the cold early part of the Sub-Boreal phase, bracketing the date of 4500 B.P. After the subsequent recession, a further readvance set in at about 2700 B.P., close to the end of Sub-Boreal times; and the next recession was followed by a third readvance, that began in about the year 1550 and reached its maximum in about the year 1850.

Clearly, the vegetational record of the standard Scandinavian succession is coarser, at least in its later part, than the record of change – thought to be mainly change of temperature – displayed by the history of mountain glaciers.

19.4d Tree-ring analysis

Trees put on thick growth rings in favourable years and thin rings in unfavourable years. Most tree-ring analysis so far performed concerns the dry regions of the U.S. Southwest, where the record has been reconstructed for the last 1550 years. Its main effect is to demonstrate the alternation of dry and wet runs of years. There is some faint suggestion that distinctly wet spells come at an average interval of 200 years. Droughts are both more frequent, and less evenly spread. Although the tree-ring record cannot be related to happenings in other regions, it does at least serve to demonstrate step-functional switch from one kind of extreme to another.

19.5 Archaeology

Interpretation of an ancient archaeological record is beset with difficulties. The longest records come from the Middle East, where at least one town, on the site of the modern Jericho, existed as early as 10 000 B.P. City civilisation also arose early in pre-history in the northwest of the Indian subcontinent and in northern China. The food surplus needed to feed city dwellers depended on irrigation farming. Hence the term hydraulic civilisations – city-centred cultures dependent on river floods and on irrigation works.

For any given hydraulic civilisation, the typical record is one of a rise to prosperity, followed eventually by decline and destruction. For a given major city centre, the sequence can have been run through more than once. Three choices are open to explain decline and destruction: conquest from outside, social disintegration within, or change of climate. According to the training and inclinations of an individual commentator, emphasis is likely to fall on one or other of these. Political historians tend to write in terms of the competition for territory and power. Social historians stress the internal difficulties of the ancient city societies, including the problem of disposing of waste. Those whose interest lies less in the founding and management of cities than in the relationship of events to environmental changes in the last 10 000 years are apt to see climatic change as a principal control.

19.5a The climatic lead-in

Furthermore, numerous efforts have been made to interpret the origins of cropping and herding, and the later development of irrigation farming, in terms of climatic shift. The general argument is that increasingly dry climate reduced the vegetation cover and the game supply of the lands bordering on great floodable valleys, and that, in consequence, hunting and gathering were perforce replaced by the deliberate cultivation of crops, and the keeping of domesticated food animals.

It is true that the existing desert that borders the inner valley of the Nile was moister than it is now during the approximate intervals 9500–7500 and 6500–5300 B.P. So much is clear from rock drawings of animals typical of the humid part of the trade-wind belt of the subtropical west coast. But the first beginnings of planting and cropping, the domestication of farm animals, and the development of city cultures simply do not match the climatic record. Village settlement and the cultivation of grain crops are known from ancient Egypt as early as 6500 B.P., and a city on the site of the classical Memphis, close to the Cairo of today, had been built by 5000 B.P.

19.5b Two episodes in the social record of ancient Egypt

The moist climates of 9500–7500 and 6500–5300 B.P. were associated with high flood levels on the Nile. After about 5000 B.P., flood levels fell dramatically. The change resulted from an abrupt decline in summer monsoon rain in the upper basin. Riots, looting, revolt and famine were concentrated in two brief episodes dating from about 4000 and 3000

B.P., each lasting only about twenty or twenty-five years. But a generation of persistent drought can bring down the social order of an hydraulic civilisation. Each episode is a dark age in the Egyptian record. During each dry spell, search parties went up and down the river in attempts to find grain. Some communities survived by means of damming the surviving trickles of river water. Some barricaded themselves in against outsiders. Some degenerated into cannibalism. Plunderers from Asia came raiding in; but the preparatory social weakening was caused by the failure of the monsoon. In the lowermost Nile valley, the modest winter rains of marginal subtropical west-coast climate also failed: potentially rain-bearing northerly winds were replaced by southerlies from the desert.

The two episodes of persistent drought serve to demonstrate not only step-functional change of the regional circulation of the atmosphere, but also the difficulty or impossibility of relating brief but significant climatic spells to the general pattern of change during the last 10 000 years.

19.5c Episodes in the social record of Mesopotamia

Ancient Mesopotamia, the Land Between The Rivers, is the broad alluvial valley of present-day Iraq. The rivers are the Tigris and Euphrates. Episodes from the early history of Mesopotamia are introduced here as an illustration of the dangers of always interpreting the archaeological record in terms of climatic change. In direct contrast to what has been said about spells of famine in ancient Egypt, the events now to be discussed exemplify the effect of man-induced positive feedback.

In the alluvial country between modern Baghdad and ancient Babylon, irrigation works existed as early as 5700 B.P. They stretched more than 100 km down-valley, and as much as 50 km across the alluvium between the two rivers. Babylon itself was fed from the irrigated fields. When the record of clay-baked inscriptions begins, at about 5500 B.P., wheat and barley were grown in about equal proportions. Shortly after 4000 B.P., records of wheat cultivation vanish. The more salt-resistant barley is the only grain (Fig. 19.3).

Substitution of barley for wheat kept yields up for more than a thousand years. Even as late as 4000 B.P., yields were comparable to those of the extensively cultivated small-grain farms of North America today. But two hundred years more reduced yields by half; and the fall continued. The cause – progressive salting of farmland soils – was recorded at the time, although its effect may not have been understood. Its mechanism was certainly not understood.

Irrigation water carries solutes into the soil. Plants and evaporation take water out, leaving the salts behind. The effect is especially strong in regions of dry climate, where the solute load of river water is typically high to begin with. Increasing salt storage in the soils of irrigated farmland begins by limiting the choice of crops, continues by reducing fertility and then, in the extreme case, ends with sterilisation. Between 3700 and 3200 B.P., four in every five settlements in middle Mesopotamia were abandoned.

Ordered conditions began to return in about 2500 B.P., after which resettlement and an effective centralised social order renewed productivity. The seven centuries of disuse had mitigated the salt damage to some extent; but the chief new measure was the extension of irrigation canals for long distances, to land that had not been previously watered by artificial means. Almost all cultivable land was brought under the plough. But the irrigation works contained the cause of their own failure. The extended canals suffered badly from silting, partly because they were poorly designed as sediment-delivering channels, and partly because logging in the headwater basins increase soil erosion and sediment delivery. The longer a distibutary canal, the greater the difficulty of keeping it clear of sediment. The farther the spread of irrigated farmland, the greater the use of labour in clearing canals, as opposed to cultivating fields. Positive feedback was at work. The measures designed to increase farm production proved automatically self-defeating. Eventually, in about 800 B.P., the systems of irrigation, farming and social organisation collapsed. They have not yet been properly restored. It was a full century after the collapse that Mongol conquerors from inner Asia appeared on the Mesopotamian scene.

Neither political history nor climatic shift need be appealed to in order to explain the two successive failures of an irrigation culture. All the same, these failures are instructive for systems behaviour. The earlier culture increased its technical stresses – progressive failure of the wheat crop and falling crop yields – over a total interval of 1850 years. Then, 500 years sufficed for something close to destruction. The later culture developed over a total interval of 1600 years, and collapsed in only a century. Societies, like climates, can obviously display step-functional behaviour.

19.6 The historical record

Relevant historical records come mainly from western Europe, relating chiefly to the last thousand years or less. In the fairly near future, analyses of

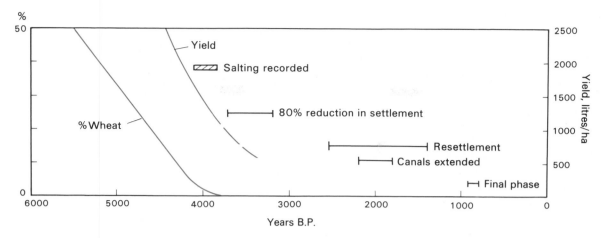

records from eastern Asia may provide comparable information. For the early part of the period, documentary evidence is in part fragmentary. It is read in association with the findings of archaeology.

The interpretation of vegetation change is especially difficult. In Britain, for example, pioneer farming settlers cleared much forest land, greatly altering the pollen record, and also greatly increasing the supply of sediment to streams and lakes. Some climatic shifts can nevertheless be firmly identified. They involved spells – and on the historical time scale prolonged spells – of generally adverse or generally favourable years. For some areas and for some time intervals, it is possible to disentangle seasonal change from change affecting the whole year.

In addition, for some climatic episodes it is possible to contrast one area to another. Climatic improvement in northwestern Europe between about 400 and 1200 A.D. seems to have been accompanied by increased rainfall and increased frost hazard in the Mediterranean region. Here we are probably dealing with a displacement of the circulation of the westerlies. In other connections, recognisable change in annual or seasonal means may have been less important than change in variability. A climatic spell that in a number of years brings unusually late

Fig. 19.3 Some trends and events in the history of ancient Mesopotamia: drawn from the data of T. Jacobsen and R.M. Adams.

spring frosts or unusually early winter frosts can be as disruptive to subsistence farming as can a persistent but slight lowering of temperature.

19.6a Variations within the Sub-Atlantic phase

We are dealing throughout with portions of Sub-Atlantic time, which began with increased rainfall and probably also with cool summers, although winters may well have been chiefly mild. Not only in the blanket bogs of the west European uplands, but also on low ground, peat markedly renewed its growth. In places it overwhelmed the forest floor and killed the trees (Fig. 19.4). In the Somerset Levels in England, trackways across the formerly dry peat had to be turned into corduroy (log-paved) roads. On the European mainland, where houses on stilts stood on lakeside peat, some lakes rose high enough to flood the dwellings and cause whole settlements to be abandoned. Alpine passes were closed, some of them for a thousand years, by the readvancing glaciers.

Fig. 19.4 Stump of birch tree, one member of a forest killed by growing peat: northeast Ireland.

19.6b Amelioration, 400–1200 A.D.

Widespread evidence testifies to a spell of climatic improvement, that may have peaked during 800–1000 A.D. with mean temperatures as much as 1.5°C above those of today. The difference may not sound much: but it meant a great deal in marginal areas, and its total effect may have increased with increasing latitude. Storm frequency and storm intensity in the North Atlantic area greatly diminished. Settlements were established in Iceland and Greenland – some Greenland settlers lived, and buried their dead, on ground which is now locked in permafrost. From Greenland, Viking ships crossed the Atlantic to the North American mainland.

The most comprehensive single document is the Domesday book of 1086, an assessment of real property and tax potential made for most of England by the conquering Normans. It records vineyards 500 km north of the existing limits of effective growing. We can safely infer a much reduced risk of late frosts in spring and early frosts in autumn, and also a decrease in summer rainfall. In the North Atlantic, winter sea ice along the shores of Iceland was in some years not at all troublesome.

19.6c Renewed climatic instability

The two and a half centuries between 1200 and 1450 A.D. brought widespread climatic, agricultural and social disaster. Mean temperatures fell to about 1°C below present levels. The change seems to have been chiefly a result of reduction in summer temperatures, as indicated by repeated crop failures, for instance in Scotland. Assessment of what actually happened is complicated by one event and one trend.

During the later 1300s, the pandemic Black Death struck western Europe. If it were related to climatic shift, however indirectly, the relationship has not been established. Between about 1300 and about 1600, many village settlements in the English lowlands were partly or wholly destroyed by the conversion from subsistence arable farming to commercial wool production. How much of the change was facilitated by worsening of climate is impossible to judge, although the wasted settlements in the southern English Midlands were certainly located on farmland that has persistently proved less than secure in terms of management practices – the area is highly liable to agricultural change as market conditions alter.

It is however certain that climate shifted. Western Europe was afflicted by numerous wet and stormy autumns, that prevented grain crops from ripening, or that lodged the stalks and obstructed the harvest.

Ice returned to trouble the harbours and fisheries of Iceland. The expansion of settlement in lowland England and its hill borders slowed to a halt; and eventually the practical limits of farming were driven 300 m down the slopes of the uplands.

19.6d The Little Ice Age

A slight and partial climatic recovery may have affected western Europe between about 1450 and 1550. Subsequently, summer weather in England became more stable than before. Winters at first were often severe. Frequent easterly winds from the mainland brought intense cold which caused the Thames to freeze over eight times during the 1600s. After about 1700, English winters tended to become less bitter; but elsewhere the climatic deterioration continued strongly enough to earn for the period 1550–1850 the name of Little Ice Age. Alpine glaciers reached their farthest limits since last full-glacial times. Glaciers overran some farmland and farm settlements in Iceland, where the northern coast remained frozen for seven months in the year.

19.6e 1850 to the present day

A general rise in temperature from about 1880 to about 1940 or 1950 is well documented for the North Atlantic area, and may have been worldwide (Fig. 19.5). It is demonstrated by instrumental records. Such records, available for some stations for a century or more, and supplemented by historical evidence, permit weather and climate to be analysed in numerical terms for the past century or two.

Much attention has been given to cycles. Some workers have concentrated mainly on indices of solar output – including, although not confined to, sunspot frequency – identifying cycles of about 11, 22, 88 and 176 years in length. The 44-year harmonic seems to be missing. Other studies have dealt with variations in the input of solar radiation to the earth-atmosphere system, taking account of periodic variations in the ellipticity of the earth's orbit, and of the probably periodic variations of the angle between the orbital plane and the spin axis.

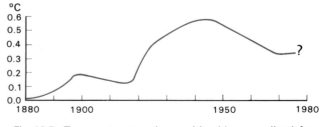

Fig. 19.5   Temperature trends, considerably generalised, for the North Atlantic area (various sources).

Non-periodic variations include those of the emission of carbon dioxide, volcanic dust, smoke dust, and industrial dust – atmospheric inputs which act to change the characteristics of the atmospheric filter. Variations that may or may not be periodic include changes in the earth's albedo (= reflectivity), in the extent of polar ice and snow (a factor overlapping with albedo), in the amount of heat stored in the oceans, and in the strength of particular ocean currents.

Because in the immediate context we are considering changes in the fairly short term, it is natural that short-term causes seem attractive. On one view, the rise in low-level air temperature up to the mid-twentieth century corresponds to the greatly increased use of fossil fuels. Burning of coal, petroleum and natural gas produces carbon dioxide. An increase in the carbon dioxide content of the troposphere acts to seal in earth radiation. However, the actual increase in temperature is only half that calculated from the rate of fuel consumption. A possible inference is that half the excess CO_2 has been absorbed by plants. If this is so, then air pollution by the burning of fossil fuels is good for cropping.

According to this same line of reasoning, the fall of temperature since the mid-century is also due to air pollution – in this case, to dust, which blocks solar radiation out. It seems unlikely that slash-and-burn cultivation in the humid tropics has greatly changed its output of atmospheric dust during the last hundred years. The North African desert, having been fought over four times in World War II, is still yielding unusually high amounts of dust today. But the chief variable is industrial dust, emitted in the dirty smoke of factory chimneys. Its rate of emission has been rising, mainly since 1900, and at an ever-increasing rate.

Industrial dust seems a better prospect as a climatic regulator than does volcanic dust. Although the latter, injected into the stratosphere, can take more than ten years to fall out, and is capable of forming a worldwide veil, the cooling that began the mid-twentieth century was not associated with any major volcanic eruptions.

19.6f A step-functional interpretation

In the context of climatic shifts within a century of time, much attention is now being paid to jumps in the behaviour of the low-level atmospheric circulation. The question is one of step-functional change in air-mass dominance.

For a given station in the westerlies belt, weather depends largely on the proportion between zonal (west-to-east) and meridional (north-to-south or south-to-north) flow of the winds. That proportion is in turn controlled by the position and form of the upper-air waves, including the jet streams. When the upper-air waves are weakly defined, zonal flow dominates at low levels. Weather in a season dominated by zonal flow does not vary greatly. A run of seasons (for instance, winter) dominated by zonal flow will not bring extreme weather. By contrast, when the upper-air waves are strongly defined, zonal flow at the surface is partly replaced by meridional flow. Particular seasons can be affected by an increase in the frequency of southerly winds, by an increase in the frequency of northerly winds, by an increase in the frequency of blocking highs or, more generally, by increasingly variable weather.

There is mounting evidence that, in the midlatitudes of the northern hemisphere, the frequency of zonal flow increased abruptly, close to the beginning of the twentieth century. An equally abrupt decrease in zonal flow came in about 1950. The episodes of frequent meridional flow in the later part of the nineteenth century and in the second half of the twentieth (up to the time of writing) are typified by highly variable weather, especially in spring and autumn. In strong contrast, the episode of frequent zonal flow, in the first half of the twentieth century, brought much less variable conditions. It is precisely these conditions which have been used in the calculation of so-called climatic normals and in development of systems of climatic classification.

Fig. 6.20, the graph of annual rainfall for Sydney, Australia – already used to illustrate step-functional behaviour – suggests that the westerlies belt of the southern hemisphere may have been affected in the same way that the westerlies of the northern hemisphere were. Fig. 19.6 illustrates the frequency of the Westerly type of weather in the British Isles. It shows how the record can be interpreted either as variation about a wave pattern, or in terms of step functions. When additional weather types are taken into account, the step-functional interpretation becomes more and more attractive.

The matter is actually far more complicated than a simple contrast between highly zonal and highly meridional flow. For the British Isles, seven weather types are recognised. For the U.S. Midwest, the best results are obtained when five types of circulation are used, and when the data are analysed month by month. For at least some areas, the timing and severity of river floods can change appreciably with change in the atmospheric circulation without necessarily involving a change in mean precipitation.

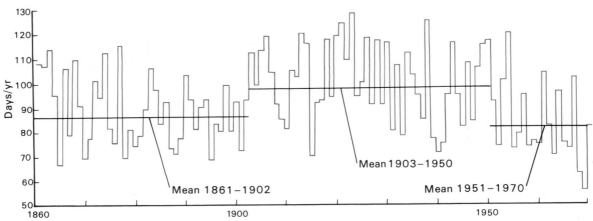

19.7 The need for prediction

Short-term prediction is a compelling need. Step-functional change of climate affects the world's food supply. The savage droughts in the Sahel region, in the years about 1970, meant starvation for many. A year in which the Indian monsoon fails brings a major food crisis. The failure of spring rains in the Russian grainlands means a failure of the grain crops. The most widely known crop hazard in the western world is that of drought on the southern U.S. Plainland, the Dust Bowl country of the 1930s (Fig. 19.7). The Dust Bowl disaster was highly publicised: not so the droughts that afflicted the same area in the 1950s. Drought struck again in the 1970s, when dust was abundantly mixed with snow in winter blizzards. The area is a major producer of

Fig. 19.6 Days per year of westerly type of circulation in the British Isles (data after H.H. Lamb) The graph suggests a sine wave variation, with a trough in about 1880–90 and a crest in about 1925–35, but can also, as here, be subjected to a step-functional interpretation.

winter wheat. Drought there threatens two-fifths of the U.S. wheat crop, an amount equivalent to one-quarter of world wheat exports.

In all these cases, drought strikes suddenly. A run of dry years may well be followed by very heavy rain and widespread flooding. Thus, even if an approximate twenty-year recurrence should be demonstrated for the southern U.S. Plains, we are still dealing with step-functional change. Predictions based on past records will never seem wholly satisfying, until the mechanisms of sudden change in the low-level circulation are understood.

Chapter Summary

The last 10 000 years or so recorded most of the *fundamental accomplishments of mankind*. The interval included *noticeable shifts of climate*. The *evidence consists of* fossil vegetation, particularly *fossil pollen*; signs of *change in surface moisture supply*, such as scoured or filled stream channels and rising or falling lake levels; the *advance or recession of glaciers*; the *tree-ring record*; and *archaeological, radiocarbon and instrumental data*.

Fossil pollen is *preserved in peat*. The *changes in vegetation* shown by the pollen strongly *imply changes in air-mass dominance*. The *Scandinavian sequence* provides a *widely applicable standard* for about 16 000 years. It *begins with an* alternating sequence of *tundra and forest climates*. The *final tundra episode* was *succeeded by drier climate*, when birch-pine forest and mixed oak woods spread. *In the Atlantic phase*, temperatures were *1 to 2°C above present means. Precipitation increased* enough to stimulate the widespread growth of peat bogs. The succeeding *Sub-Boreal phase* brought back *continental conditions*; finally *maritime climate returned in the Sub-Atlantic phase*, when peat bogs renewed their growth.

Radiocarbon dating uses the originally atmospheric radiocarbon produced by cosmic-ray bombardment of nitrogen. *Absorption* of radiocarbon *ceases when an organism dies*, but radioactive *decay continues*. *Half of a given quantity* of radiocarbon *will disintegrate in* about *6000 years*, the *half-life*. Radiocarbon *dating works* for the last *30 000 to 70 000 years*. It is especially useful for the period considered in this chapter.

Partial re-clearing of channel fills occurred in the *Atlantic phase*. Renewed *filling* took place in the *Sub-Boreal phase*. On some streams, rather *slight re-clearing and* then renewed *filling* happened during the *Sub-Atlantic phase*.

Closed lakes in the Great Basin fluctuated in sympathy with mid-latitude climatic changes. The *low-latitude lakes* were *out of phase*. In tropical Africa, there were *three main episodes of high lake stands* between *9500 and 2500 B.P.*

Mountain glaciers in the northern hemisphere *advanced* possibly *during the Younger Dryas phase*; close to *6500 B.P.*; close to *4500 B.P.* in the cold early part of the Sub-Boreal phase; close to *2700 B.P.*; and in *1550–1850 A.D.*

Tree-ring analysis defines a climatic record for the *U.S. Southwest* through the *last 1550 years*. It reveals *alternating dry and wet spells*.

Long archaeological records come from the ancient *hydraulic civilisations*. The typical sequence of growth and prosperity, decline, and destruction, can be read in more than one way. There is little evidence that increasingly arid climates encouraged the cultivation of crops and the domestication of animals in the first place.

Two sharp episodes of *low river levels* and local *drought* afflicted *Egypt* in about *4000 and 3000 B.P.* Each lasted only about twenty to twenty-five years. Each brought widespread distress and social disorder.

A *contrasted interpretation* is placed on the record of *Mesopotamia*. Irrigation works existed as early as 5700 B.P. *Progressive salting of the soil*, because of irrigation, first *caused barley* (more salt-resistant) *to be substituted for wheat*, then caused *yields to fall*, then led to the *wholesale abandonment of settlements*. Local *populations* remained very *low between 3200 and 2500 B.P. Subsequent* reorganisation included the *extension of irrigation canals* to new land, but the *demand on labour to clear out silt* eventually became *impossibly great*, and the *second hydraulic civilisation collapsed* in about *800 B.P.*

Climate in NW Europe improved between about *400 and 1200 A.D.* Viking *settlements* were *established in Iceland and Greenland. Vines* grew *500 km north of* their present limits. A renewed *deterioration between 1200 and 1450 A.D.* brought much *ice to* the shores of *Iceland, halted the spread of settlement* in lowland England and drove the *limits of cultivation 300 m downslope*. This part of the historical *record is complicated by the Black Death* and by the *spread of sheep farming* as opposed to subsistence farming.

The interval called the *Little Ice Age* dates from about *1550 to 1850 A.D. Glaciers grew* both in the Alps and in Iceland. *Winters in NW Europe* were generally *severe*.

Attempts to explain the climatic changes recorded *since about 1850* (and in some cases earlier) have included *appeals to cyclic analysis*, including variations in solar emission and/or the receipt of isolation; calculations of the production of *carbon dioxide and industrial dust*; and studies of the emission of *volcanic dust*. One main advance is the recognition of *jumps in the circulation of the westerlies* – at the simplest, the calculation of the proportion of zonal to meridional flow. *One jump* occurred at about the *end of the nineteenth century*; before it, *weather* in humid mid-latitudes was *highly variable. Far less variable conditions* obtained in the following interval, *up to* about 1950. Subsequently, more *variable conditions* have *returned*.

Short-term prediction, including prediction of jumps in the mid-latitude circulation, is *urgently needed in relation to world food supply*.

Fig. 19.7 (opposite)   Blown soil on the southern U.S. Plains, 1937: weeds trapped along the fence line of a deserted farm aid in the piling up of loose material (U.S. Dept. of Agriculture Soil Conservation Service).

20 Ecosystems I: Soil

Soil as an entity is a fine instance of a cascading system. It receives complex inputs, and makes complex outputs, of matter and energy. For many individual soils, inputs and outputs are balanced: some kind of regulating mechanism, or set of mechanisms, is at work. Where a soil fails to change through time, the rate of throughput is by definition constant. But throughput is not instantaneous; soils therefore can be regarded as storage systems. In another sense, they are process-response systems: they develop internal characteristics that can be related to external influences. In two senses they are morphological systems: they vary in respect of grain size, and also in the vertical distribution of certain attributes. They are ecosystems in the sense that they involve the interacting and mutually supporting activities of soil-dependent plants and soil-dwelling animals. Finally, where they have not been much altered by man – whether deliberately or not – they can be considered to be control systems.

Soil classification developed before the various kinds of system were identified. Nevertheless, contrasting approaches to classification reveal quite clearly that schemes of classification imply contrasting kinds of systems listing.

20.1 Definition of soil

One proposed definition of soil is the stuff that plants grow in and that at least 80% of human feed depends on. At once we come up against the kind of difficulty that bedevils all discussions of soil – too many exceptions. Some plants live in water; some live in, or on, other plants; others live simply in the air, the noteworthy example here being Spanish moss, which can grow on overhead cables. Another definition, and a better one, is that soil is the uppermost part of the waste mantle, that has been transformed by the action of organisms. Such a definition, however, has to allow for end-members of its series. One end-member is desert sand, in which transformation by organisms has not occurred at all. The other end-member is entirely organic

peat, which contains no fraction whatever of the waste mantle. Most soils come between the two extremes. A very rough average mix is 50% air + water, the proportions between the two being highly variable, and 50% mineral matter + organic matter, in the average ratio of 45:5.

20.2 Soils as morphological systems: particle size

Particle size of mineral grains in soil exerts strong controls on the ability of water to soak in and to percolate through. Broadly speaking, the coarser the particle grade, the more freely does water penetrate; but complications set in where much of the soil consists of clay particles.

The basic contrasts are those among sand particles, silt particles, and sub-microscopic flakes of clay. As noticed in Chapter 4, sand particles range mainly between 2 mm and $\frac{1}{16}$ mm in diameter. In all usual soils, they consist of 90% or more of quartz sand, SiO_2. Silt particles, $\frac{1}{16}$ to $\frac{1}{256}$ mm in diameter, can be either extremely small grains of quartz, or – as is more usual – tiny aggregates of clay. Clay particles, less than $\frac{1}{256}$ mm in diameter, always consist of clay minerals. They are compounds of silicon, aluminium, oxygen and hydrogen.

A sandy soil feels gritty when a sample is rubbed between finger and thumb. The particles fall away when they are released. A moist silt soil can be smeared between the fingers without producing a gritty feeling. A moist clay soil can be buffed into a polish or rolled into threads. The differences of particle size are differences of texture. They are highly important in the short-term response of soil to weather.

Sandy soils take in water freely, but also freely release it. They are liable to be affected by crop damage in times of drought, but warm up readily in the spring. Clay soils can be difficult to drain and heavy to cultivate; they warm up slowly. The intermediate silt soils are generally the most easily manageable.

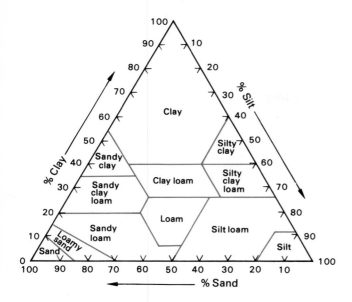

Fig. 20.1 Soil classification by particle size.

Classification according to particle size is a morphological classification, in which the particles are the components. In actual practice, twelve texture classes are recognised (Fig. 20.1). An experienced worker can distinguish all of them by feel. The boundary between one class and another, drawn as a line on the diagram, is sharp. It is also arbitrary. Nevertheless, the contrasts have real meaning, and practical meaning at that. By way of example, consider loam and clay. The loam class is roughly centred on 20% clay, 40% silt and 40% sand. Clay soils contain at least 40% clay and not more than 60% silt and sand combined. In addition, the sand content may not exceed 45%, or the silt content 40%.

20.2a Contrasted types of clays

The behaviour of soil clays, both as agricultural raw materials and as foundations for buildings, varies with chemical character and crystal structure. Three sets of clay minerals exist. One, the amorphous clays, is non-crystalline. It occurs in raw soils that have not been developing for a long time. Eventually, amorphous clays will convert to members of the other sets.

In the set of 1:1 clays, layers of silica (= compound of silicon with oxygen) alternate with layers of alumina (= compound of aluminium with oxygen). These are frequently called the kaolinitic clays, after the clay mineral kaolinite that is often present. In the other set, that of the 2:1 clays, each layer of alumina is sandwiched between two layers of silica. These are often called bentonitic clays, after the

mineral name bentonite; their detailed chemistry is highly diverse.

The main importance of the mineral contrast between the 1:1 and the 2:1 clays is that it is extremely difficult for water to invade the former. The 2:1 clays on the other hand freely take up water in wet conditions, and as freely release it in dry conditions. It is these clays that, on sloping ground, easily fail when they are wetted, and that crack badly when they dry out. Drought in the London area always means foundation and wall damage to some of the buildings resting on the London Clay formation, which is highly bentonitic.

20.3 Soils as morphological systems: soil structures

A parallel classification to that of texture uses the units known as peds. These, the system components, are the aggregates into which a given soil breaks down, especially when it dries. For instance, many clay soils crack into prisms when they dry out. For present purposes, classification by structure is less important than the general fact that the spaces around peds provide channels for the circulation – mainly downward – of water, and also provide habitats for many soil-dwelling animals. Prismatic, columnar, angular blocky, sub-angular blocky, platy and crumb structures are widely observed in nature. Soils with crumb structure, such as are common in western Europe, are resistant to rainsplash. When ploughs of European design were introduced to North America, and used on less resistant soils, severe erosion, which included gullying on slopes, resulted.

20.4 Soils as morphological systems: soil profiles

For purposes of soil description, the profile – the vertical succession down through the soil – is highly significant. However, it can scarcely be discussed without reference to process-response, which is always implicit in any profile description.

In all humid climates, soil material tends to be moved downward. Mineral matter is provided by the processes of rock weathering; but weathering does not cease when rock material is incorporated in the soil. Some of the water that falls as rain is lost by direct evaporation from the soil surface; another fraction is taken up by plants, most of it being transpired into the air. But a third fraction percolates downward, carrying with it soluble materials, both inorganic and organic. The action of downward-moving water is called leaching.

Not all the downward-moving solutions escape from the soil profile into the groundwater table. Some dissolved material will be precipitated in the lower parts of the profile. In addition, clay particles which become dispersed downward can also lodge in the lower soil. We thus need to separate eluviation (= washing out) from illuviation (= washing in).

The mineral input into soil comes from the parent material, the bedrock or superficial sedimentary deposits in which soil is formed. The organic input is concentrated on and near the soil surface, in the form of plant litter, decaying roots, decaying organisms, and the excreta of soil animals.

The net result of mineral input, organic input, eluviation in the upper soil, and illuviation in the lower soil is to produce vertical differentiation. At the clearest, the soil profile is separated into bands. These cannot be called layers, because they do not result from successive deposition, but from redistribution of, and chemical changes in, the soil constituents. Instead, the bands are called soil horizons. These are the profile components. A distinctly banded soil is said to be horizonated.

Three horizons were initially recognised (Fig. 20.2). The uppermost, the *A* horizon, consists of thoroughly weathered rock material and of partially decayed organic matter – humus. It has been strongly affected by the washing-out of soluble materials and by the downward dispersion of clay particles. It is an eluvial horizon. The underlying *B* horizon consists of less thoroughly weathered parent material, a minor proportion of organic matter, and of materials carried down from above. It is an illuvial horizon. The *C* horizon consists wholly of more or less decomposed parent material, little or not at all affected by organic action, eluviation or illuviation.

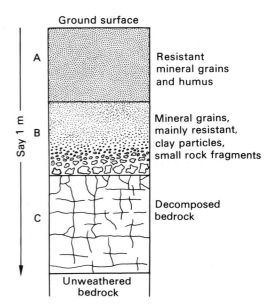

Fig. 20.2   The fundamental subdivision of a soil profile into horizons.

Fig. 20.3   Two contrasted soil profiles: a, a Spodosol (Ferrod); b, an Oxisol (Orthox). The classifications relate to the Seventh Approximation, discussed later in the text. The Spodosol profile is based on a field example from the heathlands of the North German Plain, in a region of transitional midlatitude climate. The Orthox profile is from West Africa, in an area of humid tropical climate. Stippling, etc., on this and comparable later diagrams symbolises variation of texture, without attempting to represent it exactly. Pecked horizontal lines indicate merging boundaries.

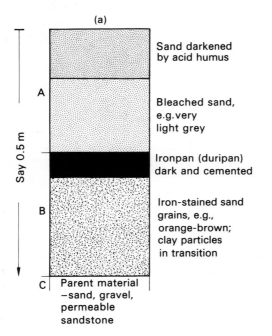

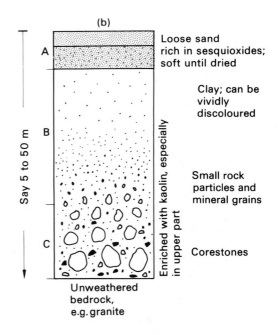

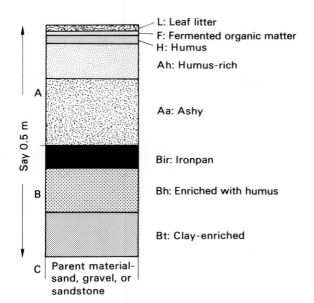

Fig. 20.4 Elaborated subdivision of a profile on the A-B-C system.

20.4a Some simple examples

Two simple fairly extreme cases will illustrate the development of horizonated profiles in leaching conditions. Under boreal forest on well-drained sites, the upper part of the *A* horizon becomes charged with dark acid litter and humus. The lower part is strongly eluviated, consisting in a typical profile of bleached sand. The *B* horizon undergoes illuviation with clay particles, and often also with iron compounds. It may contain a strongly cemented subhorizon, the ironpan, that greatly obstructs soil drainage (Fig. 20.3a).

Under humid-tropical forest, clay minerals break down in the upper part of the profile, but sesquioxides (Al_2O_3, Fe_2O_3: see Chapter 18, section 18.3) selectively accumulate. They can come to compose almost all of the *A* horizon. The *B* horizon is progressively enriched with clay minerals, especially the 1:1 kaolinitic clays (Fig. 20.3b).

Where rainfall is not enough to promote net downward movement of water through the soil and to the groundwater table, soluble material, and humus if available, progressively accumulate in the profile. The chief accumulating mineral is calcium carbonate in the soils of subhumid and subarid regions, and sodium chloride in the soils of arid regions. Wherever the soil texture is fine enough, dissolved matter is raised toward the surface by capillary action, and is concentrated at shallow depths when the soil moisture is evaporated.

20.4b Elaboration of profile descriptions

Just as a textural subdivision into sand, silt and clay is too simple, so is a subdivision of the profile into *A*, *B* and *C* horizons. More than one system of elaboration exists. Each uses number or letter systems in addition to *A*, *B* and *C* (Fig. 20.4). For example, the surface layer of plant litter can be sub-classified as *L*, *F* or *H*, according to the dominance of litter proper, fermented organic material, or humus. Part of the *A* horizon may be labelled *Aa*, *Ah* or *Ae*, according to whether it is bleached (*a* = ashy), humus-rich, or eluviated. Part of the *B* horizon may be labelled *Bir*, *Bh* or *Bt*, to show richness in iron compounds, humus, or textural (= illuvial) clays. For regions of dry climate, the added symbol *ca* shows enrichment with calcium carbonate, as in *Bca* or *Cca*.

This listing is not comprehensive. Additional symbols exist and are in use. No comprehensive review is proposed here. It will be enough to discuss the significance of the humus symbol *h*, which has already been met, and of the gley symbols *G* and *g*, which have not.

Humus is extremely finely divided organic matter. Its particles are less than 0.001 mm in diameter. Its chemistry is so complex that it has so far defeated complete analysis. Earthworms play a chief part in mixing it with the mineral fraction of the soil, and thus in creating what is known as the clay–humus complex. As will appear, the clay–humus complex is highly important in the circulation of soil nutrients.

A gley (pronounced gly) horizon results from waterlogging in the lower part of the profile. Shortage of oxygen in the waterlogged part produces grey-coloured iron compounds, so that a well-developed gley horizon is uniformly pale. The letter *G* in the profile description connotes strong gleying. Where the soil dries from time to time, the pale grey colouration is mottled with yellows, red and purples; its moderate gleying is connoted by a small *g*.

20.5 Soils as process-response systems

Soil profiles were orginally recognised for mid-latitude regions. They were related to major climatic types, implying a relation to major types of vegetation. Soils which clearly relate to climate/vegetation types are called zonal soils. Soils so dominated by local drainage conditions that they differ strongly from nearby zonal soils are named intrazonal soils. Soils as yet incompletely formed – because of very great resistance of the parent material to weathering, because of lack of time for complete development, or because of slope instability – are azonal soils.

The very recognition of zonal, intrazonal and

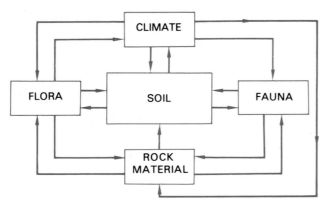

Fig. 20.5 Abstract model of process-response in soil formation.

azonal soils, and the treatment of profiles in terms of leaching/evaporation and eluviation/illuviation, mean treatment in terms of process-response. Fig. 20.5 illustrates one of several possible schemes of process-response behaviour. Rock material, acted on by plant and animal life and especially by climatic agents, provides the mineral material. The chemistry of the rock material exerts a certain influence on the plant and animal life of the soil, and on the above-ground vegetation that the soil supports. Climate affects the soil, notably in respect of temperature and moisture, but in the very lowermost air is also affected by the soil, mainly perhaps in respect of temperature. Plants and animals exchange their substance with the soil; they also act to redistribute soil constituents.

The process-response system in Fig. 20.5 must, furthermore, be imagined as acting at particular

sites, where profile drainage varies not only with climate but also with soil texture and surface slope. In addition, the system operates through time. The short-term response is as rapid as a single cycle of wetting and drying. The long-term response – that of the full development of a soil profile – may take from 50 to 500 years in well-weathered rock waste. The very fact that existing zonal soils at the surface correspond closely to the existing belts of climate/vegetation guarantees that a significant change in process is followed, in a very brief interval on the geologic time-scale, by a related change in response. But where the soil-forming process has to begin all over again, as on resistant rock scoured bare by ice, the 10 000 years or so of postglacial time have produced imperceptible effects.

20.5a Some effects of climate

Fig. 20.6 shows the influence of precipitation on the contents of clay and nitrogen, and on the depth at which calcium carbonate can become concentrated in the soil. Notice that there are three different vertical scales. Up to the indicated limit of precipitation, content of clay and nitrogen increases as precipitation increases. Calcium carbonate becomes concentrated at and near the surface where mean annual precipitation is about 250 to 400 mm, but with increasing precipitation the depth of concentration increases. No concentration occurs when precipitation rises above 1000 mm.

Temperature is also influential in the process-response behaviour of soils, both indirectly through vegetation and directly through its control over the supply of energy. In addition, the higher the temper-

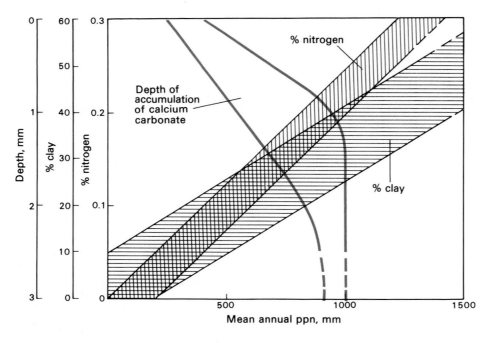

Fig. 20.6 Influence of precipitation, as a control of soil moisture content and of the leaching process, on the accumulation of clay, nitrogen, and soil carbonates.

ature, the greater the dissociation of water molecules into ions. Dissociation in this case means the breakdown of a water molecule, H_2O, into an hydrogen ion with a positive charge, H^+, and an hydroxyl ion with a negative charge, OH^-. The hydrogen-ion concentration, written pH, controls the acidity, neutrality or alkalinity of the soil. It also exerts a powerful influence on weathering processes. A pH of 7–8 is a neutral value. Lower values, down to a possible zero, indicate acidity. Higher values – usually 9 or 10, although readings up to 14 are possible – mean alkalinity.

20.5b Cation exchange

Ions with positive charges are cations; those with negative charges are anions. Clay–humus particles act as highly charged anions. Adjacent cations with opposing positive charges become locked on by the electrical forces. In addition to the already mentioned hydrogen cation, H^+, they include cations of calcium, Ca^{++}; magnesium, Mg^{++}; potassium, K^+; sodium, Na^+; and ammonium, NH_4^+. The tightness of the electrical bond is greatest for the hydrogen cation. For the metal cations, it decreases through the listed order. Variation in the strength of the bond means that some cations can be exchanged for others.

Leaching involves the loss of calcium, magnesium, potassium and sodium ions to the groundwater table. Intense leaching also involves the loss of the nitrogen-bearing ammonium. Nitrogen, potassium and calcium – and, to a lesser extent, sodium and magnesium – are important as plant nutrients. In consequence, both the pH of a soil and the character of its clay component are significant for plant growth. Acidity in itself inhibits leaching, because the abundance of negative charges retains the metallic and ammonium cations. On the other hand, if humic acids cause the breakdown of clay minerals, then the cations are released. The characteristic 2:1 soil clays of middle latitudes have a far greater capacity to lock on cations than do the 1:1 clays of the humid tropics, where plant nutrients in the root zone are concentrated in the uppermost 25 cm.

20.6 Soil classification

The great complexity of soils, the fact that one kind can grade into another – for instance, downslope – and the wide range of potential purpose in classifying at all, have combined to lead to intense confusion. Rival classifications have developed in individual countries; indeed, the *Soil Map of the World* produced by the International Food and Agriculture

Organisation of UNESCO employs a classification of its own. Many soil names are folk-terms, like the names of many rocks and minerals. But whereas an igneous rock type such as granite can be precisely defined in terms of its texture and its range of chemical and mineral composition, no comparable precision is possible with respect to kinds of soil that bear folk names. Some names have proved difficult or impossible to translate, or only too easy to mistranslate, from one language to another.

The difficulties of language are readily illustrated by reference to colour. Even in a single language, colour names may mean one thing to one observer, and something else to another. In any event, there are not enough colour names to spread through the subtly varying range of soil tints in the real world. Soil scientists, therefore, use a colour code, in which for instance 7.5YR 3/1 connotes dark reddish grey, and 7.5YR 4/3 connotes dark olive. No parallel scale of comparison exists for certain other attributes of soil.

The historical development of soil science has involved a great deal of attention to soil/climate relationships. There is however no question whatever of a one-to-one match. Chapter 15 distinguishes thirteen kinds of lowland climate. On the F.A.O. *Soil Map of the World* twenty-six kinds of soil unit appear; but it just does not happen that two kinds of soil unit go with one type of climate. In other classifications, the primary subdivision of soils runs to anything from two to twenty-three classes.

20.6a Resolving the confusion

A project team of the U.S. Department of Agriculture set out to cut right through the confusion by means of setting up a whole new technical vocabulary. At first sight, most of the terms involved look barbaric. On a closer look, they prove not only to make sense, but also to be manageable. The resulting classificatory system is known as the Seventh Approximation, the authors having concluded that six major attempts at classification had gone before.

20.6b The hierarchy of subdivision

The primary subdivision is into soil orders. There are ten of these. They are distinguished by the presence or absence of diagnostic horizons. Orders are themselves subdivided into suborders, with reference to wetness and soil moisture regime, climatic regime and (where relevant) the extent of the decomposition of plant fibre. That is, a given suborder relates to a given set of genetic processes; suborders number forty-seven. Suborders are subdivided into great groups, for the recognition of which profile characteristics are again important; great groups number more than two hundred. Great groups subdivide into subgroups, these into families – each family with its own set of conditions for plant growth – and families into soil series. A series consists of two or more pedons, the smallest units (e.g. 1 to $10\,m^2$) for which horizons can be recognised and studied. The progression from ten orders through forty-seven suborders and two hundred plus great groups suggests part of a semilogarithmic sequence. For the U.S.A., the total of soil series is estimated to be at least 10 000; projection of the semilogarithmic graph suggests that the world total is somewhere near 20 000.

Our immediate purposes call for no such detailed scrutiny. The following passages are intended to show how the Seventh Approximation works, down to the level of suborders. Moreover, it will by no means be necessary to illustrate or discuss every suborder in the list.

20.6c Diagnostic horizons and soil order

In the Seventh Approximation, surface horizons are separated, under the name epipedons, from diagnostic subsurface horizons. There are six kinds of epipedon (roughly = topsoil), of which two are strongly affected by cultivation (Table 20.1). Distinctions among the other four take account of tinting, base status and content of organic carbon. The diagnostic subsurface horizons have names referring to some combination of colour, texture, mineral composition and chemistry (Table 20.2). The way in which horizons – chiefly those of the subsurface – are used to separate out the ten soil orders is shown in the first three columns of Table 20.3.

20.7 Coping with the new vocabulary

Names of orders, suborders and so on in the Seventh Approximation are composed of word elements. Combination of elements to form actual names is like a word game. Thus, each order name ends in -sol, for soil. Its first element chiefly indicates some leading profile characteristic. Oxisols, for example, by definition possesses an oxic horizon; and an oxic horizon, again by definition, is one with a very low content of weatherable minerals and low exchange capacity, wherein the clays are kaolinitic. For purposes of rigorous soil science, fixed upper limits are set for the content of weatherable minerals and for exchange capacity, and a fixed lower limit for proportion of kaolinitic clays; terms such as high, low, rich and poor in the verbal descriptions thus have

Table 20.1   The Six Kinds of Epipedon

Anthropic	dark coloured, thick, high in exchangeable bases, with much phosphate in consequence of long-continued farming
Histic	thin, containing much organic carbon, wet most of the time
Mollic	dark coloured, thick, high in exchangeable bases, but lacking the phosphate of the anthropic epipedon
Ochric	too light in colour, too thin, or too low in organic carbon to belong elsewhere
Plaggen	man-made, thick, but with textural and chemical characteristics depending on the original natural soil
Umbric	dark in colour, but poor in exchangeable bases

Table 20.2   Diagnostic Subsurface Horizons

Albic	colour determined by sand and silt particles, clay and free oxides having been removed
Argillic	illuvial horizon with significant clay accumulation
Agric	formed by cultivation, being enriched with clay, humus, or both
Calcic	enriched with secondary concretions of calcium carbonate
Cambic	changed or altered, for instance by obliteration of the structure of parent material, by liberation of free oxides, or by clay formation
Gypsic	enriched with calcium sulfate
Natric	rich in exchangeable sodium; argillic
Oxic	content of weatherable minerals very low; clay kaolinitic, with low exchange capacity
Salic	enriched with salts more soluble than gypsum
Spodic	with accumulation of free sesquioxides, or organic carbon, or both, but not of clays
Duripan	an horizon cemented by silica or silicates
Fragipan	partly cemented loamy horizon
Petrocalcic	strong cemented by calcium carbonate
Plinthite	very sesquioxide-rich horizon

precise connotations attached to them. Histosols (Greek *histos*, tissue) are rich in organic matter, for which we may understand plant matter; and the term rich indicates an organic matter content of at least 30%.

The vocabulary is completely new, since the Seventh Approximation deliberately excludes all previously existing soil names. Its derivation largely from Greek and Latin roots has the advantage that Greek and Latin, being dead languages, are no longer evolving; modern languages, by contrast, not only develop their vocabularies, but also change the connotations of many words. Soil science has suffered as much as any part of earth study, and more than most, from the uncontrollable nature of usage. Because soils vary so greatly in nature, the full vocabulary of the Seventh Approximation is colossal; names of soil families run to between 3000 and 4000 terms, all of them brand new when first encountered. The total is far greater than that of the

TABLE 20.3   Orders and Suborders of Soils, According to the Seventh Approximation

ORDER	OUTLINE DESCRIPTION	RAPID RECOGNITION CHARACTERISTICS	SUBORDERS	
ALFISOLS	argillic horizon present; base content moderate to high	all mineral horizons present except oxic	with gleying	Aqualfs
			others in cold climates	Boralfs
			others in humid climates	Udalfs
			others in subhumid climates	
				Ustalfs
			others in subarid climates	Xeralfs
ARIDISOLS	semi-desert and desert soils	ochric or argillic horizon present, no oxic or spodic horizon; usually dry	with argillic horizon	Argids
			others	Orthids
ENTISOLS	weakly developed, usually azonal	no diagnostic horizon except ochric, anthropic, albic, or agric	with gleying	Aquents
			with strong artificial disturbance	Arents
			on alluvial deposits	Fluvents
			with sandy or loamy texture	Psamments
			others	Orthents
HISTOSOLS	developed in organic materials	30% or more organic matter	rarely saturated, 75% fibric	Folists
			usually saturated, 75% fibric	Fibrists
			usually saturated, partly decomposed	Hemists
			usually saturated, highly decomposed	Saprists
INCEPTISOLS	moderately developed; not listed elsewhere	cambic or histic horizon present; no argillic, natric, oxic, or petrocalcic horizon, no plinthite	with gleying	Aquepts
			on volcanic ash	Andepts
			in tropical climates	Tropepts
			with umbric epipedon	Umbrepts
			with plaggen epipedon	Plaggepts
			others	Ochrepts
MOLLISOLS	with dark A horizon and high base status	mollic horizon present; no oxic horizon	with albic argillic horizon	Albolls
			with gleying	Aquolls
			on highly calcareous materials	Rendolls
			others in cold climates	Borolls
			others in humid climates	Udolls
			others in subhumid climates	Ustolls
			others in subarid climates	Xerolls
OXISOLS	with oxic horizon	oxic horizon present	with gleying	Aquox
			with humic A horizon	Humox
			others in humid climates	Orthox
			others in drier climates	Ustox
			usually dry	Torrox
SPODOSOLS	with spodic horizon	spodic horizon present	with gleying	Aquods
			with little humus in spodic horizon	Ferrods
			with little iron in spodic horizon	Humids
			with iron and humus	Orthods
ULTISOLS	argillic horizon present; base status low	mean annual temperature 8°C or above; soils not listed elsewhere	with gleying	Aquults
			with humic A horizon	Humults
			in humid climates	Udults
			others in subhumid climates	Ustults
			others in subarid climates	Xerults
VERTISOLS	cracking clay soils	30% or more clay, with gilgai or other signs of up-and-down movement	usually moist	Uderts
			dry for short periods	Usterts
			dry for long periods	Xererts
			usually dry	Torrerts
MOUNTAIN SOILS	these vary greatly over short distances; they involve much steep slope			

Table 20.4 Examples of Derivation:
elements used in naming orders in block capitals

ELE-MENT	ROOT	ORIGINAL LANGUAGE AND MEANING	FAMILIAR WORDS WITH SAME DERIVATION
Agric	agri-cultura	Latin: agriculture	agriculture, etc.
Alb	albus	Latin: white	
ALF	Al,Fe	international western symbols for aluminium and iron	
Anthrop	anthropos	Greek: man	anthropology, etc.
ARID	aridus	Latin: dry	arid
Aqu	aqua	Latin: water	aquatic, aqua-sports
Arg	argilla	Latin: clay	argillaceous
Bor	boreas	Greek: north wind	boreal
Calc	calx	Latin: lime	calcite, calciferous
Camb	cambium	Latin: change	
ENT		this synthetic element is meant to suggest attributes of intermediate sorts	
Gyps	gypsum	Latin (and English): calcium sulfate	gypsum
HIST	histos	Greek: tissue	
Hum	humus	Latin: earth	humus
INCEPT	inceptum	Latin: beginning	inception
MOLL	mollis	Latin: soft	
Natr	natron	Arabic: sodium carbonate	
Ochr	ochras	Greek: pale	ochre
Orth	orthos	Greek: straight/correct/ordinary: used to group suborders not otherwise distinguished	orthodox
OX	oxide	French (and English): oxide	oxide
Psamm	psammas	Greek: sand	
Rend	rzedzic	Polish: noise; specifically, the noise of a plough in shallow soils, whence English rendzina, soil with surface horizon rich in organic matter, developed on calcareous material	
Sal	sal	Latin: salt	saline
SPOD	spodos	Greek: wood ash (reference is to colouring)	
Torr	torridus	Latin: hot and dry	torrid
Ud	udus	Latin: humid	
ULT	ultimus	Latin: last	ultimate
Umbr	umbra	Latin: shade	umbrage
Ust	ustus	Latin: burnt	
VERT	vertere	Latin: turn	inversion
Xer	xeros	Greek: dry	xerography (dry copying); xerox machine

complete working vocabulary of a non-scientist of no particular education. Some workers appear to have been put off by the extent of the lists of new words, others rather by their completely unfamiliar character. We may suspect also that some dislike the language of the Seventh Approximation simply because it is artificial: nationals of all countries cling obstinately to their own idiomatic tongues and complex grammars, despite all the logical advantages of the artificial Esperanto. Natural languages just are, to varying extents, deficient in logic. The names in the Seventh Approximation are logically constructed throughout.

20.7a Names of orders and distinctions among orders

A partial list of the elements used in composing soil names in the Seventh Approximation is given in Table 20.4: see, in alphabetical order, the derivations of ALF, ARID, ENT, HIST, INCEPT, MOLL, OX, SPOD, ULT and VERT, which are used in the construction of order names. Sorting out one order from another is like the puzzles that are usually stated in words, but are actually examples of mathematical logic: A has black hair and blue eyes, B has red hair, the one with brown hair and green eyes is not C ... and so on, ending with some such question as, Who has fair hair and brown eyes? From Table 20.3, we can see at once that a soil with an argillic horizon cannot belong to the orders Inceptisols or Entisols. One with an oxic horizon cannot be an Alfisol, Aridisol, Entisol, or Inceptisol. The sorting works by a combination of necessary inclusions and mandatory exclusions.

In a very general way, the interrelationships of the soil orders in time, and to some extent in climate, can be represented as in Fig. 20.7. Soil development begins in many instances with an Inceptisol, although two-way development between Inceptisol and Entisol is conceivable in some circumstances. By definition, profiles in these two orders are nowhere more than moderately developed. Although it is of

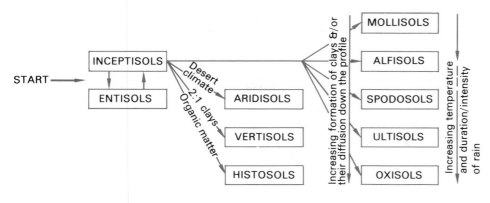

Fig. 20.7 A concept of the interrelationship of the ten soil orders in the Seventh Approximation.

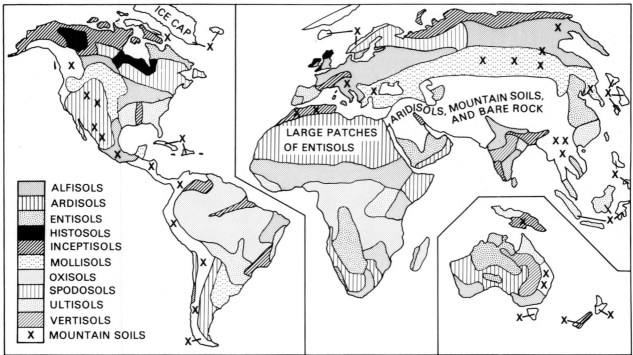

Fig. 20.8 World distribution of the ten soil orders: very highly generalised indeed.

course possible for development to go no further, if it should go further, then changes in the profile shift the soil into some other order. Accumulation of peat, as in a lowland swamp or a cool rainy moorland, leads to the development of a Histosol. Accumulation of 2:1 clays, as in a wide, shallow, floodable depression in a region of semi-arid climate, promotes the development of Vertisols. Site conditions largely apart, dry climates lead to the formation of Aridisols.

Among the other orders, there is a generally increasing sequence of the influence of soil moisture, reflecting in the degree of leaching, and at the other end of the range in the extent (always below the 30% content mark) of the accumulation of plant matter. Spodosols fit somewhat uneasily into the range, being considerably influenced in their development by the coarse texture and permeable character of

parent materials. Fig. 20.8 suggests that, in some respects, there is a very rough correspondence between the distribution of certain soil orders and the distribution of certain climates; but an overlay of Fig. 20.8 on Fig. 15.2 would show many instances of gross mismatch. Spodosols are most widespread in the northern parts of the northern landmasses, in climates that include cool continental and sub-polar. The distribution of Aridisols (with noteworthy extents of Entisols) expectably produces some kind of a match to the distribution of low-latitude and interior deserts and semideserts, while Mollisols are extensively distributed through the natural grasslands of the northern continents. Ultisols and Oxisols have some very loose connection, respectively with subtropical east-coast and with tropical or equatorial climate; but against all this, Alfisols occur widely, both in some midlatitude climates (e.g. Transitional) and within the tropics, where they are partly associated with the drier variants of low-latitude tropical climate.

20.7b Names of suborders and distinctions among suborders

Names of suborders are composed of two elements, one to indicate the soil order, the other to identify the suborder. For convenience in speaking and in writing, the element indicating the soil order is always a single syllable and always starts with a vowel. The order Alfisols readily provides the element –alf, Entisols give –ent, Oxisols give –ox, and Ultisols give –ult; but Aridisols are made to give –id, Histosols –ist, Inceptisols –ept, Mollisols –oll, Spodosols –od, and Vertisols –ert. One soon gets used both to this part of the practice, and to having the order element come second in the suborder name.

The first element of the suborder name usually says something about the profile. The element Aqu– always indicates gleying. This can be observed for soils of seven of the ten orders: hence Aqualf, Aquents, Aquepts, Aquolls, Aquox, Aquods, and Aquults, in the final column of Table 20.3. There are only twenty-four first elements, which combine with one of the ten order elements to give the forty-seven suborder names. Particular applications will appear in the profile descriptions which now follow.

20.8 Profile descriptions

The examples presented are all based on actual profiles recorded and measured in the field; but, for ease of comparison, profiles are presented wherever possible as if they were about one metre thick. The conceptual sequence follows that of Fig. 20.7

20.8a An Inceptisol profile

The selected profile (Fig. 20.9) is not well horizonated. Developed on limestone bedrock, it is sufficiently leached to be moderately acid throughout.

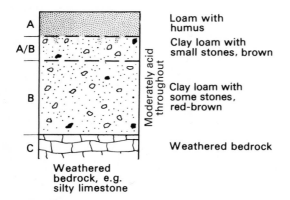

Fig. 20.9   Sample profile: an Inceptisol (Ochrept); based on a profile from the U.K., in west-coast midlatitude climate.

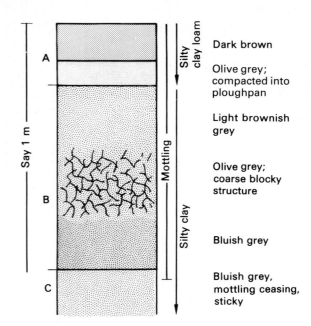

20.10   Sample profile: an Entisol (Aquent); based on a profile from an area of wet rice cultivation, Japan, in a region of subtropical east-coast climate.

Its lower part consists of clay loam in the interstices between fragments of bedrock: this is the cambic horizon. Too little water resides in the lower profile to promote gleying; the soil is developed neither on volcanic ash nor in tropical climate – it comes in fact from northwest Europe; too little plant matter is present to darken the epipedon enough to attract the name umbric, while the soil is certainly not man-made. Therefore, within the order Inceptisols, this soil is allocated to the suborder Ochrepts.

20.8b An Entisol profile

Here is a very different case (Fig. 20.10). Although the soil has developed in seasonally rainy climate in East Asia, it is azonal – that is, not related primarily to climate, but instead to site conditions. It comes from a paddy, where ploughing and planting with rice takes place twice a year. Profile development is weak, except for the agric horizon of the ploughpan and the regularly stirred soil above. Gleying is pronounced; hence, the soil is allocated to the suborder Aquents.

20.8c Histosol: Vertisol

These two orders will not be illustrated by profile diagrams. Many Histosols lack horizonation, although peats proper can be horizontally marked by surfaces that record major fires and/or shifts of climate. The subordering of Histosols depends not

primarily on vertical change in profile characteristics, but on duration of wetting and of degree of decomposition (Table 20.3).

Cracking clays in warm climates with alternate vigorous wetting and drying develop soils in which up-and-down movement is noteworthy. The profile is repeatedly churned over. In tropical climates, Vertisols are often black, as for example the vlei soils of Africa. Similar soils can form on silica-poor rocks on gentle slopes. The expression of vertical movement within the profile includes the polished surfaces of mini-faults against which individual peds move. The most obvious result, however, is produced on the surface, which under the influence of the crack/swell alternation can become patterned. The general name for the pattern, for the components responsible for it, or for a single component, is gilgai, the aboriginal word from Australia. This word has already been met in connection with mass movement (Chapter 9: Fig. 9.20b). The patterning broadly resembles that produced by periglacial action, fingerprint gilgai having visual affinities with solifluction terraces, and other forms (Fig. 20.11) looking, in the mass, like the hexagonal arrays of patterned ground in periglacial regions (Chapter 13). Authentic, if somewhat crudely defined, stone polygons can develop in stony deserts, provided that the fine fraction in the soil consists mainly of 2:1 clays.

In the African tropics especially, the vlei soils are among the most promising for cultivation, in respect of mineral content; but a surface with gilgai, after having been ploughed flat, will develop gilgai afresh within a few years.

20.8d An Aridisol profile

Midlatitude observers, a class that includes most of the users of this book, are possibly inclined to lump all tundra soils together as dominated by frost action, and equally to lump all desert soils together as saline. In actuality, soils both of cold and of dry

Fig. 20.11   Gilgai of the melon-hole type: observer stands in a depression, about 4 m across, one of hundreds in the local complex. (Tara district, Qld., Australia: T. Langford-Smith). Climate is subtropical east-coast. Contrast Fig. 9.20b.

regions are as highly varied as are those of the humid midlatitudes. It is however clear that soils of deserts and semi-deserts will as a class not undergo effective leaching, and that, again as a class, they will tend to accumulate chemical constituents that would be soluble, and removable, elsewhere. The selected profile, from an area of warm continental climate and marked soil moisture deficit in southern inner

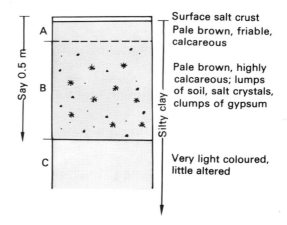

Fig. 20.12   Sample profile: an Aridisol (Orthid): based on a profile from the SW Asiatic U.S.S.R.

Asia (Fig. 20.12), has no argillic horizon in the subsurface: hence it is to be classed as an Orthid. A thin salt crust at the surface is associated both with calcium carbonate and with gypsum and sodium chloride below ground.

20.8e Two Mollisol profiles

Mollisols are by definition lime-rich; their *A* horizons are typically darkened with plant matter, being in some cases truly black. The percentage content of organics (below 30%) serves in the first instance to separate them from the Histosols, some of which, such as acid peat soils, are in any event low in exchangeable bases. Mollisols are best known from the midlatitude grasslands. There, the production and retention of grassy matter are promoted by warm springs and summers, with moisture enough to favour vigorous grass growth, but not enough to bring in forest, and by dry weather in late summer and deep frosts in winter, which act to check the decomposition of root material and surface litter. Deficiency of soil moisture in the total balance inhibits the loss of metallic cations, especially those of calcium; calcium carbonate accordingly becomes concentrated in the lower part of the profile (Fig. 20.13a). Because downward leaching is absent from the net sequence, no *B* horizon develops.

It is partly an accident of landscape history that, in both northern landmasses, conditions conducive to the formation of Mollisol profiles, of the general kind illustrated in Fig. 20.13a, have prevailed widely in regions where the soil parent material is loess. Really extensive and continuous spreads of loess occur only where relief is low. In respect of texture, loess soils cannot be improved on as agricultural material, while the Mollisols in question, with their high content of neutral to only slightly acid humus in the *A* horizon, and their charge of lime in the *C* horizon, are exceptionally well adapted to grain farming.

The most extensive suborder in the relevant areas consists of Borolls. Greater warmth and longer duration of summer reduces the humus content, in consequence of a change from the natural cover of long-grass prairie to that of short-grass prairie. Alternatively, the change can be effected simply by a reduction of precipitation. As humus content goes down, soil colour changes from black to chestnut, and lime concentration in the *C* horizon increases. Borolls pass into Ustolls.

Site conditions and bedrock character can act to develop Mollisols, in climates very different from those of the interior grasslands of the northern midlatitudes. Fig. 20.13b illustrates a Rendoll, a Mollisol profile with abundant lime in the *C* horizon, a significant lime content in the *A* horizon which here again is blackened by humus, and no *B* horizon. The controlling factors here are rock type and slope. The climate is that of midlatitude west-coast regions, where on some materials leaching can be intensive, and where on some sites peat accumulates abundantly enough to build Histosols. In the example in question, the supply of limestone is constantly replenished by the weathering and breakup of the bedrock. The development of a completely horizonated profile is prevented by creep; weathered rock is on balance moved downslope as fast as it is detached, but a time lag enables it to enter, for a time, into storage in the soil profile. Thus, morphologically and chemically, a Mollisol profile forms, displaying strong affinities with profiles that, belonging to the same order, are controlled chiefly by climate and vegetation.

Fig. 20.13 Two contrasted Mollisolls: a, a Boroll: based on a profile from the steppes of the U.S.S.R.; b, a Rendoll: based on a profile from the Chalkland on southern England. Respective climates are midlatitude cool continental and midlatitude west coast.

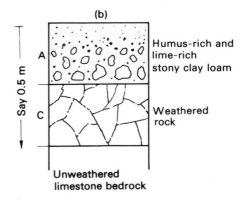

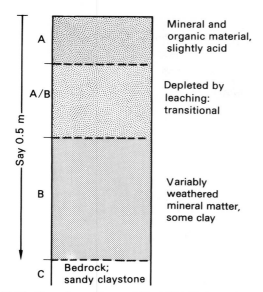

Fig. 20.14 Sample profile: an Alfisol (Udalf): based on a profile from the English Midlands, in midlatitude west-coast climate.

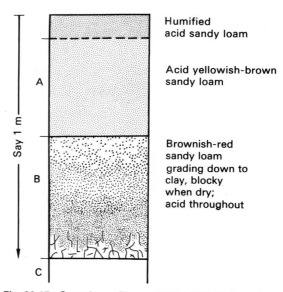

Fig. 20.15 Sample profile: an Ultisol (Udult): based on a profile from the SE United States, in subtropical east-coast climate.

20.8f An Alfisol profile

Widely occurring in humid midlatitudes, Alfisols proved somewhat puzzling in the early days of soil science. Their attributes seemed largely negative. Thus, they are by no means as clearly horizonated as are spodosols – on sight, in fact, some of them scarcely seem to be horizonated at all. There can however be no question of regarding them as immature soils, as numbers of Inceptisols are. They are neither distinctly acid, like Spodosols, nor distinctly rich in exchangeable bases like representative Mollisols. They display the effects of leaching, but not in drastic form. Because leaching has occurred, they do contain selective concentrations of sesquioxides, but never in so pronounced a way as to match what happens in some parts of the humid tropics. When however Alfisols are recognised as constituting an order in their own right, it becomes easy to see how they fit into the general scheme, and also to accommodate those of low latitudes, in addition to the middle-latitude cases that were originally studied mainly with reference to soil development under deciduous forest. Modest net leaching does not suffice to bleach the lower part of the *A* horizon. Clay particles are washed down through the profile, rather than being broken down and dispersed. Because the leaf litter is less acid than that of the boreal forest, Alfisols in middle latitudes run to only slightly acid, or even neutral; but persistent leaching does mean a progressive loss of calcium, magnesium, sodium and potassium.

The profile in Fig. 20.14 is of an Udalf, formed in humid but not cold climate. On many gentle slopes and on lowland sites, gleying is common, calling for classification in the suborder Aqualf.

20.8g A Spodosol profile

The diagram (Fig. 20.3a) has already been presented; the typical conditions for the development of many spodosols have also been outlined (section 20.4a). Nothing need be added here, except that Ferrods, Humods and Orthods are all widespread in nature, with Aquods developed wherever profile drainage is seriously impeded in the lower part. Although as climate-related soils Spodosols are particularly associated with the boreal forests, they develop readily elsewhere, as on the sandy and gravelly heathlands of western Europe, within a climate capable – if the soil were also capable – of sustaining thick hardwood forest. In this connection, as in numerous others, the Seventh Approximation well displays the advantage of discarding the zonal concept. Spodosols are spodosols wherever they occur, as opposed to being zonal in some areas and intrazonal or azonal in others. It is particularly useful to be able to recognise tropical spodosols for what they are, instead of some kind of aberrant form which belongs properly in boreal regions. Perhaps this is particularly so in respect of the unusually deep profiles – often those of Aquods, with gleying in the base – which can form on spreads of alluvial sand under impoverished rainforest; by contrast with the 1 m or so of a midlatitude Spodosol, those of the tropics can reach 2 m, 3 m or even 5 m in thickness.

20.8h An Ultisol profile

Common in the humid subtropics and parts of the tropics, Ultisols undergo persistent leaching, are typically acid close to the surface, and are enriched with clay in the lower part of the profile (Fig. 20.15). In a way, they are transitional between the less strongly leached Alfisols, and the very strongly leached, sesquioxide-rich and kaolin-rich Oxisols. It is iron compounds that impart to them their distinctive red and yellow colouration. Some of them are transitional in time rather than in space, incorporating the effects of former deep weathering, but undergoing renewed profile development in their upper parts in changed climatic conditions. Rapid breakdown of plant matter combines with widespread raininess, and in places with waterlogging at depth, to make Udults and Aquults especially frequent among the suborders. Among the Udults themselves, a particularly important subset is that of tropical and subtropical soils developed on basalt. Their profiles are not horizonated: the content of hydrated iron oxides acts to prevent much separation of constituents within the profile, which accordingly looks earthy and red to reddish-brown throughout its vertical extent.

20.8i An Oxisol profile

Finally, as has been seen, Oxisols are soils with oxic horizons. They are best and most widely developed in conditions of humid-tropical deep weathering, such as have been discussed previously in relation to the relict profiles of middle latitudes (Chapter 18, section 18.3). Oxisols range from poorly to very well horizonated, according to the selective concentration of sesquioxides in the upper part of the profile.

The thick layer of surface litter under rainforest is not clearly separable from the lowermost storey of the forest itself. Rapid decay and heavy rainfall ensure that the organic waste becomes swiftly available to ground-rooting plants. Oxisols are seldom more than moderately acid, because the decaying litter constantly supplies metallic cations. On the other hand, movement of soil moisture is copious and downward. Descending organic acids and dissociated H^+ ions not only break down the clays of the A horizon, but also mobilize silica. Clays formed in the B horizon are typically 1:1 clays – kaolinite and related types – with low capacity to hold cations. On sites with free profile drainage, cations of ammonium, silicon, calcium, magnesium, sodium and potassium escape through the C horizon to the drainage system.

All observers are impressed by the huge biomass of tropical rainforest: many have wrongly concluded that the rainforest soils must be naturally fertile. This is not so. The plant nutrients are constantly recycled from plants to litter and back to plants. If the forest is cleared, the nutrient supply is immediately cut off. The cutting-off will suddenly impoverish a soil that can reach very great thicknesses (Fig. 20.3b). If the A horizon is rich in sesquioxides, the strong probability is that it will harden within a few years into duricrust. Precisely this has happened in some western parts of the Amazon basin, where, climate notwithstanding, rainforest has been succeeded by scrub savanna – a leading instance of the overriding effect of soil on vegetation, instead of the other way round.

20.9 Soils as ecosystems

The basis of all life on the world's lands has been defined as the process whereby plants use chemical energy (derived from light energy) to convert atmospheric carbon dioxide and water vapour into sugar and oxygen. The oxygen enters the low atmosphere: sugars are converted to complex plant substances, especially cellulose.

Some such view is implicit in the often-described sequence of the initial formation of soil. Lichens, which are symbiotic (= co-operatively living) or parasitic (= one living off the other) associations of fungi and algae, can fix themselves and live upon bare rock. They extract nitrogen from the air, retain a film of water and extract mineral ions from the rock surface. Dead lichen material can provide a kind of elementary organic litter on which higher forms of plants can grow. Such a sequence is certainly being followed today, wherever bare rock is exposed for a long period in damp conditions. But no true soil will develop until bacteria have gone to work in breaking down the vegetable matter into humus.

20.9a Soil micro-organisms

This set of soil organisms includes protists and microfauna. The name protists indicates that the organisms in the subset – bacteria, fungi and actinomycetes – cannot always be successfully distinguished either as animals or as plants. They could possibly represent a very early stage in organic evolution when the clear plant/animal difference had not yet been developed. The acticomycetes have certain close resemblances both to bacteria and to fungi, and thus may refer back to a time when even these two elementary life forms had not become clearly separated from one another.

Some scientists, however, recognise the protists as plants, classing them as soil microflora. In any case, algae are included in the subset. The microfauna consists mainly of single-celled animals such as amoebae, but higher forms such as microscopic worms are also present.

The protists, and especially the bacteria, are vital to the breaking down of the complex organic molecules of plant material. The chemistry of the processes involved is still largely obscure, but the net results are clear. They appear in the recycling of sulfur, carbon and nitrogen.

There are a number of stages in the sulfur cycle: the essential outcome is the production of sulfates, inorganic SO_4 compounds of calcium, magnesium, sodium and potassium, which plants can take up.

Breakdown of organic compounds of carbon, contained both in plants and in plant-eating animals, involves the release of other materials also – most notably, compounds of nitrogen and sulfur. The released carbon combines with oxygen to form carbon dioxide, CO_2. This either returns to the atmosphere, replenishing the stock on which plants draw, or combines with soil moisture to form carbonic acid, H_2CO_3, which is an agent of weathering.

Nitrogen in plant and animal tissues is locked in highly complex compounds such as proteins. These are broken down, largely by bacterial action, into simple compounds which can be taken up by plants – ammonium, NH_4; nitrites, compounds of NO_2; and nitrates, compounds of NO_3. Not all bacteria involved in the nitrogen cycle act to fix nitrogen. Those that do are called nitrifiers. Those that do not are the denitrifiers: they convert nitrogen to gas and release it back into the atmosphere. Among the bacteria that fix atmospheric nitrogen, an important group has developed a symbiotic relationship with certain plants, including clover and alfalfa. The bacteria cause the formation of root nodules which in turn produce substances that promote the multiplication of the bacteria and the fixation of nitrogen.

The organisms of the microfauna also act in the reduction of organic debris to inorganic products, and in the promotion of ion exchange. Some feed on microflora. Their main ecological function, however, seems to be that of a link in the food web between plant material or other micro-organisms on the one hand, and the mesofauna on the other.

20.9b Soil mesofauna

By definition, this set of soil animals is of middle size. But very many of the organisms contained in the set are too small to be visible to the naked eye: they are only middle sized in comparison with the microfauna. They include nematodes (eelworms) and mites. The animals of the mesofauna feed on decaying organic matter, consuming micro-organisms as they do so. Some directly attack living plants. There is also some predation within the set, for certain members eat others.

20.9c Soil macrofauna and megafauna

Once again, the definitions are relative. The set of macrofauna (large animals) includes potworms and centipedes. The megafauna (very large animals) includes burrowing backboned animals such as moles and rabbits. As could be expected from the mere increase in size, there is a great variety of behaviour and ecological function in these two sets. The potworms eat micro-organisms and decayed plant material. Centipedes are predators. Earthworms, ingesting mineral soil and organic matter, are of leading importance. They reduce mineral substances to the silt or clay grade; they produce humus in their intestines and neutralise its acid with carbonate solutions; and they disseminate the humus by moving it through the profile. In particular, they recycle the soil material through the profile, dragging down leaves, feeding underground, and producing casts at the surface. In moderately moist midlatitude soils, they can cycle a depth of 3 cm of soil material through the profile in ten years or less. The indicated rate is equivalent to a complete turnover of an average soil in 25 to 50 years. In regions of tropical and allied climate, the soil-turning function is taken over by termites.

20.10 Soil ecosystems and dynamic equilibrium

Quite apart from visible plants – the macroflora – soils contain impressive numbers of organisms. For a representative midlatitude soil, the weight of soil organisms, aside from plant roots, runs between 2500 and 3000 kg/hectare. The total animal population in the mesofauna, macrofauna and megafauna is 10 000 000 per square metre. The total of micro-organisms is 1 000 000 000 000 per square metre. All these life forms interact with one another and with the humic and mineral constituents. In considering soil, we have to do with ecosystems of extreme complexity.

At the same time, unless they are being rapidly modified by erosion (including mass movement) or by deposition, soil ecosystems tend to achieve dynamic equilibrium. Many backboneless animals of the soil fauna move upward or downward, according to daily, seasonal or irregular changes in temperature and in moisture content. At the extreme, the protists respond to adversity by converting to spores, reverting back again when conditions change. A forbiddingly intricate array of inverweaving action goes into the life of the soil. But, simply because the life forms are interdependent and interacting, soils must be classed as ecosystems. Because rapid modifications are the exception, but also because the soil organisms range between certain limits in their behaviour, most soil ecosystems must be regarded as in dynamic equilibrium.

All the more is this so because the behaviour of soil ecosystems cannot be separated from that of other interlocking systems: ecosystems are open. The plants that grow in soil sustain, directly or indirectly, the animals that live above ground. The waste products and the dead substances of such animals return to the soil. Some aspect of the exchange of gases among atmosphere, plants and soil have been noticed in the foregoing paragraphs. In the long term, all the transfers that are here at issue are balanced out, or at most concern minor imbalances that only become significant during very long periods of geologic time. Negative feedback works to sustain ecosystem equilibrium.

Chapter Summary

Soils can be classified on the basis of particle size: that is, *texture*. The texture grades take account of the relative *proportions of clay, silt and sand*.

The behaviour of soil clays varies with chemistry and structure. Amorphous clay, typical of raw soils, will eventually convert to *crystalline clays*; these *are either 1:1* (kaolinitic) *or 2:1* (bentonitic) clays, according to the arrangement of layers of alumina and silica. The *2:1 clays take in and release water*.

Peds are the *aggregates* into which a given soil breaks down. They *define the soil structure*.

A *soil profile* is the *vertical succession* of soil *horizons. At the simplest*, it consists of *an A horizon*, enriched with humus, *at the top*; *a B horizon next beneath*, which is often enriched with material carried down from above; and a *C horizon* composed of *weathered parent material*. A profile with distinct horizons is said to be horizonated.

The *downward movement of soil materials*, such as occurs in humid climates, is the process of *leaching*. *Depleted horizons* undergo *eluviation*; *enriched horizons* undergo *illuviation. Where rainfall is low, calcium carbonate or sodium chloride can accumulate* in the profile, at depths controlled by the moisture supply.

Profile descriptions can be more elaborate than the *A−B−C subdivision. Among additional symbols are h* for a humified horizon, or part of one, and *G or g* for a highly or moderately gleyed horizon or part, where waterlogging affects the character and stability of iron compounds, producing pale or mottled tints.

Zonal soils relate to major climate/vegetation belts. Intrazonal soils are dominated by *local drainage conditions. Azonal soils are incompletely developed*, either because of their brief histories or because of slope instability.

Process-response behaviour of soils involves the *interaction* of parent material, weathering processes, climate, plants and animals. It operates both *in space and through time*. The role of *climate is* particularly *expressed* in the contents of *clay and nitrogen* and in the depth of *calcium carbonate accumulation. Temperature* especially *affects* the *dissociation of water* molecules, which controls soil *acidity/alkalinity*.

Clay-humus particles act as anions, that *lock on cations* of ammonium and mobile metals, all included in the series of *plant nutrients*. The 2:1 clays have a far greater capacity to attach cations than have the 1:1 clays. *Soil classification developed initially with reference to climate and vegetation*: hence the concept of *zonal soils*. But a highly effective modern classification, the *Seventh Approximation*, distinguishes *ten soil orders* which are only in part climate-related in any way. This classification is independent of the *zonal concept*; it *rejects all previously existing soil names*, using a *completely artificial*, and *completely new, vocabulary*. Although at first glimpse the array of unfamiliar terms is bound to prove daunting, the technique of name construction is *logical throughout*. Classification depends heavily on *profile characteristics*, and especially on the use of *diagnostic horizons*, the present or absence of which is *important* alike for the recognition of *orders* and for that of *suborders*. The listing and descriptions, from the text and especially from *Table 20.3*, will not be repeated here.

Soil organisms include protists (for example, bacteria); *microfauna*, especially single-celled animals; *mesofauna*, many of which are invisible to the eye; and *macrofauna−megafauna*, ranging from small worms to burrowing backboned animals. The protists and microfauna, collectively grouped as soil *microorganisms*, are intimately involved in the *cycling of sulfur, carbon and nitrogen*. The *general effect of soil organisms* in general is to *break down complex organic compounds into simple inorganic compounds*, or even elements, which are returned to plants, or released into the air or the soil moisture.

The *complexity of soil populations*, and the *numbers* of individuals *involved*, demonstrates that a *soil must be regarded as an ecosystem*. In most cases, the ecosystem is *in a state of dynamic equilibrium* or of very slow progressive change, *kept under control by negative feedback*.

21 Ecosystems II: Light, Air, Soil, Plants and Animals

Many participants in the so-called ecological movements of the 1970s took to using the words environment and ecology as if they were interchangeable. Of course they are not. Ecology is the science of ecosystems. The word is formed from the Greek *oikos* = household/home/dwelling place and *logos* = reckoning.

An ecosystem is a system wherein living things interact with their environment and with another (Chapter 20, section 20.9) Very detailed work, such as the famous studies of great apes in the wild, can deal with the interactions of individual animals. More broadly-based studies deal with the interactions of sets of biosystems with one another and with their environment (Chapter 1).

The 'environment' (or, mistakenly, the 'ecology') seems to be vaguely conceived as being natural, beautiful, serene, pure, sweet-smelling and unpolluted. The term has assumed emotional overtones. It is true that we all respond emotionally to scenery that we think beautiful; but a mangrove swamp on a tropical shoreline, handsome though it may seem from a distance, is by human standards polluted and stinks in human nostrils. Again, all ecosystems that are at all complex include parasites – plants and animals that live on or in others – and predators, animals that eat other animals. We human beings tend to disregard the fact that wild birds or beetles can be hosts to lice, but to object strongly if we are bitten by fleas or chiggers. Similarly, we think nothing of wild fungi and soil bacteria, but greatly disapprove of the fungal-bacterial succession of athlete's foot. We are apt to feel revulsion against the fact that the wild relatives of domestic cats – lions, tigers, jaguars and the like – prey on other warm-blooded animals. Somehow it does not seem so serious that domestic cats and wild owls will eat mice. Domestic dogs eat meat; but feeding them upon canned meat, which may well be kangaroo meat, does not seem offensive. After all, our dogs do not themselves bring down the kangaroo. By contrast, the flesh-eating wild relatives of domestic dogs – wolves, coyotes, jackals and the like – are generally distrusted and even hated. Hatred is only

partly based on the fact that these animals sometimes prey on farm livestock. Most importantly of all, people in technically advanced societies tend to think of themselves as somehow separate from, beyond, or above the workings of nature. This of course is a mistake. Humankind is just as much a part of the world ecosystem as is any other life form. We constitute a single animal species. Like all other animal species, we are interdependent on other species. If by becoming wholly vegetarian we could free ourselves of direct dependence on other animals, we should still compete against many of them for our plant food supply. This theme will be taken further in Chapter 23.

21.1 The basic ecosystem model

Components of the basic model include producers, plant eaters (herbivores), flesh eaters (carnivores) decomposers and a nutrient pool (Fig. 21.1). The succession producers→ plant eaters→ flesh eaters →decomposers is a food chain. So is the succession producers→plant eaters→ decomposers. So again is the succession producers→ decomposers. As stated in Chapter 1, food webs with cross-linkages give a more accurate picture of reality than do food chains. The diagram in Fig. 1.16 can be redrawn as a network of hexagons, in which only four pairs of sides in contact represent closed boundaries. But the abstract model of the food chain, with a nutrient pool as a storage subsystem, has great advantages in the discussion of the flows of matter and energy.

It would be possible to assume that the boundary of the basic model is closed against the transfer of matter, and that nutrients are recycled wholly within the system. To do this, however, would make it difficult to discuss the actual flow of organic material through the model. Accordingly, the model is shown as receiving two inputs, one of matter (water, carbon dioxide and mineral nutrients) and one of energy. The energy is radiant energy in the form of sunlight. It provides the primary power drive.

The input of light energy is converted to chemical

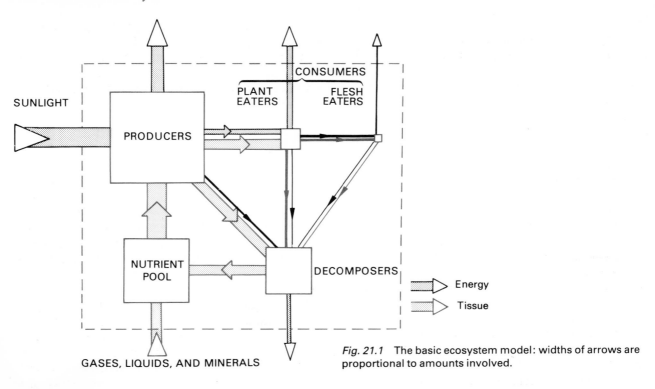

Fig. 21.1 The basic ecosystem model: widths of arrows are proportional to amounts involved.

energy, which can be stored, by plants that contain the green pigment chlorophyll. It is true that a set of purple bacteria assimilates carbon dioxide; but it does this only in the presence of organic compounds, which are by definition formed by life processes. Another set of bacteria oxidises simple organic compounds, but once again these compounds have to be formed in the first place. The fundamental process at the start of the great food chains is the use of sunlight by chlorophyll-bearing plants. These use sunlight to produce complex organic compounds.

Plants are eaten by grazing animals (herbivores), which can be classed as primary consumers. Grazing animals supply the prey of flesh-eating animals (carnivores), which can be classed as secondary consumers. If flesh-eating animals are themselves preyed on, a class of tertiary consumers exists. For example, if moles eat beetles and are themselves eaten by foxes, the foxes rank as tertiary consumers. Additional complications, and extended rankings, are easily possible.

Producers and consumers are liable to attack during their lifetimes by decomposers, which invariably take over at the time of death. Both in western Europe and in North America, elm trees during the 1970s were widely attacked by Dutch elm disease, a fungus with a beetle vector. By attacking living trees, the fungus anticipated the fate of trees that would have died eventually of old age. The decomposers convert organic matter back to an inorganic (= mineral or gaseous) state, supplying the nutrient pool on which producers draw.

21.1a Trophic levels and biomass ratios

The concept of food chains enables us to trace the cycling of nutrients and the consumption of energy through successive trophic levels. These are successive steps of feeding, named from the Greek *trophe* (= nourishment). In the basic model, the producers occupy the first trophic level. In the basic food chain, the plant-eaters occupy the second trophic level and the flesh-eaters occupy the third. Flesh-eaters that eat other flesh-eaters elaborate the pattern. When human beings eat fish that feed on other fish, a still higher trophic level appears. And when we eat fish that feed on other fish that feed on still other fish . . . it is easy to see that more than three, or more than four, trophic levels are possible. Moreover, the total pattern is complicated by the fact that some animals feed at more than one trophic level. Bears, coyotes, foxes and people are all capable of eating plants, nuts, seeds, eggs and the flesh of prey, whether cold-blooded or warm-blooded. A delicacy among Australian aborigines is the witchetty grub. Oriental delicacies include chocolate-covered ants. Wolves are reputed, on little evidence, to devour the weakest members of a pack in times of hunger stress. Human beings certainly do so. Cannibalism in some times and in some ecosystems is well documented in history. It occurs in our own time, even in westernised societies. It has been practised by survivors of air crashes. For some life forms, therefore, the concepts of the food chain and of trophic level needs to be qualified. Nevertheless, the basic concept of chain and level makes it possible to discuss the total working of an ecosystem, particularly in respect of the ratio of energy between one trophic level and the next.

Biomass is the total content of organic matter. The biomass ratio is the proportion between the biomass at a given trophic level and that at the next higher level. Calculations are made in terms of dry weight. As symbolised in Fig. 21.1, the higher the trophic level in a food chain, the less the biomass. Biomass ratios vary considerably from ecosystem to ecosystem, and within an ecosystem, from one pair of trophic levels to another. Some studies of tropical rainforest, for instance, show that plant biomass can easily be 3000 times as great as animal biomass, whereas the biomass ratio between plant-eaters and flesh-eaters is about 2:1. Dead plant tissue alone can be twenty-five times as bulky as live animal tissue. Some 70% of live animal tissue is contained in soil fauna, the animal subset of the decomposers.

Food chains and biomass ratios are keenly debated in connection with the food supply of humankind. Assume human food to be exclusively supplied by a commercial grazing economy, and that people live exclusively on farm-raised meat. Then the chain from producers through plant eaters to flesh eaters runs from fodder plants through sheep and cattle to human beings. To grow and maintain an adult human weighing 70 kg would take a tonne to a tonne and a half of farm animal. To rear this much farm animal would take from ten to fifteen tonnes of plant material.

Strictly speaking, we are looking not at biomass ratios but at a food pyramid. The biomass ratio, as stated, indicates the proportions of organic content between trophic levels. The ratios in the food pyramid are a measure of the efficiency of the consumption process in producing biomass at a next-higher trophic level. In the simplified example under discussion, the ratio of biomass production ranges roughly 10–20:1. The eating of 1 kg of plant tissue produces nothing like 1 kg of farm animal tissue; and the eating of 1 kg of animal meat produces nothing like 1 kg of human being. As a means of producing biomass at the next-higher trophic level, consumption is a most inefficient process. When the chain is lengthened to include more than two trophic levels, inefficiency is multiplied. To produce and maintain an adult human in the indicated system takes some 200 times a human's weight of plant tissue.

Some people in western societies have become vegetarians for this reason, allying themselves in practice with those vegetarians who argue that eating flesh is barbaric. Industrial interest in improved trophic efficiency is demonstrated by the production of meat substitutes made from soybeans. Two factors, however, come in here. In some systems of human nourishment, it can be physically impossible to eat enough plant tissue to supply the body with enough protein. The manufacture of meat substitutes demands energy; and the energy budget of an ecosystem is no less important than is the biomass budget.

21.1b Energy flow

Of incoming solar energy at the earth's surface, about 1% goes to drive ecosystems. At each step along a food chain, energy is dissipated. It is used by organisms in their life processes, eventually being returned to the atmosphere. Blow on your hand. Your breath feels warm. You are giving off heat energy in the process of maintaining and operating your life system.

The progressive loss of energy with progression through an ecosystem is symbolised in Fig. 21.1. Of the energy stored in plant tissue that is consumed by a large plant-eater – say, an antelope – about 90% will be dissipated in keeping the antelope alive and in supplying power for its muscular movement. If a lioness kills the antelope, further energy will be dissipated by the life processes of the lioness and her mate, their cubs if any, and of the hangers-on – vultures, jackals and hyenas. The amount of stored energy decreases very markedly from one trophic level to the next higher level. Furthermore, energy is also needed by the decomposers in their work of breaking down complex organic substances into simple inorganic forms.

21.1c Overlap of the tissue cycle and the energy throughput

Although it is convenient for purposes of discussion to separate the cycling of tissue from the throughput of energy, even the abstract model of Fig. 21.1 shows that much tissue is lost during the cycling process. This is because tissue can be, and must be, converted to energy.

Think of coal or of mineral oil. Coal consists essentially of carbon, oil of the hydrogen–carbon compounds of hydrocarbons. When we burn coal or oil, we are releasing the energy stored in the carbon. Organic compounds that contain nitrogen also store energy; the inorganic ammonia, NH_3, can be burned. In order to understand the power drive of an ecosystem, then, we need to know how energy can be stored in the first place.

21.2 Photosynthesis

Photosynthesis means construction by means of light. The process is effected by chlorophyll, an organic chemical compound where linked arrays of hydrogen atoms (H), carbon atoms (C), oxygen atoms (O), nitrogen atoms (N) and molecules of carbon dioxide (CO_2) and hydrocarbons (for instance, CH_2) are attached to atoms of magnesium (Mg). Chemically speaking, chlorophyll is simple enough; but structurally, as is usual with organic compounds, it is complex. The basic reaction, given light as the power source and chlorophyll as the operator, can be expressed as

$$6CO_2 + 12H_2O \rightarrow C_6H_{12}O_6 + 6O_2 + 6H_2O$$

carbon   water   glucose,   oxygen water
dioxide   a sugar

This equation is a symbolic statement of the fact that green plants take in carbon dioxide and water, form substances in which carbon is stored, and give off oxygen and some water – less water than was taken in to begin with. It is impossible to exaggerate the importance of this basic reaction for life on earth. Here is where the food chains begin. Here is where power enters every ecosystem, and where the construction of complex organic compounds starts.

A second set of basic reactions, equally important, is the formation of amino acids, proteins, fats and pigments. Both sets of reactions involve the production of polymers.

21.2a Polymerisation

Polymers are long chains of molecules. They are named from Greek words that mean many parts. The glucose units in the foregoing equation can be linked together as starches, which of course are carbohydrates. Bread consists essentially of carbohydrates – compounds of carbon, hydrogen and oxygen, a description which reduces itself to a basic description of the compounding of carbon with water. If we divide the specification of the formula for glucose by a factor of six, we arrive at CH_2O. So simple a specification is however not an accurate description of what actually happens.

A protein is a polymer of some twenty-four similar building blocks of amino acid. We recognise protein most easily as the substance of meat, including the meat of our own bodies. An adult human body contains from 5 to 6 kg of muscle protein. Plant proteins also exist, as advertisements for bread and breakfast cereals constantly inform us. Not only are

proteins structural materials; some are defences – the antibodies that ward off bacterial and viral attack – while others are catalysts. These latter are substances which speed up and intensify chemical reactions, without themselves undergoing change. They include the enzymes (from the Greek, = fermenters) that perform digestive processes.

Some twenty amino acids are essential to life processes in general. Each contains one or more amino groups (NH_2) and a carboxyl group (COOH), along with uncombined carbon (C) and hydrogen (H). Two include sulfur (S) or iodine (I). In terms of bulk chemistry, the amino acids, and the proteins polymerised from them, are simple, but in terms of structure they can be quite complicated. The simplest of all is glycine, the structure of which is

$$H_2N - \underset{\underset{\displaystyle H}{|}}{\overset{\overset{\displaystyle H}{|}}{C}} - COOH$$

Animals can synthesise most amino acids for themselves, but some can only be obtained from plant tissue. Among the latter is lycine, the structure of which is

$$H_2N - \underset{\underset{\displaystyle CH_2}{|}}{\overset{\overset{\displaystyle H}{|}}{C}} - COOH$$
$$|$$
$$CH_2$$
$$|$$
$$CH_2$$
$$|$$
$$CH_2$$
$$|$$
$$NH_2$$

We could write these descriptions respectively as $C_2NH_5O_2$ and $(C_3NH_7O)_2$, just as meaninglessly as glucose can be described as CH_2O_6. The arithmetic is not, however, entirely pointless. It does serve to emphasise that amino acids and proteins, unlike sugars and starches, include nitrogen. Most green plants are unable to use atmospheric nitrogen. Consequently, the activity of soil bacteria in altering nitrogen compounds, already summarised in Chapter 20, is one of the fundamental processes in the working of ecosystems. As will now be seen, phosphorus is also essential.

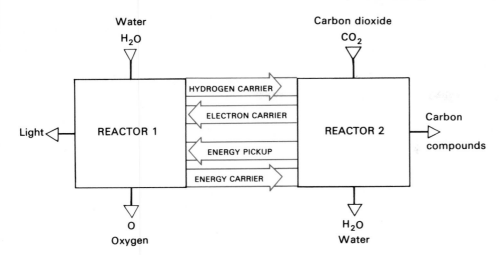

Fig. 21.2 Photosynthesis: three inputs, three outputs, four exchange paths and two black boxes.

21.2b Photosynthesis as a two-stage reaction

The complete details of photosynthetic reactions are too complex for our purposes. It will be sufficient to recognise a two-stage reaction (Fig. 21.2), and to identify each reactor as a cascading subsystem or black box. Light energy is used in Reactor 1 to break the bond between the hydrogen and oxygen of the water input, and to form a hydrogen carrier and an energy carrier by catalytic reactions. These carriers are both phosphates, PO_4 compounds. The second of them includes nitrogen. As shown in the diagram, Reactor 1 supplies an oxygen output. Here is the source of uncombined oxygen in the air that we breathe. Reactor 2 takes in atmospheric carbon dioxide. It uses its input of energy to break the carbon–oxygen bond of carbon dioxide, to combine oxygen with hydrogen for the output of water, and to synthesise the carbon compounds of sugars, starches, amino acids, proteins, fats and pigments. Stripped of its hydrogen but charged with electrons, the incoming hydrogen carrier becomes the return flow of the electron carrier. Stripped of some of its phosphate, the incoming energy carrier becomes the return flow of the energy pickup. This whole complex array of photochemical and catalytic reactions takes place within a single cell of a chlorophyll-bearing plant.

21.3 Cycling of nutrients

The cycling of nutrients through an ecosystem, already symbolised in bulk form in Fig. 21.1, can be discussed in respect of individual chemical elements. A full account would be long and complex indeed, because no fewer than seventeen elements, or com-

pounds of them, are essential for plant growth. Hydrogen, oxygen, carbon, nitrogen and sulfur have been intensively studied, because they are soluble in water and also exist as individual gases or as gaseous compounds. That is to say, they are highly mobile. Along with phosphorus, they constitute the basic elements in mineral cycling. The carbon–oxygen and the nitrogen cycles will now be used as examples of nutrient circulation.

21.3a The carbon–oxygen cycle

Fig 21.3 models the combined cycle of carbon and oxygen, which to some extent includes the hydrogen cycle. The water inputs to the plant sub-system come from the atmosphere, soil moisture, surface water, lakes and seas. Water outputs by land plants go mainly back into the atmosphere in the form of water vapour. Carbon enters the plant subsystem in the form of carbonates drawn from the soil, and also in the form of atmospheric carbon dioxide. Plants give off oxygen to the atmosphere. Breathing animals eat the carbon–hydrogen–oxygen compounds of plants, or of one another. They also assimilate atmospheric oxygen and respire carbon dioxide. For them, the oxygen/carbon dioxide exchange with the atmosphere is vital. If your oxygen supply fails, you will die. In a completely closed space, your breathing would progressively enrich the available air with carbon dioxide at the expense of oxygen. Carbon dioxide carried to the extreme would kill.

The way in which you use atmospheric oxygen and carbohydrate tissue to release energy and to produce carbon dioxide and water vapour can by symbolised by the equation

$$6O + 2(C_6H_{12}O_6) \rightarrow 6CO_2 + 12H_2O$$

oxygen   glucose,   carbon   water,

          a sugar   dioxide   water vapour

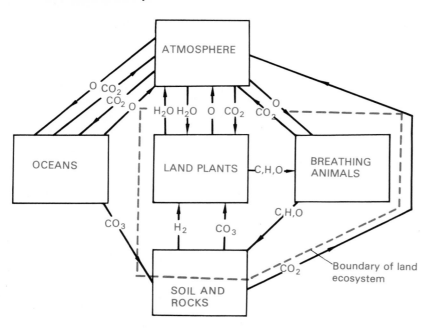

Fig. 21.3 Abstract model of the carbon and oxygen cycles: similar models can be constructed for the cycling of all other chemicals involved in the life process.

In addition to being warm on account of released energy and potentially poisonous on account of its concentration of carbon dioxide, your breath also feels damp because of its water content.

The ocean water subsystem of the carbon–oxygen cycle performs a three-way transfer. Carbon dioxide can be dissolved by seawater, and released by seawater directly back into the atmosphere. Dissolved carbon dioxide can also be assimilated by marine plant life, which stores the carbon and releases the oxygen. Dissolved oxygen can be returned directly, or can be assimilated by fish and other forms of marine life. Dissolved carbon dioxide and oxygen can be combined into carbonate (CO_3) compounds, as in the hard parts of shellfish. These hard parts, in the sedimentary cycling system (Chapter 4), go to form carbonate rocks. At a further remove, carbonate rock material in the soil can supply nutrients to the plant sub-system. At a still further remove, rock carbonates supply the carbon dioxide which is discharged into the atmosphere from volcanic vents.

If at this point, you, the reader, conclude that the cycling of chemical substances through an ecosystem is complicated, you are entirely correct.

21.3b The nitrogen cycle

Cycling of nitrogen has strongly caught the imagination of ecologists, biochemists, geochemists and evolutionary biologists. That is to say, it deeply concerns researchers into ecosystems, biology, nutrient cycling in general, earth chemistry/life chemistry relationships, the chemical makeup of the earth's

crust, and the means by which life forms take up the chemical elements and compounds on which they depend. The reason for the interest, and its very wide spread, is that the source of nitrogen–essential to the construction of amino acids and proteins – is the atmosphere, but few organisms can take up atmospheric nitrogen directly. Chlorophyll-bearing plants have no mechanism for using atmospheric nitrogen, in strict contrast to their adaptations for using atmospheric carbon dioxide, water vapour, and rainwater.

Life forms as we know them, therefore, depend on the fixing (= binding) of atmospheric nitrogen into some liquid or solid state where it can be drawn on by plants. To the operation of nitrogen-fixing bacteria, noticed in Chapter 20, we may add the probably less important operation of nitrogen-fixing algae. Nitrogen fixers convert atmospheric nitrogen, N, into compounds of ammonium, NH_4, which plants can take up (Fig. 21.4). Plants can also take up their essential nitrogen as ammonia, NH_3, as nitrates (NO_2 compounds) and as nitrates (NO_3 compounds). Decay bacteria convert ammonia and ammonium compounds to nitrites and nitrates.

Nitrogen in amino acids and proteins moves along the food chain from plants through plant-eaters to flesh-eaters. Return flow from the food chain to the subsystem of decay bacteria includes dead and rotting plant and animal tissue, along with the urine and dung of animals. Anyone familiar with the smell of a stable will know that horse dung is ammonia-rich. To the return paths through the decay processes must be added a path of output. The set of denitrifying bacteria breaks down nitrogen compounds and releases the nitrogen back into the atmosphere. In this way the circle of the nitrogen cycle is closed, except that lightning strikes make minor contributions to the formation of complex nitrogen compounds, and that fires release gaseous nitrogen by breaking down organic compounds.

A background consideration here is the original production of complex compounds on a previously lifeless planet. Some workers think that lightning strikes may have been responsible. The weight of opinion, however, lies with ultraviolet irradiation.

21.4 A basic food web model

Fig. 21.5 generalises the cycling of matter through an ecosystem on land. The sequence of soil–live plants– plant eaters–flesh eaters is taken as the basic food

chain. All of the four subsystems in the chain exchange gases with the atmosphere and water with the hydrosphere. A loop between live plants and plant eaters runs through the dead plant subsystem, a second through this same subsystem from plants back to the soil. Two-way exchanges connect the atmosphere/hydrosphere subsystem with soil and microflora, and microflora with soil and soil fauna. One-way flows, in addition to those already noticed, connect the dead plant subsystem with microflora and soil fauna, animal tissue with microflora, micro-flora with soil fauna, and the mineral content of bedrock with the soil.

It is abundantly clear that the model food web is an open system. It is also clear that changes in the operation of any one subsystem inside the ecosystem boundary can potentially affect the operations of all other subsystems.

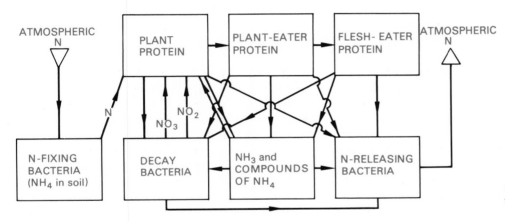

Fig. 21.4 Abstract model of the cycling of nitrogen.

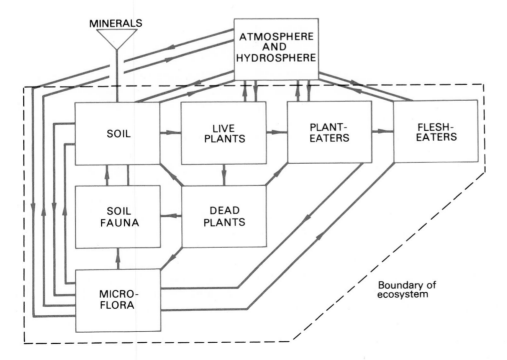

Boundary of ecosystem

Fig. 21.5 Abstract model of the cycling of matter through a land ecosystem.

21.5 Ecosystem productivity

Biomass, biomass production and productive efficiency differ from one kind of ecosystem to another. This fact raises the question of the classification of ecosystems – that is, of the subdivision of the world biosphere. An obvious separation is that of aquatic (= water) ecosystems from land ecosystems. Aquatic ecosystems include those of the world ocean or some subdivision of it – particularly water masses or enclosed seas, for example – and inland waters, which include rivers and lakes, the latter ranging from freshwater to saline. Major land ecosystems correspond to the major vegetation belts. This circumstance is in no way surprising, in view of the significance of photosynthesis in injecting energy into an ecosystem, of the fact that plants supply food to plant eaters, and of the further fact that biomass dwindles sharply along every food chain. Ultimately, therefore, the gross definition of major land ecosystems depends on a gross determination of plant assemblages in relation to climate. Within major climatic belts, account must be taken of the local impact of slope, soil parent material and soil drainage.

Measurement of the operation of major ecosystems must rely on sampling. It is simply impossible to compute the dry weight of every tree in low-latitude rainforests, or of every blade of grass in a natural mid-latitude grassland. Individual small ecosystems can be measured by complete census, or at least by very intensive sampling. Small lakes or salt marshes, for instance, can be closely studied by a

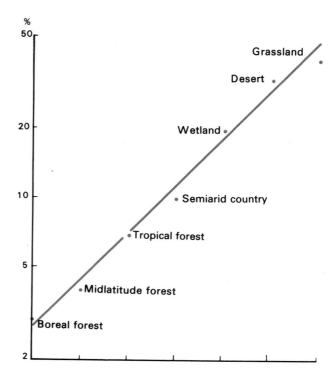

Fig. 21.7 Percentage rates of annual throughput of biomass: straight-line graph gives a 98% fit.

single researcher. It is precisely because some workers have studied small ecosystems, whereas others have sampled major ecosystems, that we know more about very great and very small ecosystems than we do of ecosystems of intermediate size.

Sampling of major ecosystems on land has progressed far enough to permit certain generalisations. Fig. 21.6 graphs biomass and rate of biomass production per unit area against major ecostystem type. Vegetation type is classified by arithmetic steps. The distance on the horizontal scale of the graph between tropical forest and midlatitude forest is equal to the distance between midlatitude forest and boreal forest, and so on down the line. The percentage contribution to world biomass is plotted against a logarithmic scale. For each right-to-left step down the sequence of ecosystem types, biomass percentage is divided. It decreases by a factor of about three from one major ecosystem type to the next.

The sequence reflects the facts that forest ecosystems contain more biomass than do grasslands, and that the forests of low latitudes contain more biomass than do forests of middle or high latitudes. Production rates for major natural ecosystems follow a similar sequence to that of biomass. Tropical forest ecosystems are the most productive, tundra ecosystems the least. Something of this kind would certainly be expected, on account of the greater input of solar energy in low latitudes than in high. Measured in dry tonnes/ha/yr, tropical forests produce some 20 tonnes of biomass, against some 2

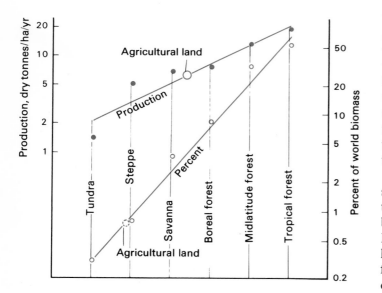

Fig. 21.6 Biomass content and biomass productivity, for major types of ecosystem: agricultural land is plotted as an extra. Straight-line graphs give a 90% fit for production rate and a 97% fit for content.

tonnes for tundra. For a step down the scale, rate of production divides by a factor of about one and a half.

Agricultural land is plotted on both graphs. On the graph of percentage contribution to world biomass it is plotted as an extra. As shown, it is estimated to be about equal to midlatitude grasslands. On the independently calculated graph of production rates, it is plotted against its place on the production scale. As a producer of biomass, farmland on the world scale ranks above natural grasslands but below forests, and this despite the deliberate injection of nutrients and energy into farming systems. If people could eat trees instead of grains, fruit, and meat, food shortages would presumably lie far in the future.

21.5a Rates of throughput

Rates of throughput, or turnover, of organic material also vary greatly from one kind of ecosystem to another. (Fig. 21.7). The information in this graph is not precisely comparable to that in Fig. 21.6, because deserts have been added and tundra has been included with wetlands in general. Once again a semilogarithmic relationship appears. For a shift down the scale of throughput, rate of throughput divides by a factor of about one and a half.

The slower rates of throughput in forests than in other ecosystems is easily understood. Some grasses are annual; some trees can live for centuries. The differences among forest ecosystems relate to differences in ecosystem energy – once again, to the contrast in solar input between low and high latitudes.

21.6 Population dynamics

Ecosystem populations are studied mainly in respect of animal components, because vegetal components tend to achieve a steady-state condition, whereas animal components can undergo drastic changes in total. Nevertheless, competition, symbiosis and parasitism all affect the vegetal components, which are also liable to change through time and space along ecosystem margins. Fig. 21.8 illustrates the succession of dominant plants in two coastal environments. As is typical of plant successions, the dominants become generally larger and more complex, and/or more capable of covering the ground entirely as the successions progress, culminating in forest, woodland, or shrubland according to the amount of moisture available. The actual diagram in Fig. 21.8 is drawn for southern Australia, but is readily adaptable to corresponding conditions elsewhere. The

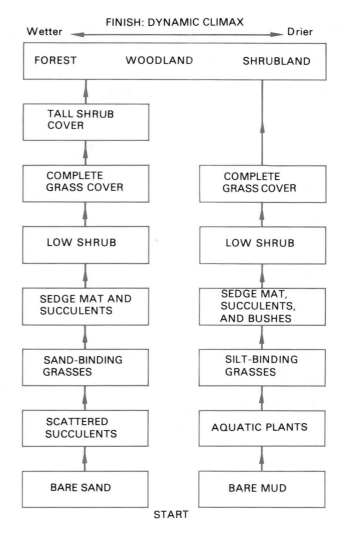

Fig. 21.8 Succession of dominant plants in two coastal environments: based on data for a region of Subtropical East-coast climate.

end-product is a dynamic vegetation climax, which can resist regression unless climatic or geologic conditions of the environment undergo significant change.

If a given animal species (say, the rabbit) fails to produce enough young to maintain its numbers, it will eventually become extinct. In actuality, a mated couple of rabbits will produce far more than just two offspring. Say they produce twenty. Then these twenty will produce two hundred. In six generations the original couple will be the great-great-great-great-great-grandparents of two million little rabbits, to say nothing of the 222 220 descendants in between. We might wonder why the world's lands are not completely overrun with rabbits. We might equally wonder why the planet is not submerged in a sea of bacteria. One single bacterium can divide into two bacteria in about twenty minutes. These two can divide into four bacteria in another twenty minutes. Growth at this rate, if continued, could cover the whole earth with bacteria to a depth of a metre in little more than a day and a half.

Some kind of check must operate. If it did not, every species and every ecosystem would be in a runaway condition. The main checks – forms of environmental resistance – are food supply, living space and shelter, disease, predation, and genetic constraints. The list recalls the conclusions reached by Malthus in his famous *Essay on Population*. The 1798 edition gives war, famine, pestilence, misery, and vice as factors preventing human populations from outrunning their food supplies. The 1803 edition adds moral restraint – deliberate refraining from the production of children.

As already noticed, many people react violently and emotionally against the competition and predation that are built into ecosystems. Plants compete against one another for access to space, light and mineral nutrients. Animals compete for living space and foodstuffs. Many years ago, plant competition was identified as a struggle for existence; and animal predation became stylised as Nature Red in Tooth and Claw. Nevertheless, unless the ultimate check of environmental resistance is opposed to population growth, positive feedback will ensure species growth by multiplication. To be and remain viable, each ecosystem must develop and impose a carrying capacity for each of its contained species.

21.6a Population trends through time

Fig. 21.9 presents six models of the change through time of population total for a given species in a given ecosystem. The two graphs in Fig. 21.9a are random walks, generated by card-flipping. The lower graph,

with an inbuilt bias toward decrease, illustrates the proposition that failure to reproduce in sufficient numbers will lead eventually to extinction. The upper graph, with an inbuilt bias toward increase, indicates that, unless some mechanism of negative feedback comes into play, population total will eventually reach infinity.

Fig. 21.9b generalises the proposition that population total will be stabilised by increase in environmental resistance – that is, that increasing numbers in time multiply the effect of resistant factors, preventing increase beyond a given level of population. Many instances could be cited. If an animal population increases in density, the mortality rate rises. Birds produce fewer eggs per nest, rate of fish growth declines. Intra-species competition often acts to check the increase in numbers. A well-known experiment with a species of hermit crab – the individuals being confined in a tank of fixed size – showed that the number of fights between pairs of crabs increased approximately with the square of the population total (Fig. 21.10). Fighting means wounding, weakening, liability to infection and even death.

Figs. 21.9c and 21.9d elaborate the model of Fig. 21.9b by providing for fluctuations about the mean population eventually attained. Thus, whereas Fig. 21.9b is an equilibrium model, Figs. 21.9c and 21.9d are steady-state models. They can be matched from studies of animal species deliberately introduced

Fig. 21.9 Models of population trends: totals on the *y* (upright) axis, time on the *x* (horizontal) axis. Although these models are usually discussed with reference to animal populations, they also apply to plant populations.

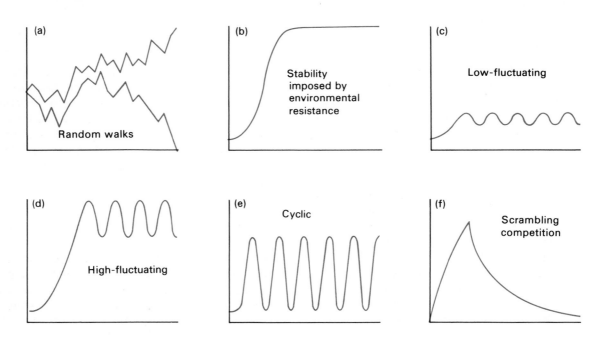

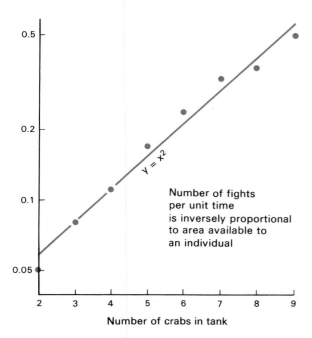

Fig. 21.10 Fighting rate against numbers, for hermit crabs confined in a tank: basic data after B.A. Hazlett. Fit of graph to data points is 98% accurate.

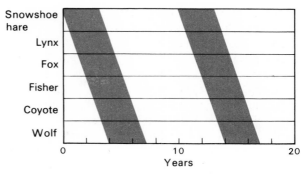

Fig. 21.11 Highly generalised relationship of abundance (dark bands) to time, in a north-Canadian food web.

into island ecosystems. Negative feedback comes into operation earlier in the situation represented by Fig. 21.9c than in that represented by Fig. 21.8d; and there is more room for large fluctuations of numbers in the latter case than in the former. But the lengthiest studies relate to populations already in being, such as those of the Canadian Arctic, for which the records of the Hudson's Bay company supply indicators of fur-bearing populations stretching back for many years. These records strongly suggest cyclic fluctuations such as the one generalised in Fig. 21.9e.

It is very tempting to suppose that cycles of abundance, such as those illustrated in Fig. 1.5 reflect simple prey – predator relationships. As the prey population increases, so does that of the predator species. But increase in number of predators leads eventually to a check on, and then to a decline in, the prey population, so that the predator population will itself decline. Some support for this idea comes from the analysis of Canadian records, which show that cycles of abundance in the snowshoe hare are followed by cycles of abundance in predators (Fig. 21.11), with time lags that at least in part depend upon breeding habits. Against this, no such corresponding cycles have been identified for the Siberian north, while in Canada, in-phase oscillations of numbers in lynx and in fish suggest that perhaps some general perturbations of the environment – and possibly random perturbations – are in question. Since, moreover, predators tend to take only the weaker prey – the young, the old and the ill

– they act to improve the gene pool and to enhance the breeding strength of the survivors. Lions in Africa seem to have little effect on the totals of the cattle populations on which they prey.

The final model of population growth, that in Fig. 21.9f, is a model of scrambling competition. Insofar as cycles of predator abundance are determined by the size of prey populations, it represents a variation on the model of Fig. 21.9e. In laboratory conditions it has been thoroughly tested for blowfly and similar populations. Given a food supply and no competition or predation, the larval population increases very rapidly to begin with, up to the point where few larvae get enough to eat. Subsequently, reproduction rate falls off, even if the food supply holds out, because the numbers surviving the pupal stage will decrease. Here, in an oscillation between positive feedback (exponential increase of numbers) and negative feedback (damage to, and inhibition of, breeding) is one possible mechanism of apparently cyclic fluctuation of population total.

21.7 Ecosystem examples

This and the following section will use rainforest ecosystems and river ecosystems to illustrate the component composition and internal interaction of well-defined and well-studied ecosystem situations. Because the examples differ so powerfully, both in content and in operation, the treatment accorded to one must differ strongly from that accorded to the other.

21.7a Rainforest ecosystems

Low-latitude rainforests attain their fullest development in regions of equatorial climate (Chapter 15). Although occasional dry spells or cold waves affect some parts of some forests, monthly averages reveal no cold season or dry season. This is not to say that

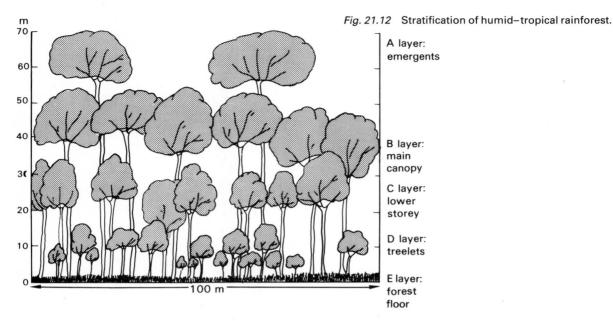

Fig. 21.12   Stratification of humid–tropical rainforest.

A layer: emergents

B layer: main canopy

C layer: lower storey

D layer: treelets

E layer: forest floor

there is no seasonal rhythm of flowering and fruiting: on the contrary, such a rhythm is widely reported. Nevertheless, plant growth is possible the year round.

No midlatitude observer can fail to be struck by the height of the tallest trees, some of which grow to 80 m, or by the enormous variety of kinds of plant. An area of only one hectare can easily contain more than a hundred species of trees alone. The number of plant species in an entire forest runs into the tens of thousands. With the richness of numbers goes a scarcely credible range of plant form, what seems to be savage competition for light among light-seeking plants, and equally savage competition for plant nutrients. Like plant life, animal life too is highly varied. A tropical rainforest can contain 200 species of reptiles and amphibia, 200 species of mammals, 200 species of visiting birds, and 600 species of nesting birds.

Ultimately, the supply of mineral nutrients must be obtained from weathered rock. But in mature forest, the nutrient pool is contained largely in the decaying organic matter of the forest floor. The pool's throughput of phosphorus and potassium is especially important. These essential elements are cycled almost entirely through the biomass – litter – biomass path. Some trees do send down deep tap roots, but such roots are believed to be rare. Most of the root biomass lies at depths of less than half a metre; indeed, some tree species send out roots along the ground surface. In compensation for the lack of anchorage, many trees have developed buttresses – common and striking features of tropical forests. Buttresses grow preferentially toward the danger side – uphill on slopes, to windward where a prevailing wind blows. They are ties, as opposed to props.

21.7b Stratification of plants

For purposes of description, the forest can be looked on as stratified. It can be described in terms of strata, layers, or storeys (Fig. 21.12). The uppermost stratum, the *A* layer, includes the tops of trees that rise above the main mass. These are the emergents, the giants of the forest. Next below comes the *B* layer, which is the main canopy. To a considerable extent this canopy blocks off sunlight from lower levels, but may not be capable of suppressing the growth of a lower treetop stratum, the *C* layer. Rising from the forest floor are the woody treelets of the *D* layer, while the *E* layer on the forest floor itself includes herbs and small seedlings.

The classification by layers is, however, somewhat misleading, since every patch of forest is a dynamic subsystem. Some small seedlings will grow into treelets, and some treelets will grow into trees, and some trees will become forest giants. Open areas where trees have died of old age will eventually revert to tall forest.

Because not every researcher has adopted a five-fold stratification, which indeed may not be suitable in all cases, it is difficult to use this stratification in locating the various contributions to plant biomass. A reasonable estimate is that some 80% of the total is contained in the trunks and branches of large trees, and another 15% in roots, leaving perhaps 2.5% for leaves in the canopy, 1.0% for smaller plants, and 1.5% for litter.

21.7c Plants growing on plants

Some plants use other plants as supports. Such is the case with midlatitude ivies which climb up trees. The tropical rainforests are as rich in climbing plants as they are in other plant forms. Big woody climbers, among which climbing palms are prominent, have freely hanging stems, or stems which spiral round the trunks of host trees. They respond rapidly to an increased input of sunlight, in clearings and along forest margins. Their dense growth in such situations is responsible for the myth that tropical rainforests are impassable jungles. Many other climbers are fixed tightly to the host trunks.

Plants which use other plants as physical environments are called epiphytes (Fig. 21.13). Some, including for instance the algae and lichens that grow on the lower parts of tree trunks, are shade-loving. Others are sun-loving. They include the many orchid species that grow in the canopy. For epiphytes of any size, and especially for those with root and stem systems, some kind of lodgement is necessary. It is widely supplied by holes in bark and by angles of branching. Tiny epiphytes need nothing more than a plant surface. Really wet forests always include epiphytes that grow on leaves.

Epiphytes that actually invade the tissue of their hosts, and draw nourishment from it, are parasites. These too are numerous and common in tropical rainforests.

Strangling plants – notably, the strangling figs – begin life as epiphytes in the lower branching angles of tree crowns. They send down roots which enter the ground, grafting onto one another, and ultimately enclosing the host tree in a strong tight web. Many host trees die. The root network preserves the shape of the trunk, and the fig system includes a branching top equal in extent to the former top of the host.

Fig. 21.13 Spike-leaved epiphytes on a tree in humid subtropical forest, coastal slope of South Brazil: see especially high on the trunk and at the base of the crown.

21.7d Stratification of animals

Animal forms tend to occupy particular forest levels, according to type of foodstuff. Above the canopy fly insect-eating birds and bats. Other species of these forms also inhabit the middle of the canopy. The top of the canopy provides a habitat for birds, such as hornbills and pigeons, and for mammals – apes, some monkeys, and flying foxes – that live on leaves, fruit, and insects. Next beneath come animals that climb back and forth between trees and ground – bears, other monkeys, and small cats. Large ground animals include deer and their relatives, various forms of wild pig, and additions according to forest region; along with the plant-eaters go large flesh-eating cats. Small animals living on the ground or in undergrowth include thrushes and pheasants, porcupines, mice, and small carnivores.

Above the ground, most animals are active either by day or by night. Apes and monkeys move and feed by day, flying foxes, tarsiers, lemurs, bats and cats do so by night. Ground animals are about equally divided into those that browse, forage or prey by day, those that do so by night, and those that do so both by day and by night.

21.7e Contents and distribution of animal biomass

Animal biomass in tropical rainforests may be less than plant biomass by a factor of thousands. Despite all the signs of vigorous evolution of plant and animal communities, and of plant–animal interaction, the actual carrying capacity of the forests appears to run far below its potential level. The rainforests remain in very large part the domain of plants.

Plant-eaters, including fruit-eaters, contribute perhaps 15% to the total animal biomass, flesh-eaters perhaps 10%. We could conclude from this that flesh-eaters in rainforests are notably successful in preying on plant-eaters. Flying insects perhaps make a 5% contribution. But the really successful animal forms are the arthropods, particularly those which feed in the litter and soil layers. In terms of dry weight, they contribute perhaps 70% to the total animal biomass. When the arthropods which feed on plants and other animal forms are added, the contribution rises to some 75%.

21.7f Plant-animal interactions

Modes of interaction between animals and plants are considered to have evolved together. Many are beneficial to plants and animals alike; some however are not. Like any forests, tropical rainforests are liable to attack by tree pests, chiefly various types of caterpillar that, by stripping trees clean, can kill them. Strong suggestions have been made that the wide dispersal of individuals of a single tree species serves as a defence. Caterpillar swarms of the female moths which lay the eggs that produce caterpillars find it impracticable to locate twenty trees of a single species, when these are scattered through an hectare of forest. Trees which have evolved poor seed dispersal systems, and which therefore occur in clusters or even stands, are usually protected by gum or resin content.

21.7g Pollination

Some plants are self-pollinating. Those which are not are pollinated by birds, bees, butterflies, moths, bats, beetles and wasps. Some of the arrangements are highly specialised: the wasps that pollinate fig trees have evolved a single wasp species for a narrow group of related fig species, or even for a single fig species. In every case, the flowers have developed attractions for the pollinators. Butterflies and moths drink nectar from tube flowers too narrow for bees. Birds visit scentless but brightly-coloured flowers with tints at and near the red end of the visible spectrum. Bees select moderately scented flowers with bright tints in the blue half of the visible spectrum, and with complex structures which defeat other potential pollinators.

Beetle pollination seems to be far more ancient than does pollination by bees, butterflies and moths. The direct ancestors of these had still to appear when, as much as 150 million years ago, beetles were already highly evolved. It is understandable that beetle pollination is in some ways rather crude. The creatures crawl over simply constructed flowers to eat the pollen. In another sense the process is beautiful. The beetles work at night. To attract them, the flowers give out, also at night, strong and fragrant scents. There is nothing in midlatitude forests to match this.

Bats also work at night, so that bat-pollinated flowers also open at night. These flowers give off sour smells. So do bats, as anyone lucky enough to have visited a large bat colony will know. In fact, the smell of bat-pollinated flowers is described as recalling the smell of bats.

21.7h Seed dispersal

Trees in the tropical forest depend mainly on animals for dispersal of their seeds. Any fruit-eating animal contributes to seed dispersal; and few of the tree-dwelling browsers are limited to leaves. Some small carnivores, and all the bears, are notable fruit-eaters. However, not much has been established as yet about their total contribution to the fruit-eating/seed dispersal process, or even about the precise way in which certain fruit-eaters are attracted to certain fruits. Matters are clearer in respect of birds and bats. Birds are attracted to fruit with shiny surface texture and bright or contrasting colours, including reds. Bats select duller fruits with musty scents. It is reasonable to expect that comparable mechanisms operate throughout the dispersal process.

21.8 River ecosystems

Like ecosystems on land, aquatic ecosystems depend ultimately on photosynthesis for their input of food energy. At sea, the important plants are those of the phytoplankton – the drifting microscopic plants which live at and near the surface. In inland waters, algae support some grazers. They and their successor mosses are responsible for the slipperiness of immobile boulders in midstream, the physical attributes of which impress many an incautious crossing human. But the main primary food source is perhaps dead vegetable matter that is washed in from the land or that falls in as leaves. The word perhaps is necessary, since the real food source, or a major part of it, may be the fungi and bacteria which are engaged in breaking down the vegetable matter.

As in land ecosystems, bacteria and fungi act as decomposers and green plants act as primary producers. Protozoa, larvae, worms and molluscs feed on green plants and are consumed by small carnivores, but no rigid linkage applies to the freshwater food web. Cannibalism becomes increasingly common as distance through the web increases: tadpoles and fish fry eat one another. Cannibalism, parasitism, predation and self-regulation – such as territoriality – all act to oppose the explosion of aquatic populations.

Life forms in rivers must be somehow adapted to resist the flow of water, to take advantage of it, or to do both. Midstream velocities of from 1 to 2 m/sec, which are by no means excessive, are capable of washing away many kinds of small aquatic animal. Against this, very low velocities are typical of a boundary layer, not more than a few millimetres thick, immediately next to the bed and banks. Adaptations to flow include shelter-seeking behaviour, attachment, casing, flattening, streamlining, strengthening of cases and arrays of powerful muscles. Other adaptations include the scraping devices of animals (including some fish) that feed on bottom vegetation. Detritus-feeders seize, filter, or strain their prey. Eggs laid in running water are commonly adhesive, or – as with some fish species – buried in soft channel sediment.

Adaptations to highly irregular flow include for instance the delay of egg hatching until flow resumes. Alternatively, or in addition, hatching can be restricted, delayed, or staggered, so that the first flood wave does not make impossible demands on the available food supply.

Extraction of carbon dioxide by plants and of oxygen by animals demands adaptations which vary with position in the ecosystem. At depths beyond about 1 m, marsh plants such as reeds and rushes are replaced by floating-leaf plants such as water lilies. Beyond about 3 m of depth, plants must either float entirely, or must develop a system of aeration which allows them to live beneath the water surface. Floating vegetation can only persist in streams with slight currents, but there, especially in the form of the water hyacinth, can prove remarkably successful, and can sustain a rich and varied fauna, especially of larvae, on the underside. Plants that live wholly underwater are eaten by animals that burrow into, and breathe from, them. Most animal forms, however, extract oxygen directly from the water. Many larval forms possess gills.

Plants growing on stony channel beds are firmly attached. Insects and their relatives are flattened, and capable of clinging or of otherwise adhering – for instance, by throwing out networks of filament, Limpets also stick on tight. Finer bed material permits plants to root, some breathing at the surface, others underwater. The plants themselves supply air and food to animal communities. Animals inhabiting sandy channel beds must be agile and able to burrow rapidly. Communities of siltbed streams often include vastly abundant larvae, such as those of craneflies, that are adapted to poor oxygen supply. Beds and banks of clay are inhabited chiefly by burrowing organisms such as mayfly larvae. High organic contents in silts and clays favour the multiplication of anaerobic bacteria and fungi. In some lime-rich streams, calcareous algae and associated mosses actually construct foundations, in the form of carbonate accumulations on the bed, and of tufa screens at waterfalls.

21.8a Combined effects of bottom and current conditions

The attributes of the environmental component that consists of channel bed material are controlled partly be sediment sources, partly by distance of transport, and partly by current velocity. Some source areas fail to supply anything below the pebble grade, or even below the cobble or boulder grade, as in the headwater reaches of many mountain streams. Rock fragments become rapidly rounded during downstream transportation, and also become smaller. Where material below the pebble grade exists, its calibre is controlled largely by flow velocity. Pebbles are moved at velocities of 1.2 m/sec and above; clay particles remain unaffected by velocities of 0.2 m/sec and below. Silt and sand typify the intermediate velocity range.

21.8b Habitat zonation

Cross-channel zonation includes differences among the plants and animals that inhabit, respectively, the channel banks, the channel bed, the midstream water and the water surface. The cross-channel zones extend along the complete length of the channel, unless they are in some way interrupted; at the same time, the species present – although not their adaptations – may well differ between midstream reaches and upstream reaches, and again between midstream and downstream reaches. Cross-channel zonation is most highly developed in large, slow-flowing streams.

Along-channel zonation is usually discussed in terms of the contrasts between upland streams with irregular beds, high levels of available oxygen, and abundant populations of limited numbers of species with limited temperature tolerance, and lowland streams with regular beds of finer materials, lower levels of available oxygen and great species diversity that includes forms with considerable tolerance of temperature change. The extreme instance of limited temperature tolerance is a species of flatworm, which lives close to actual springs and fails to breed if the water temperature rises above 12°C.

21.8c Biomass productivity

Productivity per unit area of channel bottom is probably higher in upstream reaches than in downstream reaches, although generalisation is made difficult by the ecological discontinuities imposed on many streams by major breaks of channel gradient in the forms of rapids and waterfalls. Productivity in limestone risers compares with that of boreal forests, while some mildly polluted streams match the productivity of tropical forests. In respect of production efficiency, however, freshwater streams rank low. Even the carefully managed carp of monsoon Asia average no more than about 0.2 tonnes/ha of fish a year.

21.9 The special case of humid-tropical rivers

Rivers of the humid tropics, ranging among the clearwater, whitewater and blackwater conditions, illustrate extremes of the structural–productivity –adaptive range of river ecosystems. The turbulent whitewater rivers are rich in inorganic nutrients, but lack facilities for photosynthesis except where floating vegetation occupies channel margins. Primary production of organic substances by algae is restricted by lack of light, both within the stream and on the stream surface under the forest cover. Clearwater rivers, dumping their sediment into estuarine passes, are faunally rich. Blackwater rivers, deeply stained by dissolved humic substances, poor in inorganic ions and acid in reaction, are typically poor in algae. They may lack higher plants altogether. They record few insect forms, whether those that eat micro-organisms or that live off plants. Their fish need to rise to the surface to gulp in oxygen.

21.9a Land-water interchange

River ecosystems are remarkably open, experiencing extensive interchange of matter between land and water. Rain on the floodplain drains into the stream channel. Many channels exchange water with the groundwater table, experiencing inputs when the groundwater table is high and outputs when it is low. The extreme exchange condition is that where a whole stream vanishes down a limestone sink, or when another emerges full-grown from a limestone riser; but important exchanges, varying with the seasons, are known for streams flowing in gravel. Bank erosion and point bar construction amount to interchange of sediment, involving also the destruction and re-establishment of land plant communities. Nutrients are also exchanged. To the inflow and infall of plant matter correspond the outputs effected by the taking of fish, the harvesting of commercial plants, and the cycling of organic matter through the water–land–air systems that involve insects, plants, and animals including birds. Free seasonal exchange occurs when fish and aquatic or water-breeding insects are eaten by birds. Insects and their relatives, along with decomposable plant matter, fall into the channel; and birds and fish are taken by man and by other predators.

21.10 Adaptation: an example

Structural and behavioural changes in life forms prove fascinating to outside observers. The general assumption is that such changes bring some kind of advantage to the changing species, although this assumption could be argued in connection with the fossil record. After all, the gross record is, on heavy balance, one of extinction. Against this, the history of humankind includes prominent examples of behavioural changes that have permitted particular groups to penetrate particular environments, without competition by other human groups. Leading examples include the behavioural adaptation of Eskimos to the cold rigours of Arctic regions, and the corresponding but contrasting dry rigours of the deserts inhabited by nomadic Bedouin tribespeople.

The non-human world supplies parallel examples. This section will deal with the emperor penguins, which, by evolving furthest of all penguins toward cold adaptation, have opened for themselves an ecological niche that includes breeding grounds on the margins of the Antarctic continent. King penguins, a possible rival species, have not evolved extreme adaptations to match those of the emperors; the former breed between 60° and 45°S.

21.10a The emperor penguin year

In common with their penguin relatives, emperor penguins alternate between feeding in the sea – mainly on fish, squids and small crabs – and fasting on land during the moulting and breeding seasons. Because the breeding grounds are established on thick sea ice between the mainland and offshore islands, the breeding cycle must be so placed as to avoid the seasonal melting. The penguin year, however, can be regarded as starting with the moulting season, which for a colony of emperors runs from November into January. For a single bird, the shedding of old feathers and the growth of new takes thirty days or more. The process is energy-consuming, reducing the body mass of up to 40 kg by about half. Part of the loss is heat drain to the

Fig. 21.14 Emperor penguins walking back to the rookery: one bird has elected, for a change, to toboggan (Yvon le Maho).

surrounding air, which averages a temperature of about freezing point, and part is the production of feathers. With their new plumage grown, all birds enter the sea in December or January, to feed and to restore the lost body mass.

By the end of March the sea has frozen over, and the penguins are walking in single file back to the rookeries, often for distances of 100 km (Fig. 21.14). Courting and copulating take place in late April, egg-laying at the beginning of May. By this time the sea ice is beginning to break up, and the females, having fasted for fifty days, return to sea to feed. The males take complete charge of incubation, resting the eggs (one per couple) on the tops of their feet, and covering them with brood pouches folded from stomach skin. By the time that hatching occurs, in mid-July, the males have fasted for 115 days – nearly four months – and have lost 40% of their initial body mass. The females return at the time of hatching, locating their mates by means of coded calls, and releasing the males to depart for the sea and food. It takes about four weeks for the males to renew their reserves of fat and protein, after which they return, again take over, and release the females. Change of shift in this way permits the adult birds to feed, and on their return to feed the chicks by regurgitation. Then comes the next moulting season, when the adults replace their last year's feathers and the chicks replace their down by adult plumage.

21.10b Strategies of energy efficiency

Two obvious strategies for the efficient use of energy stores are the use of fuel with high energy value per unit mass, and the limitation of energy expenditure. The emperor penguins employ both strategies in noteworthy fashion.

Birds and mammals store energy in the form of proteins, carbohydrates and fats, this last form being the most efficient on an energy/mass basis. Emperor penguins run to about 25% fat, the main energy store drawn upon during moulting and breeding, and also during the walks back to the sea which each consume about 15% of the total energy store.

During periods of fasting, the rate of breakdown of body substances and their conversion to energy is probably reduced. Down to the critical temperature of −10°C, body temperatures can be maintained without increasing the demand rate of breakdown. Feathers, fat next beneath the skin, and low rate of blood flow at the surface all promote a high level of insulation and a low rate of heat transmission from bird to surrounding air. Emperor penguins also have a low rate of heat loss because of a low body surface area in proportion to body mass. The large body mass reduces the surface/volume ratio, while the smallness of extremities also contributes to reducing the loss rate. Low rates of substance/energy exchange, as measured by the ratio of energy expenditure to mass, ensure a capacity for lengthy fasting. The plumage alone accounts for 85% of the resistance to heat transfer from inside an emperor penguin to the cold surrounding air; the feathers are dense and double-layered. In addition, the birds contain heat exchange mechanisms, wherein warm arterial blood heats the return flow through the veins. Parallel systems operate in the nostrils. An opposite adaptation for warm days is the opening of the plumage to let heat escape, the increase of surface temperatures in beaks, feet and flippers, and the by-passing of the heat exchange mechanisms by marginal veins.

In conditions of extreme cold, which for emperor penguins mean air temperatures below − 10°C, additional strategies come into play, some of them physiological and some behavioural. Shivering, as with ourselves, increases the rate of substance/heat conversion. A behavioural strategy is inactivity. An isolated emperor penguin, feeling really cold, will stand on its heels and tail, hold its flippers close, pull in its head, and if necessary remain still and silent for days on end. But in groups, cold birds form huddles. One huddle can contain thousands of birds, the close packing reducing daily loss of body mass by up to half. The huddle advances very slowly in the downwind direction, as the most exposed birds at the rear – that is, the upwind end – slowly move forward along the sides, at greater than the average huddle speed. All get to take turns at being most exposed and being most sheltered.

Here is a trade-off situation, increasingly familiar as the workings of ecosystems become increasingly better understood. Incubating males huddle. They could not do this if they were programmed to stake out, and to defend, individual nesting territories. Other penguin species do claim individual territories

– hence the wild quarrelling, snapping and flipper-slapping of the incubation season. But emperor penguins have surrendered territoriality in exchange for increased thermal efficiency.

Chapter Summary

The *basic ecosystem model* includes *producers, consumers* and *decomposers*. The *nutrient pool* is a storage subsystem. Inputs from outside include sunlight, gases, liquids and minerals. Outputs include energy.

Successive *steps of feeding* are called *trophic levels*. The *biomass ratio* is the ratio of organic matter content *between one trophic level and the next*. As a means of producing biomass, consumption is inefficient.

About *1% of incoming solar energy* goes to *drive ecosystems*. The amount of *stored energy decreases* greatly *from one trophic level to another. Energy input depends* very largely *on photosynthesis* whereby green plants take in carbon dioxide and water, construct organic compounds, and give off oxygen and some water. The *organic compounds include carbohydrates, amino acids, proteins, fats and pigments*. They are typically polymers.

Nutrient cycling involves seventeen elements and compounds of them. The cycling of hydrogen, oxygen, carbon, nitrogen and sulfur has been intensively studied: all these substances are soluble in water and exist as gases or gaseous compounds. *The basic equation for the carbon–oxygen cycle is a variant on the basic equation for* the *photosynthesis* of carbohydrates. Oxygen-breathing animals, including ourselves, use oxygen and carbohydrates to release energy, giving off carbon dioxide and water.

The nitrogen cycle is especially important, because *few organisms can take up atmospheric nitrogen directly. We depend* mainly *on bacteria*, which fix atmospheric nitrogen into ammonium compounds, or convert ammonia and ammonium compounds to nitrates and nitrites.

Like the basic ecosystem model, the slightly more complicated *basic food web model is an open system*.

Fraction of world *biomass content, for ecosystems on land, is greatest for tropical forest*, declining progressively through midlatitude forest, boreal forest, savanna, steppe and finally to tundra. The *sequence of production rates resembles* that of *biomass content*. Agricultural land rates with steppes for content, but between boreal forest and savanna for production. *Production rate is reduced by a factor of about 1.5, between one* gross *ecosystem* type *and the next* downward. *Rate of throughput is similarly reduced*, although the array of throughput rates does not match the array of production rates.

Analysis of *population dynamics* reveals a *tendency* for population totals to *increase through time*. The *alternative is extinction. Regardless of actual trend, environmental resistance* eventually *acts to control totals*. Several models of trends in numbers are available: all can be matched in actual cases.

Tropical rainforests supply one general *illustration* of the *working of actual ecosystems*. They are noteworthy for their *abundance of species*. The *nutrient pool* is *contained* largely in the decaying organic matter of the *forest floor*. For purposes of description, a tropical forest can be regarded as *stratified* into *emergents*, *main canopy*, *lower treetops*, *woody treelets* and the *forest floor* vegetation. Plant forms notably include *epiphytes*, including parasitic epiphytes. There are also *stranglers*.

Animals in tropical forests *tend to occupy particular* forest levels, varying from birds and bats above the main canopy to ground-feeders on the floor. The carrying capacity for animals runs well below its potential level.

Plant–animal interactions have probably *evolved together*. They include pollination by birds and bats, in addition to insects, and the production of fruits attractive to dispersers of seeds.

River ecosystems depend on algae, mosses, dead vegetable matter that is washed or that falls in, and on decomposing *fungi and bacteria*. *Cannibalism* is prominent. Life forms are *adapted to resist the flow of water* or to *take advantage* of it. There are also adaptations to highly irregular flow. *Adaptations to the extraction of carbon dioxide* by plants *and* of oxygen by animals *vary with position in the ecosystem*. Life *habits and adaptations vary* markedly according to the *nature of the channel bottom*. *Habitat zonation* occurs both *across the channel* and *along the channel*. *Biomass productivity varies* greatly, reaching its maximum in mildly polluted streams. *Humid–tropical rivers* include some *special cases*.

Land–water interchange is extensive. River ecosystems are remarkably open.

Emperor penguins provide an example of animal *adaptation to extreme cold*. Annual growth of feathers after moulting, breeding, incubation of eggs, and feeding of chicks amount to *drastic expenditures of energy in the form of stored fat*. The birds have evolved *sub-cycles of feeding*, some involving all adults, some males and females in alternation. *Structural adaptations* include a high *fat fraction*, effective *insulation, small extremities* and *low surface/volume ratio*. *Functional adaptations*, themselves involving structural adaptations, include *low rate of blood flow near* the body *surface, heat exchange* mechanisms, and heat exchange *bypasses*. *Behavioural adaptations*, some of which involve structural and functional adaptations, include *shivering, inactivity* and *huddling*. Huddling trades off increased thermal efficiency against territoriality.

22 Ecosystems of the Past

Application of the ecosystem concept to the record of past life is, in formal guise, fairly new, although it is implicit in all studies of the fossil record. Absolute dates, and the means of absolute dating, are discussed in Chapter 24, which also explains the reason for the division of the succession into periods. This chapter reviews, in highly compressed fashion, the appearance, evolution, and in some case the extinction, of major lines, with partial reference to the development and occupancy of major ecological niches. It samples past ecosystems with reference to early marine life and to the evolution of our own kind. Finally, it offers one possible answer to the question *Why are there so many different species?*

22.1 The frame of reference: fossils

The record of life on earth is worked out from plant and animal fossils contained in sedimentary rocks. In order to trace the courses of evolution, and to relate descendants to their ancestors, we need to classify. Classification and naming of plant forms is far better organised and controlled by the scientific community than is the classification and naming of animal forms. Against this, the treatment of animal forms is in general easier than that of plants – except, as we shall see, in the case of our own direct and immediate ancestors.

Classification depends on division into *phyla* (Greek, = stocks), which are subdivided into classes, these being subdivided into orders, orders being subdivided into families, which are subdivided into genera, and genera being subdivided into species. Further subdivision of species into subspecies and varieties is also possible. You, the reader, and I, the writer, belong to the species *Homo sapiens*, which belongs to the genus *Homo*, which belongs to the family *Hominidae*. At this level a suborder, that of the *Anthropoidea*, is distinguished; it includes ourselves and the great apes. This suborder belongs to the order *Primates*, which belongs to the class *Mammalia*, one of the eight classes in the phylum *Vertebrata*.

For purposes of a general survey of fossil life

forms, it will suffice to discuss seven phyla of invertebrates (= animals without backbones) and the phylum of vertebrates (= animals with backbones), taking notice of the three classes of the phylum Mollusca, three of the five classes of the phylum *Anthropoda*, and four classes of *Vertebrata* plus the four fish classes as a group (Table 22.1, Fig. 22.1). Three phyla of worms and one of wheel animals will be ignored.

Table 22.1 A Partial Classification of Animal Life Forms

Phylum	Leading examples
Protozoa	Single-celled animals, such as those of the plankton
Porifera	Sponges
Coelenterata	Corals and their relatives, such as sea anemones
Brachiopoda	Shellfish which construct lampshells
Bryozoa	Moss animals
Echinodermata	Starfish, sea urchins, sea lilies, and their relatives
Molluscs	Molluscs
Class pelecypoda	Clams and their relatives
Class gastropoda	Snails and their relatives
Class cephalopoda	Squids, octopuses, nautili, and their relatives
Arthropoda	Invertebrates with jointed legs
Class trilobitoidea	Trilobites
Class crustacea	Lobsters and crabs
Class insecta	Insects
Vertebrata	Animals with backbones
Four pisces classes	Fish
Class amphibia	Amphibians: frogs and their relatives
Class reptilia	Crocodiles, turtles, snakes, dinosaurs, and their relatives
Class aves	Birds
Class mammalia	Mammals

This table omits one phylum of wheel animals (rotifers) that is never found in fossil form: two phyla of worms – flatworms and threadworms – that are also never found fossil: and another phylum of worms, the segmented worms, that is very rarely represented by fossils but is fairly well known from fossil tracks.

Fig. 22.1 (pp. 319–20) Examples of marine invertebrate fossils (scales in cm) (R. O'Brien).

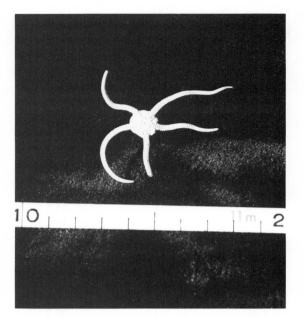

Fig. 22.1 – continued

22.2 The frame of reference: time scales

Fig. 22.2 and Table 22.2 present and compare some scales of time which are useful in the discussion of cosmic and terrestrial history. About three-quarters of universe time elapsed before the solar system formed and life first appeared on earth. On the scale of a twenty-four hour day, these events would have occurred at about 1800 hours. Of the remaining six hours, five would have been consumed by Precambrian time, during which the development of sexual reproduction introduced great biological advantages, and the development of photosynthesis led eventually to the enrichment of the earth's atmosphere with oxygen. Quite complex organisms had evolved before the Cambrian period opened, but their lack of hard parts means that few traces remain today. It was in the Cambrian that there took place an explosive multiplication of marine invertebrates which did possess hard parts. With Cambrian times begins the abundant fossil record of life forms, by means of which the courses of evolution can be traced through the last 600 years – that is, through the last hour of our universe day.

22.3 Outline of the evolutionary record

During the Cambrian period, the oceans widely invaded the world's lands in an event known as a marine transgression. The resulting shallow seas were copiously inhabited by trilobites. This arthropod class varied so greatly that it accounts for some 70% of all the known species of those Cambrian marine animals that were more complex than the single-celled Protozoa (Fig. 22.3). The next most numerous group was that of the brachiopods, shellfish which lived attached to the sea bed. As Fig. 22.3 shows, all other phyla were represented in Cambrian times, with the noteworthy exception of the vertebrates. Oceanic plankton were also present.

In the succeeding Ordovician period the proportionate contribution of trilobites to shallow-marine life was drastically reduced. No simple cause-and-effect explanation suggests itself. It is true that the contribution of brachiopods increased, and that the mollusc line produced clams, but neither clams nor brachiopods seem likely to have been direct competitors for food with the mobile trilobites. Nor would the graptolites, floating colonial organisms possibly related to corals, have competed, because the trilobites were bottom-feeders. The major evolutionary event of the Ordovician period – if one such event can be thought more important than another – was the production of vertebrates (animals with backbones) in the form of air-breathing lampreys and hagfish. Our own basic design is already evident in their remains.

The earliest known coral reefs date from the Silurian period. But the most impressive change of the time was the beginning of the invasion of the world's lands by plants. Silurian plants included forms such as the horsetail, *Equisetum*, adapted to moist conditions. It is reasonable to assume that the invasion began as penetration along river banks. The

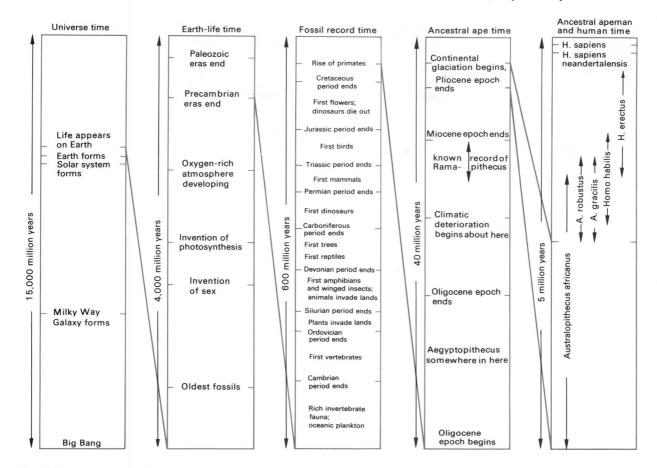

Fig. 22.2   Some time-scales.

Table 22.2   Comparative Lengths of Time-scales

	Universe time	Earth–life time	Fossil record time	Ancestral ape time	Ancestral apeman and human time	Homo sapiens sapiens time
Universe time	1	3.75	25	375	3000	150 000
Earth–life time	0.27	1	6.7	100	800	40 000
Fossil record time	0.04	0.15	1	15	120	6000
Ancestral ape time	0.0027	0.01	0.07	1	8	400
Ancestral apeman and human time	0.0003	0.00125	0.008	0.125	1	50
Homo sapiens sapiens time	0.000007	0.000025	0.00017	0.0025	0.02	1

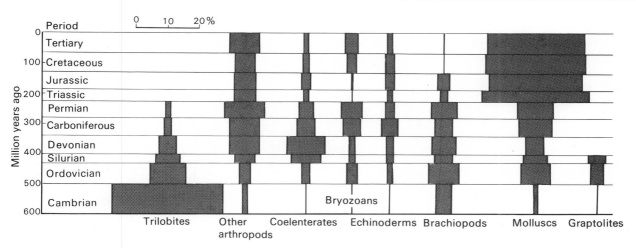

Fig. 22.3   Relative abundance of marine invertebrate animals during the last 600 000 000 years: protozoa and some numerically minor groups excepted.

ecological niches thus opened were soon occupied by arthropod species.

Corals reached their peak of success in the Devonian period, and brachiopods their peak of diversity. On land there sprang up the earliest forests. Ferns and related plants, at the opening of the period leafless and small, had by the end of the Devonian times lost their immediate dependence on damp ground, and grew in dense leaf-bearing stands, attaining heights as great as 10 m. Insects, initially wingless but later including winged forms, diverged from the previous arthropod line. The vertebrate line also split. One branch became gill-breathing fish, striking in the fossil record because many possessed external plating. The other, still including air-breathing fish, also diverged into amphibians. These, such as modern frogs, are forms that breed in water but can spend time on land. Some of the early Devonian amphibians seem intermediate between recognisable fish and recognisable amphibians.

A second major marine transgression took place in the Carboniferous period, when huge rivers constructed deltas at the invading coastlines. The largest deltas of all underwent intermittent subsidence in response to their accumulating sedimentary loads, the peat of swamp forest going to form the chief coal reserves of our own day. Tree forms multiplied, scale trees being added to the earlier ferns and rushes. Land insects developed some very large forms, land snails multiplied, and land-dwelling amphibians evolved the first reptilian line, later to enjoy huge ecological success. In the shallow marginal seas, among the echinoderms, sea-lilies underwent explosive increase of numbers and of species (Fig. 22.3): they have never been so prominent since.

To the Permian period belong the first known fossils of dinosaurs (from Greek, = terrible lizards). The first evolved forms were, however, small, in no way comparable to forms to come. Also during this period, the cephalopod class of the mollusc phylum produced ammonites, coiled-shell relatives of the nautilus, but with complex patterns of the juncture of interior partitions with the outer shell. Interest in Permian life is focused partly on the massive extinctions with which the period ended. Some 75% of amphibian families and 80% of reptilian families died off. Similarly drastic reductions affected bryozoans, echinoderms, ammonites and brachiopods. The bryozoans barely escaped complete extinction; the brachiopods were never to recover; and trilobites finally vanished from the record (Fig. 22.3).

One suggested cause for extinction is the desalting of the world's oceans. During Permian times, huge salt deposits were formed. In all probability, these were precipitated in land-marginal basins, where evaporation concentrated brine but where shallow sills prevented free mixing with the outside sea. As much as 500 000 km³ of seawater may have been evaporated; oceanic salinity may have been driven down below the level of tolerance. It is a fact that, by comparison with modern forms, the doomed species of late Permian times appear to have been those least adapted to withstand large/long-term fluctuations of changes of salinity. However, considerable extinction of land plant forms also marked the end of the Permian period. Apart from complex arguments involving salt in rainwater, oceanic desalting cannot be thought to apply here.

Surviving stocks evolved anew during the Triassic period. Coniferous forests established themselves on land, inhabited by reptiles which could run down their prey. Kangaroo-like in general design, with small forelimbs and large hind limbs, the Triassic dinosaurs did not however hop, but raced like emus and other flightless birds. The Triassic dinosaurs were partly flesh-eaters, partly plant-eaters. At sea, renewed explosive speciation of ammonites and of their squid-like relatives the belemnites supplied food stocks for sea-going reptiles. The most dramatic evolutionary and ecological event of all was the development of strengthened shells and lengthened siphons on the part of clams and their relatives, members of the Pelecypod class of the mollusc phylum. The evolutionary changes made it possible for the species concerned to burrow into the sea bed, and thus to lodge in an ecological niche that had previously been closed except to the soft-bodied worms (Fig. 22.4). Fig. 22.3 shows how, in consequence, the molluscs became numerically dominant among marine invertebrates.

Progression from reptiles to mammals, begun in the Permian period, was continued in the Triassic period. It was completed in the Jurassic period, but with no foreshadowing of what was eventually to come – complete mammal dominance. On land, the Jurassic period saw the increased spread of evergreen forests, and the marked variation of insect life,

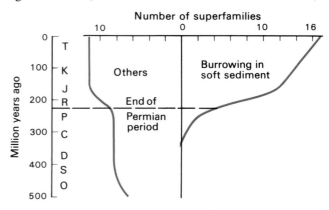

Fig. 22.4 Invasion of an ecological niche, the muddy bed of the sea, by burrowing pelecypods: very freely adapted from the data of S. Starkey.

including the development of moths and flies, which are highly specialised life forms. At sea, ammonites became remarkably abundant, evolving so rapidly that their fossils are used to subdivide the geologic succession. Belemnites were similarly abundant, becoming a preferred food source for the porpoise-like icthyosaurs (Greek, = fish-lizard), sea-going reptiles which bore live young. The ecological niche now occupied by turtles was, in Jurassic times, occupied by the plesiosaurs (Greek, = near-lizard), reptiles generally turtle-like in form except that their early species possessed long necks. The flesh-eating crocodile line, still with us today, also appeared during the Jurassic period.

Most of the large reptile tribes of Jurassic times, in contrast to the ichthyosaurs, laid eggs. Fossil eggs have been found with the vertebrate young still inside. The wholly plant-eating line diverged to produce the stegosaurs (= armoured lizards), great land animals with rows of bony plates along the back. The originally flesh-eating line continued to produce flesh-eaters, but also gave rise to the brontosaurs (Greek, = thunder-lizard), colossal creatures up to 20 m long that browsed on swamp plants. These are the lovable dinosaurs of comic-strip fiction.

Two kinds of adaptation to the air were made during the Jurassic period. The pterosaurs (Greek, = winged lizard) flew, or more probably glided, with the aid of membranous wings, rather as bats do today. But the pterosaur wing was attached to a single digit, rather as a yacht sail is attached between mast and boom. It was useful for soaring, but not for flapping. But true birds also appeared, their originally scaly covering becoming feathers, and five-digit wings being adapted for takeoff and for powered flight. Recent studies have served to emphasise the similarities between bird design on the one hand, and one kind of reptile design on the other.

The third major marine transgression took place during the Cretaceous period, when minute calcium-fixing organisms showered their remains down on the sea bed, in the sediment that now constitutes chalk rock. Echinoderms in the shallow transgressive seas varied so rapidly that they provide the means of subdividing the rock succession in minute degrees, just as the ammonites do for Jurassic strata. On land appeared, for the first time, flowering plants, which have subsequently driven non-flowering plants into marginal situations. We are bound to infer a parallel development of pollinating insects and birds. Also on land, the dinosaur lines produced the horned and armoured triceratops among the plant-eaters, and the massive and savage tyrannosaurus among the flesh-eaters. Primates with opposable thumbs appeared as the forerunners of humankind, while marsupial (= pouched) mammals

appeared as the forerunners of kangaroos, opossums and their kindred.

The Cretaceous period ended with extinctions, at least as impressive in some respects as the extinctions of the Permian. Ammonites and belemnites flourished vastly through Triassic, Jurassic and Cretaceous times, despite the drastic end-Permian reduction, only to vanish entirely with the closing of the Cretaceous. The dinosaurs also died out, leaving only crocodiles, lizards, turtles and snakes to represent the reptilian line.

Some writers consider the evolution of flowering plants a response to an increased climatic contrast between the high-sun and the low-sun seasons, reasoning further that the same environmental change was adverse to the dinosaurs. The argument depends on the view that dinosaurs were cold-blooded – a view that has been challenged. At present, we can only look on the total dinosaur extinction, after 160 million years of dominance, as mysterious. Equally mysterious is the disappearance of the ammonites, especially in the light of the complete reorganisation of the marine plant biomass during Jurassic and Cretaceous times. The phytoplankton (assemblage of floating/swimming plants) became greatly elaborated, becoming the basis of a marine food web in much like the present form. Whatever destroyed the ammonites, it can scarcely have been starvation. During the Cretaceous period, numbers of ammonite species developed so-called degenerate forms, forsaking the earlier tight coiling for loose coiling, screw-like twisting, or even straightness. One is tempted to conclude that the whole line was going genetically wrong.

Finally, the last 65 million years of earth history, the Tertiary period, records the rise to dominance of mammals. There can be no doubt that the multiplication of grazing individuals and of certain grazing species was a response to the evolution and spread of grasses, which previously had been unknown. We know far more of the land animals of the Tertiary than of their precursors. Some evolutionary lines have been traced in detail. Numbers of mammal lines display a strong tendency to produce giant forms, some of which survive today. The common ancestors of all living horses, including the largest of workhorses, were only as big as moderate-sized dogs. Our own ancestry, along with that of gorillas, is probably to be sought among the tarsiers of sixty million years ago. These small creatures, about the size of mice, had the opposable thumbs that are common to all primates, but in addition had evolved forward-directed eyes that provided full stereoscopic vision.

With universe time scaled down to the length of one day, Tertiary time reduces itself to the last six minutes, 23.54 hrs to midnight. Ancestral ape time is

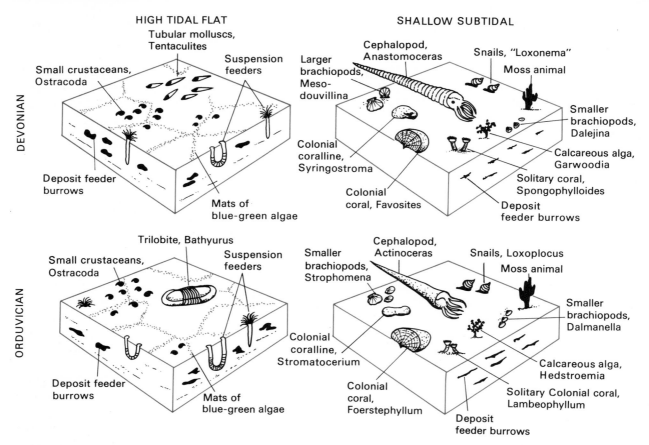

HIGH TIDAL FLAT

Tubular molluscs, Tentaculites
Small crustaceans, Ostracoda
Suspension feeders
Deposit feeder burrows
Mats of blue-green algae

DEVONIAN

SHALLOW SUBTIDAL

Cephalopod, Anastomoceras
Larger brachiopods, Meso-douvillina
Snails, "Loxonema"
Moss animal
Smaller brachiopods, Dalejina
Colonial coralline, Syringostroma
Calcareous alga, Garwoodia
Colonial coral, Favosites
Solitary coral, Spongophylloides
Deposit feeder burrows

Trilobite, Bathyurus
Small crustaceans, Ostracoda
Suspension feeders
Deposit feeder burrows
Mats of blue-green algae

ORDUVICIAN

Cephalopod, Actinoceras
Smaller brachiopods, Strophomena
Snails, Loxoplocus
Moss animal
Smaller brachiopods, Dalmanella
Colonial coralline, Stromatocerium
Calcareous alga, Hedstroemia
Colonial coral, Foerstephyllum
Solitary Colonial coral, Lambeophyllum
Deposit feeder burrows

Fig. 22.5 Pairs of shallow-marine ecosystems of differing geologic age: simplified, by permission, from a diagram of K.R. Walker.

even shorter, occupying only the final four minutes. Ancestral apeman time amounts to no more than 30 seconds, starting at 23:59:30 hrs. Ancestral humans came on the scene later still, at 23:59:45 hrs, while our own sub-species, *Homo sapiens sapiens*, accounts for only the final half-second. In terms of real time, we have existed for only 100 000 out of the universe total of 15 000 000 000 years.

22.4 Reconstructing past environments

The various animals and plants of the past were necessarily components of the ecosystems of their times. Physical aspects of past environments can be reconstructed from the rock record, and particularly from records of sedimentation. Rock type not only reveals the nature of sediment supply; it can show modes of sediment transportation, and not infrequently the weathering conditions in the source region. Internal rock structures can permit reconstruction of the environments of deposition.

The sedimentary rock type arkose is composed of quartz grains and of feldspars, the latter showing no more than modest signs of weathering. A source region for arkose is interpreted as being granitic

terrain on a steeply-plunging coastline. Weathering of the granite supplies quartz grains, feldspar crystals, and mica flakes. Steep slopes and short distances of transportation permit the feldspars to be carried into the sea before weathering can destroy them. Whereas the coarser rock waste settles immediately to the sea bed, the mica flakes are more widely dispersed.

Sand deposited in shallow water, as along a gently shelving shore, is liable to be deformed into ripples as tiny waves pulse to and fro on the bottom. If net deposition is taking place, then sets of ripples will be buried, becoming preserved in sandstone rock when the sediment hardens. Fossil ripples are known from sandstones 2000 million and more years old. In addition to indicating deposition of sand in shallow-water conditions, they can be used to show the direction of wave approach.

22.5 Reconstructing past ecosystems

The examples now to be given of the reconstruction of past environments are only two of the many available. When life forms are added to the environmental record, then whole ecosystems can be reconstructed. Although it is not always possible to determine just what ate what, a great deal can be done with modes of feeding.

22.5a Some comparative marine ecosystems

Fig. 22.5 illustrates the leading ecosystem components for the high tidal flat zone and for the shallow subtidal zone, respectively in part of Ordovician and in part of Devonian times. The evidence is presented in limestone rocks in what is now upper New York State. The Ordovician rocks in question are about 450 million years old, the Devonian rocks about 70 million years younger. Despite this gap, there are close ecological similarities between the two sets.

In each period, the inshore was lime-rich, with a mean temperature not less than 20°C. So much can be read from the calcareous nature of the sediments themselves, and from the presence of corals. The sea bed consisted of lime mud, soft enough to be penetrated by soft-bodied burrowers. If the burrowers had possessed hard parts, it is inconceivable that none of these would have been preserved; but in actuality, infilled burrows only are found. These, however, make it possible to distinguish two kinds of burrowers – mobile forms which fed on the mud itself, as earthworms feed on soil, and immobile forms of the high tidal flat, which formed vertical or U-shaped burrows and must be taken for suspension feeders.

In each period, the high tidal flat carried reticular mats of blue–green algae. Small ostracods scavenged on surface deposits. The only signal differences for the high tidal flat ecosystem are the presence of the trilobite *Bathyurus* in the Ordovician and its absence from the Devonian, and the presence of the tubular molluscs *Tentaculites* in the Devonian and their absence from the Ordovician. It may be over-simple to suggest that the apparent switch from one life form to another represents successful interspecies competition by the molluscs, unless these and *Bathyurus* were similar scavengers.

For the shallow subtidal zone, the main components of the ecosystem remain unchanged in function and also essentially in form, although there is a complete changeover from the Ordovician to the Devonian record at the genus level. Calcareous algae, moss animals, solitary corals, colonial corals and colonial corallines occur in each period. Each ecosystem includes one larger and one smaller genus of brachiopods, living attached to the sea bed and feeding by filtering water through themselves. Each system includes snails, which are herbivores, grazing on the plant life of the floor. Each, again, includes predatory cephalopods, distant relatives of the modern nautilus. This particular record represents remarkable ecosystem stability that does not, however, exclude evolutionary change.

22.5b Ecosystems of human evolution

Ancestors of modern apes, and of modern people, lived in East Africa twenty million years ago. At about this time there began a climatic deterioration which brought average midlatitude temperatures down by 10°C. The drop is likely to have been less in low latitudes, but nevertheless sufficient to produce an increase in grassland at the expense of forest. Questions of converting from a tree-feeding to a ground-feeding habit, and possibly back again, along with questions of the influence of feeding locations on modes of walking on four feet as against two, are keenly argued. But ecosystem reconstruction shows that all the ancestral forms in question, the ramapithecines, lived in complex patchwork environments of forest, scrub, grasslands, and swamp. Environmental complexity should have encouraged complex adaptive evolution.

Evidence for the approximate period from twelve to five million years ago is so far fragmentary. When it becomes more abundant, the ancestral line is represented by *Australopithecus*, the so-called great southern ape. Lines of evolutionary divergence, being largely uncertain, are omitted from Fig. 22.2. Descendants of the original stock, members of the species *Australopithecus africanus*, were still living as late as 1.5 million years ago, but by that time as contemporaries of the sturdy *Australopithecus robustus* and the smaller and more slender *Australopithecus gracilis*. Although usually distinguished as separate species, *A. robustus* and *A. gracilis* were possibly no more than male and female creatures of the same species; a similar contrast in size and build typifies the ground-dwelling apes of the present day.

Australopithecines coexisted in common territory. On the known evidence it is impossible to say how the territory was subdivided, if indeed it was. But another evolutionary line had also appeared – that of our own direct ancestors, represented by the species *Homo habilis*, who was already a maker of stone tools.

It is important here to distinguish among tool using, tool making, and tool making to a pattern. Sea otters carry pebbles which, rested on their stomachs as they float on their backs, provide anvils for the bashing open of shellfish. Some apes will throw stones. Chimpanzees make tools, drawing off the leaves from straight twigs, and using the twigs to poke into termite mounds. Termites clinging to the withdrawn twigs constitute a protein source. Some researchers interpret the lack of enlarged canine teeth in the australopithecines of four million years ago to mean that these creatures used defensive hand-held weapons, such as pointed sticks. But our nearer ancestors, doubtless starting with throwing stones and progressing to using broken stones with sharp edges, took eventually to making stone tools to a pattern.

Homo habilis lived from about 2.5 to about 1.5 million years ago, becoming superseded by, or evolving into, *Homo erectus*, who by 500 000 years ago had spread widely through Africa, Europe and Asia. *Homo erectus* developed a remarkably effective toolkit, including choppers, pounders, scrapers, cutters and piercers, wherein hand axes and cleavers are prominent. It may by no means be coincidental that the emergence and spread of this species coincides more or less with the extinction of the australopithecines. If the various species were competitive, then *Homo erectus* had all the advantages. It is highly likely that *Homo erectus* lived in groups larger than those of modern apes, covering considerable territory and possibly migrating with the seasons. It is certain that large game animals were hunted. Here is the first record of the general invasion of the world's grasslands by human beings.

Homo erectus also domesticated fire. The importance of this cultural step, ultimately basic to metallurgy, cannot be exaggerated. It lengthened the working day, if lengthening were necessary, by permitting tool-makers to go on working after the sun had set. It made possible animal drives, far beyond the scope of drives set off by shouted alarms; and it eventually helped signally in the penetration of regions of cold climate, such as the northern tundras. Perhaps most importantly of all, it helped human digestion. Raw meat is less easy to digest – that is, has a lower input/power production ratio – than has cooked meat. With all due respect to convinced vegetarians, humankind is not really designed to subsist on plants alone. Part of the protein intake should be flesh. Cooked flesh, weight for weight, is more efficient an intake than is raw flesh.

With tools, with group hunting – implying the use of some kind of language – and with invasion of formerly hostile and even impenetrable ecosystems, *Homo erectus* dispersed the human stock throughout much of the world. Ecological adaptation, for the genus *Homo*, depended no longer on evolutionary change alone, but also on behavioural adaptations. At the same time, *Homo erectus* was less brainy than is our own species, *Homo sapiens*. So much is shown by a comparative survey of brainpan size (Fig. 22.6).

As in so many other connections, the sequence arranges itself into an exponential progression. Brainpan size, and presumably also brainpower level, rises from the australopithecines through *Homo habilis* and *Homo erectus* to *Homo sapiens*, our own species. The multiplication rate from one order of size to another is about 1.275. We have no reason to suppose that human evolution has ceased.

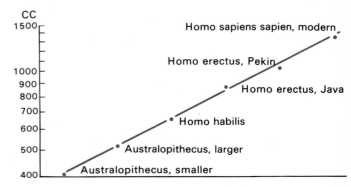

Fig. 22.6 Brainpan size in hominids: the sequence from the larger Australopithecus to Homo sapiens is the sequence of appearance in the fossil record. Points are plotted at constant intervals along the *x*-axis: fit of the straight-line graph is 98% perfect.

If it has not, and if the sequence in Fig. 22.6 is any guide to the future, then our successors will have brainpans of some 1720 cc capacity, proving as skilful to us as we have proved to *Homo erectus*. The prospect is exciting or frightening, according to one's point of view. *Homo erectus* produced little technological change in a million and a half years. *Homo sapiens*, in only 100 000 years, has developed the cultures and technologies in and with which we live. The record suggests an accelerating rate of change. What might be done by a human species more brainy still defeats the imagination.

22.5c Our extinct close relatives

Somewhat earlier than 100 000 years ago, the human line split into two subspecies, *Homo sapiens sapiens* and *Homo sapiens neandertalensis*, the latter being familiarly known as Neandertal man. Brainier than *Homo erectus*, the Neandertals took to wearing animal skins, developed tools more refined and specialised than those of their predecessors, and learned how to make tent-dwellings in their seasonal hunting camps. Their remains and their tools occur chiefly in Europe. Whether or not their separate evolution resulted from physical and biological isolation is uncertain. What is known is that they had large noses and prominent lower jaws. What is likely is that they carried considerable body hair on chest and back. These are, of course, attributes of the stereotype caveman. But the famous animal pictures in European caves were produced by *Homo sapiens sapiens*, who suddenly replaced the Neandertals about 37 000 years ago, when our nearest relatives became mysteriously extinct.

The history of Neandertal tool-making shows that the rate of innovation was more than twenty times as slow as that of the succeeding *Homo sapiens sapiens*.

Analysis of trash-heaps shows that the Neandertals had not properly learned to kill dangerous and/or swift-moving game. Evidence that they were liable to sinusitis, rheumatism and abcesses of the teeth is not in itself enough to prove that the descent into a new glacial, with which their extinction coincided, was the cause of their destruction. We can perhaps guess at genetic deterioration. An obvious alternative possibility is that our own ancestors, engaging in subspecies competition, killed them off.

22.6 Some partial information on past climates

Paleoecology, the study of ecosystems of the past, can be approached not only by the reconstruction of past environments and the analysis of their known life forms, but also by the reconstruction of past climates.

22.6a Oxygen–isotope analysis

This technique applies to former marine environments. Oxygen possesses three isotopes (see Chapter 24) – ^{16}O, which is by far the most common, ^{17}O, which is unimportant for our purpose, and the heaviest, ^{18}O. For every thousand atoms of ^{16}O in sea-water, there are about two atoms of ^{18}O. Both isotopes, ^{16}O and ^{18}O, are built into the hard parts of marine organisms, for example the shells of brachiopods. Neither isotope is radioactive. The observed $^{18}O:^{16}O$ ratio, therefore, relates to conditions at the time of growth. Now the ratio, in the indicated form, rises in value as ocean salt content increases, because ^{16}O is preferentially removed by evaporation. The ratio also rises as water temperature increases. If one assumes that the $^{18}O:^{16}O$ ratio is controlled primarily by temperature, then isotopic analysis of fossil marine organisms should indicate water temperatures when the organisms were alive. Work has been directed very largely to surface-swimming organisms such as the cephalopods, and to shallow-water forms such as brachiopods and molluscs. Results indicate considerable changes of water temperature through time (Fig. 22.7), although precise interpretation is impossible unless samples are closely spaced through the sequence. There is moreover some doubt that, for a given temperature and salinity, the isotope ratio of the distant past was identical with that of today. Nevertheless, isotope analysis offers, at the very least, the means of exploring changes in one important attribute of former marine ecosystems. It suggests, and very strongly, that faunistic shifts on the part of shallow-marine organisms are due to the supersession of one life form by

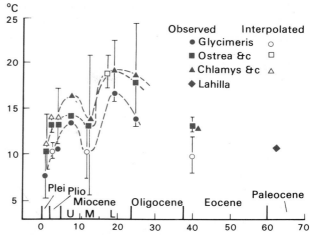

Fig. 22.7 Paleotemperatures determined on shallow-marine fossils from southeast Australia: original data from F.H. Dorman and E.D. Gill, interpolations by G.H. Dury and G.M. Habermann.

another, as opposed to progressive adaptation to a widened range of climate.

22.6b Pollen analysis

Comparative counting of fossil pollen grains has been discussed in Chapter 19, section 19.2. The floristic changes that provide the basis for Table 19.1 imply drastic changes in the complete ecosystem. If the vegetation changes from tundra to boreal forest, or back from boreal forest to tundra, then a corresponding change in the assemblage of animal life is inescapably implied.

There is more. Duly instructed, computers can take in the pollen record and print out reconstructions of temperature, surface wetness and even conditions of air-mass dominance. In this way, the floristic record is translated into a numerical climatic record. In due course, these records will doubtless be integrated with the record of animal and with the record of paleosols (= former soils), so that the major components of land ecosystems of the past 10 000 to 15 000 years will be identified in relation to one another.

22.6c Leaf margins

Most tree species in tropical rain-forests have leaves like those of laurels – oval or elliptical in outline, with the margins broken only by the drip-tips at the extremity. Leaves of this type are called entire-margined. They contrast for instance with the highly indented margins of the leaves on midlatitude oaks and maples.

There is in fact a fairly systematic relationship between forest type and percentage of species

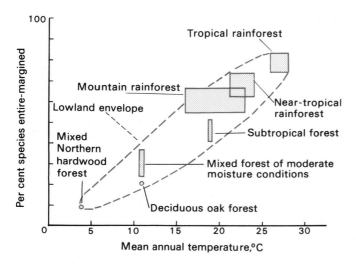

Fig. 22.8 Forest types and per cent of species with entire-margined leaves: basic data adapted from E.A. Wolfe.

entire-margined (Fig. 22.8). Because there is also a fairly systematic relationship between forest type and mean annual temperature, analysis of leaf margins can lead to reconstruction, not only of forest type, with all its ecological implications, but also of climatic type. For a given range of temperature, the minimum degree of wetness needed to sustain a forest can also be stated. Leaf margin analysis promises richly for the reconstruction of Tertiary forest ecosystems in terms of climate and vegetation. No doubt, in due course, the larger forms of forest-dwelling animals will be added in.

22.7 Past ecosystems and evolutionary change

Two general questions arise from the earlier sections of this chapter: why are there so many species? and, how does a given species diverge into two separate species?

A first-approximate answer to the first question is that species diversity makes for ecosystem stability. The more complex an ecosystem, the more capable it is of resisting damage. A second-approximate answer is that species diversity enables the species concerned to occupy distinct ecological niches. For instance, about fifty million years ago there existed some five lines of flesh-eating mammals. In the next ten million years three of these lines became extinct, while a fourth continued a somewhat uncertain existence until five million years ago. The remaining fifth line produced all the five families, and two extinct families, of later times. The common ancestors of fifty million years in the past were small unspecialised creatures about half the size of a small domestic cat, which fed perhaps upon rodents, in-

sects, and birds if they could catch them. The extinct families include the huge bear-dogs which failed to survive the oncoming of Pleistocene glaciation, and a branch of the dog family that vanished somewhere during that glaciation. The surviving five families – cats, dogs, bears, raccoons, seals and their closest relatives – have feeding habits which are in some respects highly distinct, and occupy distinctive ecological niches. Living cats all belong to the genus *Felis*. All members of the genus are biting cats. A branched-off genus of stabbing cats, with enormous incisor teeth and including the so-called sabre-toothed tiger, failed, like the one branch of the dog family, to survive the Ice Age. If the stabbing cats and the biting cats competed directly, the latter were, by demonstration, the better fitted to survive. But despite extinctions, the array of flesh-eating mammals of the ultimately successful line increased from the original one family to our contemporary five.

Part of the success related to behavioural adaptations. Domestic dogs and cats have opened ecological niches for themselves in the homes of human-kind, while many raccoons successfully use North American city suburbs as scavenging territories. In strict contrast, the fossil record is one of structural changes. Why these should occur, and how they can occur, is central to the issue of the history of life forms on earth.

22.7a Genetic variation

Heredity in organisms that reproduce sexually, as opposed to undergoing the simple division of single total-organism cells, depends on the transmission of genes. These contain programmed instructions for the development of the organism. They come in pairs, one of each pair from each parent. They copy themselves as cells subdivide (= as the organism develops after fertilisation), except that sudden mutations take place from time to time. For a wide range of life forms, mutation (= sudden change) in the message transmitted by a given gene pair occurs once in 500 000 individuals, once in a generation. That is, the chance of a mutation in a gene pair combination is two in a million. Mutation in a gene complex is far rarer still.

Since one gene pair controls a single characteristic, it is clear that the branching of a genus into more than one species involves changes of whole gene complexes. Among the biting cats, tigers are striped, leopards, jaguars and cheetahs are spotted, lions much the same colour all over as adults, although spotted as kittens, and some panthers are completely black. Sizes, habitats, hunting habits and type of prey vary from species to species. Inheritance of mutated gene complexes is responsible for the total contrasts among the various species of the cat genus *Felis*.

Mutation constitutes a change in the gene pool, which, if persistent, amounts to a long-term statistical change. General adaptations such as the development of biting teeth, claws, keen sense of sight and smell, speed in running and powers of springing extend, with the biting cats of today, right down to the species level. Adaptive pressures at the species level ensure for example a range of size. Small size makes for concealment from larger predators and for rapid dissipation of heat in hot weather. Large size makes for conservation of heat in cold weather, an increased range of possible prey, and the ability to fast for a time if necessary, because more energy in proportion can be stored in a large frame than in a small one. Every species represents a trade-off situation. If the effective trade is adverse, the species fails. Here is the process of natural selection. This process can be regarded as the suppression of ill-adapted mutant strains.

Ill-adaptation, then, can be looked on as a genetic mistake, a failed experiment. It can also be looked on as a variation which has had too little time to become fixed in the succession. If mutation is too rapid, evolution of variant forms cannot occur: these forms will be destroyed by the pressures of the ecosystems in which they appear.

22.7b Some qualifications

In a given gene pair, one gene may be more influential than another – one dominant, one recessive. A particular hereditary line may carry a whole dominant gene complex: it has been claimed, for example, that free-breeding cats in the London metropolis are becoming dominantly black and white and of moderate size. Recessive genes however are transmitted down the hereditary line, retaining their coded instructions in storage for possible use if the environment changes – as it certainly will. The grasses ancestral to the small grains of modern farming all retain recessive genes in their total gene pools. A grave concern with some bioscientists is that the wild gene pools may be destroyed as the Middle Eastern lands concerned become increasingly exploited for commercial grazing and for cropland, so that the potentially useful – indeed, vital – store of recessive genes will vanish.

Changes in ecosystem attributes can, in some cases, show that inheritance of characteristics is not irreversible. Fish evolved paired eyes, as opposed to a single eye, as long as 400 million years ago. One fish genus still grows the rudiments of a third eye, which however never becomes sighted. But if developing fish are raised in brine enriched with magnesium chloride, $MgCl_2$, they will develop only a single eye, centrally placed. Four hundred million years has not been enough to eliminate the influence of environment over the response to genetic programming.

22.7c Evolution as random variation

It has been known for a considerable time that the random mutation, segregation and recombination of genes offers a sound explanation, in the statistical sense, of the control of directions of evolution by natural selection. Fig. 22.9 is a wholly synthetic family tree, produced by random means. In this case

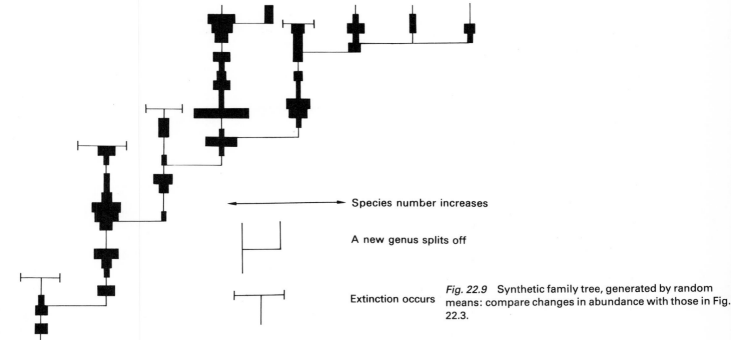

Species number increases

A new genus splits off

Extinction occurs

Fig. 22.9 Synthetic family tree, generated by random means: compare changes in abundance with those in Fig. 22.3.

the means was that of card-turning, with on each throw an approximately 77% chance of increase in species number, a 20% chance of further evolution, a 2% change of species divergence and a 1% chance of extinction. This diagram is remarkably similar to diagrams drawn for the time/abundance/diversity of known fossil forms, as can be seen from a comparison with the highly generalised information in Fig, 22.3. Thus, for life forms as for stream nets (Fig. 7.5), random variation provides an intellectually satisfying explanation of what actually happens. With the aid of computers, random variation of life forms can be extended into variation of individual characteristics such as size and shape, again with convincing results.

Purely random variation of genetic instruction must involve a great deal of waste, in the form of unsuccessful, because ecologically ill-adapted, mutants and recombinants. But it must also lead to successful adaptation and recombinations. Some of us may resent the proposition that we have orginated through random but adaptively successful alterations in our genetic coding. The total fossil record however is one of major adaptive successes, usually followed by major – and presumably non-adaptive – failures. Only our own species has the potential to manipulate whole ecosystems for its own purpose; all others are obliged to fit into what is there. This theme will be taken further in Chapter 23.

Chapter Summary

Life forms are classified into phyla. The sequence of subdivision then runs class, order, family, genus, and species. Suborders, super-families, subspecies and varieties can also be recognized.

The *record* of fossil life *before 600 million years ago* is *scanty*, because few animals possessed hard parts. Highlights of the succession are the abundance of *trilobites in the Cambrian period*, the appearance of *vertebrates in the Silurian period*, the beginning of the *invasion of the lands by plants in the Silurian period*, the spread of *forests*, appearance of *insects, abundance of gill-breathing fish*, and development of *amphibians in the Devonian period*, the spread of *deltaic swamp forests* and evolution of *reptiles in the Carboniferous period* and the appearance of the *first dinosaurs* and of *ammonites in the Permian period*. This period *ended with massive extinctions*, both of animals and of plants.

During the *Triassic, Jurassic and Cretaceous periods* animal life on *land* was *dominated by the dinosaurs*, whose *sea-going reptile relatives* also *flourished. One pelecypod line invaded a new ecological niche* in the *Triassic* period *by becoming adapted* for burrowing in soft sediment. *Gliding reptiles* and *true birds* evolved in the *Jurassic* period, when *progression from reptiles to mammals was completed.*

Ammonites and belemnites evolved rapidly in *Triassic* and *Jurassic* times, *sea urchins* in *Cretaceous* times. Great changes occurred in plant life. The *marine plant biomass was reorganised* by the *elaboration of the phytoplankton* in the *Jurassic–Cretaceous* periods, while *flowering land plants* appeared in the *Cretaceous*, which also recorded the *first primates. Cretaceous times ended* with another series of *massive extinctions*, including those of the *dinosaurs*, the great *marine reptiles, ammonites* and *belemnites.*

Mammals dominate the Tertiary record. The *evolution and spread of grasses opened whole new ecosystems* to grazers and their predators. Numbers of evolutionary lines show *tendencies to gigantism.*

Analysis of *sedimentary rocks* permits past *depositional environments* to be reconstructed. Where *fossil assemblages* are known, then *past ecosystems* can be reconstructed. *Comparison of shallow marine ecosystems, separated* by seventy million years in time, *reveals striking similarities* of function, despite complete turnover of species and in many cases of genus.

Our own ancestors of twenty million years ago lived in a spatially *complex environment* of *forest, scrub, grassland* and swamp, which in all probability *encouraged* complex *adaptive evolution*. The *means of manipulating the environment* in a large way *began* to develop *with Homo habilis* within the interval from 2.5 to 1.5 million years ago; this species *made stone tools*. It was *superseded· by, or evolved into, Homo erectus*, who *domesticated fire* and was therefore able to *spread widely* through Africa, Europe and Asia by half a million years ago, invading ecosystems that had formerly been hostile or even impenetrable. *About 100 000 years ago* the ancestral line split into *Homo sapiens sapiens* and *Homo sapiens neandertalensis*, the latter possibly resulting from physical and biological isolation. In Europe the *Neandertals were suddenly replaced by our own species* about 37 000 years ago.

Information about the *climatic attributes of past ecosystems* can in some contexts be obtained. *Oxygen–isotope analysis* provides *paleotemperatures* for sea surfaces or for shallow sea beds. *Pollen analysis* permits *past floras* to be reconstructed; for the past 16 000 years in western Europe, floristic change can be used to reconstruct temperature change (Chapter 19). *Leaf margin analysis* is applied to the fossil record in the reconstruction of *past forest types*, which can be interpreted in terms of *mean annual temperature.*

Increasing species *diversity* in a given genus, or of genus diversity in a given family, *means increased ecological adaptation*, as is well illustrated by the Tertiary evolution of flesh-eating mammals. *Evolution*, including evolutionary divergence, of any line of descent involves a *statistical change in the gene pool, brought about by* the *mutation* of genes and of gene complexes. *Ill-adapted mutant forms are eliminated by natural selection.*

Evolutionary trees can be simulated by random process. Increase in species number in given time brackets, evolutionary change in a given line, division of a line into two lines of descendants and extinction of lines can all be handled by card-flipping. Where single characteristics are to be treated, such as size and shape, random instructions are provided by a duly programmed computer.

23 Humans as Ecosystems Controllers

All species other than our own are compelled to adapt to the ecosystems which contain them. It is true that many behavioural adaptations and ecosystem effects on the part of other animals invite direct comparison with the behaviour of humans. Burrowing creatures excavate their dwellings. Among nest-builders, weaver birds can sew. Some species, in constructing habitations for themselves, unwittingly construct habitats for others. A beaver dam, built solely for the purposes of beavers, greatly modifies the whole ecosystem of the channel reach next upstream. Ventilation chimneys in a large termite mound in Africa are likely to be occupied by jackals. Complete transformation of large ecosystems by species other than our own is, however, never deliberate. If it does occur, as when a plague of caterpillars devastates an entire forest, it is likely to destroy the food supply.

In time, however, the forest will recover. Not so, or not so easily, the land, rivers, lakes, oceans and air into which harmful wastes from towns and factories are dumped. But environmental pollution, outside the scope of this chapter and of the book itself, merely involves destructive inputs. Ecologically speaking, it is a far simpler matter than the transformation of entire ecosystems with the object of increasing food supply for the human species.

Agriculture in its modern form aims, with partial success, at replacing wild ecosystems by control systems with deliberately selected, and few, components. Commercial–industrial forestry works in the same way. This chapter will eventually deal with the ecological implications that are involved, including those of world population and world food supply. But in order to understand how our species has arrived at its present situation, we need to review the development of our ancestors, in their roles of ecosystem components, in more detail than was offered in Chapter 22. In the agricultural context, the present situation is partly summarised by the statement that, in less than ten thousand years, one-third of the world's land area has been brought into farms.

23.1 Pre-agricultural adaptations

Some two million years of human prehistory belong to Paleolithic (Old Stone Age) time. During the whole of this interval our ancestors were hunters and gatherers. Gathering of fruit, berries, nuts and roots can be inferred from their rather unspecialised teeth. Hunting is indicated by the basic kit of stone implements, which, including choppers, pounders, and scrapers, was suited to the butchering of large game. Contrary to what is often imagined, our Paleolithic ancestors may have fed better than half the world does today. At the same time, their life expectancy was low – possibly an even chance of survival until age 20, and a 90% probability of death by 40. One factor in life expectancy may have been brain size. Among mammals that, like ourselves, maintain constant body temperatures, there is a close relationship between size of brain and capacity to live a long time.

Two million years is equivalent to about 100 000 generations. The toolkit of stone underwent very little change in all this time, either in range or in technique of manufacture. We are bound to conclude that way of life also changed very little, except for behavioural responses to unfamiliar ecosystems, as these were successively penetrated. The highly conservative nature of Paleolithic technology is all the more surprising, in view of the supersession of *Homo habilis* by *Homo erectus*, the associated increase in brain size, and the domestication of fire by the latter species.

Our remote ancestors were spread thinly. In good country, they needed perhaps 250 ha of territory a head, or 5000 ha for a group of twenty. The 125 000 direct ancestors who may have existed two million years ago could have been contained in as little as 1000×1000 km of eastern Africa. In colder or drier regions, more territory would have been needed – say, ten times as much, or 50 000 ha for a group of twenty. An entire continent could have supported a few million Paleolithic people – but only if these were able to penetrate, and to adapt to, every

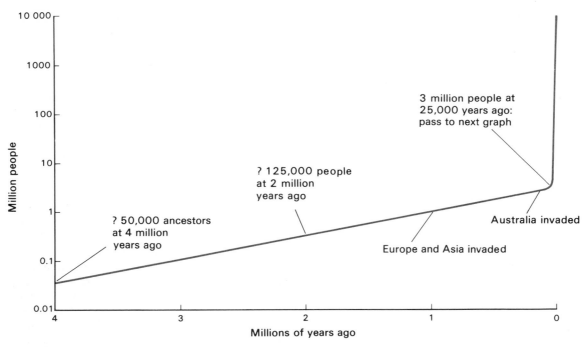

Fig. 23.1a Possible world total of humans during four million years.

regional ecosystem, and to overcome the competition of their non-human enemies.

It seems unlikely that our hunting-and-gathering ancestors made any great environmental effect in the areas which they inhabited; and whole continents, or large parts of continents, long remained unoccupied by humans. The unchanging technology strongly suggests no basic change in the fundamental relationships between our ancestors and other ecosystem components. Any group that outran its food supply must have perished. Some groups, doubtless, did exactly this. But the long-term balance was slightly in favour of the genus *Homo*. The total world population of hunters and gatherers is estimated to have risen from about one million, a million years ago, to three million at 25 000 years ago (Fig. 23.1a). Such an increase reduces itself to about 2% a century, with a doubling time of half a million years. It is difficult to recognise population pressure – that is, intraspecies competition – as the factor responsible for the invasion of Europe and Asia, from about 1 000 000 B.P. onwards, and of Australia, from somewhat earlier than 25 000 B.P. Rather, it seems that ecological niches empty of humans were attractive simply because they were empty.

In the interval between about 25 000 and 15 000 B.P., world population is estimated to have risen from three to five million (Fig. 23.1b). Doubling time had come down to 15 000 years. The change in the rate of increase probably resulted from a combination of causes, including the emergence of our own subspecies. This, *Homo sapiens sapiens*, with a brain larger than earlier human brains, revolutio-

nised the technique of tool-making, producing better, more varied, and more specialised implements than had hitherto been known. First appearing about 100 000 B.P., our subspecies took over Europe from the neandertals within 65 000 years, spread throughout Australia within the next few thousand years, and not long after 15 000 B.P. entered the Americas by the Bering Strait route, spreading within a few thousand years thereafter throughout both of the American continents. The pace of radiation far exceeded anything previously known. Radiation was necessarily accompanied by the effective behavioural adaptation: our subspecies lodged itself, adjusted, survived, and increased, in environments as diverse as the Australian desert and the Arctic tundra. Some structural or functional adaptations – by no means easy to distinguish from one another – also occurred. These are recognisable as bodily characteristics, such for example as skin pigmentation (or lack of it), amount and distribution of body fat, and nose design. Genetic drift and natural selection can be looked on as responsible for these. In the present context, they are chiefly important as demonstrations that our pioneering ancestors were still adapting to their containing ecosystems. Attempts at control had yet to come.

23.2 The beginnings of agriculture

Neolithic (New Stone Age) times brought in the use of tools and weapons of polished stone. They saw the development of deliberate herding, of cultivation, and of food storage. They also saw the establishment of the first fixed settlements.

The dating of the changes varies from one part of the world to another. Australia escaped altogether the technological evolution from paleolithic to neolithic culture, whereas the latter became firmly established on the neighbouring New Guinea. Similarly, the timing of the development of agriculture, and of the building of the first cities, ranges widely according to the area concerned. The estimated world total of fifteen million people at 6000 B.P. (Fig. 23.1b) is meant to be fairly representative of full Neolithic times and of the beginnings of civilisation.

There can be little doubt that the first beginnings of farming were accidental. Domestication of animals may well have begun with the keeping of captured young as pets. Very young lambs, kids, calves and piglets will adapt happily to life with humans, even though the responses to the earliest domesticators may have been less amiable than those observable today. Some fledgling birds, including goslings and chickens, will fixate on humans as surrogate parents.

Collection, storing, and planting of seeds appear likely to have begun with the observation that grain, accidentally spilled on disturbed soil, would grow. Seed-grinding, and therefore seed collection, is well documented for the paleolithic culture of the Australian aborigines; but it was left to our neolithic forebears to save seed, to plant it, to harvest the crop and to store part of the harvest. Storage is highly important, for it signifies that production can exceed immediate demand.

Much effort has been expended in the search for the original home of agriculture. That effort now appears to have been misdirected. The strong probability is that, wholly or largely free from the influence of cultural diffusion, deliberate cultivation developed independently in Hither Asia, in Egypt, in SE Asia in the area that is now northern Thailand, in Middle America, and possibly also in the Indus basin in the Indian subcontinent and in the Huangho basin of northern China. Grain farming, with or without work animals and meat animals, evolved in all of these areas.

We can easily distinguish an ecosystem relationship for these agricultural beginnings. Few, if any, of the originally cultivated plants and originally domesticated animals seem to have been derived from simple natural ecosystems. Neither the reindeer of the boreal tundra nor the horse of subhumid midlatitude grasslands has provided the basis for settled farming in its native region. The most favourable areas for the early development of agriculture appear to have resembled those where our human ancestors first evolved – that is, ecosystem boundary zones or steep ecosystem gradients on the foothills and flanks of mountain belts. In such places there could be expected high rates of productivity, marked seasonal changes in the availability of food, and marked spatial variation in the availability of particular foodstuffs.

23.2a The special case of the humid tropics

In contrast to the grain-dependence of midlatitude farming, the farming systems of the humid tropics – aside from wet-rice cultivation – depend basically on the raising of root and stem crops such as yams and manioc. They also depend on shifting cultivation, known as slash-and-burn, or swidden, agriculture. Marked variations occur from region to region. But the central practice is that of clearing part of the forest floor, burning the felled plant debris, and planting a diversity of crops in the burnt-over patch. In as little as a year, and at most within a few years, soil exhaustion sets in and the forest reclaims the cropland. The farmers move to another patch. Inter-

Fig. 23.1b Possible world total of humans during 25 000 years.

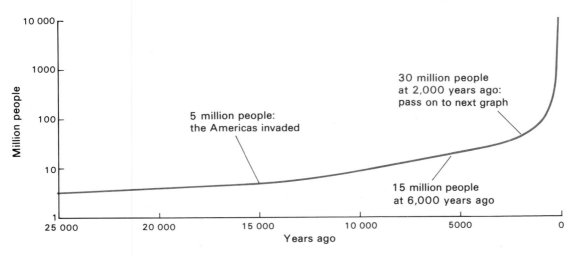

vals between successive croppings on a given patch can be as long as twenty-five years. It is for this reason that many communities of swidden cultivators construct new huts at every planting site.

Against the fact that swidden cultivation can maintain few people per unit area, and the additional fact that it produces starches, leaving proteins to come from game animals and fish, must be set the circumstance that it is ecologically stable. It probably originated in humid–tropical climates with alternating wet and dry seasons. Although it has spread to some extent into regions of rainforest climate, it is quite unsuited to extension beyond the humid tropics.

Grain cultivation, which does supply protein in the diet, frees its farmers from heavy dependence on game and fish, and makes possible the invasion of ecosystems other than those where it first developed. Not surprisingly, it has affected swidden cultivation to some extent, adding grains to root and stem crops. By demonstration, it has massively invaded the humid to subhumid mid-latitudes.

23.2b Choice or necessity?

Attempts to explain the origins of agriculture in terms of response to declining rainfall, and with especial reference to ancient Egypt, have been summarised and rejected in Chapter 19 (section 19.5). When the only full and reliable archaeological record was that of Egypt and Hither Asia, a climatic interpretation remained possible. With additional information, we can regard the early farming process as efforts – and, in the main, successful efforts – at ecological adaptation. We can also see how the development of agriculture involves a shift from the role of predator to that of ecosystem manager.

If, in the pre-urban stage of human culture, farming could support ten times as many people as could hunting and gathering, then 25 ha per head would have sufficed. There are, however, two important qualifications. Not all terrain that is suited to hunting and gathering is suited to agriculture. Much is too cold, too dry, too steeply sloping, too poorly endowed with soil, or too thickly covered with unconquerable forest to be put to crops. Fixed agricultural settlements probably offered strong temptations of pillage and slaughter to the hunting bands of uncultivated areas. History, when it begins, certainly shows that urban cultures, from the China Sea to the Black Sea and the Caspian, were liable to attack and overthrow by organised nomads. There is no reason to suppose that things were different in prehistoric times. Intracultural strife, then as later, seems likely to have been based to some extent on contrasting modes of ecological adaptation.

23.3 Early control systems with large agricultural surpluses

Between 6000 and 2000 B.P., world population is estimated to have doubled, rising from 15 to 30 million (Fig. 23.1b). Doubling time has come down to 4000 years. Part of the increase was presumably due to the random upward drift of the total. Part was certainly due to the development of controlled irrigation, as practised in Egypt, Mesopotamia, the Indus basin, and northern China. Another part, which cannot be wholly disentangled from the increase made possible by irrigation, reflects the rise of cities, with all that this implies in respect of city life and activity, specialisation, technology, culture, economics, and trade.

Our own city-centred culture inclines us to ignore the cultural achievements of pre-urban times. Politics and government, understood as involving social duties and taboos, we share with all other social forms of life – among which bees, wasps, termites and wolves, to name only a few, are genetically programmed for advantageous social behaviour. Burial of the human dead, surely indicating some kind of religious belief, goes back at least 70 000 years. The magnificent cave art and accompanying sculpture of western Europe dates from not long after 35 000 B.P. Speech, ritual, music, song, dancing, and religion all date from before the times of the first fixed settlements. These settlements, made possible by farming, gave rise to such accomplishments as potting: specialisation, inevitably implying subdivision of labour, had set in.

Specialisation was to go fast and far when cities grew up. It notably included writing, literature, building in masonry, and the working of metals, all of which eventually led to modern science and technology. None of this would have been possible if cities could not have been fed from outside.

Despite the close attention focused on the cities of the hydraulic civilisations, it would be overly simple to conclude that irrigation agriculture was alone capable of producing the necessary agricultural surplus. Cities independent of irrigation farming arose in Middle America, and what may be the oldest of all, the recognisable city that existed on the site of Jericho as early as 10 000 B.P., compares poorly in its regional setting with the early Egyptian and Mesopotamian cities, surrounded as these were by extensive alluvial plains. We can only infer that the transition from a control system managed from farming villages, to a much more complex system managed by cities, was likely to happen. The great ecological advantage is that, like a tropical rainforest, a city offers a multitude of niches. But the larger the city, the greater the agricultural surplus on which it needs to draw.

This fact appears to have been well understood in the early days of urbanisation. There is good archaeological evidence, in the form of abundant crude jars, that manual workers in the early Mesopotamian cities were paid in grain. Rationing in times of flood and famine is well documented from ancient Egypt. City managers and regional managers were completely aware that their cities and regions were integrated with supply areas in unified control systems. A factor in the expansion of the ancient Roman empire was the need to command distant supplies of grain.

23.3a Agricultural surpluses today

Nothing fundamental in the city–farm relationship has changed during the last two thousand years. Farming has tended to become progressively more intensive, both in subsistence systems and in commercial systems. Intensive rice cultivation can feed one person on as little as 0.1 ha of land. Intensive grain and livestock farming, as practised in much of the U.S.A., can feed one person from as little as 0.5 ha. An additional 2 ha is needed for dwellings, factories, roads, railroads, airports and the like; but even so, highly developed commercial agriculture can sustain population densities a hundred times as great as those possible, even in the most favourable circumstances, with hunting and gathering. Because agriculture was the chief life support of 1000 million people in 1850, and of 4000 million in 1975, and because it is the prospective chief support of the 8000 million expected for the year 2000 (Fig. 23.1c), we might well incline to accept its history as a tale of fabulous success, even despite the fact that nutritional levels run low in much of today's world. But when we pause to consider the ecological characteristics of farming, a very different view appears.

Fig. 23.1c Likely world total of humans during 2 000 years.

23.4 Agriculture as ecologically regressive

A well-known tabulation of ecosystem attributes contrasts the quality or status of these attributes in developing ecosystems to their quality or status in mature systems. In agricultural systems, about three in four of the attribute conditions are appropriate to development rather than to mature stages. For control systems to be maintained and operated, it is necessary to arrest development.

23.4a Simplification

Not only does agriculture replace wild vegetation by deliberately selected components: it also aims to keep the number of components down. In some of its most successful manifestations – in the commercial sense – agriculture is a one-crop operation (Fig. 23.2). Similarly with commercial forestry: a plantation frequently consists of a single tree species and of trees of a single age (Fig. 23.3). Nothing less like the diversity of wild mature ecosystems can be imagined.

In ecosystem terms, cultivated lands are low in respect of species diversity, biochemical diversity, and diversity of pattern. Here are some leading marks of ecological immaturity. With lack of diversity go broad niche specialisations. A wheatfield is intended to provide a single floristic niche, or two niches if clover or lucerne are undersown. A wild ecosystem, even at field size, would provide multiple niches for the wild plants that the wheat-grower identifies as weeds.

Life cycles are simplified in agricultural systems, and in the main are also foreshortened by comparison with the cycles of wild ecosystems. The life cycles of cultivated plants, except for tree crops, are mainly annual and at the most biennial. Even tree crops will be felled for timber, long before the trees'

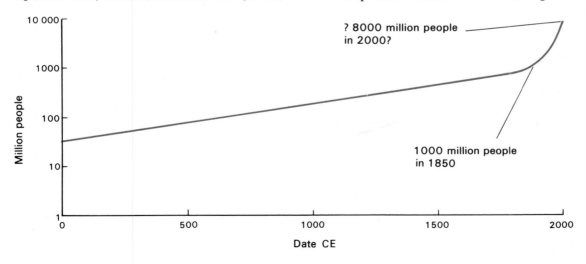

Fig. 23.2 (left) Wheat in southern Saskatchewan (Canadian Government Travel Bureau).

Fig. 23.3 (below left) Commercial forest: part of the Thetford district, East Anglia (Aerofilms). All the trees in a given block are a single type of conifer, all planted as seedlings of a single age.

life cycles have been completed, or in the case of fruit trees will be grubbed up as soon as signs of age set in. The fifty-year term of commercial forests and the twenty-year term of commercial orchards are in direct contrast to the centuries required for the cycling of trees in the wild.

In addition, agriculture as a destroyer of natural habitats shares the responsibility for the extinction of animal species. As Fig. 23.4 shows, the rate of extinction of mammal species has risen during the last five hundred years, from about twenty species per century to about two hundred. Extinction by direct means includes extinction by hunting, an increasing menace in our own day to whales and to numbers of non-mammalian fish species. Extinction by indirect means includes extinction by destruction of habitats; and this means of extinction, for mammals, has been by far the most important since about 1750. Like the suppression of plant species, the extinction of animal species acts to reverse the development of natural ecosystems toward increased diversity.

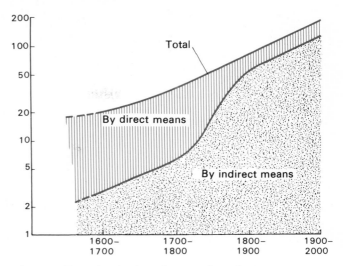

Fig. 23.4 Rate of extinction of mammal species.

23.4b Nutrient cycling

In mature wild ecosystems, nutrient cycles are largely to entirely closed. Nutrients are recycled through a system – as, notably, through tropical forest ecosystems. In such systems, the detritus component of the system is of great importance, as is internal symbiosis, the effective co-operation of one kind of organism with another. Resistance to perturbation from outside in a complex wild ecosystem is high, partly because of the very complexity and partly because of the internal cycling of nutrients.

Agricultural systems work in directly opposite ways. Nutrient cycles, including mineral cycles, are open. Nutrient conservation is poor. Unless fertilisers are supplied from outside the system, decline, and possibly breakdown, will ensue.

There are catches here. The Chinese practice of spreading human excrement on farmland is chemically sound, but has dire bacteriological implications. Spreading of farmyard manure or of deactivated sewage sludge on the farmland of western countries restores only part of what is taken out; it is also expensive. Composting, beloved of many western gardeners, is far more beneficial to soil texture than to soil nutrient supply. Commercial agriculture depends heavily on fertilisation with industrial chem-

icals. Inorganic nutrients are supplied from outside the system – again a sign of ecosystem immaturity. Furthermore, agriculture is designed on the principle of the food chain, producing grains and vegetables for humans, or grains and grasses for livestock and so meat and milk for humans, as opposed to the food web principle that applies to large natural ecosystems.

23.4c Production

In mature ecosystems, total production and total loss tend to equal one another. In the early stages of ecosystem development, the production/loss ratio is typically far from unity. If production exceeds loss, biomass accumulates within the system, but an increasing loss rate eventually leads to the steady-state situation where the two attributes balance out. Agricultural systems are designed to maintain high production rates, but also to deliver large outputs of biomass. The ratio between biomass supported and energy flow through the system is typically high for mature ecosystems and low for immature systems. Although the energy demands of agricultural systems vary widely, for instance with crop type and with climate, modern commercial agriculture records a very low ratio of biomass to energy flow.

23.4d Energy demands of agricultural systems

Only when ecosystem theory came to include the analysis of energy inputs and energy flows did it become possible to examine the use of energy in agricultural systems. It has been obvious for some time that total energy use varies widely from country to country, both on a national and on a per capita

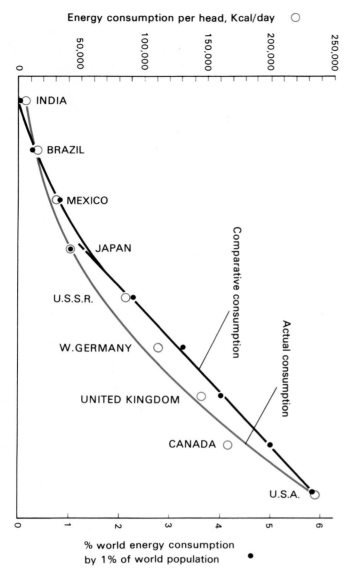

Fig. 23.5 Rates of energy consumption for selected nations: comparative consumption reckoned as per cent of world consumption by 1% of world's people.

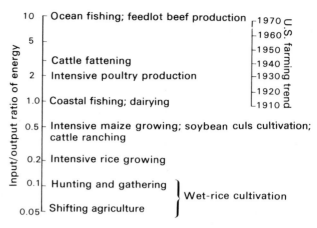

Fig. 23.6 Ratio of energy input to energy output for selected food-producing activities: very freely adapted from a diagram by J.S. Steinhart and C.E. Steinhart.

basis. For the nine countries graphed in Fig. 23.5 – which include the world's leading users of energy – there is not very much difference between the actual consumption, which makes no allowance for size of population, and the comparative consumption, which does allow for size of population. For the most part we are once again faced with a semilogarithmic array. Up the range from Japan, through the U.S.S.R., West Germany, the U.K. and Canada to the U.S.A., the rate of consumption of energy increases geometrically with rank order. The basis of comparison used in scaling the diagram is the amount of consumption by 1% of the world's people; use of this criterion compensates for differing sizes of national populations. It is tempting to assume that the relationships illustrated are wholly due to industrial and domestic use of energy. This,

however, is not necessarily so. The results of analysis of the use of energy in agriculture are highly disturbing.

Commercial agriculture is most highly developed in the U.S.A., where it is capable of producing huge national surpluses of food and of feed grains. But these surpluses are made possible only with the aid of huge subsidies of energy from outside the agricultural system. Parts of the subsidies come from solar radiation; but other parts come from the expenditure of the energy stored in fossil fuels.

Fig. 23.6 supplies the basis of discussion. With the proviso that inputs of solar energy are taken as free, swidden agriculture comes off very well, in a comparison of energy input with food energy output. One Kcal (= kilocalorie) of muscular effort yields 20 Kcal of food energy for swidden cultivators. A hunting-and-gathering economy is about half as energy-efficient as swidden cultivation, partly because it can range into climates where energy is needed for keeping warm, and partly because energy must be freely expended in the actual chase. Nevertheless, in input/output terms it still does well, as also does wet-rice cultivation. Swidden agriculture, hunting-and-gathering, and wet-rice cultivation are all energy-efficient enough to support dependants of the farmers and hunters – but, needless to say, only on subsistence terms.

Intensive rice growing on the Japanese model perhaps breaks even, when farmers' dependants are allowed for. The 1:5 ratio of input to output is enough to provide for aged parents and for children too young to work in the fields. Beyond this the picture darkens. The intensive growing of maize, the cultivation of soybeans, and the raising of ranch cattle demand 0.5 Kcal of energy input for every 1 Kcal of energy output. If a cultivator or rancher has more than one other family member who contributes nothing to the process of food production, then energy subsidies become necessary. The situation with small grains, notably wheat, is somewhat worse still.

Coastal fishing and dairying achieve an approximate balance between energy output and energy input. The production of broiler fowl and the fattening of cattle require twice to three times as much fuel energy as they produce in food energy. Feedlot beef production is still more energy-intensive, with input ten times as much as output. Ocean fishing comes at this same place on the scale, chiefly because so much of it is carried on in distant waters. The very heavy energy subsidy needed by distant fishing shows that exploitation of the antarctic Krill is no simple solution to the world's food supply problem.

The data presented for energy subsidies to agriculture in the U.S.A. from 1910 onwards reveal the irony of mechanisation. Planting and harvesting machinery go back to ancient times, respectively in Mesopotamia and in Rome, but in their modern form descend from the horse-drawn drills and hoes developed from the eighteenth century onward. Steam-powered tractors of the nineteenth century were replaced by petroleum-powered machines in the twentieth century, which also produced the self-propelled combine grain harvester. So far has mechanisation gone, that 1 ha of corn can be planted, grown, and harvested with the expenditure of 15 man-hours of labour. The social benefits are the feeding of cities, and the freeing of farm workers for other employment; the paid farm workforce in the U.S.A. fell from 20% of the country's population in the late 1800s to 3% in 1970. But, as Fig. 23.6 shows, these results have been possible only with the aid of increasingly heavy subsidies of energy. Shortly after 1910, agriculture in the U.S. no longer delivered more energy as food than it consumed as fuel. The subsidy has increased at a geometric rate, until the national agricultural system is now as energy-inefficient as distant fishing, and 200 times less energy-efficient than shifting cultivation.

23.5 The struggle against unwanted components

Weeds, insects, and plant diseases are estimated to destroy 20% of the crops grown on farms in the U.S.A. This figure is probably fairly representative of modern commercial agriculture in general. Precisely because of its ecological simplicity, farmland is under constant threat of competition, parasitism, and predation. The threat takes two forms, the continuing tendency for the control systems to evolve toward ecological complexity, and sudden calamities such as plagues of locusts or the disastrous spread of blight.

23.5a Weed control

Not only do weeds compete with crops and grasses for space, light, and nutrients; they also provide storage systems for destructive viruses, bacteria, and insects. Some are actually poisonous to farm livestock. The prospects of successful control of weeds in grainfields go back no more than 300 years. They depended to begin with on the conversion from broadcast sowing to drilling in rows, and the use of the horse-drawn hoe already mentioned. The weeding implements of the early eighteenth century have subsequently been improved almost out of recognition; but mechanical weeding is falling into some disfavour because it tends to damage the shallow portions of the root systems of crops.

Chemical weed killers have been in quite widespread use for about a hundred years. They provide one example of the exchange flow between industry and farmland – food from the country to industrial cities, synthetic fertilisers and pesticides back from industry to the farms. The subscience of agricultural chemistry was established in the early 1800s, a leading initial concern being fertilisation. As early as 1820, Europe began to import Peruvian guano; this, aged seabird dung, was an effective form of nitrate. But the nineteenth century rate of exploitation was so much greater than the rate of accumulation that guano was reduced, in effect, to the status of a non-renewable resource. Industrial chemistry came to the rescue, producing synthetic fertilisers from the 1840s onward. During the late 1800s it was discovered that broad-leaved weeds in grainfields could be killed by sprays, the solutions used being those of sulfates and nitrates of copper and iron. Already before 1900, organic chemicals were recognised as potential herbicides.

Chemical weed control advanced rapidly during the 1940s, especially with the introduction of the substance 2, 4–D. This and generally related toxins, highly effective and needing application only in very low doses, have encouraged some enthusiasts to forecast completely weed-free farming. Such farming is certainly within the range of theoretical possibility.

23.5b Pest control

Poisons have been in use against pests at least since the eighteenth century. One noteworthy early success was the large-scale application of sulfur and sulfur compounds to European vines afflicted by mildew. Coming very shortly after the establishment of the synthetic fertiliser industry, this event raised high hopes of major contributions of chemistry to the elimination of pests. Some of the substances that came into use are truly alarming in nature – compounds of arsenic, used for poisoning rodents (and also in weedkillers); mustard gas, used to fumigate glasshouses; and potassium cyanide, employed in the destruction of the European wasp. Like weed control, pest control eventually became increasingly dependent on organic substances, and, within this range, on substances originally synthesised in the laboratory and subsequently manufactured in the factory.

Again like weed control, pest control advanced rapidly during the 1940s – or appeared to do so. Almost seventy years after it had first been synthesised, the insect-killing properties of the substance DDT were recognised. The full chemical name, dichlorodiphenyltrichloroethane, hints at its chemical complexity. Here seemed to be a miracle applicant. Additional chlorinated compounds followed, soon to be accompanied by organic phosphorus compounds. Use of these, and above all of DDT itself, so reduced pest damage that, in the twenty years 1945–65, crop yields in large areas of the world rose by 50%.

Already by 1965, however, national governments had begun to ban the use of DDT and similar poisons. These proved to be scarcely at all biodegradable. Once they are injected into an ecosystem, they work their way along food chains and through food webs, becoming increasingly concentrated at increasingly higher trophic levels. Stored in body fats, DDT can cause birds to suffer reduced breeding rates, to lay fragile eggs, or even to die. Commercial poultry farmers have suffered loss of whole flocks, when the feeding rate is seasonally reduced, and when the birds draw on their body fats for energy. DDT has worked its way through all the world's ecosystems, from arctic seals to Hawaiian birds and to antarctic penguins. Your own body fat contains DDT. Whether it is dangerous to humans remains undecided.

Quite aside from passive contamination of ecosystems, the use of pesticides is apt to prove self-defeating. The risks are of two kinds. The poisons may kill off desirable species in addition to pests, as for example when DDT came into use during the 1940s in the citrus orchards of California. It destroyed the vedalia beetle – a wholly unintended target – permitting an explosive outburst of citrus blight, which previously the beetle had kept within bounds. Then, in the second place, attack on a particular pest may leave alive a few genetically resistant individuals. These can multiply into a population on which normal pesticide has no effect. If competitors and predators have been killed off, the inevitable result is a destructive wave of attack. On some U.S. farmlands, pesticides are now being applied at twenty-five times the original rate. A

particular case of genetic resistance is that of some rats to warfarin. This substance, an anti-coagulant useful in the treatment of arterial disease, will give hemophilia to nearly all of the rats that eat it. They bleed to death. But the few survivors then have the rat niche to themselves, rapidly filling it in the absence of competition within the species, and probably with the aid of an increased rate of reproduction.

Unless a given pest can be wiped out entirely, a resistant population can be expected to develop. Such appears to be the normal situation. Ecologically, but not agriculturally, speaking, the situation appears wholly sound. The steady escalation of attack and resistance recalls the parallel evolution of prey and predators. The total prospect is, surely, that insect pests will not be obliterated by chemical means.

23.5c Biological control

Deliberate introduction of predators and parasites has a mixed history, both of success and of esteem. The so-called common sparrow, now likely to be reclassified as a species of finch, was introduced from Europe into the northeastern U.S.A. during the 1850s. The purpose of introduction was to control inchworms, caterpillars that, at times of population peaks, can denude trees entirely of their leaves. The sparrow proved highly successful in checking inchworm damage, but also proved highly adaptable in their choice of food. Spreading widely across the North American continent, they concentrated on standing grain for the vegetable fraction of their diet; and it turned out that the insect fraction decreased sharply at times other than those of breeding. Here is a leading case of a non-specific predator.

If specific predators can be identified, then effective control can be achieved. Such was the case with the citrus blight of California. The predator selected was the vedalia beetle, imported from Australia in the 1880s. The great distance between Australia and California points up the likely difficulty in the identification of an appropriate predator; and even when a predator or parasite has been identified, it may be difficult or even impossible to establish in the desired area. Even when established, it may be overtaken by disaster, as the vedalia beetle was when DDT came into use. The odds on successful introduction, even in the best circumstances, are no better than 25%.

The unfortunate results of many deliberate introductions, and of many escapes, combined in the late 1800s and the early 1900s with the increasing attraction of chemicals to divert attention from the general possibilities of biological control. But biological possibilities have by no means been fully explored. At least in some cases, the unwanted insects can themselves be manipulated, being sterilised by chemicals or by radiation. The most noteworthy agricultural success to date is probably the control of the screwworm, a cattle parasite very troublesome in the U.S. South. Sterilised insects, released in large numbers into the breeding stock, caused a dramatic and highly profitable drop in screwworm parasitism.

Reduction of the breeding rate in this manner could in time be widely achieved. It eliminates the need to search for specific parasites or predators, which must inevitably come from a ecosystem so distant, and so different, that the treated system has no defences. It is far more promising mathematically than is the development of chemical pesticides, where the probability that an effective compound will eventually be adopted for commercial use is 300 to 1 against. It bypasses the risk of genetic resistance to poisons and the subsequent emergence of a large and totally resistant population. It could potentially be used in all situations where a particular insect pest can be singled out. Here we may include carriers of disease, such as the tsetse flies which exclude domestic cattle from huge areas of Africa, the fleas that carry plague, and the mosquitoes that carry malaria. On the other hand, the total of insect species that are undesirable is about 10 000; and of these, some 500 are so highly destructive as to demand control in some measure. Control by sterilisation would mean attack species by species, and ecosystem by ecosystem.

23.6 Control of agricultural systems by governments

Efforts by national governments to control agricultural systems go back at least 5000 years, to the national system of irrigation agriculture in ancient Egypt. In modern historical times, such efforts go back to the 1400s. They differ from those of ancient Egypt in being applied to commercial agriculture in a money economy, and in being designed to control crop output, crop prices and farm income. None has been really successful. Less successful still have been efforts at international control. But all measures intended to influence crop output, crop prices, and farm incomes impinge both on national agricultural systems and on the practice of individual farms. Modern commercial agriculture, therefore, is engaged in the control of its own simplified and immature ecosystems, while at the same time being obliged to respond to external processes.

The central difficulty is that, in the absence of control, farm prices fluctuate more widely than do other prices, and that farm incomes fluctuate more widely still. Within fairly narrow limits, farm costs in commercial agriculture are fixed. Prices for farm products are notoriously variable, largely because production and demand are themselves variable. Production of the world's chief food crops varies significantly with the weather of growing seasons. The extreme possibilities are that all major producing areas, say of wheat, will have a run of simultaneously good seasons, or that they will all have a run of simultaneously bad seasons. The greatest probability is that, in a given year, the growing season will be good in some producing areas and bad in others; but there is still plenty of room for a distinct balance toward high or low total production. With production and demand varying, at times wildly, and with farm prices varying in consequence, fixed costs ensure great variation of farm incomes.

23.7 Ecological dilemmas

It is ironical that government measures intended to stabilise farm incomes include limitations on cropping, for world food supply runs persistently below world need. There are presumably limits to what can be done by way of breeding farm plants and farm livestock. There are certainly limits on the extent of land that could conceivably be brought into cultivation or range; and the limits on the exploitation of the oceans have already in some respects been exceeded. Industrial pollution of the environment can be checked: whether or not it will be is another matter altogether. The central dilemma remains that of food supply for the human species; and this dilemma reduces itself in large part to the dilemma of species population, quite independently of the dilemma of finite mineral resources.

Models for the past and future total of the world's human population are contained in Fig. 21.9. The net increase in numbers since the appearance of our first recognisable human ancestors is represented by the rising random walk in Fig. 21.9a. To the low-fluctuating model in Fig. 21.9c corresponds the situation in western Europe in the late Middle Ages, when the tendency of upward drift was held back by recurrent famine and plague. Options for the future seem to include only the situations portrayed in Figs. 21.9b and 21.9f – that is, the conditions of stability imposed by environmental resistance, and drastic reduction of numbers from a sharp peak as a result of scrambling competition.

That group of writers known popularly as the doomsayers prefer the scrambling-competition model of Fig. 21.9f. They predict that overpopulation will bring our species to disaster. Fig. 23.7, however, suggests that environmental resistance may already be setting in. During the last million years, the

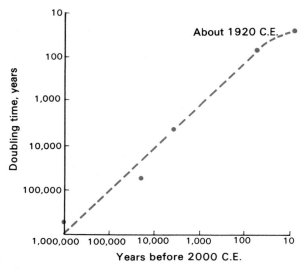

Fig. 23.7 World population: possible change in the trend of doubling time.

doubling time for the numbers of people on our planet has roughly equalled the number of years before 2000 C.E. (= Common Era: 2000 in the Christian calendar). The doubling times illustrated are those listed in the foregoing text. Despite the so-called population explosion that began in about 1850 C.E., the rate of increase in the doubling time has declined during the twentieth century. It just could possibly be that Fig. 21.9b is the more likely model for the future. Our species population may stabilise at some 8–10 000 000 000 on account of environmental resistance. But if environmental resistance is to be the controlling factor, its control will not be pleasant.

Chapter Summary

Agriculture and commercial–industrial *forestry* aim to operate *control systems with few components. They are late developments* in human history; *for two million years our ancestors were hunters and gatherers, using stone implements* to butcher large game. *Thinly spread,* they needed perhaps 250 ha of territory per head in good country. They are thought to have numbered about *one million, a million years ago.* The species *Homo habilis was replaced by Homo erectus,* and *Homo erectus by Homo sapiens.* Our own subspecies, *Homo sapiens sapiens,* first *appeared about*

100 000 years ago, taking over Europe from Homo sapiens neandertalensis about 37 000 years ago.

Our direct ancestors *revolutionised the working of stone,* producing improved and specialised tools. Perhaps *fifteen million* of them existed *by 6000 B.P.,* by which time *fixed village settlements* had come into being and *city life had started.* Fixed villages mean occupied farmland: cities demand agricultural surpluses.

Farming seems to have developed independently in more than one area using plants and animals derived from complex ecosystems and being located *originally in ecosystem boundary zones.* Its first beginnings were probably accidental. The *humid tropics developed swidden cultivation* with its slash-and-burn technique and with root and stem crops. This form of cultivation is *ecologically stable,* but does not itself produce much protein. The *grain-and-livestock farming* of middle latitudes, which does supply protein, has *widely invaded ecosystems outside* its areas of origin.

Irrigation and the widespread rise of cities contributed to the increase of world population to an estimated *thirty million in 2000 B.P.* Cities have the ecological advantage of offering *multitudes of niches,* including specialised occupations. The *agricultural systems* on which they depend, on the other hand, can be called *ecologically regressive,* being *low in diversity, simplified* as to *life cycles,* and *broad in niche specialisation.* Their *nutrient cycles* are *open,* their *production rates* and *biomass outputs* are *high.* The *simplification* of system components has *contributed to the extinction of animal species.* The *high energy demands* of modern commercial agriculture *require energy subsidies* in the form of fossil fuels.

Crop losses on account of *weeds, insects* and *diseases* run at *20% or more. Weed-free farming is theoretically possible* with the aid of chemical herbicides. *Insect control* by chemical means *is more difficult.* Quite apart from the entry of poisons into ecosystems, and their storage therein, it is only too *easy to kill off predators* in addition to pests, while *pests* have a great capacity to *develop resistance. Biological control* of pests, although complex, *has recorded some successes.*

Many governments attempt to control national agricultural systems, in order to damp down variations in farm incomes, but so far have had but limited success. International efforts at control have proved less successful still. Ironically, government measures often include *restrictions on cropping, even though world food supply runs below world need.* Because world *population is still increasing,* the *future options* seem to be *stability imposed by environmental resistance* and *sudden collapse* of numbers after a certain peak is reached. *Signs* that the rate of increase in doubling time is declining might *suggest that environmental resistance is already setting in,* and that our numbers may stabilise at 8–10 000 000 000.

24 Appendix I: Some Geological and Geochemical Data

This chapter traces the successive formation of the chief minerals that go to make up igneous rocks, with special reference to the reaction series. It explains the two geologically important types of bonding, whereby elements are combined into compounds. It uses the compounds that are the chief rock-forming minerals to classify igneous rocks, and the radioactive decay of certain elements to date the geologic past.

Names of minerals prominent in igneous rocks, and of types of igneous rocks, will of necessity be freely employed. It is certainly possible to comprehend, on an intellectual basis, the chemical nature of the chief rock-forming minerals, and similarly to comprehend the way in which certain minerals combine to produce certain types of rock. There can, however, be no substitute for visual and tactile experience. For preference, rock and mineral types should be examined both in hand specimen and in microscope sections. By no means all readers will have access to microscopes and microscope slides; but rocks and minerals are quite readily accessible, either in the field, or in museum or school collections, or in commercially available sets. This text deals with only a few of the hundreds of known mineral species, and with only the best-defined and commonest igneous rocks; but first-hand acquaintance with the essential kinds means at least a good start in the understanding of geology.

24.1 The reaction series

When a rock melt starts to cool, the first minerals to crystallise out are those which are stable at high temperatures. As cooling continues, new mineral species appear. The succession is called the reaction series (Fig. 24.1).

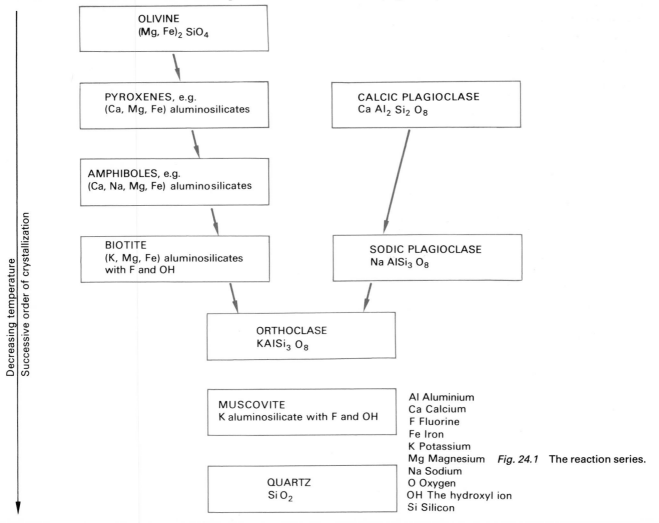

Decreasing temperature

Successive order of crystallization

OLIVINE
(Mg, Fe)$_2$ SiO$_4$

PYROXENES, e.g.
(Ca, Mg, Fe) aluminosilicates

AMPHIBOLES, e.g.
(Ca, Na, Mg, Fe) aluminosilicates

BIOTITE
(K, Mg, Fe) aluminosilicates
with F and OH

CALCIC PLAGIOCLASE
Ca Al$_2$ Si$_2$ O$_8$

SODIC PLAGIOCLASE
Na AlSi$_3$ O$_8$

ORTHOCLASE
KAlSi$_3$ O$_8$

MUSCOVITE
K aluminosilicate with F and OH

QUARTZ
Si O$_2$

Al Aluminium
Ca Calcium
F Fluorine
Fe Iron
K Potassium
Mg Magnesium
Na Sodium
O Oxygen
OH The hydroxyl ion
Si Silicon

Fig. 24.1 The reaction series.

All the minerals but olivine at one extreme and quartz at the other are aluminosilicates – compounds of metals other than aluminium with aluminium, silicon and oxygen. The first part of the sequence is double. Olivine probably comes out first of all; but, as soon as compounds including calcium appear, the plagioclase feldspars develop from calcium-rich (calcic) to sodium-rich (sodic) in parallel with the successive production of pyroxenes, amphiboles and biotite (= black mica).

Magnesium and iron are included throughout that portion of the sequence that runs from olivine to biotite. The minerals involved are, by definition, ferromagnesian minerals, often called mafic minerals for short.

The sequence from wholly calcic to wholly sodic plagioclase feldspars in a continuous one. It does, however, serve to show that calcium compounds appear before sodium compounds; orthoclase feldspars, aluminosilicates of potassium, appear later still. Similarly in the ferromagnesian sequence, calcium enters into the combinations first, sodium next and potassium last. We can assume that, by the time that orthoclase forms, the available magnesium, iron, calcium and sodium have been taken up. Muscovite (= white mica) differs from biotite mainly in its lack of magnesium and iron. Finally, surplus silicon and oxygen combine as silica, SiO_2, which freezes in the interstices among earlier-formed crystals, or is injected as liquid into the surrounding rock, where it sets solid as quartz veins.

24.1a How compounds form

Compounds are combinations of two or more elements. The substances listed at the side of Fig. 24.1, fluorine and the hydroxyl ion excluded, make up about 98.5% of the earth's crust by weight. The minerals named in the body of Fig. 24.1 make up 95% of the crust. In view of the fact that oxygen and silicon together account for nearly 75% of the crust, and that aluminium supplies another 8%, the prominence of silicate minerals, and of aluminosilicates except at the extremes, is in no way surprising. But the chemistry of the minerals actually produced entails the bonding of certain elements to certain others.

Two kinds of bonding, ionic bonding and covalent bonding, are geologically important. Both involve the behaviour and action of sub-atomic particles. Some forty such particles have been identified. For our purpose, it is enough to recognise protons, which carry positive electrical charges, electrons, which carry negative charges, and neutrons, which are uncharged. Protons and neutrons together form the atomic nucleus, round which the electrons pur-

sue orbital paths. This description seems at first to recall the solar system, but the orbital arrangements of electrons are more complex than those of the sun's planets. In the first place, the orbiting electrons sweep out spherical rather than circular (or elliptical) paths. Secondly, there is usually more than one electron to each orbital sphere. The subsets of electrons, with their spherical paths, are called electron shells.

The number of protons in the nucleus is the atomic number of the element, running from 1 for hydrogen to 92 for the heaviest naturally occurring element, uranium. A series of man-made elements has more protons than has uranium. Atomic number is written as a subscript in front of the symbol for the element – $_1H$ in the case of hydrogen and $_{92}U$ in the case of uranium. Number of protons + number of neutrons = mass number, also called atomic weight. It is written as a superscript; uranium, with 146 neutrons, has a mass number of $92+146 = 238$, symbolised as ^{238}U and read as uranium 238. If desired, the atomic number can also be shown, as in $^{238}_{92}U$ (for discussion of isotopes, see below).

In atoms, the number of protons is matched by the number of electrons, so that the atom is electrically neutral. The most stable state is attained where the outermost electron shell contains eight electrons, as with the inert gases radon, xenon, krypton, argon and neon. But sodium, with eleven protons and eleven electrons, has two electrons in its innermost shell, eight in an intermediate shell, and a lone one in the outermost shell (Fig. 24.2, upper left). Chlorine, with seventeen protons and seventeen electrons, has two electrons in its innermost shell, eight in an intermediate shell, and seven in the outermost shell, (Fig. 24.2, upper right). Here are the makings of a trade-off situation. The sodium atom releases one electron to the chlorine atom. The former loses one negative charge, the latter gains one. They are no

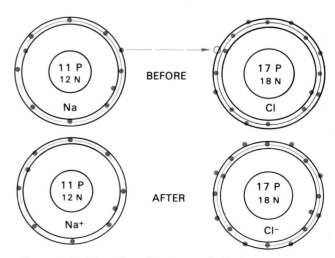

Fig. 24.2 Ionic bonding of sodium and chlorine.

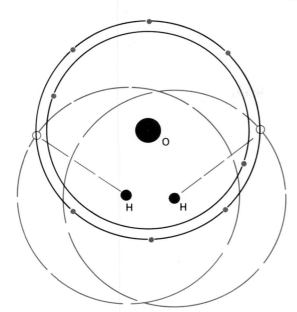

Fig. 24.3   Covalent bonding of hydrogen and oxygen.

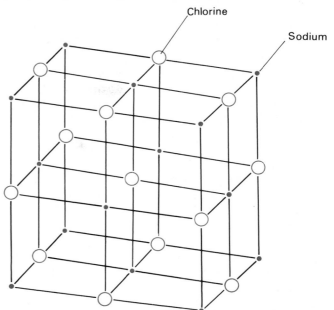

Fig. 24.4   Part of an atom lattice for common salt, NaCl: compare Fig. 24.2.

longer properly called simply atoms but ions (or charged atoms) (Fig. 24.2, lower). The opposite charges attract one another, so that the two ions are locked together by the process of ionic bonding. They form a molecule of sodium chloride, common salt – an excellent illustration of the fact that the properties of compounds can be greatly different from the properties of the elements that compose them.

In covalent bonding, electrons are shared. Oxygen, with eight protons in the nucleus, possesses an inner shell of two electrons and an outer shell of six. If two atoms of hydrogen both share their electrons with an oxygen atom, the stable number 8 is attained (Fig. 24.3), the resulting compound being the familiar H_2O, water. Here again, the properties of the compound are strikingly different from those of the elements: oxygen will burn, hydrogen can explode.

24.1b Atom lattices

In crystalline minerals, ions are combined in distinctive three-dimensional patterns. It is these patterns which impart the properties of hardness, cohesion, crystal structure as seen under the microscope, and crystal shape where this is developed.

Fig. 24.4 is a diagram of the crystal structure of common salt. The formula, NaCl, merely states that for each atom of sodium, the compound contains one atom of chlorine. The diagram shows how, in this case, the bonding works. The chlorine ion in the very centre is bonded to six ions of sodium. Imagine the array reversed, with a sodium ion in the centre; then it would be shown bonded to six ions of

chlorine. Extend the diagram, and the same thing happens. Bonding is six-way throughout.

Fig. 24.5 shows the relative sizes of the ions of the metals, and of silicon and oxygen, that figure prominently in rock-forming minerals. The unit of measurement is the Ångstrom, one ten-millionth of a millimetre. Because the ions of magnesium and iron are closely similar in size, and because each carries two positive charges, they can substitute for one another in the atom lattices of the mafic minerals. The calcium and sodium ions are also closely similar in size, and do effectively substitute for one another in the plagioclase feldspars. But because calcium carries two positive charges whereas sodium carries only one, the electrical balance is assured by the replacement, in sodic plagioclase, of one silicon (4+) for one aluminium ion (3+). Here again, the substitute is close to the size of the original.

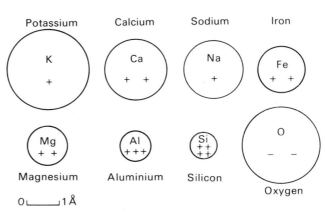

Fig. 24.5   Relative sizes of the very common ions contained in rock-forming minerals.

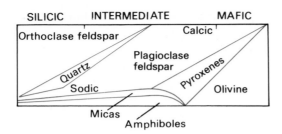

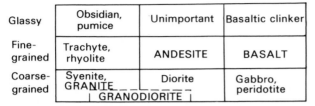

Fig. 24.6 Fundamental classification of igneous rocks.

24.2 Mineral content and the classification of igneous rocks

Fig, 24.6 outlines the two-way classification of igneous rocks. A two-way classification is necessary, because names vary not only with mineral composition but also with texture. The faster a rock melt cools, the finer its texture will be. Lavas range in texture from fine-grained to glassy. Slow cooling at depth, on the other hand, permits the growth of large crystals and of coarse texture. Exceedingly coarse texture (not represented in the diagram) typifies the set of igneous rocks known as pegmatites. These can contain feldspar crystals up to 1 m or more across, set in a matrix of quartz.

Gross mineral content ranges from silica-poor to silica-rich. At the extreme silica-rich (silicic) end of the spectrum, igneous rocks consist chiefly of orthoclase (K) feldspars, with fairly minor additions of quartz, soda feldspars, micas and amphiboles. These rocks are syenites when coarse-grained, trachytes when fine-grained lavas. In some classifications, because of their low quartz content, they are regarded as somewhat less silicic than granites, the chief representatives of silicic igneous rocks. As shown, granites have a noticeable quartz content; sodic plagioclases bulk quite largely in some of them, but are usually outweighed by orthoclase. If granite magma reaches the surface, it flows as rhyolite lava. Both trachyte and rhyolite, very rapidly cooled, can freeze into glass, the rock name for which is obsidian. Silicic lavas are highly viscous, which means that they can become much distended by gas bubbles. Frozen lava froth, with its bubbles drawn out in the direction of flow, constitutes the rock type pumice.

At the other end of the spectrum come silica-poor rocks, containing no free quartz and typified by high contents of olivine, pyroxenes and plagioclase feldspar, mainly calcic. These rocks are all dark-coloured and dense. The deep-seated, coarse-grained set is represented mainly by gabbro, with peridotite, consisting essentially of olivine, being a specialised variant. The fine-grained set is represented chiefly by crystalline basaltic lavas, although some volcanic vents blow out blobs of lava which cool in the air and set into clinker.

Rocks of intermediate composition include diorite in the coarse-grained slot. In this rock type, plagioclase feldspars intermediate between the calcic and sodic extremes are characteristic, but orthoclase feldspar is also present. Content of free quartz is low to zero. Mafic minerals include biotite, amphiboles and pyroxenes. As Fig. 24.6 indicates, there is a continuous range in some areas from granites to diorites, with the result that some large deep-seated masses are classified as granodiorites. Fine-grained lavas are andesites, prominent in part of the circum-Pacific volcanic belt, and thought to represent the mixing of mafic magma with silica-rich rocks of continental origin (Chapter 3, section 3.4b).

24.3 The geological column

Attempts at dating the geologic past were being made long before the application of the radiometric dating methods discussed below. Radiometric dating has so far been applied mainly to igneous rocks. Relative dating, and the establishment of a total geologic sequence, relies mainly upon sedimentary rocks. The leading principle is that a sedimentary unit is younger than the unit upon which it rests, and older than the unit that rests on it. So much is self-evident. But the exposures available for study can rarely be traced for long distances across country; and sedimentary formations incorporated in orogens may be highly deformed and even overturned. Some means is required, therefore, of correlating units across gaps in space, and of telling which way is up. This means is supplied by the fossil succession, which is used both to close gaps in space and to trace the change in life forms through time. A further principle implicit here is that the evolutionary process is irreversible. A given life form, once having vanished from the record, does not reappear later.

When a relative sequence of sedimentary rocks has been worked out, it can be used for the relative dating of igneous rocks. These latter are necessarily younger than any sedimentary rocks which they invade, and are older than any sedimentary rocks which rest on their eroded remnants.

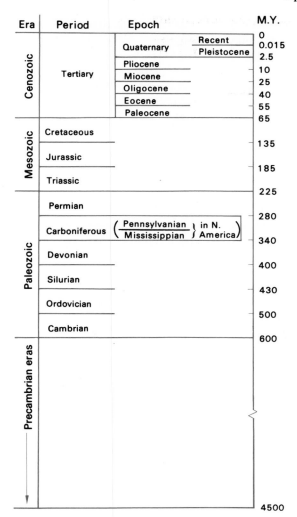

Fig. 24.7 The geologic column.

Systems and period of the Mesozoic (middle life) era are named partly for rock types and partly for areas. The Triassic system of southern Germany is clearly subdivisable into three major subsets: hence the name. The Jurassic is named for the Jura mountains on the Franco–Swiss border. The Cretaceous system includes a very thick and widespread development of chalk rock (Latin *creta* = chalk).

Periods are subdivisible into epochs (not shown for pre-Cenozoic time), epochs subdivisible into stages, and so on. The subdivision of the Cenozoic (recent life) era is controversial, partly because we know more about it than about earlier eras, partly because it contains the time of Pleistocene glaciation, and especially since it includes the evolutionary history of mankind. In many classifications, the Cenozoic era is made to include Tertiary and Quaternary periods, names reflecting an antique and discredited gross classification of rock types. But the radiometric dates listed in Fig. 24.7, added after the total sequence had been established, show that a Quaternary period would amount to less than 4% of Cenozoic time, while a Recent epoch would amount to only about 0.00025%. On the geologic time-scale, these durations are derisorily short. The average duration of a period during Paleozoic and Mesozoic time is 59 000 000 years (Europe) to 53 500 000 years (North America). The 65 000 000 years allocated to the Tertiary period in Fig. 24.7 is exceeded by the durations of the Cambrian, Ordocician and Cretaceous periods. We have no means of knowing whether or not the Quaternary has ended; but on available evidence it does not seem to deserve period status. Its sub-epochs are the Pleistocene, the time of alternating expansion and retraction of the continental ice sheets – also not known to have ended – and the Recent, the very short bracket of geologic time since the last great glacial retraction.

The names for subdivision of Cenozoic time that end in '-cene' indicate, in approximate fashion, the percentage increase through time of life forms closely resembling, or even identical with, those of today. Any good dictionary will supply the derivations from Greek roots. There is one catch. The Eocene epoch (dawn of modern forms) originally embraced all of pre-Oligocene Tertiary time. Further work made it desirable to split off a Paleocene (ancient dawn) epoch.

24.3a Drawing period boundaries

The initial idea in separating one period from another was to use the break between one system and another. Such a break could be a pronounced change in fossil content of sedimentary rocks, a pronounced change in rock type, or both together.

The first task in establishing a total sedimentary sequence was to divide the total succession into major sets. These, still obscure for Precambrian time, are called systems. The corresponding divisions of geologic time are periods (Fig. 24.7). As is usual with geologic names, those of sedimentary systems and of the time periods form a mixed bag. The systems and periods now allocated to the Paleozoic (ancient life) era were, with one exception, named for areas where the systems were first recognised. The Cambrian takes its name from the ancient Roman title of Wales; the Ordovician and Silurian systems take their names from corresponding titles of Welsh tribes. The Devonian is named for a southwestern English county, the Permian for a province in Czarist Russia, investigated geologically by a Scot. The Carboniferous of Europe is named for the coal measures which its rock succession includes, but in North America two systems and two periods are recognised, named after individual states.

Alternatively, it could be a pronounced discontinuity in the record – a gap representing a prolonged episode of erosion, after which newly-appearing life forms would necessarily contrast to those of former times. The most widespread break of all is that between Precambrian and later rocks: Cambrian rocks typically overlie Precambrian rocks with strong erosional discontinuity. But because erosion in one area must be accompanied by sedimentation somewhere else, the total record of sedimentation must be continuous. Furthermore, a change of sediment type in one area need not be replicated in another area. Much must depend on the environment of deposition. As a result, although system names were established first – and in a piecemeal fashion – period names, with their basis in the fossil record, the superset era names, similarly based, and the subset epoch names, similarly based again, have become the main frame of reference for the 600 000 000 years of Cambrian and later times.

It might seem, by comparison with subsequent averages, that there is room in the Precambrian for ten eras and forty periods. In terms of crustal history, this may well be so; but in terms of the history of life, it is not. Fossils of bacteria and of blue–green algae are known from rocks 3 000 000 000 or more years old. Sponges and protozoa – single-celled animals with shells – had evolved before the end of the Precambrian, as also had worms. But, whatever allowance must be made for the very widespread development of hard parts in Cambrian times, and thus for the greatly increased chance of preservation, it seems highly probable that evolution during the first 2400 million years or more of life on earth was far less rapid than it proved subsequently to be. Indeed, some ecological niches that we take for granted today did not exist, or remained unfilled, for much or all of Paleozoic and/or Mesozoic time.

24.4 Dating the past

At various places in the main text, dates are given for events in the remote geologic past. These dates depend on the analysis of exponential decay, the opposite to exponential growth. They relate to the disintegration of radioactive substances.

24.4a Processes of radioactive decay

Radioactive elements can change into other elements by emitting alpha particles (each composed of two protons and two neutrons); by emitting beta particles (electrons) from the nucleus, thus converting neutrons into protons; or by electron capture, which converts protons into neutrons.

Among the best-known radioactive elements are isotopes of uranium – that is, forms of uranium with different totals of neutrons in the nucleus. Recall that uranium was symbolised above as $^{238}_{92}U$, indicating a nucleus containing 92 protons and 146 neutrons. Most uranium does in fact come in this form; but there is also an isotope, uranium-235, symbolised as $^{235}_{92}U$ to indicate the same atomic number, 92 (total of protons), but a mass number of 235: the nucleus is composed of 92 protons and 143 neutrons. Most elements, whether radioactive or not, possess two or more isotopes.

The stable end-products of the radioactive decay of uranium-238 and uranium-235 are two isotopes of lead, respectively $^{206}_{82}Pb$ and $^{207}_{82}Pb$. In each case, ten protons and twenty-two neutrons have been lost. Radioactive decay of thorium, $^{232}_{90}Th$, leads to the formation of another lead isotope, $^{208}_{82}Pb$: eight of the original protons and sixteen of the neutrons have been lost.

Decay time is measured in terms of the half-life. For example, the half-life of U-235 is 713 000 000 years. Half of a starting amount, supplied by the solidification of a rock melt, will have decayed into lead-207 after 713 000 000 years; half the remainder will decay in the next 713 000 000 years, that is, 1 426 000 000 years in total, and so on (Fig. 24.8). We can guess that the process continues until just two atoms are left, of which one will decay in the next half-life, and that in the final half-life, the odds are even that the sole remaining atom will decay either before or after the midway point of the half-life.

If the Pb-207/U-235 ratio is 3:1, this means that two half-lives have elapsed since the original U-235 was formed, and that the age of formation is 1 426 000 000 years ago. Similar calculations can be made for the decay of U-238, with a half-life of some 4 500 000 000 years, and of Th-232, with a half-life of some 14 000 000 000 years. In favourable circum-

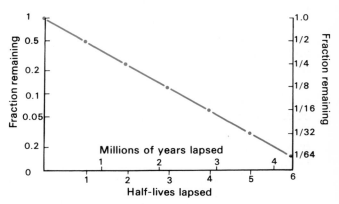

Fig. 24.8 The timing of the radioactive decay of Uranium-235.

stances, two of these calculations, or even all three, can be checked against one another. The ratio Pb-207/Pb-206 provides a further check: it increases exponentially with mineral age.

The radioactive isotope of rubidium, $^{87}_{37}$Rb, decays by electron emission from the nucleus. Loss of an electron converts a neutron to a proton, forming strontium, $^{87}_{38}$Sr. With a half-life of some 50 000 000 000 years, rubidium, analysed for the rubidium/strontium ratio, is important in the radiometric dating not only of igneous but also of sedimentary rocks.

The radioactive isotope of potassium, $^{40}_{19}$K, is also attractive to analysts because of its presence in micas and in sedimentary rocks, and also because of the length of its half-life, short in comparison with those already noticed. It would be more accurate to say half-lives, because radiopotassium decays both into calcium and into argon. About 90% of radiopotassium decays into the commonest isotope of calcium, losing an electron from the nucleus and so increasing the number of protons by one, converting $^{40}_{19}$K to $^{40}_{20}$Ca, which is useless for purposes of analysis, even though the half-life of the radioactive decay is an attractive 1 500 000 000 years. The remaining approximate 10% of the radiopotassium decays in opposite fashion, capturing an electron, converting a proton to a neutron, and changing to argon, $^{40}_{19}$Ar. The difficulty is that argon, being a gas, is apt to escape from the mineral system, so that dates determined from the K/Ar ratio may tend to run too low. Where possible, they are checked against Rb/Sr dates. The half-life of decay from potassium to argon, about 12 000 000 years, makes the K/Ar method applicable to the dating of a considerable time range of igneous materials. The method is applied especially to silicic rocks, with their high contents of micas and orthoclase.

24.4b Carbon-14

Radiocarbon dating relies on the formation and decay of a carbon isotope, ^{14}C, radiocarbon. Its half-life, about 5750 years, is highly relevant to archaeology. As total half-lives mount up, the problems of precise analysis multiply. Radiocarbon dating, given carbon samples of 10 gm each (for preference), or even of 5 gm, can be dated as far back as 36 000 years ago, or within somewhat imprecise limits to as far back as 72 000 years ago. These spans, however, include all the history and protohistory of mankind.

Radiocarbon is produced by the bombardment of atmospheric nitrogen by cosmic rays. A hit by a neutron on $^{14}_{7}$N dislodges a proton, converting the atom to $^{14}_{6}$C, carbon-14, radiocarbon. This combines

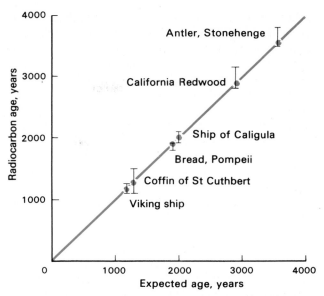

Fig. 24.9 Some of the early radiocarbon determinations of W.F. Libby.

with atmospheric oxygen as carbon dioxide, which is built in carbon form into the organic structures of plants and animals. When a plant or animal dies, the carbon dioxide intake ceases. Radioactive decay takes over, converting radiocarbon, by means of electron loss, back to gaseous nitrogen. Calculations of the relative proportion of radiocarbon to other carbon isotopes permits age to be determined.

Fig. 24.9 plots some of the first analyses made. All points are plotted on the line for radiocarbon age = expected age, with vertical bars showing the range of uncertainty involved in the analysis. All ranges overlap with, or at least end on, the graph of equality. Correlation of central y, x values gives $r = +0.999$, which is significant at better than the 0.001 level (see Chapter 25). The value of r^2 is 0.998, indicating that variation in expected age accounts statistically for 99.8% of the variation in radiocarbon age, within the qualification that the ranges of uncertainty of radiocarbon ages apply.

Subsequent work, however, has shown that the radiocarbon content of the world's lower atmosphere is not strictly constant. If the rate of production of ^{14}C increases, as in response to an increase in cosmic radiation, then the starting ratio between radiocarbon and the common isotope of nonradioactive carbon, ^{14}C/^{12}C, will increase. If the rate of production of inert ^{12}C increases, as in response to greatly increased emittion of CO_2 by volcanoes, then the ^{14}C/^{12}C ratio will decrease. The changes will be reflected in the ratios in the carbon contents of plants and animals.

Tree rings older than 8000 years, in particular those of the bristlecone pine, have been analysed both for radiocarbon dates and for tree-ring dates. The results show that, for a somewhat indefinite interval before 5000 years ago, the ^{14}C content of the atmosphere was as high as 8% above the level of today. In consequence, wood with a real age of 8000 years would yield a radiocarbon age of some 7300 years. Because of this fact, the radiocarbon chronologies of western Europe are being completely overhauled. One powerful group of archaeologists concludes that the traditional view of cultural diffusion from Egypt through the Balkans and the Danube Valley to western Europe may be the reverse of what actually happened. Alternatively, roughly parallel cultural advances, similarly involving the making of large stone structures, may have occurred independently.

25 Appendix II: Analytical Techniques and their Extension

Portions of the main text depend on the relationships among sets of variables, others on methods of dating, and others again upon models in the form of flow-charts. The present chapter will describe some of the analytical methods employed, and will show how the mathematical parts of the work lead toward significance testing and work with computers.

25.1 Modes of growth

Diagrams previously presented include those where values for one variable are plotted against values for another – for instance, the planetary arrays in Figs. 2.6 and 2.7, stream network geometry in Fig. 7.4, the relationship of meander wavelength to bedwidth in Fig. 8.9, and the relationships of channel characteristics to discharge in Figs. 8.19 and 8.20. The modes of growth principally involved in the working of natural systems are exponential and power-functional. But in order to understand these, and particularly in order to understand how graphs are fitted to arrays of plotted points, we need to begin with the simplest relationships of all.

25.1a Linear relationships

Table 25.1 lists the condition of three bank balances at selected points in a fifty-year period. No balance receives any interest. Its amount is controlled wholly by the inputs of deposits and the outputs of withdrawals.

In the first case, the starting balance is 500 units (dollars, pounds, marks, francs, whatever). At the end of each year another 100 units are added. By the end of fifty years the account will contain 5500 units. The data plot on a straight line on arithmetic graph paper (that is, paper with plain scales on both axes) (Fig. 25.1, graph A).

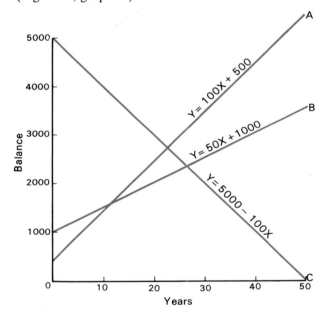

Fig. 25.1   Some linear relationships: two of increase, one of decrease.

Table 25.1   Status of three Bank Balances, not earning interest

Year	Case 1: initial deposit 500 units, annual addition 100 units	Case 2: initial deposit 1000 units, annual addition 50 units	Case 3: initial deposit 5000 units, annual withdrawal 100 units
0	500	1000	5000
1	600	1050	4900
2	700	1100	4800
3	800	1150	4700
4	900	1200	4600
5	1000	1250	4500
10	1500	1500	4000
20	2500	2000	3000
30	3500	2500	2000
40	4500	3000	1000
50	5500	3500	0

By convention, the horizontal axis of the diagram is called the *x*-axis, the vertical axis, the *y*-axis. Also by convention, when one variable can be termed independent and the other dependent, the independent variable is identified as *x* and the dependent variable as *y*. In Table 25.1, time is the independent variable: time values are therefore *x* values. The amount of a bank balance is the dependent variable: its values are *y* values.

A graph of the type shown can be expressed by an equation in the form

$$Y = mX + c$$

where *c* is the starting balance and *m* is the amount by which *Y* increases for every unit increase in *X*. Note that where $X = 0$, the graph intersects the *y*-axis at $Y = 500$. We already know that the balance increases by 100 units each year; therefore, in numerical terms, the equation is

$$Y = 100X + 500.$$

Check: when $X = 30$, $Y = (100 \times 30) + 500 = 3500$, which is correct. The constant *c*, then, controls the intersection of the graph with the *y*-axis where $X = 0$, and the coefficient *m* controls the slope of the graph (Fig. 25.1, graph A).

The second case is one where the starting balance is 1000 units, and where another 50 units are added at the end of each year. The equation for this case is

$$Y = 50X + 1000$$

and, as can be seen from Fig. 25.1, graph B, the slope of the plotted line for $m = 50$ is less than that for $m = 100$. This second graph intersects the *y*-axis where $X = 0$ and $Y = 1000$.

In the third case, where the starting balance is 5000 units, no additional deposits are made. Instead, money is withdrawn at the rate of 100 units a year. The coefficient *m* has a negative value. The graph (Fig. 25.1, graph C) slopes downward to the right, showing that, as time increases, the amount of money left decreases. The equation becomes

$$Y = -mX + c$$

or, as it is usually written,

$$Y = c - mX$$

which in numerical terms would be

$$Y = 5000 - 100X.$$

Check: in fifty years there is nothing left; $Y = 5000 - (50 \times 100) = 0$.

Table 25.2   Status of a Bank Balance with initial Investment of 100 units, earning 6% compound interest

Year	Balance	Year	Balance
0	100	10	179
1	106	20	321
2	112	30	574
3	119	40	1 029
4	126	50	1 842
		⋮	⋮
5	134	100	33 930

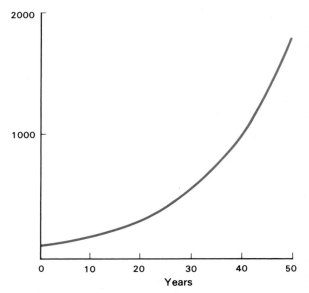

Fig. 25.2   A non-linear relationship: graph of the data in Table 25.2.

If there were no starting balance, but 100 units were added each year, the equation would be $Y = mX$, and the graph would run through the common zero point of $X = 0$, $Y = 0$. Sometimes a relationship of this kind is plotted on double-logarithmic paper, as with the meander wavelength –channel width relationship in Fig. 8.9, but only because the paper spreads out the many points in the lower end of the range, and makes it possible to include other points in the upper range. Sometimes, too, it is appropriate to use logarithmic values rather than arithmetic values in calculation, in order to prevent the results from being distorted by a few very high values in the upper range. But in all linear relationships between two sets of variables, each set grows by addition or diminishes by subtraction.

25.1b Exponential relationships

Interest-bearing accounts serve to illustrate exponential growth – the situation in which one variable increases by addition (or decreases by subtraction) and the other increases by multiplication (or decreases by division). Table 25.2 lists the increase in a starting balance of 100 units, where the interest rate is 6% a year, and the interest is added to the account. In fifty years the starting balance will be multiplied by more than eighteen times (in 100 years it will be multiplied by nearly 340 times!). Fig. 25.2 graphs the data in Table 25.2 against arithmetic scales. If the scale on the *y*-axis is large enough to show the starting balance and the growth through the first few decades, then high values simply run off the diagram – in this case, the 33 930 monetary units which would be in the account at the end of 100 years. If the *y*-scale is small enough to contain very large quantities, then it is useless in the lower range.

The solution, of course, is to use semilogarithmic graph paper. Fig. 25.3 graphs the data of Table 25.2 against an arithmetic scale on the *x*-axis and a logarithmic scale on the *y*-axis. This latter scale permits large *Y* values to be plotted. As years are added, the balance multiplies, at the constant rate of 6% a year. In consequence, the graph on semilogarithmic paper is a straight line.

The general equation for an exponential relationship is

$$Y = ab^x$$

or, in words, the value of *Y* equals the product of some constant *a* and another constant *b* raised to the *x*th power. The constant *a*, in this case, is the starting balance of 100 monetary units. The constant *b* is the annual multiplier: each year's entry of 6% interest means that the balance is multiplied by a factor of 1.06. Thus, in numerical terms, our equation becomes

$$Y = 100 \times 1.06^x.$$

There is, however, an easier way of doing things. The straight-line graph of Fig. 25.3 resembles the straight-line graphs of Fig. 25.1 in everything but the logarithmic scale of the *y*-axis. Taking this scale into account, we can rewrite the general equation for an exponential relationship as

$$\log Y = X \log b + \log a$$

or, in numerical terms in the case in question, as

$$\log Y = X \log 1.06 + \log a$$

which reduces to

$$\log Y = X(0.025306) + 2$$

Check: $X = 50$, $\log Y = 3.2653$, $Y = 1842$.

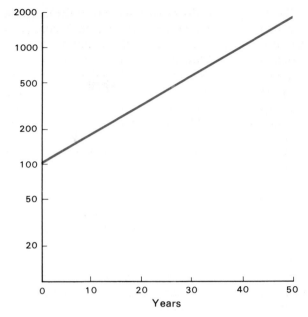

Fig. 25.3 Semilogarithmic graph of the data in Table 25.2.

Straightening out the graph by semilogarithmic plotting has great advantages when the work runs in the other direction. In the examples given, there can be no doubt about the relationships that exist. Money has been added to, or taken from, bank accounts at constant rates; alternatively, interest has been compounded at a constant rate. But in the analysis of the natural world, when relationships must be sought after instead of being defined in advance, any technique that produces approximately straight-line graphs tends to simplify matters. In particular, it greatly simplifies the analysis of how closely the variation in one quantity – for instance, stream number – matches the variation in another quantity – in this particular instance, stream order. This matter will be taken further in a later section.

25.1c Power-functional relationships

When two variables change in value by multiplication and/or division, but at different rates, their relationship is power-functional. The general equation is

$$Y = aX^b$$

which means that the value of *Y* is given by the product of the coefficient *a*, and *X* raised to the *b*th power. Here again, it proves convenient to deal in terms of logarithms. We write

$$\log Y = b \log X + \log a$$

Table 25.3 Channel Width and Meander Wavelength (metres), in a sample of fifteen stream reaches, listed in decreasing order of width: some basic calculations added

	Channel width = X	Meander wavelength = Y	X²	Y²	XY
	71.6	670.5	5 126.56	449 570.25	48 007.8
	70	888.5	4 900	789 432.25	62 195
	66.6	670.5	4 435.56	449 570.25	44 655.3
	65.5	623	4 290.25	388 129	40 806.5
	44.8	600	2 007.04	360 000	26 880
	39.6	475	1 568.16	225 625	18 810
	39.6	468	1 568.16	219 024	18 532.8
	38	588	1 444	345 744	22 344
	38	323	1 444	104 329	12 274
	33.5	239	1 122.25	57 121	8 006.5
	31.4	288	985.96	82 944	9 043.2
	23.3	239	542.89	57 121	5 568.7
	22.9	216	524.41	46 656	4 946.4
	19.4	323	376.36	104 329	6 266.2
	13.7	144	187.69	20 736	1 972.8
Sums	617.9	6 755.5	30 523.29	3 700 330.75	330 309.2
Means	41.193	450.367	2 034.886	246 688.717	22 020.613

which plots as a straight line on a double-logarithmic paper (see Figs. 8.20 and 25.5). If the graph slopes downward to the right (the value of Y divides as the value of X multiplies), then the equation becomes

$$Y = \log a - b \log X$$

as in the relation between stream size = discharge on the one hand, and channel slope on the other. Here again, an approach to a straight-line plot in observational data proves convenient in analysis.

25.2 Fitting graphs to observational data

Table 25.3 lists a sample of concurrent data on channel width and meander wavelength. Width was determined as the average for a small number of closely spaced cross-sections, and measured as width of water surface at most probable annual peak discharge. Wavelength was averaged for three up-stream bends and three downstream bends. A graph of wavelength against width (Fig. 25.4) reveals some degree of scatter. The points do not plot precisely on a straight line. A width of 71.6 m is associated with a wavelength of 670.5 m, whereas a lesser width of 70 m is associated with a distinctly greater wavelength of 888.5 m. Lower down the table, two widths of 39.6 m are associated with wavelengths of 475 and 468 m, whereas a lesser width of 38 m is associated with a wavelength of 588 m. Although in general wavelength decreases as width decreases, the match is by no means perfect. Three questions immediately arise: How close is the association between the two variables, meander wavelength and channel width? Is the association close enough to mean something? and How should a graph relating

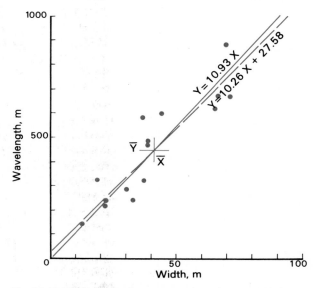

Fig. 25.4 Graph of the data on wavelength and width in Table 25.3.

wavelength to width be drawn? In order to answer these questions, we need to analyse the observed relationship by statistical means.

25.2a Closeness of fit

How closely the variation of one set of variables matches the variation of another set is tested by means of correlation analysis. Where only two sets of variables are in question, and where the numbers of pairs are not many, correlation can easily be run with the aid of a pocket computer. Where many columns (sets of variables) and many rows (groups of observations) are present, then analysis needs a large data-processing machine. The principle, however, remains the same throughout.

If you, the reader, are already familiar with correlation, this portion of the chapter can be skipped. It does, however, include some possibly useful hints on what to do when things go obviously wrong. If you are not already familiar with correlation, then the best thing by far is to use a small calculator to check the calculations made.

The first object is to calculate the statistic r, the coefficient (= measure) of correlation between the dependent variable, wavelength, and the independent variable, width. Wavelength is already plotted against the y-axis of Fig. 25.4, and width against the x-axis. Wavelength values are y values, width values are x values. An inbuilt assumption here is that wavelength depends to a considerable extent on width, as indeed it does.

Table 25.3 shows the initial calculations. Width values are listed in the first column, wavelength values in the second. The squares of these values appear in the third and fourth columns, their products in the fifth column. The two bottom rows of the table contain the sums and means of the five columns.

Here we meet a few symbols. The upper case Greek letter sigma, Σ, is used for the sum of ...; thus ΣX is the sum of all the x values, in this case 617.9. The italic letter n is used for the number of pairs of x and y values, in this case fifteen. Means are given by Σ/n, and are symbolised by a bar; thus, $\overline{X} = \Sigma X/n$, in this case 617.9/15 = 41.193, as shown. The remaining values are symbolised as $\overline{Y}$, $\overline{X^2}$, $\overline{Y^2}$ and $\overline{XY}$. These are what we want. Many small calculators in common use permit an operator to go straight to $\overline{X^2}$, $\overline{Y^2}$ and $\overline{XY}$ without writing down individual entries or sums.

In addition to the means, we need a measure of variability both for the X values and for the Y values. This measure is the standard deviation, symbolised as σ, the Greek lower case sigma. It is given by the square root of the difference between the mean square and the squared mean. Thus, $\sigma_X = \sqrt{\overline{X^2} - (\overline{X})^2}$. In words, the standard deviation of X is the square root of the difference between the mean of X–squared and the square of the mean value of X. Similarly, $\sigma_Y = \sqrt{\overline{Y^2} - (\overline{Y})^2}$.

Sometimes calculation produces a negative number under the square root sign. Many small calculators are programmed to display an error sign if you try to extract the square root of a negative number: large machines will simply spit out the whole operation. There is indeed an error somewhere in the calculation. The quantity $\overline{X^2}$ is always greater than $(\overline{X})^2$, and $\overline{Y^2}$ is always greater than $(\overline{Y})^2$. The only possible thing is to backtrack, locate the error in calculation, and correct it. With desk machines, a

fairly frequent cause of error is the confusion of $(\overline{X})^2$ with $\overline{X^2}$ and of $(\overline{Y})^2$ with $\overline{Y^2}$.

Using the values listed in Table 25.3, we compute σ_X as

$$\sqrt{2034.886 - (41.193)^2} = \sqrt{2034.866 - 1696.863}$$
$$= \sqrt{337.993} = 18.3846$$

and σ_Y as

$$\sqrt{246\,688.717 - (450.367)^2}$$
$$= \sqrt{246\,688.717 - 202\,830.435}$$
$$= \sqrt{43\,858.282} = 209.424$$

Now we can compute r. We set

$$r = \frac{\overline{XY} - \overline{X} \cdot \overline{Y}}{= \sigma_X \cdot \ \ \sigma_Y}$$

that is, r is the difference between the mean of the products of X and Y, all divided by the product of the two standard deviations. In this case,

$$r = \frac{22\,020.613 - (41.193 \times 450.367)}{18.3846 \times 209.424}$$

$$= \frac{3468.645}{3850.176} = +0.901.$$

Here is part of the answer to the first question, How close is the association between the two sets of variables? The rest of the answer is given by the square of r: $r^2 = 0.812$. This states that, statistically speaking, 0.812 = 81.2% of the variation in wavelength is explained by variation in channel width.

If the values of one variable increase, in a general fashion, as the values of the other variable increase, then the correlation coefficient always carries a + sign. A − sign would show the opposite situation, one variable decreasing in value while the other increases. In the case under consideration, a result of $r = -0.901$ (or any other negative value) would signal an error, and call for backtracking.

Sometimes calculation produces an r value greater than + 1.0 (say, +1.124), or less than −1.0 (say −1.113). Here again is the mark of error. Values beyond + 1.0, perfect positive correlation, and −1.0, perfect negative correlation, are impossible. A likely general cause of error is hitting the wrong button on the calculator, especially on one that provides printout but not readout. The usual safeguard, whether or not obvious errors occur, is to perform all calculations twice, and more than twice if the first two results of a calculation fail to match.

Table 25.4 Extract from a table relating Correlation Coefficients to Probability Levels for a Two-variable Situation:
n= number of data pairs

	Probability, *P*						
n	0.10	0.05	0.25	0.02	0.01	0.005	0.001
5	0.805	0.878	0.924	0.934	0.959	0.974	0.991
6	0.729	0.811	0.868	0.822	0.917	0.942	0.974
15	0.441	0.514	0.575	0.592	0.641	0.683	0.760

25.2b Significance of fit

Could the value $r = +0.901$ have been obtained solely as the result of chance? To this question there is no simple yes or no answer. We can, however, determine the odds against. Table 25.4 contains an extract from a probability table for a two-variable situation. The number of x, y pairs appears in the left-hand column. Cutoff values of correlation coefficients (whether positive or negative does not matter) are listed in rows against the numbers of x, y pairs. Probability values appear in the top row. As shown, the probability value increases as the value of the correlation coefficient increases. For fifteen x, y pairs, $r = 0.760$ gives a probability of 0.001, or one in a thousand, meaning that, with this number of data pairs, $r = 0.760$ would be expected to result by mere chance once in a thousand separate analyses. Our own result, $r = 0.901$, is better than this. Hence, we write $0.001 > P$, to state that the odds against a merely chance result are more than a thousand to one. We are justified in concluding that the relationship defined is statistically real. The correlation is said to be significant at better than the 0.001 level.

If we had obtained $r = \pm 0.600$, we should read P as less than 0.02 but greater than 0.01, and should write $0.02 > P > 0.01$. That is, the odds against a purely chance relationship would be assessed at better than 2% but worse than 1%. What odds are unacceptable is to some extent a matter of choice, but few workers are satisfied with anything worse than $P = 0.05$, the 95% level of significance.

25.2c Drawing the best graph

The values obtained for means of the variables, for the correlation coefficient, and for the standard deviations are used in fitting a graph to the scatter of points. The form of the equation employed varies, according to what is being fitted to what. In this particular two-variable situation, where meander wavelength (Y) is being fitted to channel width (X), the equation is

$$Y - \overline{Y} = r\,\sigma_Y/\sigma_X(X - \overline{X}).$$

With the known numerical values substituted,

$$Y - 450.367 = +0.901(209.424/18/3846)\,(X - 41.193)$$
$$= 10.2635\,(X - 41.193)$$
$$= 10.2635\,X - 422.784$$

so that $Y = 10.2635\,X + 27.583$

which is the equation of the desired graph. No harm is done if the equation is slightly simplified to

$$Y = 10.26X + 27.58;$$

this of course is the familiar form $Y = mX + c$. The graph of the equation is drawn in Fig. 25.4. Any two convenient points may be chosen to define it. Thus, where $X = 0$, $Y = +27.58$; where $X = 80$, $Y = 848.38$.

Common sense now takes over. The equation predicts that where $X = 0$ – that is, where channel width is zero and there is no stream – wavelength is 27.58 m. But where width is zero, wavelength must also be zero. The result of graph-fitting has been distorted by the actual scatter of the data, and also probably by the small size of the sample. We are justified in settling for a slightly inferior fit, and in passing the graph through the common zero mark and through the mark of $\overline{X}$, $\overline{Y}$ (Fig. 25.4). The equation for this graph is $Y = X$ $(\overline{X}/\overline{Y}) = X(450.367/41.193) = 10.93X$.

The fitted graph also passes through the point for $\overline{X}$, $\overline{Y}$. If it failed to do this, the calculation, the drawing, or both, would be in error somewhere. If the fitted graph displayed an obviously incorrect slope, again there would be an error somewhere. As before, backtracking is the remedy.

25.2d Semilogarithmic data

For correlation and graph-fitting in cases of exponential relationships, one data set must be converted to logarithmic values, and these values used in calculation (Table 25.5). It will be convenient to

Table 25.5 Basic Calculations for the data on the randomly generated Stream Net (Figs. 7.4 and 7.5)

$Y =$ stream number	$X =$ stream order	$y =$ log stream number	X^2	y^2	Xy
262	1	2.4183	1	5.8482	2.4183
60	2	1.7782	4	3.1620	3.5564
12	3	1.0792	9	1.1647	3.2376
3	4	0.4771	16	0.2276	1.9084
1	5	0.0	25	0.0	0.0
Sums	15	5.7528	55	10.4025	11.1207
Means	3	1.15056	11	2.0825	2.22414

distinguish arithmetic values as X or Y and logarithmic values as x or y.

In this case, $\sigma_x = 1.41421$ and $\sigma_y = 0.87104$, leading to $r = -0.997$ and $r^2 = 0.994$. The negative value of r reflects the fact that stream number decreases as order increases. The value of 0.994 for r^2 indicates that 99.4% of the variation in (log of) stream order is statistically accounted for by variation in stream order. Reference to the five-pair row in Table 25.4 shows that $0.001 > P$. There is less than one chance in a thousand that the observed correlation is accidental. In view of the wholly synthetic character of the network, this result is highly encouraging. The best-fit equation, graphed in Fig. 7.4, is

$$y = 2.9928 - 0.6141X$$

in the convenient form $Y = mX + c$, from which a y value can be obtained for any given X value. Remember that y values here are logarithmic; they need to be converted back to arithmetic values for purposes of plotting. In arithmetic terms, the equation would be approximately

$$Y = 2.99 - 0.61^X.$$

25.2e Double-logarithmic data

When power-functional graphs are to be fitted, both sets of data need to be converted to logarithmic form. Table 25.6 lists fifteen pairs of stream discharges and channel widths, in decreasing order of discharge. The widths are those which have already appeared in Table 25.3. Discharges are those at most probable annual peak flow, $q_{1.58}$ (annual series).

Width generally decreases as discharge decreases, although not perfectly in step; a width of 22.9 m is associated with a discharge of 85 m³/sec, whereas one of 38 m is associated with a discharge of 71 m³/sec, and another of 38 m with a discharge of 139 m³/sec. The graph of width against discharge displays

Table 25.6 Basic Calculations for data on Stream Discharge and Channel Width

$X =$ discharge, m^3 sec	$Y =$ channel width, m	$x =$ log discharge	$y =$ log width	x^2	y^2	xy
289	70	2.461	1.845	6.057	3.404	4.541
283	66.6	2.452	1.823	6.012	3.323	4.470
262	71.6	2.418	1.855	5.847	3.441	4.485
229	65.5	2.360	1.816	5.570	3.298	4.286
170	44.8	2.230	1.651	4.973	2.726	3.682
153	39.6	2.185	1.598	4.774	2.554	3.492
139	38	2.143	1.580	4.592	2.496	3.386
133	39.6	2.124	1.598	4.511	2.554	3.394
85	22.9	1.929	1.360	3.721	1.850	2.623
71	38	1.851	1.580	3.426	2.496	2.925
68	33.5	1.833	1.525	3.360	2.326	2.795
67	19.4	1.826	1.288	3.334	1.660	2.352
45	23.3	1.653	1.367	2.732	1.869	2.260
44	31.4	1.643	1.497	2.699	2.241	2.460
23	13.7	1.362	1.137	1.855	1.293	1.549
Sums		30.470	25.520	64.463	37.531	48.700
Means		2.0313	1.5680	4.2289	2.5021	3.2467

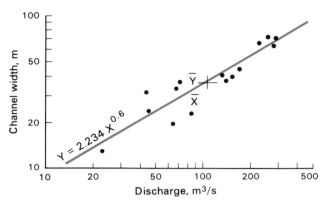

Fig. 25.5 Double-logarithmic graph of width against discharge.

some degree of scatter (Fig. 25.5).

Standard deviations are $\sigma_x = 0.3205$, $\sigma_y = 0.2805$, leading to $r = +0.922$ and $r^2 = 0.850$. The correlation coefficient is highly significant, with $0.001 > P$ (Table 25.4). Variation in (log of) discharge accounts statistically for 85% of the variation in (log of) width.

The best-fit graph is given by the equation

$$y = 0.6x + 0.349,$$

again in the convenient form $Y = mX + c$. In this case, both x and y are logarithmic, needing to be converted to arithmetic X and Y values for purposes of plotting. Where $X = 500$, $x = 2.699$, giving $y = 1.968$ and $Y = 93.0$; where $X = 20$, $x = 1.301$, giving $y = 1.130$ and $Y = 13.5$. In arithmetic terms, the equation reads

$$Y = 2.234X^{0.6}.$$

Here again the graph passes through the point defined by the means – here, $\overline{x}$, $\overline{y}$. The equation states that, in this sample, channel width varies with the 0.6 power of discharge, close to the 0.55 power derived from analysis of a much larger sample.

25.2f Difference-of-means tests

Although correlation as such is not involved, the kind of computation performed in correlation analysis is useful in further connections such as that of the rainfall time series for Sydney in Fig. 6.20. Specifically, the means and standard deviation determined for 1886–94, 1895–1948 and 1949–75 can be used to test the probabilities that means and standard deviations do not differ significantly from one interval to another. The technique employed will not be described here; but its results are instructive. The probability that the variability of mean annual rain-

fall, 1885–94, was not different from that in 1895 –1948 is less than one in a thousand: $0.001 > P$. The probability that the variability for 1895–1948 was not different from that in 1949–75 is less than one in a hundred: $0.01 > P$. We can accept the proposition that variability changed significantly from 1885–94 to 1895–1948, and then again from 1895–1948 to 1949–75. But a comparative test of variability between 1885–94 and 1949–75 produces $P > 0.10$. There is less than a 10% chance that the variability differs significantly between these two periods. Taking the cutoff point at $P = 0.05$, we reject the hypothesis that variability in 1885–94 differs from that in 1949–75.

Mean annual precipitation for 1885–94 was about 2100 mm. Mean annual precipitation for 1895–1948 was about 1720 mm. The difference between the two means is statistically significant at $0.20 > P > 0.10$ – that is, there is better than an 80% chance but less than a 90% chance that the difference is not merely accidental. Using a cutoff of $P = 0.05$, we reject the hypothesis that the difference is statistically significant. At the same time, if the series of observations up to 1894 had started in 1887, and had displayed the same variation and recorded the same mean value, they would have produced $P = 0.05$, and have defined a statistically acceptable difference.

Things are clearer for the contrast between 1895–1948 and 1949–75. Mean annual precipitation increases from about 1720 mm back to about 2100 mm, and the means differ at $P = 0.005$. There is about 1 chance in 200 (5 in 1000) that the difference is merely accidental. We conclude that the difference is real. From 1895–1948 to 1949–75, mean precipitation at Sydney increased, on average, by more than 20%.

25.3 Going on from here

As has been illustrated, reduction of sets of data pairs to the situation where analysis can be performed in terms of $Y = mX + c$, or in terms of the variant $Y = c - mX$, can be enormously useful. Possibilities are by no means limited to linear graphs. The fit of observed lake levels to a sine wave (Fig. 6.4) can be performed with the aid of the correlation technique.

At this juncture, several alarm signals go up. Correlation analysis is enormously powerful; but the technique explained, simple linear correlation of two variables, is not the only technique available, even for correlation analysis. It is used as a detailed illustration both because it is easy to use and is in widespread use, and because it is simple to apply. If some other technique is preferable, such as χ^2 (chi-square, pronounced ky-square) analysis, which

Fig. 25.6 Information encoded on a computer punchcard.

makes a statistical comparison between observed and expected values of a given variable, then it should substitute for correlation analysis or be used in conjunction with it.

It is easy – all too easy – to become over-enthusiastic about any form of statistical analysis. While completely capable of revealing levels of statistical significance, the analytical process has nothing whatever to do with causal relationships. One might, for example, find a statistically significant relationship between sunspot numbers and the numbers of motor vehicle sales, but a causal relationship would be impossible to demonstrate. Some tyro analysts gaining access to a computer program will feed in data sets regardless of the likely value of the resulting printout. Hence the GIGO principle of work with computers – Garbage In, Garbage Out.

But for all the risks, the data of geoscience are very variable data, susceptible to treatment in terms of statistical probability rather than of applied mathematics. Furthermore, the data frequently come in more than two sets at a time. For short series of data on three variables, it may not be worth using a computer as opposed to a desk calculator: the computer is quicker, but someone has to transfer the data to punchcard form (Fig. 25.6), instruct the computer which program to run, and feed in the set of punchcards. For long data series on three variables, or where more than three variables are in question, the computer offers the only realistic option.

For example, the width/depth ratio of a stream channel can be taken as a dependent variable.

Suppose that a correlation is run between the width/depth ratio and channel slope. In the general case, the w/d ratio would be expected to increase in the downstream direction while slope decreased. Assume that the r value $= -0.592$, so that $r^2 = 0.350$: 35% of the variation in the w/d ratio is statistically explained by variation in slope. Now add in variation in suspended-sediment load. Suppose that the joint correlation of the w/d ratio with slope and with suspension load leads to $r^2 = 0.510$: 51% of the variation in the w/d ratio is now explained. The addition of suspension load to the calculations has explained an additional 16% of the variation in the w/d ratio, or 24.6% of the variation left unexplained by variation in slope alone.

More follows. It would be possible to add in variation in bottom-sediment size, the size-distribution of bottom sediment, calibre of bank sediment, position in the drainage network . . . and so on until we reached the limits of useful addition. The selection of useful pollen types for the reconstruction of past climates (Chapter 22, section 22.4b) was made in exactly this manner. A computer will take in modern data on the pollen–climate relationship and identify the indicators in order of importance, according to level of explanation afforded, and will give total explanation at each step.

In addition, a computer will show what happens, in a multivariate situation, when one or more of the independent variables is held constant. It will also, when duly instructed, batch the independent variables into clustered subsets, called factors, and indicate the percentage explanation obtained from each factor. Computers can also draw maps.

Small wonder that computers are coming into ever wider and wider use in the investigation of environmental systems, including ecosystems. The number of components in all but the most limited natural systems, even when these components are already complex in themselves – for instance, individual animals – is so great that only computer analysis can handle the available and necessary information. Hence the development of the applied science called systems analysis. There can be no doubt that this will add greatly to our knowledge of how environmental systems actually work – what variables are principally significant, what variable interacts with what other variable, and where the thresholds are at which systems behaviour breaks down. But the intellectual jump from the systems idea to systems analysis is quite short, even though the technical jump may be long.

25.3a Computers and the yes/no situation

The computer flow charts now to be presented resemble, in a general way, the flow diagrams presented in the main text. The only additional item is that computers can make a yes or no choice. To some extent this choice is implicit in the working of environmental systems. An animal population falls below a critical level: the indicated result is no perpetuation. Persistent undercutting of a cliff by waves leads eventually to slope failure: the indicated result, in terms of failure, is yes. But computers can be instructed to go round again if the analysis does not work out on the first round. Because going round again means an additional input, it amounts to positive feedback, even though involving a negative choice of operation (Fig. 25.7).

Fig. 25.8 is a flow chart for a computer program of the interaction among erosion, sedimentary deposition, and isostatic rebound, applicable to situations on and near to former margins of continental ice-sheets. The chart includes a feedback loop (I), equivalent to the responses to the 'no' answers in Fig. 25.7, a connected sequence of internal components (II), inputs additional to the initial topography (III), an input of probabilistic influence (IV) and the application of the yes or no decision function, which for a 'no' answer to the question, Is the surface below sea level? means a bypass of the deposition–transportation–subsidence sequence by the erosion–transportation–elevation sequence. When the processing of data reaches the point where the machine instructs itself to print out results, these appear at a rate of tens of thousands of characters per second (Fig. 25.9).

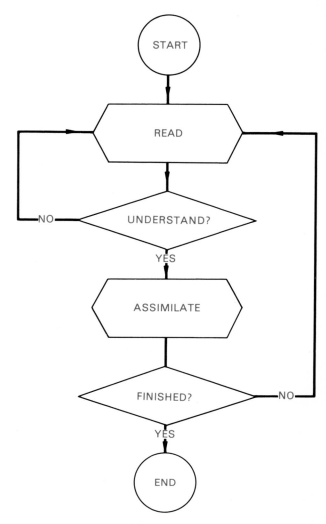

Fig. 25.7 One of the simplest possible computer models.

If numerical values were used in a computer program designed in the style of Fig. 25.8, the result would be a simulation model. This model would make it possible to predict the effect of variation on the part of any of the independent variables. These are, to begin with, initial topography, crustal warping independent of isostasy, and gross sediment supply resulting from weathering and warping. Local sediment supply depends on these three factors. The input of pseudo-random numbers reflects the manner in which events occur in nature. For purposes of local sediment supply, it constitutes another independent variable. And so on down the chain: at each step, the operative variable at each earlier step acts as an independent variable until the final crustal elevation becomes dependent upon all previous inputs.

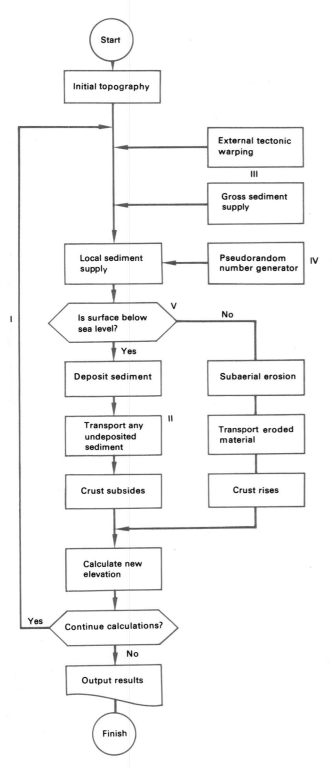

Fig. 25.8 Simplified flow chart for simulation program dealing with interaction among erosion, deposition and isostasy: redrawn, by permission of John Wiley & Sons, Inc., from a diagram in J.W. Harbaugh and D.F. Merriam *Computer Applications in Stratigraphic Analysis* (1968).

```
RUNID:   A00960   PROJECT: 07039           USER: 521863
17:31:18   A00960  MAX - PAGES
17:31:18   A00960  ERROR

ITEM                              AMOUNT        COST

CPU TIME                      00:00:07.925      $0.14
FILE I/O REQUESTS                     88        $0.02
FILE I/O WORDS                    198987        $0.02
MEMORY USAGE                       0.154        $0.08
CARDS IN                             299        $0.06
PAGES PRINTED                         31        $0.56
SOFTWARE SUPPORT                      24        $0.20
JOB CHARGE                             1        $0.05

TOTAL COST                                      $1.13

    THE ABOVE DOLLAR AMOUNTS ARE APPROXIMATE AND ARE BASED
USER BALANCE                                    $2.49

    INITIATION TIME:    17:30:24 MAY 11,1979
    TERMINATION TIME:   17:31:19 MAY 11,1979
    PREVIOUS RUN TIME:  16:50:07 MAY 11,1979
```

Fig. 25.9 Part of the final page of a computer print-out. Total time for the operation was 95 seconds, print-out time was just under 8 seconds: the machine handled 299 punchcards of instruction, printed nearly 200 000 bits of information on 31 pages, and charged U.S. $1.13 (at 1979 prices). To perform the operation manually, a single operator would need at least two years.

25.3b Final statement

The argument of this book has now come full circle. To recapitulate what has been said: systems are structured; they possess components, attributes and internal order; they receive inputs and make outputs; and they are susceptible, according to circumstances, to positive and/or negative feedback. They can be in states of static equilibrium, dynamic equilibrium, or disequilibrium. Changes of systems behaviour can be either zero, cyclic, or step-functional. But at every level of complexity, the workings of the natural world operate in systems terms, notably including the influence of random variation. Systems analysis of the almost incredibly complex operation of the natural world – insofar as we can measure it as not affected by the activities of mankind – and parallel analysis for predictive purposes of the effect of mankind on ecosystems, contain so many variables that computer analysis and computer modelling alone seem capable of producing sensible answers to any well-designed necessary question.

Meantime, an understanding of the systems idea, at once so simple that it scarcely needs explanation, and so complex that it leads into the whole history of life on earth and into the total interaction of life forms, makes a sound intellectual beginning.

Index